# Dissolved Organic Matter in Lacustrine Ecosystems: Energy Source and System Regulator

Developments in Hydrobiology 73

*Series editor*

H. J. Dumont

# Dissolved Organic Matter in Lacustrine Ecosystems: Energy Source and System Regulator

*Edited by*
K. Salonen, T. Kairesalo & R. I. Jones

*Reprinted from Hydrobiologia, vol. 229 (1992)*

Kluwer Academic Publishers
Dordrecht / Boston / London

ISBN 0-7923-1652-5

Published by Kluwer Academic Publishers,
P.O. Box 17, 3300 AA Dordrecht, The Netherlands.

Kluwer Academic Publishers incorporates
the publishing programmes of
D. Reidel, Martinus Nijhoff, Dr W. Junk and MTP Press.

Sold and distributed in the U.S.A. and Canada
by Kluwer Academic Publishers,
101 Philip Drive, Norwell, MA 02061, U.S.A.

In all other countries, sold and distributed
by Kluwer Academic Publishers Group,
P.O. Box 322, 3300 AH Dordrecht, The Netherlands.

*Printed on acid-free paper*



Printed in Belgium

# Contents

**DOM as a system regulator**



**Concluding remarks**

# Preface

Dissolved organic matter (DOM) is quantitatively the most important pool of organic matter in most lakes. Photosynthesis by the littoral and pelagic flora produces autochthonous DOM within lakes through various secretion and decomposition processes. In addition, part of the production of the terrestrial vegetation within the catchment area enters surface waters as allochthonous DOM. The two components can be widely different in composition. Allochthonous DOM is largely composed of humic substances which have already undergone a complicated series of modifications in soil. The fate of these humic substances in water is poorly understood, but they are generally considered refractory to further rapid microbial degradation. The role of DOM in the food chains of water bodies has been controversial since the beginning of the 20th century, and with few exceptions has been largely ignored. Only in the 1980's did the importance of DOM become more widely accepted with the recognition of the 'microbial loop' as a mechanism by which DOM can be utilised by higher trophic levels.

From the tropics to the northern coniferous forest zone, allochthonous DOM may produce a degree of brown colouration in surface waters. In Finland, lakes are generally highly coloured because of the high input of allochthonous humic substances from catchments with peaty soils. As a consequence, awareness of the potential importance of DOM in lakes developed relatively early in Finnish limnology. H. Järnefelt, the first professor of limnology in Finland, took account of DOM in the 1930's while developing his classification scheme for lakes. Subsequently, in the 1960's, his successor R. Ryhänen laid the foundations of Finnish research into humic waters. In 1986, the report of an international evaluation of scientific research in hydrobiology in Finland, organised by the Academy of Finland, recommended increased support for fundamental research on the limnology of humic lakes in Finland. Following that recommendation, a three year research programme entitled "Food chains of humic lakes" was established at Lammi Biological Station (University of Helsinki) with financial support from the Academy of Finland and from the Maj and Tor Nessling Foundation. The main aim of the programme was to evaluate the role of DOM in humic lakes, with detailed study of a polyhumic lake, Mekkojärvi. The end of this project provided a timely opportunity for a symposium to disseminate the results of the research and to discuss them in the wider context of international research into ecosystem effects of DOM.

This volume contains the proceedings of that symposium, held at Lammi Biological Station in May 1990. We were happy to be able to gather together many of the most distinguished scientists in the field to present their results and to discuss important recent developments, such as the role of bacteria-nanoflagellate food chains. The symposium highlighted the many kinds of effects which allochthonous DOM can have in lacustrine ecosystems, and helped to clarify the needs and directions for future research on the food chains of humic lakes. We hope that the publication of these proceedings will help to focus attention on the importance of DOM in lakes as an energy source and system regulator.

K. Salonen
T. Kairesalo
R. I. Jones

*Hydrobiologia* **229**: 1–21, 1992.
*K. Salonen, T. Kairesalo & R.I. Jones (eds.),*
*Dissolved Organic Matter in Lacustrine Ecosystems: Energy Source and System Regulator.*




# Natural and man-caused factors affecting the abundance and cycling of dissolved organic substances in precambrian shield lakes

D. W. Schindler,[1] S. E. Bayley,[1] P. J. Curtis,[1] B. R. Parker,[1] M. P. Stainton[2] & C. A. Kelly[3]
[1] *Departments of Zoology and Botany, University of Alberta, Edmonton, Alberta T6G 2E9, Canada;* [2] *Freshwater Institute, Fisheries and Oceans Canada, 501 University Crescent, Winnipeg, Manitoba R3T 2N6, Canada;* [3] *Department of Microbiology, University of Manitoba, Winnipeg R3T 2N2, Canada*

## Abstract

Effects of natural factors (drought and forest fire), and experimental perturbations (fertilization and acidification) on dissolved organic carbon (DOC) concentrations and ratios to other nutrients in lakes of the Experimental Lakes Area are examined using data obtained over a period of 20 years. DOC concentration, and the ratio of dissolved iron to DOC in lakes of the area were strongly correlated with the relative size of the catchment to the lake.

DOC in many lakes of the area declined over 20 years, due to increased water residence times caused by increasing average temperature and decreasing precipitation. Inexplicably, Lake 382 was an exception to this general observation. Acidification of Lake 302S to below pH 5.0 also caused DOC to decrease. The lesser acidification of Lake 223 (min. pH 5.0) did not significantly affect DOC. Experimental acidification of a small peatland also caused a temporary decline in DOC concentrations in bog pools. Changes in DOC appear to affect the availability of mercury for methylation. Addition of aluminum to a small acidic lake caused a two-fold decline in DOC.

Fertilization of Lake 227 caused a considerable increase in DOC, and in ratios of DOC to other carbon fractions. New stable ratios did not occur for a decade after fertilization began. Lake 226N, fertilized at a lower rate, showed similar but less pronounced increases in DOC, but the experiment was terminated after only eight years. Phosphorus fertilization caused a dramatic increase in the lability of the DOC pool in Lake 226N, where the autochthonous carbon pool was labelled with $DI^{14}C$. A large increase in autochthonous production of DOC and increased microbial utilization of allochthonous DOC appear to have occurred.

DOC concentrations in streams were higher after drought, but concentrations were unrelated to flow volume during wet periods. Due to lower streamflows in drought years, annual yields of DOC from streams were unaffected by drought.

Mesocosm experiments showed that DOC's primary effect on iron is to inhibit sedimentation, possibly by suppressing flocculation reactions that are known to control the cycles of many metals.

The changes in DOC in lakes brought about by changes in water renewal, acidification, or other perturbations can have major effects on the cycles of metals, lake transparency, and phytoplankton production and standing crop.

## Introduction

Dissolved organic matter (DOM) is an important energy source for microbially-based aquatic food webs (Wetzel *et al.*, 1972). It also has important effects on other elemental cycles (Stumm & Morgan, 1981; Drever, 1988; Weilenmann *et al.*, 1989; Curtis, 1991). In softwaters, high concentrations of dissolved organic compounds with low pKa's can be a major source of acidity (LaZerte & Dillon, 1984; Eshleman & Hemond, 1985; Kortelainen & Mannio, 1987). A number of new methods of study have rekindled interest in DOM among aquatic ecologists and geochemists.

In most cases, the index of DOM that is measured in lakes is dissolved organic carbon (DOC). Our studies are no exception, and to be strictly correct, we shall refer to DOC throughout the rest of this paper.

Curiously, methane has not been considered as part of DOC, despite the fact that it is an extremely important energy source for bacteria (reviewed by Rudd & Taylor, 1980). Recently, attention has shifted to methane because of its importance as a greenhouse gas (Lorius *et al.*, 1990). We shall follow the usual procedure, by excluding methane from our discussion of DOC, but its importance as an organic carbon pool should not be forgotten.

More typically, the DOC pool is regarded as containing humic and fulvic acids, plus smaller organic molecules that are excreted by organisms and are an important substrate for metabolism by microorganisms. In this paper, we focus instead on ecosystem-level changes in DOC in response to various natural factors and experimental perturbations at the Experimental Lakes Area. These records reveal some characteristics of DOC and its interactions with the cycles of other elements that cannot be easily studied by smaller-scale, short-term studies.

*Description of the studies*

At the Experimental Lakes Area (ELA) in northwestern Ontario, we have measured a number of physical, chemical and biological variables in lakes, streams and wetlands for over two decades, during a period of climatic warming and lower than average precipitation (Schindler, 1988; Schindler *et al.*, 1990). The lakes are small, softwater systems in catchments underlain by Precambrian gneisses or granites. Catchments have thin, sandy soils, usually under 1 m in depth. The catchments are covered with typical pristine boreal forest communities dominated by jackpine (*Pinus banksiana*) and black spruce (*Picea mariana*). Small wetlands also occur in some of the catchments, causing major changes in DOC. Brunskill & Schindler (1971) give more detailed information on the watersheds.

The period of record is one of climatic warming and drying. Average lake temperatures increased by about 2.5 °C, and water renewal times for the lakes increased several-fold (Schindler *et al.*, 1990). The warm, dry weather caused a high incidence of forest fires in the area. Of the lakes studied in their natural state, Lake 239 (L239) had over half of its watershed burned in 1974, and all of it burned in 1980. Lake 382 (L382) had only one small fire in its catchment, so any changes should reflect the effects of drought alone. These reference systems have been important in interpreting experimental results, for the increasing water renewal times and water temperatures have caused many changes in the physical, chemical and biological properties of lakes in the area (Schindler, 1988; Schindler *et al.*, 1990).

In addition, DOC and many other parameters have been monitored in lakes that have been experimentally altered by adding nutrients, acids or radiotracers. Lakes 227 (L227) and 226N (L226N) were fertilized with phosphorus and nitrogen. Carbon, as sucrose, was also added to L226N. Lake 226S (L226S) received nitrogen and carbon, but no phosphorus. The two basins of L226 were fertilized from 1973–1980, after which they were monitored during recovery. Recent summaries are provided by Schindler, 1990; Schindler *et al.*, 1987; Shearer *et al.*, 1987; Findlay & Kasian, 1987.

Lake 223 (L223) and 302S (L302S) were acidified with sulfuric acid. The former lake was acidified for eight years, reaching a pH of 5.0, after which it was allowed to recover slowly to a pH of 5.8 in 1988. Lake 302S has been acidified to a pH of 4.5, at which value it is currently being held. Further details of these lake experiments are summarized by Schindler (1988; 1990), Rudd *et al.* (1990) and Schindler *et al.* (1991).

Carbon-14 was added as carbonate to three lakes with differing DOC concentrations and nutrient inputs (Hesslein *et al.*, 1980; Bower & McCorkle, 1980; Bower *et al.*, 1987). These experiments offer unique opportunities to observe differences in the patterns of transformation of inorganic carbon into DOC, and in how DOC interacts with the cycles of nutrients and trace metals. Further opportunities to observe the behaviour of DOC under a wide range of conditions were offered by a comparative study of the iron cycle in lakes with over an order of magnitude range in DOC concentrations (Curtis, 1991), a study of streams draining upland and wetland areas (Bayley, unpubl.) and a study of the effects of artificial applications of sulfuric and nitric acids to wetland systems (Urban *et al.*, 1989; Bayley *et al.*, 1987, unpubl.). In all of the studies, chemical analyses were done as described by Stainton *et al.* (1977).

The changes in ratios of carbon, nitrogen and phosphorus in the dissolved organic pool resulting from various man-made or natural stresses may change the availability of such compounds for metabolism. Much of the dissolved organic fraction in a pristine oligotrophic lake consists of relatively refractory residuals from biological processes in the lake and its catchment. As we shall show, lake fertilization causes DOC to increase, adding both refractory and labile compounds to the DOC pool. In addition, fertilization with phosphorus seems to cause increased availability of DOC. As discussed by Münster *et al.*, 1991, DOC residuals may represent 'capacitors' that extend the nutrient reserves for algae and bacteria over longer time periods than would otherwise be possible. Their availability may also depend on the nutrient sufficiency of bacteria (see also Jones, 1991). We shall therefore compare the ratios of DOC to DON (dissolved organic nitrogen = TDN − ($NH_4$ + $NO_3$ + $NO_2$) or to DOP (dissolved organic phosphorus), to see whether the relative sizes of pools of nutrients in dissolved organic matter are affected by any of the perturbations. Because our earlier experiments have shown that for all practical purposes, DOP = measured total dissolved phosphorus, i.e., that inorganic P is below the limits of detection (Levine & Schindler, 1980; Levine *et al.*, 1986), we shall use TDP in our examination of dissolved organic nutrient ratios.

### *Observations on DOC in natural lakes and streams at the Experimental Lakes Area and elsewhere in the precambrian shield*

In preliminary surveys, Schindler (1971a) observed that water color was correlated with the ratio of terrestrial drainage area ($A_d$) to lake surface area ($A_o$), suggesting that DOC was largely of terrestrial origin. This observation was tested by Curtis (1991), who showed a close correlation between DOC concentration in lakes and $A_d/A_o$. Engstrom (1987) also found a strong relationship between DOC and $A_d/A_o$ for lakes of Labrador with small catchments, and Rasmussen *et al.* (1989) provided detailed analyses of the DOC-watershed relationships in a variety of lakes. They found that catchment slope, lake size and depth also had significant effects on DOC concentrations. Gorham *et al.* (1986) found that organic acidity accounted for a high proportion of the acidity in Nova Scotia lakes, regardless of whether or not they received acidic precipitation. Schindler (1971b) showed that the color of water was the major factor attenuating light in the oligotrophic lakes at ELA, a conclusion which had previously been reached by Birge & Juday (1934) for lakes in northern Wisconsin. Lakes with the highest concentrations of DOC (1500–1800 $\mu$M) also had maximum light penetration strongly shifted toward yellow or orange wavelengths (>600 nm), while the clearest lakes of the area (which generally have DOC concentrations of 240–400 $\mu$M) had maximum penetration by blue to green wavelengths (<500 nm) (Schindler, 1971b). Similar results were reported by Effler *et al.* (1985) and Watras & Baker (1988). Vertucci & Likens (1989) showed that DOC also had a strong effect on the reflectance of light from lakes. DOC was positively correlated with reflectance of wavelengths from 600–750 nm, but negatively correlated with visible wavelengths. Large spectral shifts were found for rather small changes in DOC concentration. The euphotic zones in the clearest lakes at ELA are >20 m while those in the most stained ones are <3 m. Thus, the attenuation of light by DOC greatly reduces the potential for phytoplankton production in the hypolimnions of stained lakes. Although there appears to be some light compensation by the plankton of such lakes, allowing higher rates of photosynthesis per unit light (E. J. Fee, pers. comm.), there is a general tendency for lakes with higher DOC to support less phytoplankton biomass per unit input of phosphorus (Janus & Vollenweider, 1981). This may also be due in part to inhibition of alkaline phosphatase activity by DOC, which seems to be most pronounced in soft water (Wetzel, 1991; Münster *et al.*, 1991). The spectral characteristics of DOC in lakes from a variety of areas seem to be very similar (Davies-Colley & Vant, 1987), so that responses observed in one area are probably widely applicable. The quantification of light-DOC relationships is discussed in more detail by Jones (1991).

DOC may also play an important role in the elemental cycles of lakes. For example, the concentration of iron per unit DOC in lakes is closely related to $A_d/A_o$ (Fig. 1). This can be partially explained by dilution of inflowing streams with precipitation on the lakes' surfaces, for precipitation has a very low Fe/DOC ratio. The same would be true for autochthonous sources, i.e., production by organisms. However, these mechanisms are not enough to account for all of the observed differences in the Fe/DOC ratios of lakes, and Fe and DOC must therefore be removed from the lakes at different rates. Removal rates are likely to be a function of water renewal time, which is shorter in lakes with high $A_d/A_o$. In turn, the cycles of phosphorus and trace metals are known to be closely linked to the aquatic iron cycle (Hutchinson, 1957).

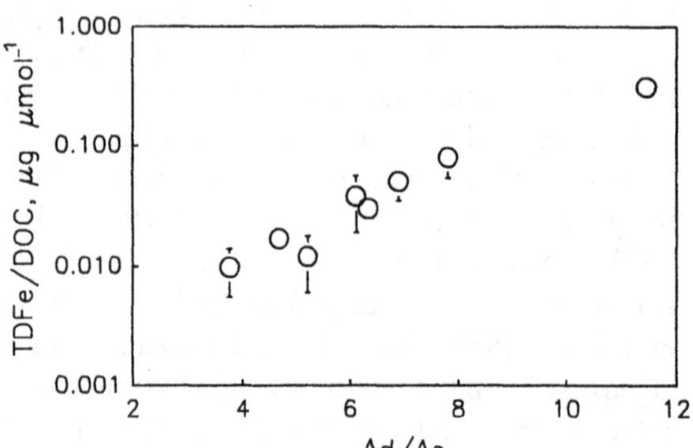


*Fig. 1.* The ratio of dissolved iron to DOC, as a function of the ratio of catchment area to lake surface area. Vertical lines are confidence limits for several years of observations. From Curtis (1991).

*Watersheds and lakes as sources and sinks for DOC*

On an areal basis, more DOC is exported from wetlands than from upland catchments (Urban *et al.*, 1989, Table 1). Within uplands, more DOC per unit area is exported from areas with thicker organic soils (McDowell & Likens, 1988). As a result of the higher losses of DOC from wetland catchments, lakes in extensive areas of peatlands have higher color than those with catchments dominated by uplands, a fact well known to boreal limnologists (Pitkänen, 1986; Heikkinen, 1989). Urban *et al.* (1989) estimated that 20–40% of the lakes in eastern Canada, northern Europe, and the northeastern and midwestern USA, all areas containing considerable areas of wetlands, have color of >50 mg Pt $l^{-1}$, indicating high con-

*Table 1.* DOC export (g $m^{-2}$ $yr^{-1}$) from various ecosystems. Data from Urban, *et al.* 1989.

| Ecosystem | Export | Reference |
|---|---|---|
| *Bogs* | | |
| Thoreaus Bog, Mass. | 8.6 | McKnight *et al.*, 1985 |
| Marcell S-2, Minn. | 21.2 | Urban *et al.*, 1989 |
| ELA 239 NE, Ont. | 23.8 | Urban *et al.*, 1989 |
| Eastern North America | 5.0–20.0 | Urban *et al.*, 1987 |
| *Uplands* | | |
| Marcell | 3.4 | |
| ELA 239 NW dand E, Ont. | 5.0 | |
| First-order streams | 0.43–2.5 | Tate & Meyer, 1983 |
| Hubbard Brook | 1.0 | Hobbie & Likens, 1973 |
| *Fens* | | |
| Marcell | 43.0 | Urban *et al.*, 1989 |
| *Southern Swamps* | | |
| | 2.5–10.4 | Muholland & Kuenzler, 1979 |

centrations of DOC. Our studies show that precipitation is also a significant source of DOC, furnishing 10–25% of total annual input for lakes at ELA. It probably originates as emissions from terrestrial vegetation, or from leaching of airborne plant particles.

Even if *in situ* sources of DOC are disregarded, retention of DOC in most lakes at ELA is high, as expected for lakes having rather long water renewal times. For example, in L239, only 15–55% of DOC inputs from watersheds and precipitation reach the outlet of the lake (Fig. 2). The concentrations of DOC in water are rather low, indicating that the retained DOC is either sedimented or decomposed. This is not surprising, considering the role of DOC in flocculation or coagulation with metals and particulate matter (Weilenmann *et al.*, 1989), the reputed photodegradation of DOC in the presence of peroxides (Cooper & Lean, 1989) and the bacterial degradation of DOC (Likens, 1985; Wetzel, 1991). Studies in Precambrian Shield lakes of eastern Canada employing stable carbon isotopes also indicate a reasonably rapid turnover for DOC (Schiff *et al.*, 1991). The retention of DOC, and associated organic anions, is an important source of in-lake buffering in many softwater lakes (P. J. Dillon, D. W. Schindler, unpubl.).

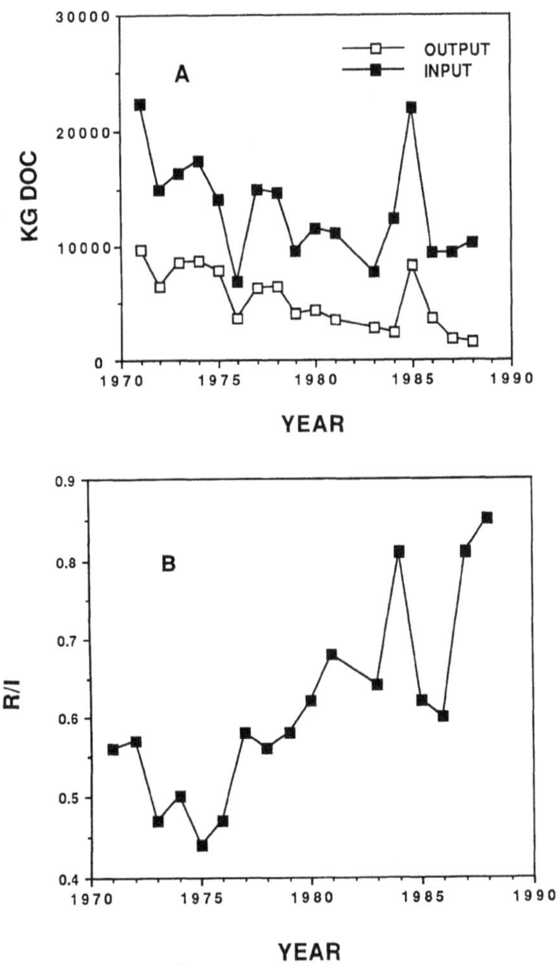


*Fig. 2.* DOC budgets for Lake 239. a – DOC inputs (streams + direct drainage + precipitation) and outflow from the lake. b – Retention (R) of DOC in the lake as the proportion of input (I).

## *The effects of climate change on DOC budgets*

The 20 years of records at ELA have been collected during a period of slow, continuous warming and increasing drought (Schindler *et al.*, 1990). The incidence of forest fires in the area has increased, including two fires in the catchment of L239. These records provide a preview of what increasing greenhouse warming may do to DOC concentrations in lakes and streams.

DOC concentrations in streams at ELA are poorly correlated with rates of water discharge, in contrast to the positive correlations generally observed (Heikkinen, 1989; Lewis & Grant, 1979; Lewis & Grant, 1979). Instead, concentration appears to depend heavily on antecedant moisture conditions. For example, at a given rate of hydraulic flow, much higher concentrations of DOC are observed after dry weather than following wet conditions (Fig. 3). In wet years, there is a pronounced summer maximum in the DOC concentrations in both upland and wetland streams, probably reflecting production of DOC by microbial activity in soils (Fig. 4a). In contrast, in drier years there is a progressive increase in concentration of DOC throughout the ice-free season (Fig. 4b). Kerekes *et al.* (1986) and Naiman (1982) reported seasonal DOC concentrations in streams of eastern Canada that have a midsummer maximum like that observed in wet years at

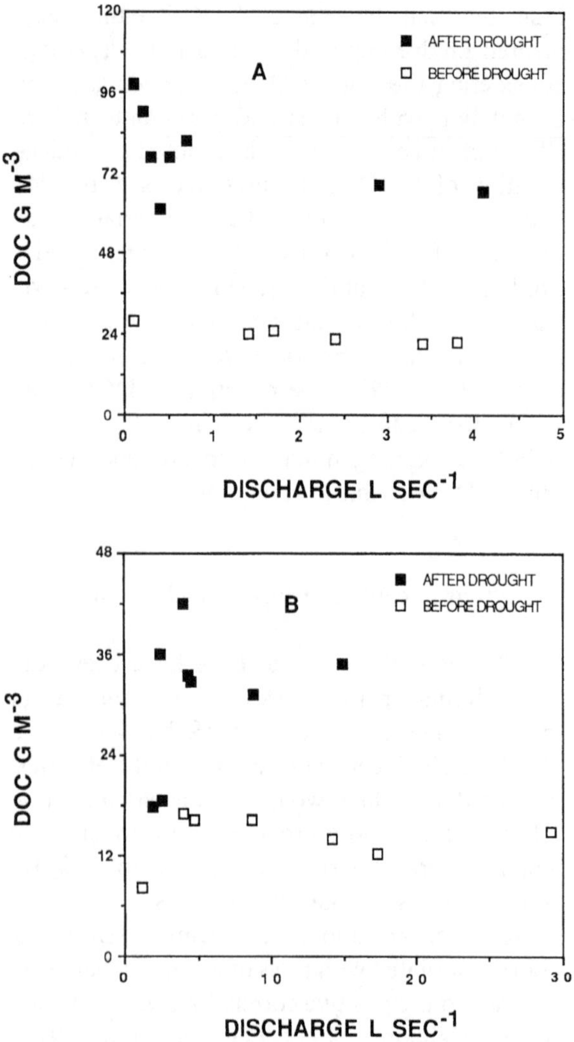


*Fig. 3.* DOC concentrations in two of the streams entering Lake 239, illustrating the general lack of relationship between discharge and concentration, but the increase in DOC following several weeks of drought. a – The NE subbasin, which contains a wetland. b – The East subbasin, which contributes 70% of the streamflow to the lake.

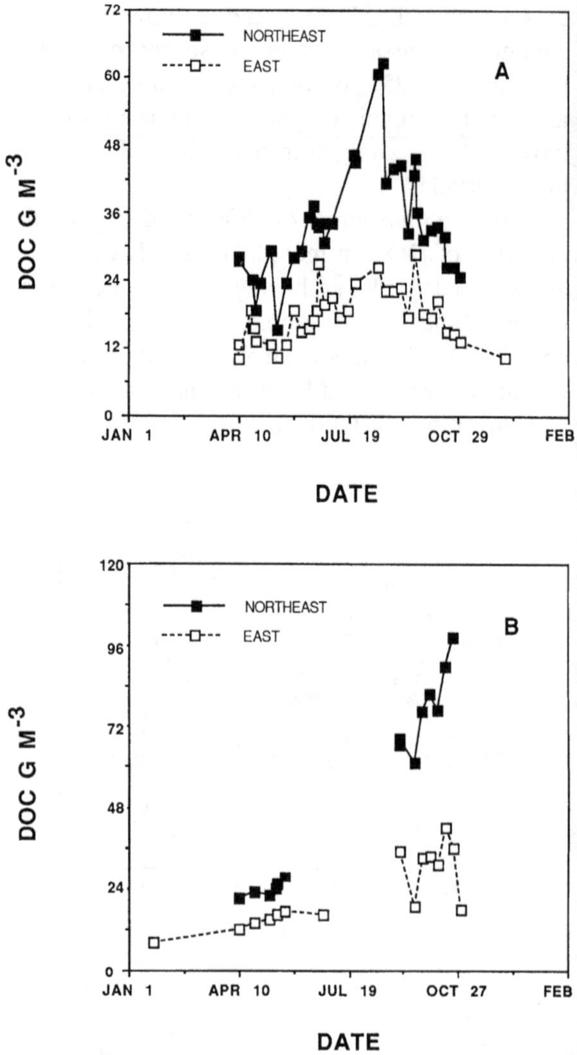


*Fig. 4.* Seasonal patterns of DOC concentrations in two of the inflow streams to Lake 239. a – 1985, a wet year. b – 1986, a dry year.

ELA, but Glover & Webb (1979) report maximum concentrations during spring in a Norwegian river. Meyer *et al.* (1987) found that highest DOC concentrations occurred in summer and autumn in streams at Coweeta. This was attributed to high rates of leaching of organic matter liberated by microbial decomposition of organic debris in the streams.

Overall, considerable differences occur in the seasonal patterns of DOC at ELA, as the result of climate and vegetation type. High concentrations of DOC and low flows in headwater streams following drought are balanced somewhat by lower concentrations at higher flows in wet periods, and annual average concentrations of DOC in streams are remarkably stable (Fig. 5).

Long-term retention of DOC in L239 is well-correlated with water renewal time (Fig. 6). A comparison of Figs. 2 and 7a shows that increased DOC retention in L239 is accompanied by decreased concentrations of DOC in the water,

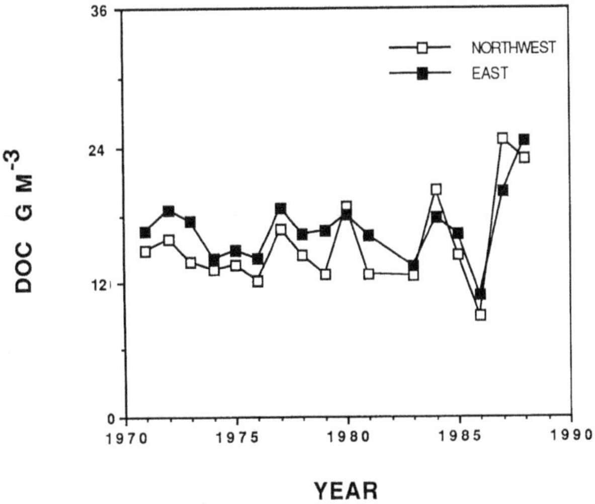


*Fig. 5.* Volume-weighted mean annual DOC concentration vs. time in two of the inflows to Lake 239, showing the overall lack of effect of weather and perturbation by windthrow and fire on annual average DOC concentration.

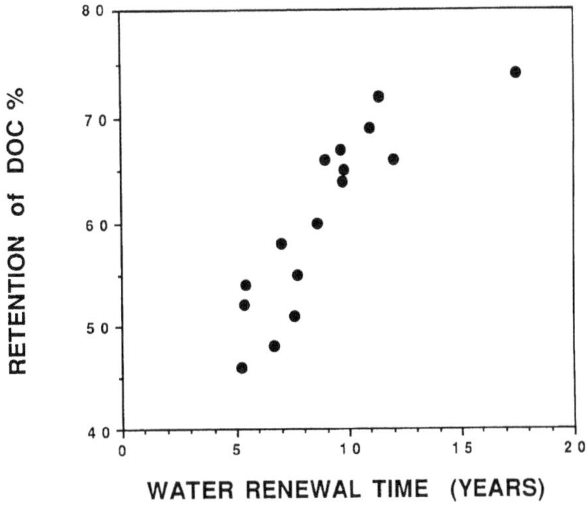


*Fig. 6.* Percent retention of DOC vs. water renewal time in Lake 239. Values are 3-yr running averages.

indicating that retention is caused by increased losses to sediments or degradation, rather than increased concentrations in water. Ali *et al.* (1984) showed that the transport and sedimentation of particulate matter in lakes was highly dependent on both DOC and water renewal time. If climate becomes warmer and drier, as is forecast for the central boreal regions of North America, decreasing DOC concentrations and lowered rates of water renewal should alter many lacustrine processes. Obviously, alterations in climate, catchment soils or vegetation and hydrological regimes can have important effects on DOC and on the transparency, temperature, productivity and elemental cycles of lakes.

### *Changes in concentrations and ratios of DOC, DON and TDP caused by natural phenomena*

In order to properly interpret results of experimental perturbations, it is necessary to know what changes in lakes of the area are caused by natural events, such as the effects of warmer, drier conditions as discussed above. For this purpose, we have chosen two lakes for which long records are available, L239 and L382. As discussed above, the former lake has had its catchment burned, some of it twice. A small fire in the L382 basin in 1979 burned $<5\%$ of the catchment area. The effects of this fire are undetectable on lake chemistry, and the response of the lake should be typical for natural lakes in the absence of forest fire. Both lakes had similar DOC concentrations at the beginning of our period of study.

DOC in L382 varied widely, but there has been no significant long-term trend, despite high concentrations in 1986–1987 (Fig. 7a). In contrast, the DOC concentration in L239 increased dramatically for two years after the 1980 fire, which burned the entire catchment. Unfortunately, there are no DOC data for L382 in those two years. As we shall discuss later, high DOC values were also found in experimental lakes in the same years, suggesting that the high DOC's had a common cause. Because dry weather would normally be expected to cause DOC in lakes to decrease due to longer retention times as discussed above, we hypothesize that the increases were caused by increased dry fallout from the large areas burned by forest fires in 1980. During windy periods of the years just after the fire, visibility in the area was often decreased by airborne 'soot', and air filters at the ELA meteorological site were black after a single day of sampling. The increased

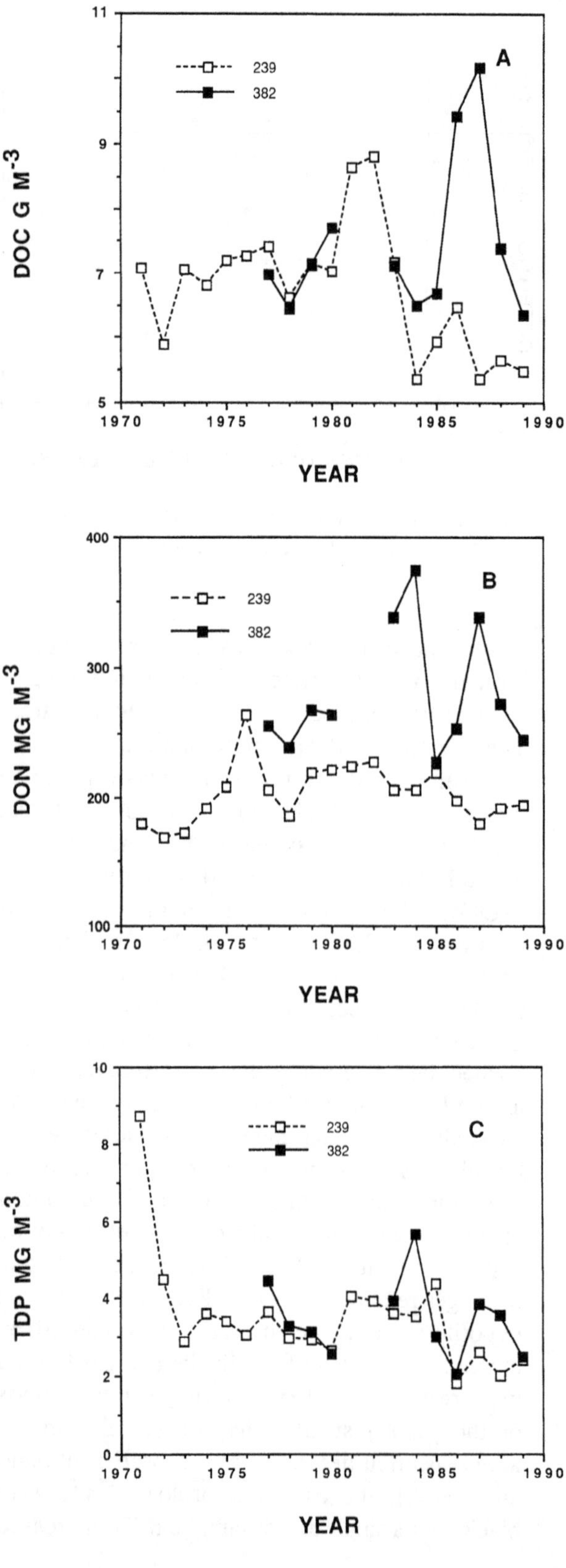


DOC in L239 was followed after two years by a rapid decline to stable concentrations about 1500 mg m$^{-3}$ lower than in prefire years. This may be related to the destruction of organic soil profiles in the East Subbasin, which contributes 70% of the inflow to the lake, after the 1980 fire. DOC is correlated with soil depth (McDowell & Likens, 1988), and is known to decline after severe soil disturbance (Meyer & Tate, 1983; Meyer *et al.*, 1987). The 1974 fire did not cause a noticeable change in DOC concentrations from the East Subbasin, but the fire did not destroy organic soils as in 1980. Trees were killed in 1974, but left standing, so that wind erosion of soils was also lower than in 1980. Similarly, Hobbie & Likens (1973) found no change in DOC lost from a denuded watershed at Hubbard Brook, New Hampshire. The extent of soil disturbance was not given.

The dissolved organic nitrogen (DON) concentration in L239 increased after both fires, but the sharp decline observed for DOC two years after the 1980 fire was absent (Fig. 7b). This may have been caused by increased nitrogen losses from burned catchments, which consisted primarily of DON (Schindler *et al.*, 1980; Bayley *et al.*, in press). In contrast, DON concentrations in L382 were highly variable, but generally followed DOC. In both lakes, there were slow, long-term declines in TDP, although if the extremely high 1971 value for L239 is omitted, the trends are not statistically significant (Fig. 7c).

Over the years, DOC in L239 declined while total carbon (TC = DOC + DIC + Particulate C) increased slightly, causing a decrease in the ratio of DOC:TC. In L382 the ratio remained constant, which supports the hypothesis that the above-mentioned degradation of organic soils in the watershed was responsible for the decline in DOC:TC in L239 (Fig. 8a). DOC:DON in L239 also declined slowly over the entire period

*Fig. 7.* a – Mean annual DOC concentration vs. time in two reference lakes, Lake 382 which is unperturbed, and Lake 239 which had its catchment partly burned by a forest fire in 1974 and totally burned in 1980. b – As above for DON. c – As above for TDP.

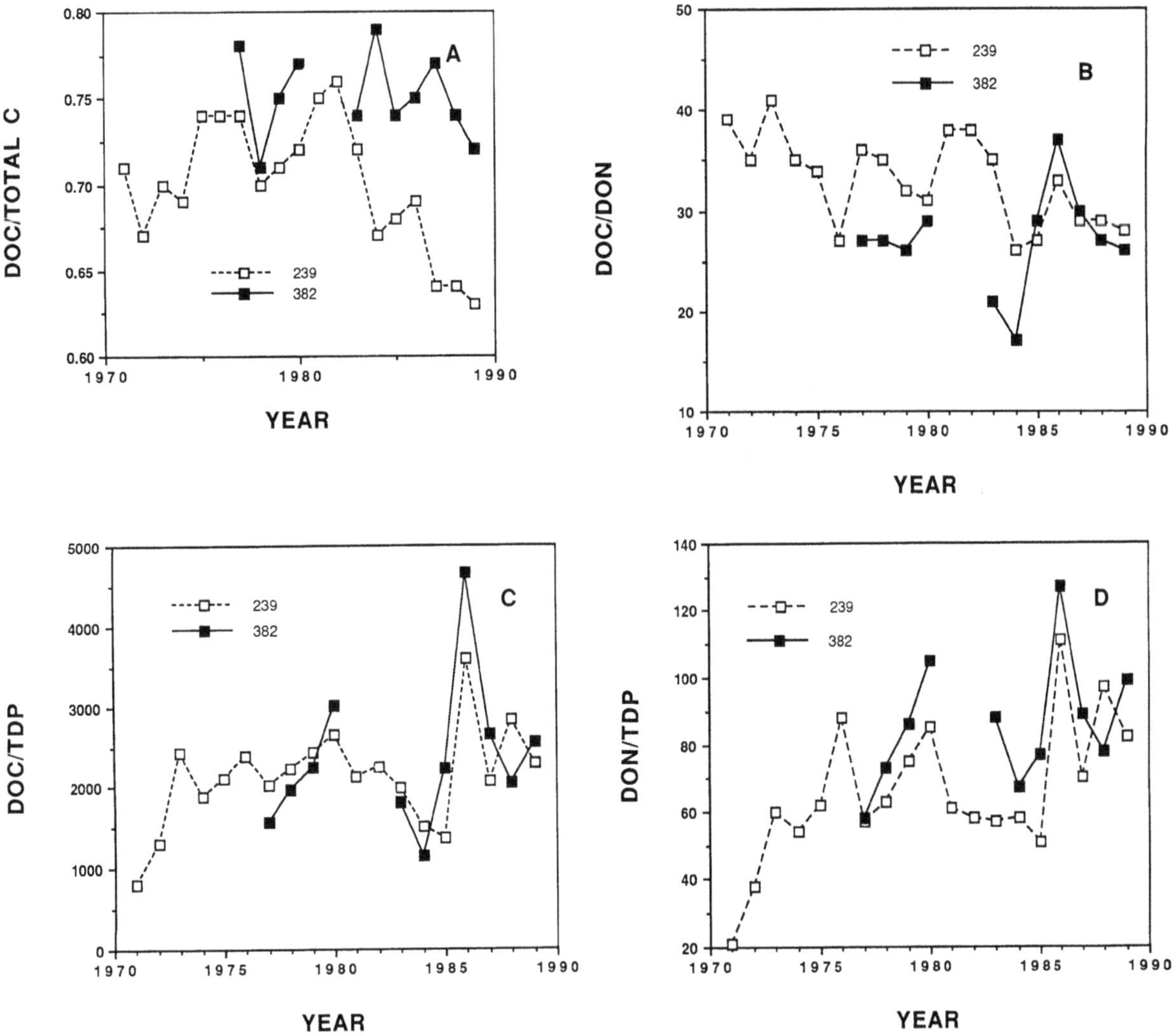


*Fig. 8.* Annual average nutrient ratios in reference lakes.

of record, while that in L382 remained quite variable, but without apparent trend (Fig. 8b). Increases in the ratios of DOC:TDP and DON:TDP occurred in both lakes (Figs. 8c–d). We hypothesize that declining ratios of DOC to DON and TDP may cause increased lability of the DOM pool, as indicated by $^{14}C$ experiments that are discussed later. This, as well as longer water residence time, may be responsible for the increased retention of DOC in L239. Further research will be needed to distinguish whether this hypothesis is correct, and which DOC removal mechanisms in L239 are important.

*Autochthonously produced DOC*

Measurements following experimental additions of radiocarbon-labelled DIC to pristine ELA lakes suggest that autochthonously-produced DOC is not a high proportion of the total DOC. For example, L224, the site of the first radiocarbon addition, has the lowest natural concentration of DOC and the smallest terrestrial catchment of all lakes at ELA. It would therefore be expected to have the highest ratio of autochthonous DOC to allochthonous DOC of any lake in the area. When $DI^{14}C$ is added to the lake, one might therefore expect a high proportion of the DOC to become labelled with radiocarbon. This was not the case. The $^{14}C/^{12}C$ ratio in the DOC pool reached a steady-state far lower than for suspended particulate matter (which consisted largely of algae) or DIC. This suggests that most of the lake's DOC pool is both refractory and of allochthonous origin. Similar results were obtained in L302S and L226S (Bower, 1981; Bower & McCorkle, 1980; Bower *et al.*, 1987; Hesslein *et al.*, 1980; Fig. 9). Likens (1985) reached similar conclusions about Mirror Lake, based on total input-output budgets. As we shall see later, this situation is dramatically altered by nutrient enrichment.

This is not to suggest that autochthonously-produced DOC is unimportant in the carbon cycle, for it represents up to 20% of the carbon fixed by phytoplankton in ELA lakes (Nalewajko & Schindler, 1976). However, much of the excreted DOC is in compounds of low molecular weight, which are rapidly consumed by bacteria and algae, and do not accumulate in the DOC pool. Sucrose added to entire experimental lakes of the area was also consumed quickly, showing that low molecular weight compounds are very labile (Thompson & Hamilton, 1973). However, some of the excreted DOC appears to quickly bond to high molecular weight substances, rendering it much less labile (Nalewajko & Lean, 1972; Nalewajko & Schindler, 1976). Some of the autochthonously-produced high molecular weight compounds are metabolized slowly, remaining in the water long enough to support approximately half of the total lake respiration under the ice (Fig. 10; C. Kelly, unpubl.). In addition, some autochthonous DOC compounds seem to be as refractory as DOC from allochthonous sources,

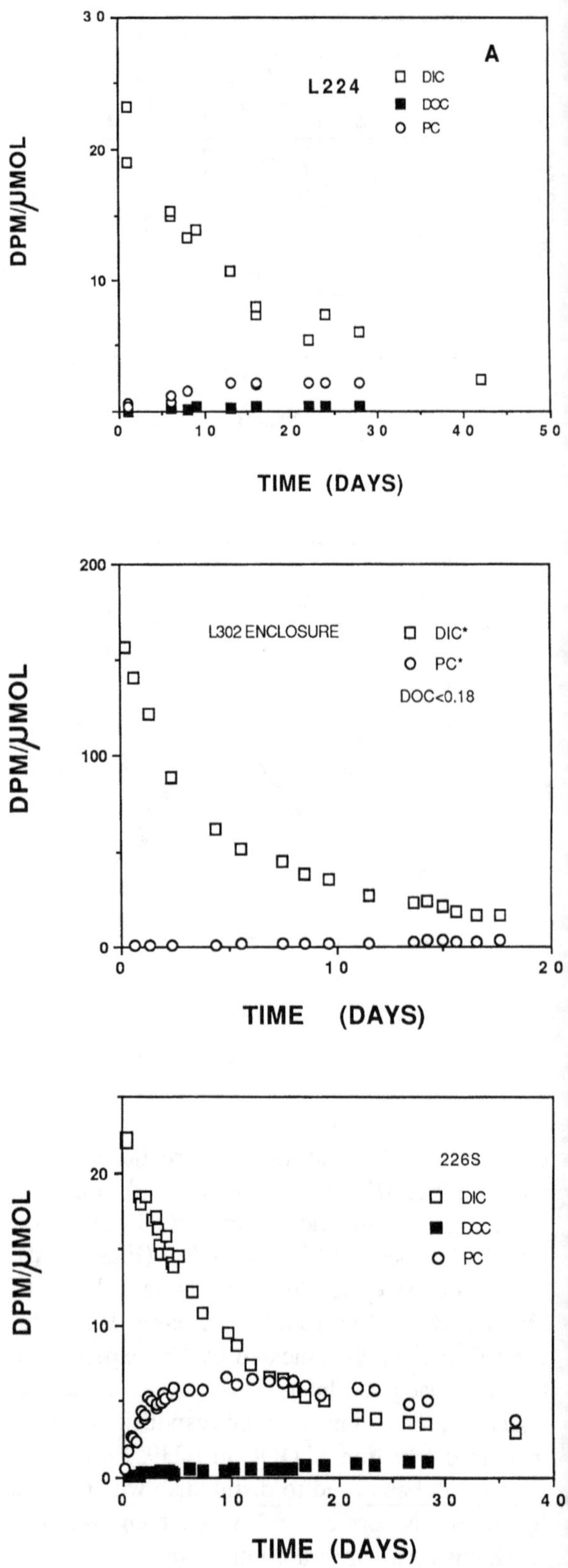


*Fig. 9.* Specific activity of carbon isotopes in 3 oligotrophic lakes spiked with $^{14}C$ in midsummer. a – Lake 224, clear (<5 Pt units) and oligotrophic. b – Lake 302S, slightly stained (20 Pt units), data from a limnocorral experiment). c – Lake 226S, about 20 Pt units, fertilized with N and C, but no P. Note that $DO^{14}C$ does not reach isotopic equilibrium with particulate carbon (PC) in any of the lakes, indicating that most DOC is refractory. (Data from Hesslein *et al.*, 1980; Bower, 1981).

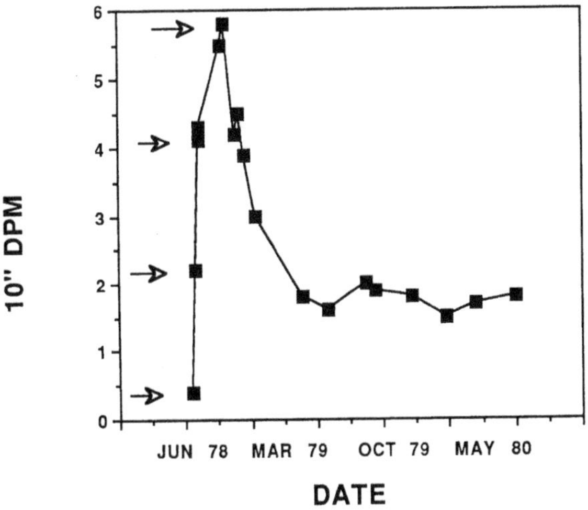


*Fig. 10.* Activity of $^{14}C$ as DOC from Lake 226S after the addition of $^{14}C$-labelled DIC on four occasions in the summer of 1978. The activity of $^{14}C$ in the lake after 1978 was corrected for outflow losses. The arrows indicate when radiocarbon was added.

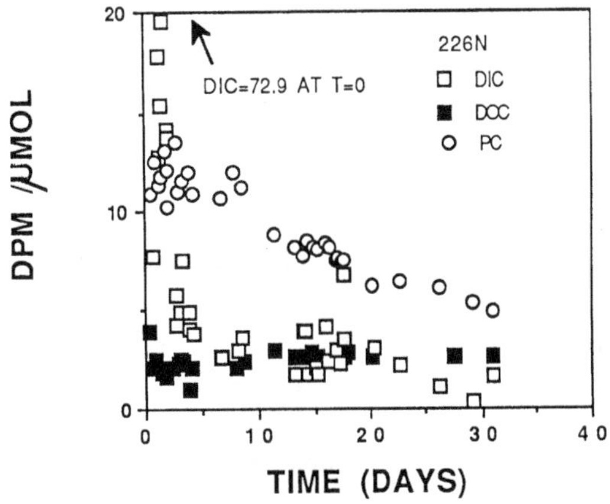


*Fig. 11.* Specific activity of carbon isotopes in Lake 226N, a eutrophic lake, after $^{14}C$ addition. Note that in contrast to the oligotrophic lakes (Fig. 9), DOC in Lake 226N reaches almost half the specific activity of PC, indicating high *in situ* DOC production or increased lability of DOC as the result of fertilization.

as evidenced by the fact that after May, 1979, the remaining DO$^{14}$C did not degrade further (Fig. 10). After this date, the loss of DO$^{14}$C occurred only by outflow.

### *Effects of eutrophication on DOC, DON and TDP*

Addition of $^{14}C$ to L226 revealed that phosphorus addition caused dramatic changes in the turnover of the DOC pool. In the south basin (L226S), which was fertilized only with nitrogen and carbon, DOC was as poorly labelled with $^{14}C$ as in other oligotrophic lakes of the area. By contrast, in L226N, which received phosphorus as well as nitrogen and carbon, the DOC pool became much more completely labelled with $^{14}C$ than in other lakes. A much higher proportion of the DOC was apparently of autochthonous origin, due to higher rates of algal growth (Fig. 11). However, the size of the DOC pool did not increase greatly after fertilization, so that increased turnover of refractory material must also have occurred.

The literature on sources of carbon for the lacustrine carbon cycle appears to be contradictory. Several investigators have concluded that the atmosphere is the most important source of carbon (for example Schindler *et al.*, 1972, 1977; Talling, 1976; Quay *et al.*, 1986), while others have concluded that allochthonous sources are more important (Wetzel, 1991; Wissmar *et al.*, 1977; Rau, 1978; Likens, 1985). The L226 experiment shows that fertilization with phosphorus is a mechanism that can explain the differences, unifying contradictions in earlier observations. In L226N, the carbon pool was apparently shifted from allochthonously dominated to autochthonously dominated by fertilization with phosphorus.

DOC concentrations in L227 increased dramatically during the years of fertilization. Oddly, only a slight similar trend is observable in L226N, as mentioned above. Despite the increased autochthonous DOC production indicated by more complete labelling with radiocarbon, DOC concentrations remained very similar to those in L226S (Fig. 12a). In some years, an increase in DOC was noticeable in L226N during the summer, but it was entirely metabolized during the ensuing winter.

In L227, the increase in DON was similar to DOC. A slight increase in DON in both L226N and L226S during the period of fertilization was also noticeable (Fig. 12b). TDP increased in both L227 and L226N, but not in L226S, where no phosphorus was added (Fig. 12c). In all three lakes, the ratio DOC:TC increased during the first several years of fertilization (Fig. 13a). The ratio levelled off in L227 after roughly a decade of fertilization. At first look, it seems peculiar that DOC would increase in L226S, but there is some long-term leakage between the basins through the curtain separating it from L226N. This would be expected to equalize concentrations of less biologically-active solutes like DOC much more than for highly reactive substances like inorganic phosphorus. In L227, the proportion DOC:TC increased from 50% to over 60% of the total carbon pool during the first 10–12 years, even though DIC and POC also increased as the result of fertilization. The only additional source of carbon to the lake was atmospheric input of $CO_2$ induced by fertilization with P and N (Schindler *et al.*, 1972). This result suggests that in similar geological settings, eutrophic lakes will have higher DOC than oligotrophic lakes, and that if all other factors are equal, DOC will constitute a higher proportion of the total carbon pool in eutrophic lakes. Very slight, non-significant increases were observed in ratios of DOC:DON and DOC:TDP in the three lakes, and in L226 the increase continues well after fertilization has ceased, perhaps indicating that it is caused by reduced water renewal rates (Fig. 13b–c). DON:TDP also increased slightly as the result of fertilization, even though the added fertilizers greatly exceeded natural inputs and had much lower N:P ratios (Fig. 13d). The nutrient ratios in these pools thus reflect the net balance between the direct effects of fertilization and compensatory in-lake processes, rather than the peculiarly phosphorus and nitrogen-rich ratios of the fertilizers applied to the lakes. DOC:TC in L226 also continued to increase after fertilization ceased

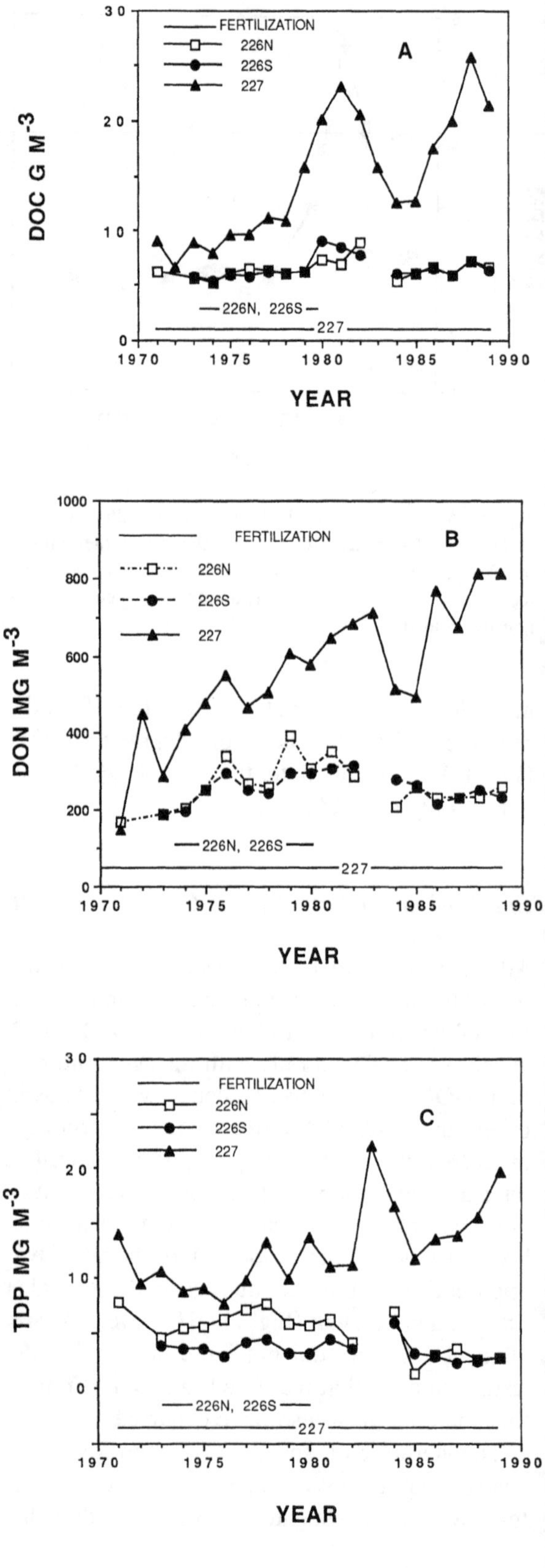


*Fig. 12.* Dissolved organic nutrients vs. time in the fertilized lakes. a – DOC. b – DON. c – TDP. Horizontal lines indicate the years that lakes were fertilized. Lake 227 was fertilized with P and N; Lake 226N with P, N and C and Lake 226S with N and C.

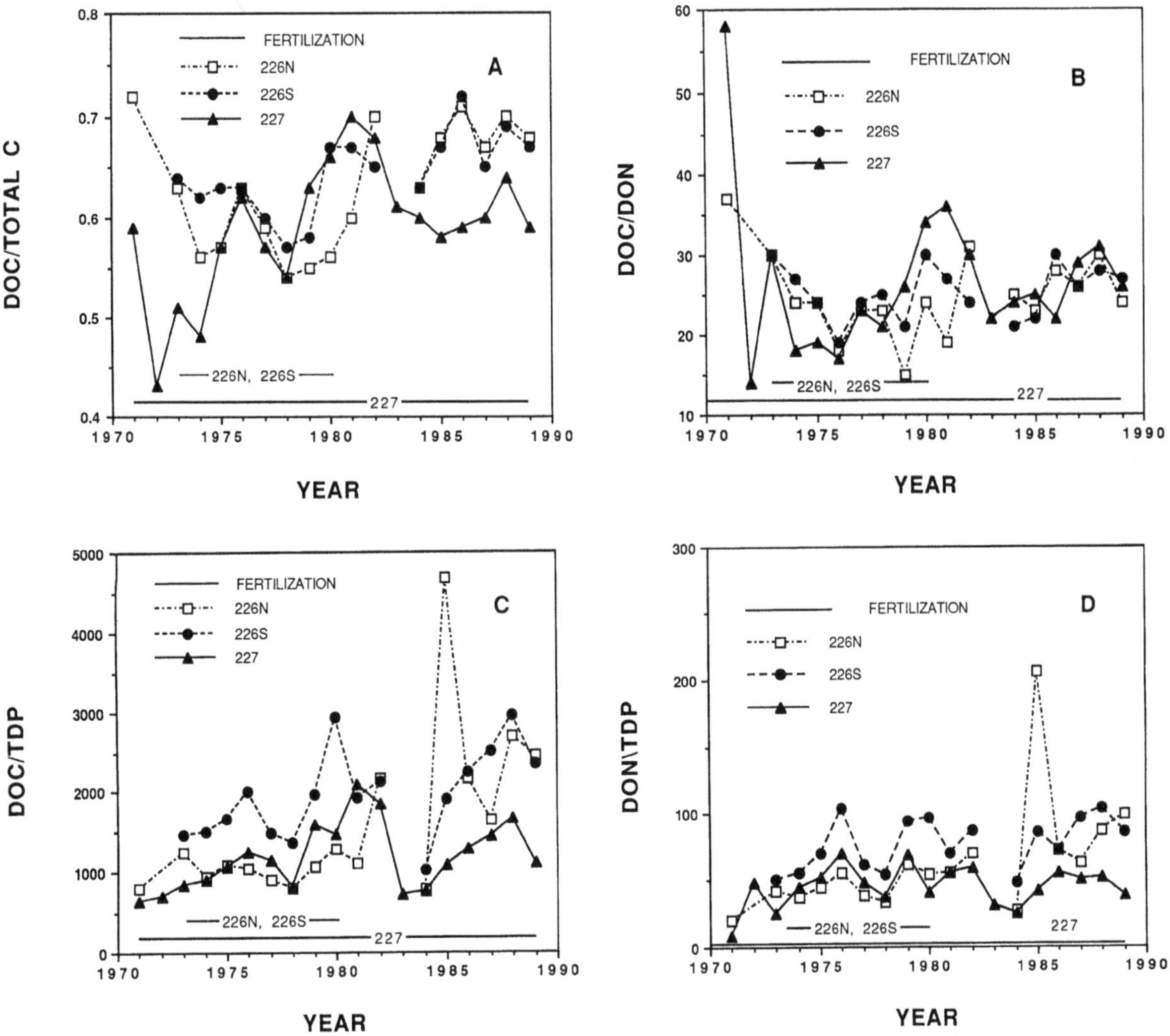


*Fig. 13.* Ratios of dissolved nutrient pools in the fertilized lakes. a – DOC:Total C. b – DOC:DON. c – DOC:TDP. d – DON:TDP.

(Fig. 13a), despite the expected opposite effects of lower water renewal during the years of record. It may reflect some long-term enrichment of sediments with organic carbon, which is slowly removed from the system. A forest fire also burned part of the catchment in 1979. Organic soils were not greatly disturbed, and a higher yield of DOC may have resulted.

The increases in DON accompanying fertilization may have resulted from increased fixation of atmospheric nitrogen as bluegreen algal blooms became dominant on both lakes, as well as from biological conversion of the inorganic nitrogen added as fertilizer. Some increased return of nitrogen from sediments may also have occurred (Schindler *et al.*, 1987).

In contrast to DOC, DON in L226 declined after fertilization ceased, suggesting that most of the DON had been autochthonous in origin, while the DOC was allochthonous. The ratio remained highly variable in L227, for reasons that are not clear. No significant long-term trends were evident.

DON:TDP trends in eutrophied lakes were

similar to those in natural and acidified lakes, indicating that low N:P in fertilizer had little effect on the nutrient ratios in DOM.

### *Effects of acidification on DOC, DON and TDP*

Previous studies have shown that DOC or color decreases and transparency increases as the result of acidification (Yan, 1983; Effler *et al.*, 1985). In lakes acidified by precipitation, terrestrial catchments are also acidified, causing increased exports of aluminum, calcium and iron and perhaps other metals to lakes and streams (Wright *et al.*, 1988; Dillon *et al.*, 1988; Cronan, 1980; van Breeman *et al.*, 1984; Nilsson, 1985). It has been hypothesized that the increased input of aluminum is responsible for the flocculation and removal of DOC, and thus the clarity of acidic lakes (Effler *et al.*, *op. cit.*). At pH > 4, DOC may also co-precipitate with $Fe(OH)_3$.

An interesting test of the aluminum hypothesis is provided by our experimentally acidified lakes, where acid has been applied only to the lakes. Because catchments were not acidified, only slight increases in dissolved aluminum, iron and manganese have occurred, resulting from small net releases of metals from lake sediments (Schindler *et al.*, 1980).

The only whole-lake experiment (Lake 114) where aluminum and hydrogen ion were both increased, by adding pulses of alum, showed the strong influence of aluminum on DOC concentrations. A single addition of aluminum caused a two-fold decline in DOC concentrations.

Acidification had little effect on the DOC concentration in L223, which was acidified to pH values of 5.0. In fact, DOC increased slightly in the early years of acidification, then declined dramatically in 1983, the third year at 5.0. Concen-

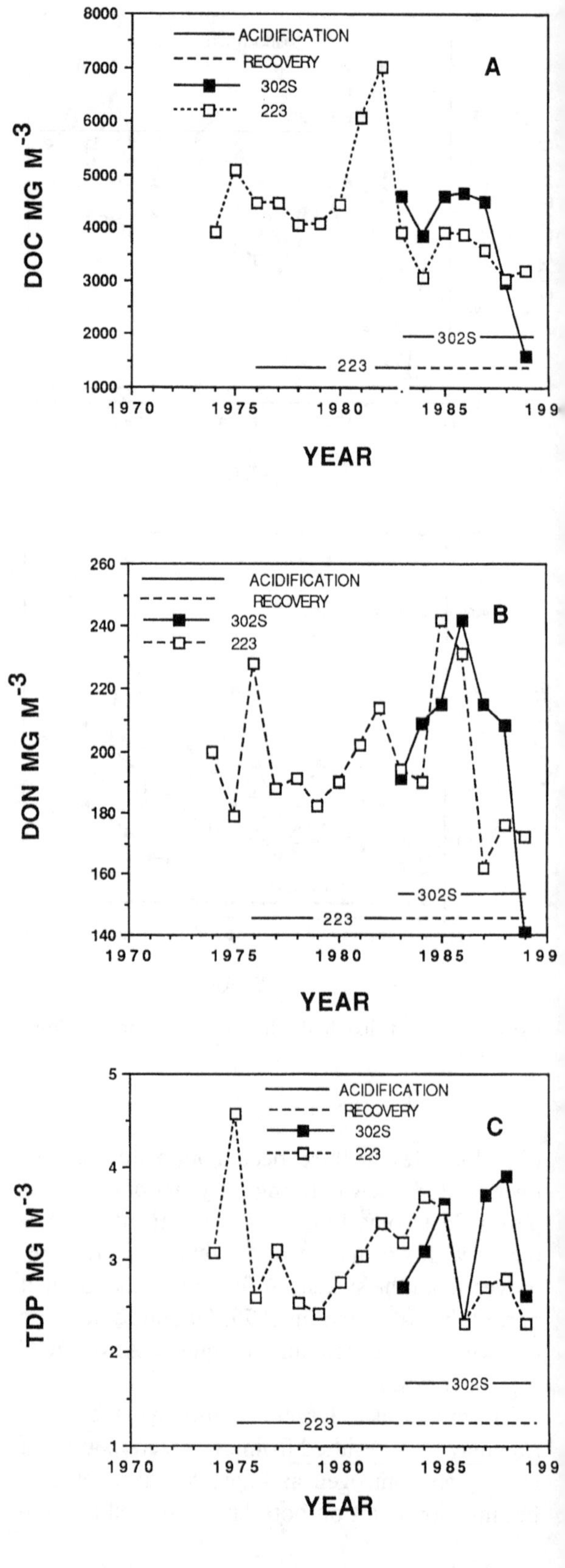


*Fig. 14.* Concentrations of dissolved organic nutrients vs. time in acidified lakes. a – DOC. b – DON. c – TDP. Horizontal solid lines indicate periods when the acidity of lakes was increased or held stable at low pH. Dashed lines indicate the period when the pH of Lake 223 was being slowly increased.

trations remained stable and low for several years during the lake's recovery, even though pH values increased to 5.8 (Fig. 14a). DOC in L302S remained stable at pH values of 5 or greater, then decreased rapidly once the pH was below 5. The decrease in DOC was accompanied by a considerable increase in lake transparency (E. Fee & E. Schindler, pers. comm.). Concentrations of aluminum, iron and manganese were only a few micrograms per liter in surface water (D. W. Schindler & M. A. Turner, unpubl.). The lower DOC and higher transparency of experimentally-acidified lakes do not therefore seem to be the result of high concentrations of metals.

Overall, DOC in L302S was inversely related to $H^+$ concentration (Fig. 15). There are reasons for expecting DOC and organic staining to be removed from water when a pH threshold is passed. If DOC is colloidal and removal is by coagulation, pH will affect the surface charge of colloids. If DOC functional groups have $pK_a$'s that are typically lower than ambient pH (hence the negative charge), acid titration will have little effect until pH approaches $pK_a$, at which point surface charges change rapidly with increasing acidity as functional groups are protonated. The change in the proportion of ionized functional groups is proportional to the intensity of electrostatic repulsion among DOC colloids, so that colliding colloids are more likely to coagulate at low pH. The acidification of L223 and L302S also revealed an apparent loss of charge per unit of DOC which normally has a net anionic charge at neutral pH.

The decline in DOC concentration caused by acidification has also been observed in our experimentally-acidified wetland, where monthly doses of sulfuric and nitric acid are applied via an irrigation system (Bayley *et al.*, 1987). The hydrogen ion content of bog pools is nearly doubled in the course of one day when irrigation is in progress. Immediate DOC declines of up to 20% have been observed (Fig. 16). As the pH recovers to pre-acidification values over several days, DOC concentrations also recover. In bog pools, aluminum concentrations are about ten-fold higher than in lake water, and some sort of flocculation reaction may be involved. Decreasing pH also increases the lipophilic nature of humic substances (De Haan, 1991), which may partly explain the decline in DOC. If these results are typical, part of the observed increase in transparency of lakes acidified by acid precipitation may also be caused by retention of DOC in acidified catchments.

DON in L223 increased during acidification,

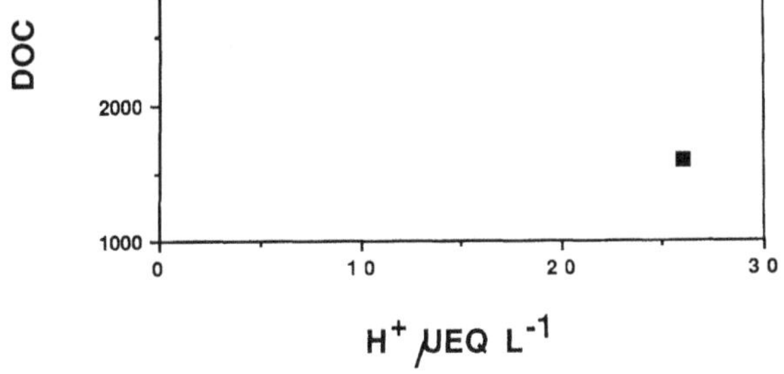


*Fig. 15.* Annual average DOC concentration vs hydrogen ion in Lake 302S.

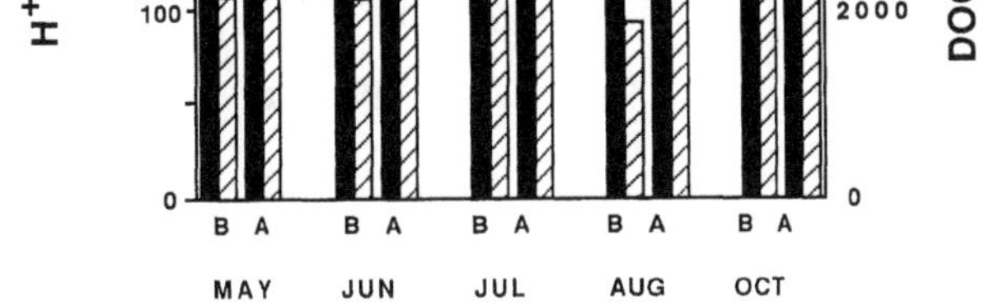


*Fig. 16.* DOC and hydrogen ion concentrations in the bog pool of the experimentally-acidified wetland, the day before (B) and the day after (A) experimental spraying with sulfuric and nitric acids. Applications were made monthly, at a rate sufficient to lower ambient rainfall from pH 5 to pH 4.

parallelling ammonium increases observed after acidification had caused nitrification to cease (Rudd *et al.*, 1988). As the pH increased during recovery, DON declined rapidly (Fig. 14b). DON in L302S also increased as the lake acidified, but it declined very suddenly in 1989, after two years at pH 4.5, similar to the decline in DOC. The sudden decline may be related to further disruptions of the nitrogen cycle, although the parallel to L223 suggests a common cause. Further observations will be necessary to resolve the possibilities. Overall, the decline in DON in L302S was not as great as that observed for DOC. There was no indication of a major change in TDP concentrations in either lake as the result of acidification or recovery (Fig. 14c).

As the result of the greater decline in DOC than in DON or TDP in L302S, the ratios DOC:TC, DOC:DON and DOC:TDP declined. DON:TDP showed no apparent trend (Fig. 17a–d).

*Interactions of DOC and the cycles of other elements*

Perturbations that affect DOC such as those discussed above can directly affect the cycles of many other elements. Changes in DOC are expected to

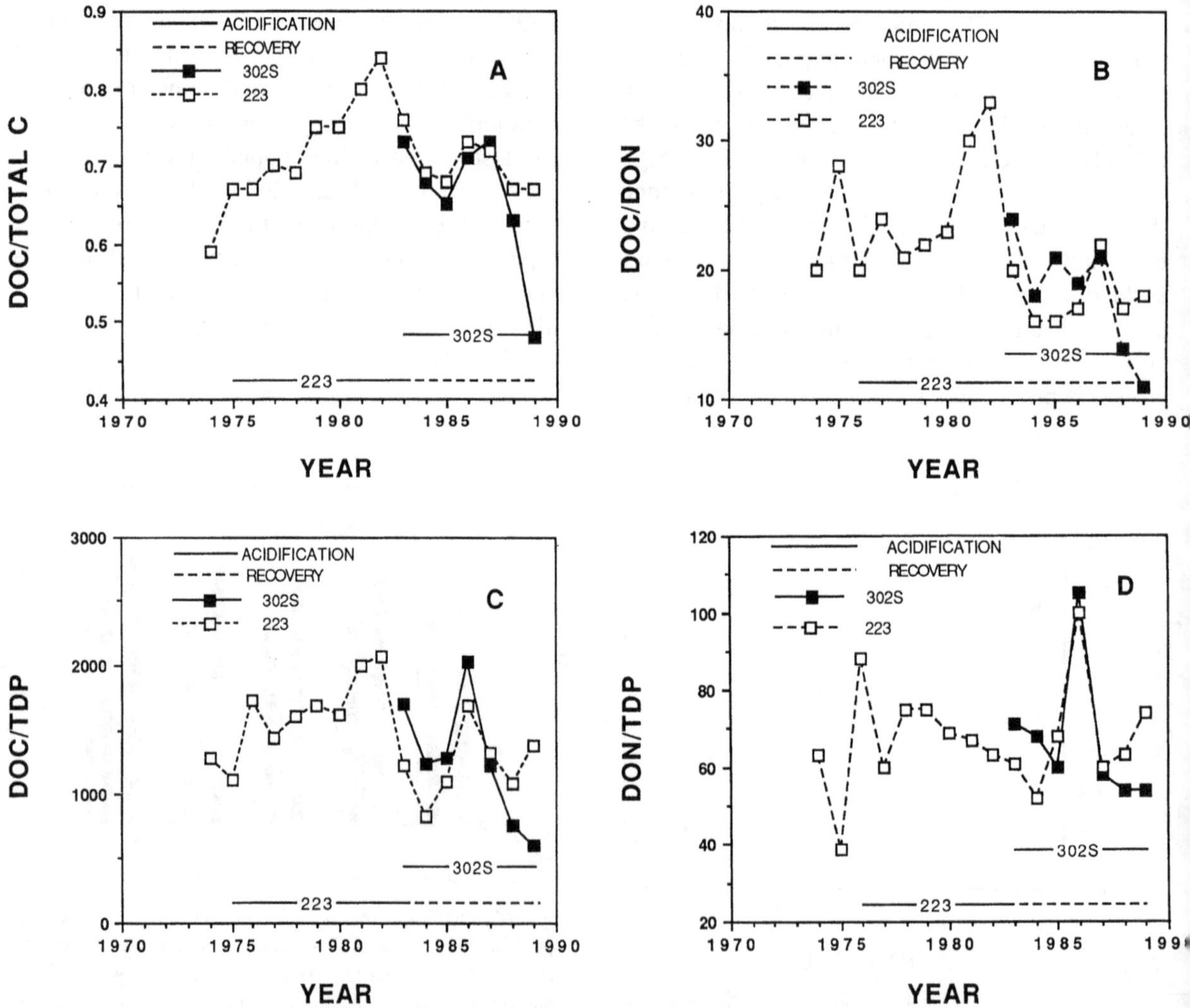


*Fig. 17.* Ratios of dissolved nutrient pools in the acidified lakes. a – DOC:Total C. b – DOC:DON. c – DOC:TDP. d – DON:TDP.

affect the cycles of many metals (Sholkovitz & Copland, 1982: Nilsson, 1985; Tipping & Cooke, 1982; Tipping & Ohnstad, 1984; Effler *et al.*, 1985). Effler *et al.* (*op. cit.*) found that DOC losses from acidic Dart's Lake in the Adirondacks were correlated with losses of aluminum. They hypothesized that coagulation or adsorption of DOC by aluminum was responsible for DOC removal. Also, aluminum bound to DOC is much less toxic to organisms than that in ionic form. As in the unperturbed ELA lakes, most of the DOC in Dart's Lake was allochthonous in origin. At ELA, the sedimentation efficiency of iron is inversely proportional to the concentration of DOC in lakes (Curtis, 1991). In general, the sedimentation of iron increases more rapidly than the increase in DOC, and DOC removal increases with increasing DOC concentration, as might be expected from a reaction that involves flocculation. DOC also appears to affect the concentrations of methyl mercury, by decreasing the rate of net methylation (Miskimmin *et al.*, 1991).

Several studies have shown that DOC causes changes in the distribution of metals between water and sediments. Curtis (1991) found that the distribution coefficient for iron between water and sediment ($K_d$) declined with increasing DOC. Similarly, Wahlgren & Orlandini (1981) found that the distribution coefficients ($K_d$'s) for thorium, plutonium and uranium were negatively related to DOC concentrations, in lakes at ELA and elsewhere. There was also a positive correlation between the concentration of plutonium (IV) in water and DOC. The $K_d$'s and concentrations were uncorrelated with other parameters tested, including mean depth, flushing time, pH, suspended solids, conductivity and alkalinity. Brunskill & Wilkinson (1987) found that the concentrations of radium, thorium, uranium, lead and polonium isotopes dissolved in stream water at ELA were proportional to DOC concentrations. Herczeg *et al.* (1985) showed that DOC altered the response of pH electrodes, causing errors of several tenths of a pH unit. The degree of response was correlated with DOC concentration, as shown by addition and removal of DOC from waters of lakes and streams at ELA. Also, Gran alkalinity titrations were consistently greater than $HCO_3$-calculated from DIC, pH and $pCO_2$. The discrepancy was found to be due to the binding of protons by weak organic acids, that are a part of DOC. Dillon *et al.* (pers. comm.) found that much of the *in situ* buffering capacity of lakes in the Precambrian Shield of Ontario was supplied by DOC. The interference of DOC with colorimetric sulfate analysis is well known (Kerekes *et al.*, 1984). The unknown charge density of DOC also makes it an obstacle to performing simple charge-balance checks on water chemistry, particularly in DOC-rich bog waters.

### *Effects of DOC on biological productivity*

Many studies at ELA and elsewhere indicate that DOC may affect the productivity of phytoplankton. Several mechanisms seem to be involved. In experiments where humic materials alone were added, Sakamoto (1971) found that phytoplankton production was inhibited. The inhibition was overcome by the addition of a trace element mixture. Sakamoto also found that addition of chelators + trace metal and nutrient mixtures had a greater effect in clear than in humic lakes. Similar inhibitory effects of organic substances on production were noted by Jackson & Hecky (1980) in lakes and reservoirs of Manitoba, and by Arvola (1984) in lakes of southern Finland. Guildford *et al.* (1987) found that addition of *Sphagnum* to limnocorrals in a northern Manitoba reservoir caused an initial surge in productivity, which was caused by enrichment with nitrogen and phosphorus. The increase was followed by a decline in productivity related to DOC, as the above authors had observed. Bioassays indicated that the depression was due to the binding of iron or some other metal by DOC. This conclusion is similar to Sakamoto's (*op. cit.*). Altogether, these conclusions are consistent with Janus & Vollenweider's (*op. cit.*) observation that colored lakes supported less phytoplankton. Weilenmann *et al.* (*op. cit.*) suggest that DOC entering lakes with wastewater effluents may prolong algal blooms by stabilizing nutrient elements in apparent solution. However, Wetzel (1991) & Münster *et al.* (1991) present evidence that DOC can interfere with alkaline phosphatase and other extracellular enzymes, which would be expected to cause results indistinguishable from those of the above authors. Investigations are needed to resolve these possibilities.

As a result of decreasing DOC, the depth of the euphotic zone increases as a lake acidifies, as we discussed above. The small increases in biomass and production in metalimnetic and hypolimnetic regions can cause areal production and biomass to increase slightly (Schindler, 1980; Findlay, 1984; Shearer & DeBruyn, 1986). The effects of DOC on photosynthesis are discussed in more detail by Jones (1991), and clearly deserve further study.

Rasmussen & Kalff (1987) also found that higher DOC was correlated with lower zoobenthos densities, which they attributed to lower hypolimnetic oxygen in humic lakes. No studies to date have examined either the response of oxygen or zoobenthos in a single lake following changes in DOC. This also seems an important area for further study.

In addition, increased light penetration appears to cause warming at depth, destabilizing thermal stratification and increasing thermocline depth (Effler *et al.*, *op. cit.*; Schindler *et al.*, 1990).

In summary, long-term observations and experiments show that DOC controls a large number of important physical, chemical and biological processes in lakes. DOC concentrations are highly susceptible to changes induced by man both directly to lakes and to their catchments, including fire, clearcutting, wetland drainage, acidic precipitation, eutrophication and climate change. A better understanding of the role of DOC in lacustrine processes is necessary to predict the effects of global change on freshwater ecosystems.

## Acknowledgements

Research at the Experimental Lakes Area was sponsored by the Canadian Department of Fisheries and Oceans. Radiocarbon experiments were funded in part by the U.S. National Science Foundation, via grants to W. S. Broecker of Lamont-Doherty Geological Observatory. The mesocosm experiments with iron and the period of writing and data analysis were supported by National Research Council of Canada operating grants to Schindler, Bayley, Curtis and Kelly and a World Wildlife Fund grant to Bayley.

## References

Ali, W., C. R. O'Melia & J. K. Edzwald, 1984. Colloidal stability of particles in lakes: Measurement and significance. Wat. Sci. Tech. 17: 701–712.

Arvola, L., 1984. Vertical distribution of primary production and phytoplankton in two small lakes with different humic concentration in southern Finland. Holarct. Ecol. 7: 390–398.

Bayley, S. E., D. H. Vitt, R. W. Newbury, K. G. Beaty, R. Behr & C. Miller, 1987. Experimental acidification of a *Sphagnum*-dominated peatland: first year results. Can. J. Fish. aquat. Sci. 44 (Suppl. 1): 194–205.

Bayley, S. E., D. W. Schindler, K. G. Beaty, B. R. Parker & M. P. Stainton, 1992. Effects of multiple fires on nutrient yields from streams draining boreal forest and fen watersheds: nitrogen and phosphorus. Can. J. Fish. Aquat. Sci. (in press).

Birge, E. A. & C. Juday, 1934. Particulate and dissolved organic matter in inland lakes. Ecol. Monogr. 4: 440–474.

Bower, P. M., 1981. Addition of radiocarbon to the mixed layers of two small lakes: Primary production, gas exchange, sedimentation, and carbon budget. Ph.D. thesis, Columbia University, New York. 238 pp.

Bower, P. M. & D. McCorkle, 1980. Limnocorral radiocarbon spike: Gas exchange, photosynthetic uptake and carbon budget. Can. J. Fish. aquat. Sci. 37: 464–471.

Bower, P. M., C. A. Kelly, E. J. Fee, J. A. Shearer, D. R. De Clercq & D. W. Schindler, 1987. Simultaneous measurement of primary production of whole-lake and bottle radiocarbon additions. Limnol. Oceanogr. 32: 299–312.

Brunskill, G. J. & D. W. Schindler, 1971. Geography and bathymetry of selected lake basins, Experimental Lakes Area, northwestern Ontario. J. Fish. Res. Bd Can. 28: 139–155.

Brunskill, G. J. & P. Wilkinson, 1987. Annual supply of 238U, 234U, 230Th, 226Ra, 210Pb, 210Po, and 232Th to Lake 239 (Experimental Lake Area, Ontario) from terrestrial and atmospheric sources. Can. J. Fish. aquat. Sci. 44 (Suppl. 1): 215–230.

Cooper, W. B. & D. R. S. Lean, 1989. Hydrogen peroxide production in a north-temperate lake. Envir. Sci. Technol. 23: 1425–1428.

Cronan, C. S., 1980. Controls on leaching from forest floor microcosms. Plant & Soil 56: 301–322.

Curtis, P. J., 1991. Biogeochemistry of iron in small Precambrian Shield lakes at the Experimental Lakes Area, northwestern Ontario. Ph.D. Thesis, University of Manitoba. 284 pp.

Davies-Colley, R. J. & W. N. Vant, 1987. Absorption of light by yellow substance in freshwater lakes. Limnol. Oceanogr. 32: 416–425.

De Haan, H., 1991. Impacts of environmental changes on the biogeochemistry of aquatic humic substances. Hydrobiologia (this volume).

Dillon, P. J., H. E. Evans & P. J. Scholer, 1988. The effects of acidification on metal budgets of lakes and catchments. Biogeochem. 5: 201–220.

Drever, J., 1988. The geochemistry of natural waters. Prentice-Hall Inc., New Jersey. 437 pp.

Effler, S. W., G. C. Schafran & C. T. Driscoll, 1985. Partitioning light attenuation in an acidic lake. Can. J. Fish. aquat. Sci. 42: 1707–1711.

Engstrom, D. R., 1987. Influence of vegetation and hydrology on the humus budgets of Labrador lakes. Can. J. Fish. aquat. Sci. 44: 1306–1314.

Eshleman, K. N. & H. F. Hemond, 1985. The role of organic acids in the acid-base status of surface waters at Bickford Watershed, Massachusetts. Wat. Resour. Res. 21: 1503–1510.

Findlay, D. L., 1984. Effects of phytoplankton biomass, succession and composition in Lake 223 as a result of lowering pH levels from 5.6 to 5.2. Data from 1980 to 1982. Can. MS Rep. Fish. aquat. Sci. 1761:iv + 16p.

Findlay, D. L. & S. E. M. Kasian, 1987. Phytoplankton community response to nutrient addition in Lake 226, Experimental Lakes Area, northwestern Ontario. Can. J. Fish. aquat. Sci. 44 (suppl. 1): 35–46.

Glover, G. M. & A. H. Webb, 1979. Weak and strong acids in the surface waters of the Tovdal region of southern Norway. Wat. Res. 13: 781–783.

Gorham, E., J. K. Underwood, F. B. Martin & J. G. Ogden III, 1986. Natural and anthropogenic causes of lake acidification in Nova Scotia. Nature 324: 451–453.

Guildford, S. J., F. P. Healey & R. E. Hecky, 1987. Depression of primary production by humic matter and suspended sediment in limnocorral experiments at Southern Indian Lake, northern Manitoba. Can. J. Fish. aquat. Sci. 44: 1408–1417.

Heikkinen, K., 1989. Organic carbon transport in an undisturbed boreal humic river in northern Finland. Arch. Hydrobiol. 117: 1–19.

Herczeg, A. L., 1985. Carbon dioxide equilibria and 13C studies in some soft water lakes. Ph.D. thesis, Columbia University, New York,. 260 pp.

Herczeg, A. L., W. S. Broecker, R. F. Anderson & S. L. Schiff, 1985. A new method for monitoring temporal trends in the acidity of fresh waters. Nature 315: 133–135.

Hesslein, R. H., W. S. Broecker, P. D. Quay & D. W. Schindler, 1980. Whole lake radiocarbon experiment in an oligotrophic lake at the Experimental Lakes Area, northwestern Ontario. Can. J. Fish. aquat. Sci. 37: 454–463.

Hobbie, J. E. & G. E. Likens, 1983. Output of phosphorus, dissolved organic carbon, and fine particulate carbon from Hubbard Brook Watersheds. Limnol. Oceanogr. 18: 734–742.

Hutchinson, G. E., 1957. A Treatise on Limnology. Vol. 1. John Wiley & Sons, New York. 1015 pp.

Jackson, T. A. & R. E. Hecky, 1980. Depression of primary production by humic matter in lake and reservoir waters of the boreal forest zone. Can. J. Fish. aquat. Sci. 37: 2300–2317.

Janus, L. L. & R. A. Vollenweider, 1981. The OECD cooperative report on eutrophication: Canadian contribution. Summ. Rep. Sci. Ser. 132.

Jones, R. I., 1991. The influence of humic substances on lacustrine planktonic food chains. Hydrobiologia (this volume).

Kerekes, J., G. Howell & T. Pollock, 1984. Problems associated with sulfate determination in colored, humic waters in Kejimkujik National Park, Nova Scotia (Canada). Verh. int. Ver. Limnol. 22: 811–817.

Kerekes, J., S. Beauchamp, R. Tordon, C. Tremblay & T. Pollock, 1986. Organic vs. anthropogenic acidity in tributaries of the Kejimkujik watersheds in western Nova Scotia. Wat. Air Soil Pollut. 31: 207–214.

Kortelainen, P. & J. Mannio, 1987. The contribution of acidic organic anions to the ion balance of lake waters, pp. 229–238. In: Acidification and Water Pathways. UNESCO conference, Bolkesjo, Norway, 4–5 May 1987. Vol. II.

LaZerte, B. D. & P. J. Dillon, 1984. Relative importance of anthrophogenic versus natural sources of acidity in lakes and streams of central Ontario. Can. J. Fish. aquat. Sci. 42: 1664–1677.

Levine, S. N. & D. W. Schindler, 1980. Radiochemical analysis of orthophosphate concentrations and seasonal changes in the flux of orthophosphate to seston in two Canadian Shield lakes. Can. J. Fish. aquat. Sci. 37: 479–487.

Levine, S. N., M. P. Stainton & D. W.Schindler, 1986. A radiotracer study of phosphorus cycling in a eutrophic Canadian Shield lake, Lake 227, northwestern Ontario. Can. J. Fish. aquat. Sci. 43: 366–378.

Lewis, W. M. Jr. & M. C. Grant, 1979. Relationships between stream discharge and dissolved substances from a Colorado mountain watershed. Soil Sci. 128: 353–363.

Likens, G. E., 1985. An ecosystem approach to aquatic ecology: Mirror Lake and its environment. Springer-Verlag, New York. 516 pp.

Lorius, C., J. Jouzel, D. Raynaud, J. Hansen & H. Le Traut, 1990. The ice-core record: climate sensitivity and future greenhouse warming. Nature 347: 139–145.

McDowell, W. H. & G. E. Likens, 1988. Origin, composition and flux of dissolved organic carbon in the Hubbard Brook Valley. Ecol. Monogr. 58: 177–195.

McKnight, D., E. M. Thurman, R. Wershaw & H. Hemond, 1985. Biogeochemistry of aquatic humic substances in Thoreau's Bog, Concord, Massachusetts. Ecology 66: 1339–1352.

Meyer, J. L. & C. M. Tate, 1983. The effects of watershed disturbance on the dissolved organic carbon dynamics of a stream. Ecology 64: 33–44.

Meyer, J. L., C. M. Tate, R. T. Edwards & M. T. Crocker, 1987. The trophic significance of dissolved organic carbon in streams, pp. 269–278. In W. T. Swank and D. A. Crossley, Jr. (eds). Forest hydrology and ecology at Coweeta. Springer-Verlag, New York.

Miskimmin, B. M., J. W. M. Rudd & C. A. Kelly, 1991. The influence of dissolved organic carbon, pH, and microbial respiration on mercury methylation and demethylation in lake water. Can. J. Fish. aquat. Sci. (in press).

Mulholland, P. J. & E. J. Kuenzler, 1979. Organic carbon export from upland and forested wetland watersheds. Limnol. Oceanogr. 24: 960–966.

Münster, U., P. Einiö, J. Nurminen & J. Overbeck, 1991. Extracellular enzymes in a polyhumic lake: important regulators in detritus processing. Hydrobiologia (this volume).

Naiman, R. J., 1982. Characteristics of sediment and organic carbon export from pristine boreal forest watersheds. Can. J. Fish. aquat. Sci. 39: 1699–1718.

Nalewajko, C. & D. R. S. Lean, 1972. Growth and excretion in planktonic algae and bacteria. J. Phycol. 8: 361–366.

Nalewajko, C. & D. W. Schindler, 1976. Primary production, extracellular release, and heterotrophy in two lakes in the ELA, Northwestern Ontario. J. Fish Res. Bd. Can. 33: 219–226.

Nilsson, J. I., 1985. Budgets of aluminium species, iron and manganese in the Lake Gardsjon catchment in SW Sweden. Ecol. Bull. (Stockholm) 37: 120–137.

Pitkänen, H., 1986. Discharges of nutrients and organic matter to the Gulf of Bothnia by Finnish Rivers in 1968–1983, pp 72–83. In Kangas, P. and M. Forsskahl (eds). Proceedings of the 3rd Finnish seminar on the Gulf of Bothnia. Water Research Inst. Publ. no. 68, National Board of Waters and Environment, Helsinki.

Quay, P. D., S. Emerson, B. M. Quay & A. H. Devol, 1986. The carbon cycle for Lake Washington – a stable isotope study. Limnol. Oceanogr. 31: 596–611.

Rasmussen, J. B. & J. Kalff, 1987. Empirical models for zoobenthos in lakes. Can. J. Fish. aquat. Sci. 44: 990–1001.

Rasmussen, J. B., L. Godbout & M. Schallenberg, 1989. The humic content of lake water and its relationship to watershed and lake morphometry. Limnol. Oceanogr. 34: 1336–1343.

Rau, G., 1978. Carbon-13 depletion in a subalpine lake: Carbon flow implications. Science 201: 901–902.

Rudd, J. W. M. & C. D. Taylor, 1980. Methane cycling in aquatic environments. Adv. Aquat. Microbiol. 2: 77–150.

Rudd, J. W. M., C. A. Kelly, D. W. Schindler & M. A. Turner, 1988. Disruption of the nitrogen cycle in acidified lakes. Science 240: 1515–1517.

Rudd, J. W. M., C. A. Kelly, D. W. Schindler & M. A. Turner, 1990. A comparison of the acidification efficiencies of nitric and sulfuric acids by two whole-lake addition experiments. Limnol. Oceanogr. 35: 663–679.

Sakamoto, M., 1971. Chemical factors involved in the control of phytoplankton production in the Experimental Lakes Area, northwestern Ontario. J. Fish Res. Bd. Can. 28: 203–213.

Schiff, S. L., R. Aravena, S. E. Trumbore & P. J. Dillon, 1991. Dissolved organic carbon cycling in forested watersheds: A carbon isotope approach. Wat. Resour. Res. 26: 2949–2957.

Schindler, D. W., 1971a. An hypothesis to explain differences and similarities among lakes in the Experimental Lakes Area, northwestern Ontario. J. Fish Res. Bd Can. 28: 295–301.

Schindler, D. W., 1971b. Light, temperature and oxygen regimes of selected lakes in the Experimental Lakes Area (ELA), northwestern Ontario. J. Fish Bd Can. 28: 157–169.

Schindler, D. W., 1977. Evolution of phosphorus limitation in lakes. Science 195: 260–262.

Schindler, D. W., 1980. Experimental acidification of a whole lake: a test of the oligotrophication hypothesis, pp. 370–373. In D. Drablos and A. Tollan (eds). Proc. Int. Conf. Imp. Acid Precip., Sandefjord, Norway. SNSF Project, Oslo, Norway.

Schindler, D. W., 1988. Experimental studies of chemical stressors on whole lake ecosystems. Edgardo Baldi memorial lecture. Ver. int. Ver. Limnol. 23: 11–41.

Schindler, D. W., 1990. Experimental perturbations of whole lakes as tests of hypotheses concerning ecosystem structure and function. Oikos 57: 25–41.

Schindler, D. W., G. J. Brunskill, S. Emerson, W. S. Broecker & T.-H. Peng, 1972. Atmospheric carbon dioxide: Its role in maintaining phytoplankton standing crops. Science 177: 1192–1194.

Schindler, D. W., R. Wagemann, R. B. Cook, T. Ruszczynski & J. Prokopowich, 1980. Experimental acidification of Lake 223, Experimental Lakes Area: background data and the first three years of acidification. Can. J. Fish. aquat. Sci. 37: 342–354.

Schindler, D. W., R. H. Hesslein & M. A. Turner, 1987. Exchange of nutrients between sediments and water after 15 years of experimental eutrophication. Can. J. Fish. aquat. Sci. 44: (suppl. 1): 26–33.

Schindler, D. W., K. Beaty, E. J. Fee, D. R. Cruikshank, E. R. DeBruyn, D. L. Findlay, G. A. Linsey, J. A. Shearer, M. P. Stainton & M. A. Turner, 1990. Effects of climatic warming on the lakes of the lakes of the central boreal forest. Science 250: 967–970.

Schindler, D. W., T. M. Frost, K. H. Mills, P. S. S. Chang, I. J. Davies, D. L. Findlay, D. F. Malley, J. A. Shearer, M. A. Turner, P. J. Garrison, C. J. Watras, K. Webster, J. M. Gunn, P. L. Brezonik & W. A. Swenson, 1991. Comparisons between experimentally- and atmosphericallyacidified lakes during stress and recovery. Proc. Roy. Soc. Edinburgh 97B: 193–226.

Shearer, J. A. & E. R. DeBruyn, 1986. Phytoplankton productivity responses to direct addition of sulfuric and nitric acids to the waters of a double basin lake. Wat. Air Soil Pollut. 30: 695–702.

Shearer, J. A., E. J. Fee, E. R. DeBruyn & D. R. DeClercq, 1987. Phytoplankton productivity changes in a small, double-basin lake in response to termination of experimental fertilization. Can. J. Fish. aquat. Sci. 44 (Suppl. 1): 47–54.

Sholkowitz, E. R. & D. Copland, 1982. The chemistry of suspended matter in Esthwaite Water, a biologically productive lake with seasonally anoxic hypolimnion. Geochim. Cosmochim. Acta 46: 393–410.

Stainton, M. P., M. Capel & F. A. J. Armstrong, 1977. The chemical analysis of freshwater. 2nd. edition. Can. Fish, Mar. Serv. Misc. Spec. Publ. 25: 166 pp.

Stumm, W. & J. Morgan, 1981. Aquatic chemistry. J. Wiley & Sons, New York. 780 pp.

Talling, J. F., 1976. The depletion of carbon dioxide from water by phytoplankton. J. Ecol. 64: 79–122.

Tate, C. M. & J. L. Meyer, 1983. The influence of hydrologic conditions and successional state on dissolved organic carbon export from forested watersheds. Ecology 64: 25–32.

Tipping, E. & D. Cooke, 1982. The effects of adsorbed humic substances on the surface charge of goethite (alpha-FeOOH) in freshwaters. Geochim. Cosmochim. Acta 46: 75–80.

Tipping, E. & M. Ohnstad, 1984. Colloid stability of iron oxide particles from a freshwater lake. Nature 308: 266–268.

Thompson, B. M. & R. D. Hamilton, 1973. Heterotrophic utilization of sucrose in an artificially enriched lake. J. Fish Res. Bd Can. 30: 1547–1552.

Urban, N. R., S. J. Eisenreich & E. Gorham, 1987. Proton cycling, in bogs: Geographic variation in eastern North America, pp. 577–598. In T. C. Hutchinson & K. M. Meema (eds). The effects of air pollutants on forests, wetlands and agricultural ecosystems. Springer-Verlag, New York.

Urban, N. R., S. E. Bayley & S. J. Eisenreich, 1989. Export of dissolved organic carbon and acidity from peatlands. Wat. Resour. Res. 25: 1619–1628.

van Breeman, N., C. T. Driscoll & J. Mulder, 1984. Acidic deposition and internal proton sources in acidification of soils and waters. Nature 307: 599–604.

Vertucci, F. A. & G. E. Likens, 1989. Spectral reflectance and water quality of Adirondack mountain region lakes. Limnol. Oceanogr. 34: 1656–1672.

Wahlgren, M. A. & K. A. Orlandini, 1981. Comparison of the geochemical behavior of plutonium, thorium and uranium in selected North American lakes. In IAEA-SM-257/ International symposium on migration in the terrestrial environment of long-lived radionuclides from the nuclear fuel cycle. Knoxville, Tenn., USA, 27–31 July 1981.

Watras, C. J. & A. L. Baker, 1988. The spectral distribution of downwelling light in northern Wisconsin lakes. Arch. Hydrobiol. 112: 481–494.

Weilenmann, U., C. R. O'Melia & W. Stumm, 1989. Particle transport in lakes: Models and measurements. Limnol. Oceanogr. 34: 1–18.

Wetzel, R. G., 1991. Gradient dominated ecosystems: sources and regulatory functions of dissolved organic matter in freshwater ecosystems. Hydrobiologia (this volume).

Wetzel, R. G., P. H. Rich, M. C. Miller & H. L. Allen, 1972. Metabolism of dissolved and particulate organic matter in the sea. Deep Sea Res. 17: 19–27.

Wissmar, R. C., J. E. Richey & D. E. Spyridakis, 1977. The importance of allochthonous particulate carbon pathways in a subalpine lake. J. Fish Res. Bd Can. 34: 1410–1418.

Wright, R. F., E. Lotse & A. Semb, 1988. Reversibility of acidification shown by whole-catchment experiments. Nature 334: 670–675.

Yan, N. D., 1983. Effects of changes in pH on transparency and thermal regimes of Lohi Lake, near Sudbury, Ontario. Can. J. Fish. aquat. Sci. 40: 621–626.

*Hydrobiologia* **229**: 23–41, 1992.
*K. Salonen, T. Kairesalo & R.I. Jones (eds.),*
*Dissolved Organic Matter in Lacustrine Ecosystems: Energy Source and System Regulator.*




# Sources, concentrations and characteristics of organic matter in softwater lakes and streams of the Swedish forest region

Markus Meili
*Institute of Limnology, Uppsala University, Box 557, S-75122 Uppsala, Sweden*

*Key words:* TOC, COD, humic, colour, Secchi, phosphorus, carbon, permanganate

## Abstract

18 Swedish forest lakes covering a wide range of dystrophy were studied in order to quantify and characterize the organic matter in the water with respect to origin (allochthonous or autochthonous), physical state (particulate or dissolved) and phosphorus content. Samples were collected repeatedly during a two-year period with unusually variable hydrological conditions. Water from three different depths and from tributaries was analysed with standard monitoring methods, including water colour, Secchi disk transparency, total organic carbon (TOC), $COD_{Cr}$, $COD_{Mn}$, total phosphorus and molybdate reactive phosphorus. Interrelationships were used to compare different methods and to assess the concentration and composition of organic matter. It is estimated that in remote softwater lakes of the Swedish forest region, autochthonous carbon is typically $< 5\ g\ m^{-3}$. Most lakes in this region receive significant amounts of humic matter originating from coniferous forest soils or peatland in the catchment area. In most humic lakes with a water colour of $\geq 50\ g\ Pt\ m^{-3}$, more than half of the organic carbon in the surface water is of allochthonous origin, and in polyhumic lakes ($> 200\ g\ Pt\ m^{-3}$) the proportion can exceed 90%. Secchi depth readings were related similarly to organic matter from both sources and provided good estimates of TOC with a single optical measurement. Water colour was used to distinguish allochthonous and autochthonous matter. High concentrations of phosphorus were found in humic waters, most of it being molybdate reactive, and probably associated with humic matter rather than as dissolved free inorganic forms. $COD_{Mn}$ yielded only 25–60% of TOC and appears to include mainly truly dissolved substances of low molecular weight.

## Introduction

Organic matter in surface waters can be classified according to origin (allochthonous or autochthonous), physical state (particulate or dissolved) and nutrient content (carbon/phosphorus ratio). In lakes of the boreal forest zone (Bruenig, 1987), two main types of organic matter can be distinguished: organic matter from primary or secondary production in the water body, and humic matter originating from coniferous forest soils, peatland or marginal rooted vegetation. However, the origin and composition of organic matter in inland freshwaters is largely unknown. It is presently not possible to identify all organic compounds in lake water, but the description of gross properties may help in differentiating organic matter from different sources (Cole *et al.*, 1984).

Swedish surface waters are typically oligotrophic and stained to a varying degree by humic substances (SNV, 1986). The bulk of humic substances in brownwater lakes and streams is dissolved, and a minor proportion is colloidal or particulate (Pennanen, 1988; Olofsson, 1989).

The dissolved fraction is dominated by fulvic acids of relatively low molecular weight, which are less coloured than heavier humic substances (Pennanen, 1988; Cole *et al.*, 1984). Organic matter from autochthonous planktonic production occurs both in particulate and dissolved forms, with dissolved components prevailing (Birge & Juday, 1934; Wetzel, 1983), but contributes little to water colour (Birge & Juday, 1934; Åberg & Rodhe, 1942; Hutchinson, 1957).

Various methods have been applied to determine the concentration of organic matter. The greatest convenience is provided by optical methods. Water colour gives a simple estimate of humic content (Åberg & Rodhe, 1942; Hutchinson, 1957; Wetzel, 1983) and is widely used in monitoring programs (e.g. SNV, 1986). Organic matter from autochthonous production causes absorption and dispersion of light, which is easily assessed by measuring water transparency with a Secchi disk. Despite subjective disadvantages, Secchi depth provides a visual index of water clarity (Preisendorfer, 1986) and has been used to describe the trophic state of lakes (e.g. Carlson, 1977). However, Secchi depth readings are also influenced by inorganic and detrital particles (e.g. resuspended sediment), by the humic content of the water, and to a minor degree by the attenuation of light by the water itself (e.g. Åberg & Rodhe, 1942; Wetzel, 1983).

Chemical methods usually include an oxidation or combustion step, where either the consumption of an oxidant or the emerging carbon dioxide is quantified. Consumption measurements (chemical oxygen demand, COD) have traditionally been common in monitoring programs, as high sensitivity combustion techniques are more expensive and have become available only in recent decades. In this study, two chemical oxidation methods were used, one based on dichromate ($COD_{Cr}$) and one on permanganate ($COD_{Mn}$). In both methods, the chemical oxygen demand can be converted into concentrations of organic carbon, assuming constant stoichiometric relationships for organic matter and little disturbance by other oxidizable compounds (Wilander, 1988). In this paper, converted $COD_{Cr}$ values are referred to as CrOC, and converted $COD_{Mn}$ values as MnOC. In natural waters, the dichromate method provides a good estimate of total organic carbon, whereas the permanganate method is less efficient in decomposing organic matter and usually yields lower values (*op. cit.*, Hutchinson, 1957).

Primary production in freshwaters is usually limited by the access to nutrients, mainly phosphorus. Concentrations of total phosphorus are therefore frequently used to estimate autochthonous biomass. As a result of an efficient retrieval by organisms, most phosphorus in lakes is usually found in the particulate form (Wetzel, 1983). However, humic substances can efficiently bind dissolved inorganic phosphate in the presence of iron (Jones *et al.*, 1988). In brown waters, a considerable proportion of phosphorus may thus occur in the dissolved or colloidal phase.

The aims of this study are (1) to quantify and characterize the types of organic matter in different natural waters of the boreal forest zone, with special attention to the impact of humic matter and the distribution of different phosphorus forms; (2) to compare and evaluate different methods commonly used to measure the concentration of organic matter; and (3) to provide estimates of organic carbon based on simple analytical procedures, partly for use in low-budget monitoring programs, and partly for conversion and comparison of available data obtained with different methods.

## Methods

The study was conducted in 18 Swedish forest lakes in two areas, 8 in central Sweden (Delsbo, Hudiksvall, 61.6–7°N 16.6–8°E), and 10 about 500 km further south in southern Sweden (Aneboda, Växjö, 57.2–2°N 14.4–7°E). The lakes were selected to represent the typical range of softwater lakes in the Swedish forest region: water colour 20–200 g Pt $m^{-3}$, total phosphorus 5–25 mg $m^{-3}$, pH 5–7 (SNV, 1986; Meili *et al.*, 1991; Meili, 1991a). The morphometrical, hydrological and chemical characteristics of the study lakes are given in Table 1.

*Table 1.* Morphological, hydrological and chemical characteristics of the 18 Swedish forest lakes included in this study.
Location: C = central Sweden (62° N 17° E), S = southern Sweden (57° N 15° E)
Dynamic ratio = $\sqrt{Lakearea}$/Mean depth.
Res. time = theoretical hydrological residence time
Cultiv. = cultivated (deforested) land.

| Lake name | Code | Location | Morphology | | | | Hydrology | | | Catchment area | | | | Surface water chemistry 1985–87 | | |
|---|---|---|---|---|---|---|---|---|---|---|---|---|---|---|---|---|
| | | | Lake area ($km^2$) | Max. depth (m) | Mean depth (m) | Dyn. ratio | Lake volume ($10^6$ $m^3$) | Res. time (yr) | Drain area ($km^2$) | Lakes (%) | Forest (%) | Mire (%) | Cultiv. (%) | pH | Water colour (gPt $m^{-3}$) | Total P (mg $m^{-3}$) |
| Blacksåstjärn | BL | C | 0.12 | 6.8 | 2.4 | 140 | 0.3 | 0.5 | 2.2 | <1 | 76 | 15 | 9 | 6.0–7.1 | 75–130 | 11–14 |
| Bottentjärn | BO | C | 0.15 | 4.9 | 2.9 | 140 | 0.4 | 0.4 | 4.2 | <1 | 83 | 17 | <1 | 5.2–6.4 | 70–130 | 8–12 |
| Loppesjön | LO | C | 0.28 | 14.0 | 5.4 | 100 | 1.5 | 1.2 | 4.4 | 2 | 73 | <1 | 25 | 6.5–7.6 | 15–80 | 15–24 |
| Stensjön | ST | C | 0.59 | 8.5 | 4.3 | 180 | 2.6 | 2.5 | 3.6 | 1 | 74 | 25 | <1 | 5.8–6.6 | 40–80 | 7–13 |
| Löjesjön | LÖ | S | 0.31 | 2.5 | 1.5 | 380 | 0.5 | 0.8 | 2.1 | <1 | 74 | 19 | 7 | 6.2–6.6 | 80–120 | 18–23 |
| Skärhultsjön | SK | S | 0.36 | 14.0 | 3.8 | 160 | 1.1 | 0.8 | 6.0 | 1 | 76 | 8 | 16 | 5.2–6.1 | 80–140 | 10–21 |
| Gårdsjön | GÅ | S | 0.15 | 8.5 | 3.2 | 120 | 0.5 | 1.9 | 0.9 | <1 | 71 | 3 | 21 | 6.3–7.3 | 20–45 | 9–20 |
| Holmeshultasjön | HO | S | 0.69 | 16.0 | 4.5 | 180 | 3.1 | 1.8 | 6.0 | 7 | 70 | 3 | 20 | 6.4–7.3 | 20–40 | 9–15 |
| Sörsjön | SÖ | C | 0.25 | 7.2 | 3.2 | 150 | 0.8 | 0.2 | 17.9 | | | | | 5.5 | 160 | 12 |
| Vålåssjön | VÅ | C | 0.38 | 11.0 | 4.8 | 130 | 1.8 | 1.3 | 4.8 | | | | | 7.1 | 40 | 12 |
| Håknorrbosjön | HÅ | C | 0.50 | 17.0 | 4.5 | 160 | 2.2 | 0.9 | 8.5 | | | | | 6.7 | 55 | 10 |
| Lillbosjön | LI | C | 0.65 | 18.0 | 5.8 | 140 | 3.8 | 4.1 | 3.2 | | | | | 7.0 | 25 | 9 |
| Fräjen | FR | S | 0.21 | 1.2 | 0.8 | 600 | 0.2 | 0.1 | 5.7 | | | | | 5.6 | 280 | 26 |
| Vrången | VR | S | 0.55 | 6.4 | 1.2 | 600 | 0.7 | 0.2 | 15.8 | | | | | 5.2 | 165 | 28 |
| Hacksjön | HA | S | 0.28 | 7.0 | 1.4 | 380 | 0.4 | 0.8 | 1.7 | | | | | 6.5 | 35 | 11 |
| Lundasjön | LU | S | 0.10 | 8.3 | 4.0 | 80 | 0.4 | 2.8 | 0.5 | | | | | 6.5 | 10 | 10 |
| Burken | BU | S | 0.21 | 7.0 | 3.1 | 150 | 0.7 | 2.0 | 1.2 | | | | | 5.5 | 65 | 12 |
| Klintsjön | KL | S | 0.11 | 17.0 | 4.4 | 70 | 0.5 | 3.6 | 0.5 | | | | | 5.8 | 5 | 7 |

Water samples were collected between August, 1985 and May, 1987. In late summer 1985, all 18 lakes were sampled once (23–28 August in central Sweden, 18–25 September in southern Sweden). In 1986, samples were taken in eight lakes during four periods: a few days after ice break-up (5–6 May in southern Sweden, 16–21 May in central Sweden), in early summer (16–30 June), in late summer (29 July–11 August), and in autumn during or (in deep lakes) shortly before overturn (2–6 September in central Sweden, 13–17 September in southern Sweden). Complementary surface samples were taken in four lakes in spring 1987 (6–7 May). Both hydrology and water chemistry were unusually variable during the study period, especially in central Sweden, where the theoretical water residence time repeatedly dropped from several months to a few days in many lakes (cf. Meili *et al.*, 1991).

Lake water samples were collected with a Ruttner sampler at three depths at the deepest point of each lake. After recording the temperature profile, epilimnion samples were taken at 0.5 m depth to represent the most productive layer of the water body, metalimnion samples were taken in the lower part of the thermocline where effects of mixing, production, decomposition and accumulation of organic matter presumably are least pronounced, and hypolimnion samples were taken 0.5 m above the sediment surface to obtain water with decomposed organic matter. Stream water was collected where possible once in late summer 1985 from a major inflow close to the lakes. Samples were kept cool, and subsamples were frozen within 2–12 hours of sampling for storage until analysis. In 1985, a portion of the samples was filtered before freezing with Whatman GF/C glassfibre filters (apparent pore size $\approx 1.2\ \mu m$).

Water transparency (Secchi depth) was recorded with a Secchi disk of 0.25 m diameter in the shadow of a boat to minimize disturbances caused by light reflections from the water surface. Water colour was measured within 24 h in untreated water using a Hellige colour comparator and expressed as g Pt $m^{-3}$ (= Pt units). Particle concentrations were usually low, and no significant colour difference was observed between unfiltered and filtered water. Total organic carbon (TOC) was measured after combustion in a Shimadzu TOC500 carbon analyser. Chemical oxygen demand ($COD_{Cr}$ and $COD_{Mn}$) was analysed from frozen samples according to standard methods (Ahlgren & Ahlgren, 1978; cf. Wilander, 1988). For the determination of $COD_{Cr}$, samples were autoclaved for 1 hour (120 °C) in 4 mN $K_2Cr_2O_7$, 0.08 N $AgNO_3$ and 50% $H_2SO_4$. For $COD_{Mn}$, samples were treated with 1.6 mN $KMnO_4$ in 1% $H_2SO_4$ in a 100 °C water bath for exactly 20 min. Care was taken to keep digestion time and temperature constant, and to avoid effects of oxidant exhaustion by diluting samples when required. The samples were then titrated with Fe(II)ammoniumsulphate (phenanthroline as indicator) and thiosulphate (iodide/starch as indicator), respectively. COD values were converted into concentrations of organic carbon (CrOC and MnOC) assuming constant stoichiometry for organic matter ($O_2/C = 32/12$, cf. Wilander, 1988). Phosphorus concentrations were determined according to Murphy & Riley (1962). Molybdate-reactive phosphorus (MRP) was determined in filtered samples, whereas total phosphorus (TP) was obtained by analysis of MRP after oxidation of unfiltered samples in 1% peroxodisulphate ($K_2S_2O_8$) in an autoclave (120 °C) for 1 hour (Menzel & Corwin, 1965).

Mathematical relationships were generated partly by means of least square regression, partly by graphically or mathematically fitting the parameters of conceptually or mathematically suitable functions with a microcomputer. Parameters were rounded to the lowest feasible number of significant digits in order to generate simple, appealing and useful conversion formulas with a high predictive power.

## Results and discussion

In surficial lake water, total organic carbon (TOC) ranged from 3 to 28 g $m^{-3}$, water colour from 5 to 280 g Pt $m^{-3}$, total phosphorus from 7 to 24 mg $m^{-3}$, and Secchi depth from 0.4 to 5.5 m

*Table 2.* Ranges of chemical and optical water characteristics at different sites in 18 Swedish forest lakes 1985–1986.

| | Inflow | Epilimnion | Metalimnion | Hypolimnion |
|---|---|---|---|---|
| pH | 4.1–6.6 | 5.2–7.6 | 5.1–7.0 | 5.0–6.9 |
| Alkalinity [meq $m^{-3}$] | 0–220 | 0–200 | 0–280 | 0–850 |
| Conductivity [mS $m^{-1}$] | 1.8–17.6 | 1.9–8.5 | 1.9–9.0 | 2.3–11.8 |
| Oxygen saturation [%] | – | 72–134 | 2–120 | 0–114 |
| Water colour [g Pt $m^{-3}$] | 130–660 | 5–280 | 10–190 | 25–500 |
| Secchi depth [m] | – | 0.4–5.5 | – | – |
| Total organic carbon [g $m^{-3}$] | 13–55 | 3–28 | 3–19 | 3–35 |
| Labile organic carbon (MnOC) [g $m^{-3}$] | 6–30 | 1–12 | 1–9 | 1–16 |
| Total phosphorus [mg $m^{-3}$] | 9–35 | 7–34 | 7–42 | 7–107 |
| Molybdate-reactive P [mg $m^{-3}$] | 3–32 | 0.4–11 | 0.4–11 | 0.7–26 |

(Tables 1 and 2). Metalimnetic water was usually quite similar to surface water. In contrast, samples from the deepest strata of the water column often differed markedly from the other samples. Due to incomplete mixing, this water layer was temporarily anoxic in some lakes, and during the whole study period in two deep lakes with a sheltered deep hole (mesotrophic Loppesjön and polyhumic Skärhultsjön). In most inflow samples, all measures of organic carbon (13 to 55 g $m^{-3}$) and phosphorus (9 to 35 mg $m^{-3}$) yielded higher values than in the surface water of the respective lake. This indicates that a significant proportion of the organic matter in the lake is produced in the drainage area and transported to the lakes with surface runoff.

*Chemical measures of organic carbon*

Results from combustion (TOC) and chemical oxidation methods ($COD_{Cr}$ and $COD_{Mn}$) were very well correlated but differed in yield. When applying strong oxidants, most estimates of organic carbon from wet oxidation (CrOC) were within 80 and 90% of TOC, which shows the high reproducibility of both methods. In undiluted samples with high concentrations of organic matter, the chemical method was less efficient due to the depletion of oxidant, but TOC can be estimated with good reliability from CrOC with a correction formula accounting for the proportion of oxidant remaining after analysis ($Ox = 0 \ldots 1$):

$$TOC = CrOC/(0.9 \cdot Ox^{0.27}) \qquad (1)$$

This relationship was also valid in anoxic waters, which indicates that the interference from inorganic reduced compounds such as Fe-hydroxides on CrOC is of minor importance. 65% of corrected CrOC values were within ±5% of TOC, and 80% within ±10%, which shows that both methods produce equivalent results, provided appropriate dilution is applied.

Estimates of organic carbon from the less efficient permanganate method (MnOC) deviated significantly and systematically both from TOC and CrOC. In the clearest lakes, MnOC yielded only 25–40% of TOC, in brownwater lakes (>100 g Pt $m^{-3}$) about 35–50% and in forest streams 40–60% (Fig. 1a, 2b). This shows that the widely used permanganate method provides poor information about the total amount of organic matter present, especially in clearwater lakes, and that the yield of the method strongly depends on water quality (cf. Wilander, 1988). On the other hand, the non-linear correlation of the MnOC/TOC ratio with water colour is fairly consistent for both stream and lake surface water (Fig. 2b). This indicates that the translation of permanganate numbers into MnOC concentrations may provide information about the character of organic matter. The MnOC/TOC ratio in surface waters can be quantitatively described as a function of water colour [g Pt $m^{-3}$]:

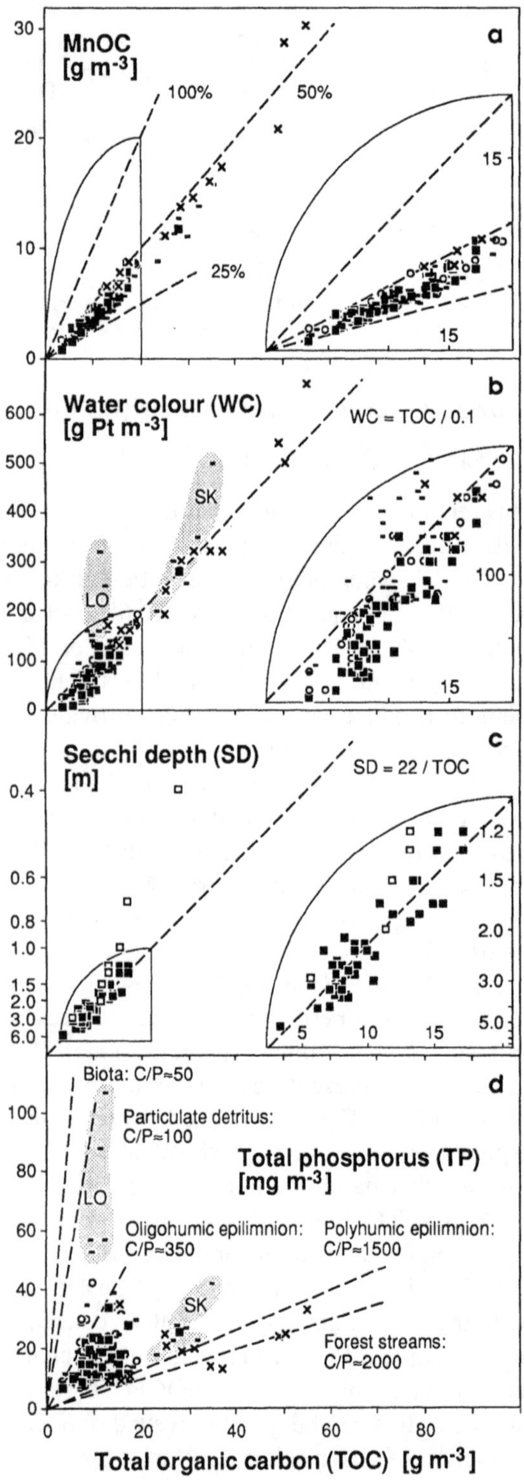


*Fig. 1.* Correlation of optical and chemical water characteristics with the concentration of total organic carbon (TOC) in 18 Swedish forest lakes and streams 1985–1986. Enlarged insets show the lake data in more detail. Data from epilimnion (squares), metalimnion (circles), hypolimnion (hatches) and inflow (crosses) are shown. Anoxic hypolimnion samples from mesotrophic Loppesjön and polyhumic Skärhultsjön are highlighted. Open squares denote shallow lakes. MnOC = labile organic carbon susceptible to permanganate oxidation.

$$\mathrm{MnOC/TOC} = 0.5\cdot(\mathrm{WC} + 25)/(\mathrm{WC} + 50) \quad (2)$$

This formula can be used to estimate TOC [g m$^{-3}$] in surface waters when both permanganate number ( = 10.5·MnOC) [g m$^{-3}$] and water colour [g Pt m$^{-3}$] are available:

$$\mathrm{TOC} = \mathrm{Perm.}\cdot 0.19\cdot(\mathrm{WC} + 50)/(\mathrm{WC} + 25) \quad (3)$$

*Optical measures of TOC*

Water colour was, as expected, correlated with TOC concentrations (Fig. 1b). However, the deviation of the values from proportionality is evident, and the scatter is considerable, especially among the lake samples. Consequently, direct estimates based on proportionality, as are sometimes applied, will be doomed to failure. Only for highly coloured (> 200 g Pt m$^{-3}$) samples of stream and lake waters, can a reasonable estimate of TOC [g m$^{-3}$] be obtained from water colour [g Pt m$^{-3}$] by applying

$$\mathrm{TOC} = 0.1\cdot\mathrm{WC} \quad (4)$$

whereas in clear waters, proportionality factors were several times higher (cf. Fig. 1b). A better correlation and proportionality with water colour was observed for MnOC (Fig. 2a). Water colour is well established as a measure for the concentration of humic substances (Birge & Juday, 1934; Åberg & Rodhe, 1942; Hutchinson, 1957; Wetzel, 1983; SNV, 1986), most of which are dissolved or colloidal (Pennanen, 1988). This is a first indication that MnOC in fact is a measure of dissolved organic carbon (DOC) rather than of TOC. Dissolved compounds are likely to be most susceptible to mild chemical oxidation, whereas particulate matter is more resistant.

Some hypolimnion samples deviated from the rest of the samples, being considerably more stained than would be expected from their TOC concentration (Fig. 1b). These samples all represent strongly hypoxic or anoxic waters from oligo-mesohumic lakes, where iron compounds are likely to accumulate. The observed water colour in these cases is caused by inorganic rather than organic matter. The strongest deviations were observed in mesotrophic lake Loppesjön, where the deepest water was anoxic on all sampling occasions due to incomplete mixing even during spring overturn. However, in other samples from hypoxic strata, high colour values corresponded fairly well to values expected from TOC, e.g. in the polyhumic lake Skärhultsjön (Fig. 1b), where anoxia in the bottom water appeared to be permanent, similar to Loppesjön. This suggests that in the polyhumic lake, settling organic matter is dominated by humic colloids, whereas in the mesotrophic lake, autochthonous matter appears to prevail (cf. C/P ratios, Fig. 1d). Colour values may be depressed when iron occurs in reduced forms, and may also be affected by differences in pH or iron content (Åberg & Rodhe, 1942; Gjessing, 1976). However, no correlation was found between the colour/TOC-ratio and pH in humic surface waters.

Another optical measure of water quality was derived from Secchi disk transparency. Inverse values of Secchi depth, which describe the light attenuation by surficial lake water, provided excellent proportionality with TOC, when disregarding very shallow lakes. The scatter was comparatively small and constant over the TOC range of 3–20 g $m^{-3}$, and systematic deviations were insignificant (Fig. 1c). A good prediction of TOC [g $m^{-3}$] is provided with a simple equation for inverse proportionality with Secchi depth [m]:

$$TOC = 22/SD \quad (5)$$

In some lakes, all rich in TOC, the Secchi disk transparency was sometimes lower than expected

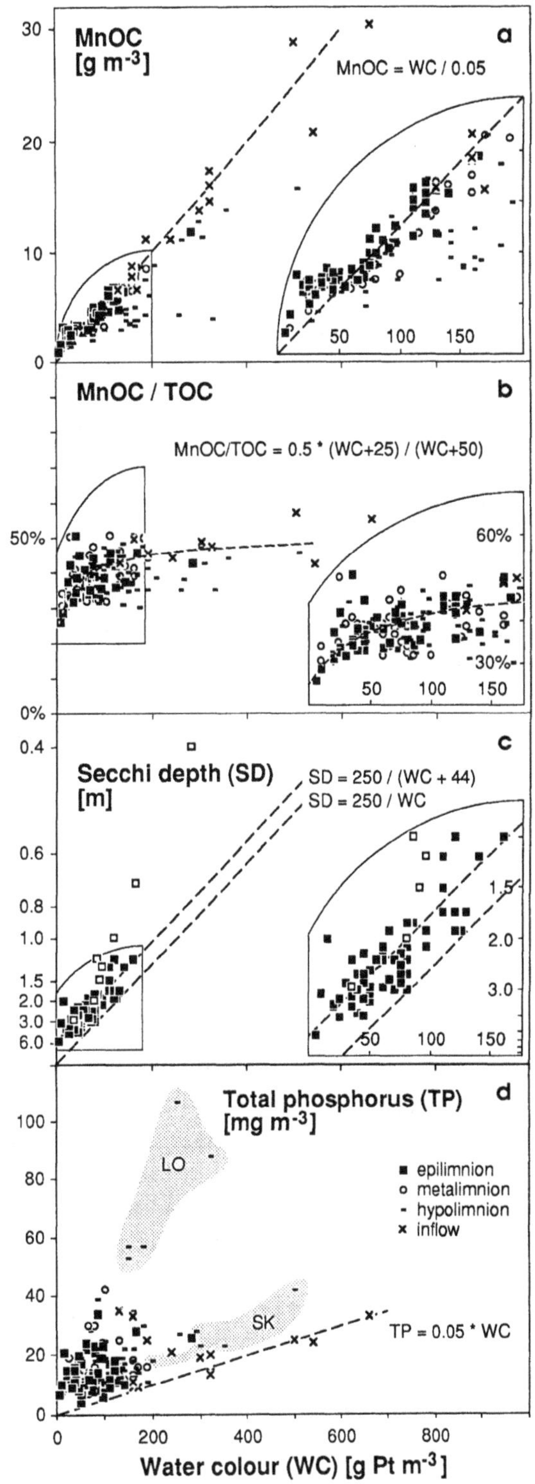


*Fig. 2.* Correlation of optical and chemical water characteristics with water colour in 18 Swedish forest lakes and streams 1985–1986. Enlarged insets show the lake data in more detail. See also Fig. 1.

from TOC (open symbols in Fig. 1c). A closer examination of the data revealed that these lakes belong to a group of four lakes with a very low mean depth (0.8–1.4 m), whereas the mean depth of the other lakes is 2.4–5.8 m. In addition, the four lakes have a surface area of 0.21–0.55 $km^2$ and are thus not among the smallest of the study lakes. This results in a high dynamic ratio, a relative measure for the potential impact of wind-induced mixing forces (Table 1). Consequently, it is highly probable that the decreased transparency is a result of periodic resuspension of sediments, a large proportion of which consists of inorganic clay particles (typically 40–80% of dry weight assuming C = 50% org. dw; Meili, 1991b). Deviations were often observed on windy days and were accompanied by unusually high concentrations of organic matter and nutrients in the water samples. Consequently, Secchi depth measurements are unsuitable for estimating TOC in lakes sufficiently large and shallow for sediment resuspension to significantly increase particle concentration in the water column. When excluding these lakes, however, predictions with this convenient method are satisfactory ($R^2 = 0.83$, $n = 43$).

*Phosphorus content of organic matter*

In contrast to MnOC, water colour and Secchi depth, the concentrations of total phosphorus were not correlated with TOC when comparing all samples. Especially the values for lake surface water show a pattern similar to random scatter (Fig. 1d). However, a more detailed analysis of the relationship between total phosphorus and TOC revealed interesting patterns about the C/P-ratio [$g\,g^{-1}$] in different types of waters. The highest C/P-ratios were found in stream waters, where most values were around 2000 (Fig. 1d). On the other hand, values on the order of 100 were found in the hypolimnion of mesotrophic lake Loppesjön. In surface lake waters, all samples except one showed an C/P ratio between 350 and 1500. Higher values were typically observed in dystrophic lakes, and lower values in lakes with low water colour and varying concentrations of total phosphorus, where most production is autochthonous. Evidently, the C/P-ratio reflects qualitative aspects of the organic matter in lakes, suggesting that in oligo-mesotrophic lakes, the C/P-ratio is basically dependent on the influence of allochthonous humic matter from the drainage area and can thus be used as a measure of dystrophy.

In humic waters, concentrations of total phosphorus (TP) were high and increased with water colour, most prominently in stream waters. In polyhumic waters, low phosphorus concentrations are very rare. A lower threshold appears to apply (cf. 987 Finnish lakes in Kämäri *et al.*, 1990), which is similar to the relationship of TP [$mg\,m^{-3}$] with TOC [$g\,m^{-3}$] (Fig. 1d) and colour [$g\,Pt\,m^{-3}$] (Fig. 2d) found in stream samples, and which may represent the proportion of recalcitrant phosphorus in decomposed organic matter:

$$TP = 0.05 \cdot WC \approx 0.5 \cdot TOC \qquad (6)$$

High phosphorus concentrations in humic lakes as well as positive correlations of TP with TOC or light attenuation have been reported previously (Åberg & Rodhe, 1942; Jackson & Hecky, 1980; Wetzel, 1983; Pennanen, 1988; Kämäri *et al.*, 1990). Other studies indicate that planktonic productivity or biomass in humic lakes is lower than expected from relationships with TP established for non-humic lakes (Chow-Fraser & Duthie, 1987; Jackson & Hecky, 1980). In humic waters, total phosphorus by itself is thus most probably of little value as an indicator of trophic state or primary productivity. Instead, it rather turns into a measure of humic content in heavily stained waters.

Phosphorus can be part of organic or inorganic compounds, which may occur in particulate or dissolved state. Roughly half of total dissolved phosphorus both in streams (Olofsson, 1989) and oligotrophic lakes (Tranvik, 1988) of the boreal forest region is typically MRP. Particulate phosphorus is roughly half of TP in forest streams (e.g. Olofsson, 1989), but is the dominant fraction in many lakes (Wetzel, 1983). In humic waters, particulate phosphorus as defined by various filtration procedures may include part of colloidal organic phosphorus (cf. Wetzel, 1983). To assess the phosphorus content in different types of organic matter, chemical analyses of molybdate-reactive phosphorus (MRP) and total phosphorus (TP) were applied and compared to different measures of organic matter.

In contrast to TP (Fig. 1d and 2d, see above), MRP was correlated with water colour in all types of water (Fig. 3a). The influence of humic substances on the chemical reactivity of phosphorus becomes obvious when considering the MRP/TP ratio (Fig. 3b). Whereas in oligohumic lakes both the concentration and proportion of MRP were very low, the vast majority of TP in polyhumic stream waters was MRP. This indicates an efficient biotic fixation of labile phosphorus in clear-water lakes, whereas in humic waters, phosphorus is strongly interacting and to a large extent associated with humic substances (cf. Jackson & Hecky, 1980; Jones *et al.*, 1988).

In the lake water samples of this study, MRP was typically 5–40% of TP, whereas in stream water, the proportion was typically 30–90% (Fig. 3b). This difference could lead to the conclusion that in runoff water, most phosphorus occurs as inorganic orthophosphate ($PO_4^{3-}$). However, $PO_4^{3-}$ has many times been reported to be almost completely absent in humic waters (cf. Pennanen, 1988), and MRP values from this study probably represent dissolved organic phosphorus (cf. Stainton, 1980). Humic matter alone has no capacity to bind $PO_4^{3-}$ (Young & Comstock, 1984) and is only interactive in the presence of ferric iron (Koenings & Hooper, 1976). In polyhumic waters, both iron and phosphorus are almost exclusively associated with high molecular weight humic substances, which are also responsible for most of the brown water colour (Pennanen, 1988). Consequently, the correlations in lake and stream water between TP and colour (Fig. 2d) as well as MRP and colour (Fig. 3a) are most probably a result of chemical interactions between phosphorus and colloidal humic matter, whereas truly dissolved low molecular weight substances, which constitute the majority of the organic matter in polyhumic waters and soil solutions (Pennanen, 1988; Cole *et al.*, 1984; Jackson & Hecky, 1980, see also below), contain little phosphorus.

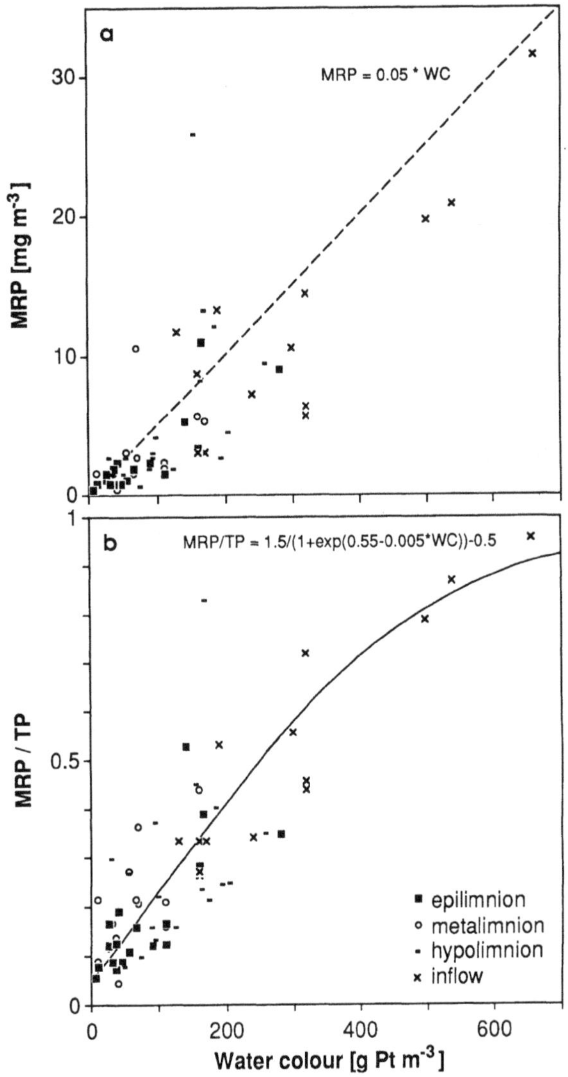


*Fig. 3.* Correlation of molybdate-reactive phosphorus (MRP) (a) and its proportion in total phosphorus (TP) (b) with water colour in 18 Swedish forest lakes in August–September 1985. The line in Fig. 3a is the same as in Fig. 2d.

## *Source-related estimates of TOC*

Oligo- and eutrophic lakes are usually little stained by brown dissolved compounds (Åberg &

Rodhe, 1942; Hutchinson, 1957; cf. Fig. 2d). Water colour is enhanced in the presence of metals, especially iron and manganese (Gjessing, 1976) which, like humic substances, are supplied to forest lakes almost exclusively by surface run-off. Consequently, water colour can be regarded as an indicator of the concentration of allochthonous carbon. This view is also supported by the inverse relationship between colour and water residence time (Fig. 4, Table 1) and by the temporary increase of colour in lakes during periods of high water flow (Meili *et al.*, 1991).

In temperate lakes, the correlation of autochthonous production with the concentration or the availability of phosphorus is well established (e.g. Carlson, 1977). Whereas most organic matter in oligohumic lakes originates from primary production in the lake, a large proportion of TOC in humic lakes is produced in the catchment and transported to the lake, where it is mixed with carbon from autochthonous production in the lake. The latter causes a lower colour/TOC ratio in most lakes when compared to the inflowing water (Fig. 1b). Thus TOC in lake water is basically composed of strongly coloured allochthonous humic matter and weakly coloured but more nutrient-rich autochthonous matter such as algae

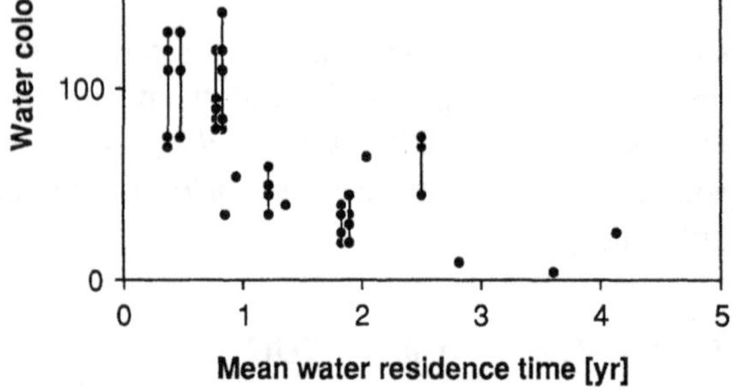


*Fig. 4.* Relationship of water colour and theoretical mean water residence time in 18 Swedish forest lakes 1985–1986.

and excreted dissolved compounds (Åberg & Rodhe, 1942; Hutchinson, 1957). Accordingly, it can be assumed as a first approximation that the concentration of TOC is proportional to both the concentration of phosphorus and to the water colour, each representing a different type of organic matter. Using only surface water data from the eight lakes which were sampled during the whole study period, empirical relationships from multiple linear regressions suggest the following TOC estimate for lake surface waters:

$$\mathrm{TOC} = 0.085 \cdot \mathrm{WC} + 0.27 \cdot \mathrm{TP} \qquad (7)$$

where TOC = total organic carbon [$\mathrm{g\,m^{-3}}$], WC = water colour [$\mathrm{g\,Pt\,m^{-3}}$] and TP = total phosphorus [$\mathrm{mg\,m^{-3}}$]. This equation was derived from several regressions (usually $p < 0.0001$, $R^2 = 0.5...0.8$), where different periods (seasons, years) and lake types (morphological, hydrological, chemical) were included or excluded, and where the parameters in the equation varied interdependently within 0.074...0.099 and 0.35...0.17, respectively. 25% of the values observed in different seasons in the surface water of the lakes from which the model was derived were within ±5% of the model values, 40% were within ±10%, and 90% were within ±30% ($R^2 = 0.71$, $n = 41$).

However, Eq. (7) resulted in systematic overestimates for polyhumic lakes and especially streams. Moreover, the second term in the equation for the contribution of autochthonous carbon is conceptually not satisfactory, as it suggests a C/P ratio of 270 in autochthonous matter, which is lower than any of the observations from the study lakes, although part of them have a very low humus concentration and a long water residence time, suggesting that most production is autochthonous. According to Fig. 1d, the C/P ratio in epilimnetic waters is typically above 350, which is in agreement with data from other Swedish lakes of the meso-eutrophic type (Meili unpubl. data). In addition, Fig. 2d indicates that in humic waters, a large proportion of phosphorus is associated with humic substances and probably contributes little to primary production. Using Eq. 6, phosphorus associated with autochthonous matter (AP, mg m$^{-3}$) can be estimated as follows:

$$AP = TP - 0.05 \cdot WC \quad (8)$$

Assuming that the C/P ratio of 350 is representative of organic matter from primary and secondary production in lakes, the concentration of autochthonous organic carbon (AOC, g m$^{-3}$) can be estimated from AP:

$$AOC = 0.35 \cdot AP \quad (9)$$

Allochthonous humic organic carbon (HOC) can be calculated by subtracting AOC from the first TOC estimate (Eq. 7) and by assuming that the remaining phosphorus term represents detrital matter (C/P ratio ≈ 1000), which can be expressed in Pt-units according to Eq. (5):

$$\begin{aligned} HOC &= TOC - AOC \\ &= 0.103 \cdot WC - 0.08 \cdot TP \\ &\approx 0.103 \cdot WC - 0.008 \cdot WC \\ &\approx 0.095 \cdot WC \quad (10) \end{aligned}$$

By including these considerations on the phosphorus content of humic matter and the C/P-ratio in oligohumic lakes, Eq. (7) can now be replaced by a new, more conceptual equation. It is based on the addition of one term representing the contribution of allochthonous humic carbon to TOC, and another term for the proportion of carbon from autochthonous primary and secondary production:

$$\begin{aligned} TOC &= HOC + AOC \\ &= 0.095 \cdot WC + 0.35 \cdot AP \\ &= 0.095 \cdot (WC + 3.7 \cdot AP) \quad (11) \end{aligned}$$

where TOC = total organic carbon [g m$^{-3}$], WC = water colour [g Pt m$^{-3}$] and AP = available phosphorus from Eq. (8) [mg m$^{-3}$]. When applying this model to the surface water data from all 18 lakes, 40% of the observed values were within ± 10% of the model values, 70% within ± 20% and 90% within ± 30% ($R^2 = 0.75$, $n = 51$). When applying the model to all lake surface and stream data, 35% of the observed values were within ± 10% of the model values, 70% within ± 20% and 90% within ± 30% ($R^2 = 0.93$, $n = 65$).

The model can be illustrated with a set of parallel lines in a nomogram for the dependence of TOC on water colour and phosphorus concentration (Fig. 5a). The model accounts for the fact that low phosphorus concentrations at high water colour are very rare, which is illustrated in Fig. 5a by the bold line derived from the left term in Eq. (11). Systematic deviations of the predicted values from the observations occurred only in some meta- and hypolimnion samples (Fig. 5b). Prediction statistics and confidence intervals were therefore calculated for each sample type separately (Fig. 6a). In lake water samples, an increasing overestimation of TOC with water depth was observed, but only in hypolimnion samples was the median ratio of predicted/observed values significantly different from one. This can be attributed to anoxia in the hypolimnion of deep lakes, where excess colour is created from the accumulation of inorganic Fe-compounds, and where decomposition of settling matter and the presence of Fe-compounds results in an accumulation of inorganic phosphorus. In some deep and oligohumic lakes, however, the ratio was largest in the metalimnion. Speciation of phosphorus (Fig. 3b) and organic carbon (Fig. 2b) both indicate that the organic matter in the metalimnion of these lakes was dissolved, either diffusing from the hypolimnion or remaining after spring overturn, whereas in the lowest layer of the hypolimnion, settling detrital particles including precipitated humic colloids were more common.

As samples were collected during a year with unusually variable conditions with respect to hydrology and water chemistry, systematic seasonal deviations from predicted values were considered separately (Fig. 6b). In the northern study area, high water fluxes occurred as a result of unusually heavy rainfall and soil erosion in autumn 1985 and 1986, and a rapid snowmelt in spring 1986, whereas the summer period was rather dry (cf. Meili *et al.*, 1991). However, seasonal effects on the TOC predictions were small in epilimnion

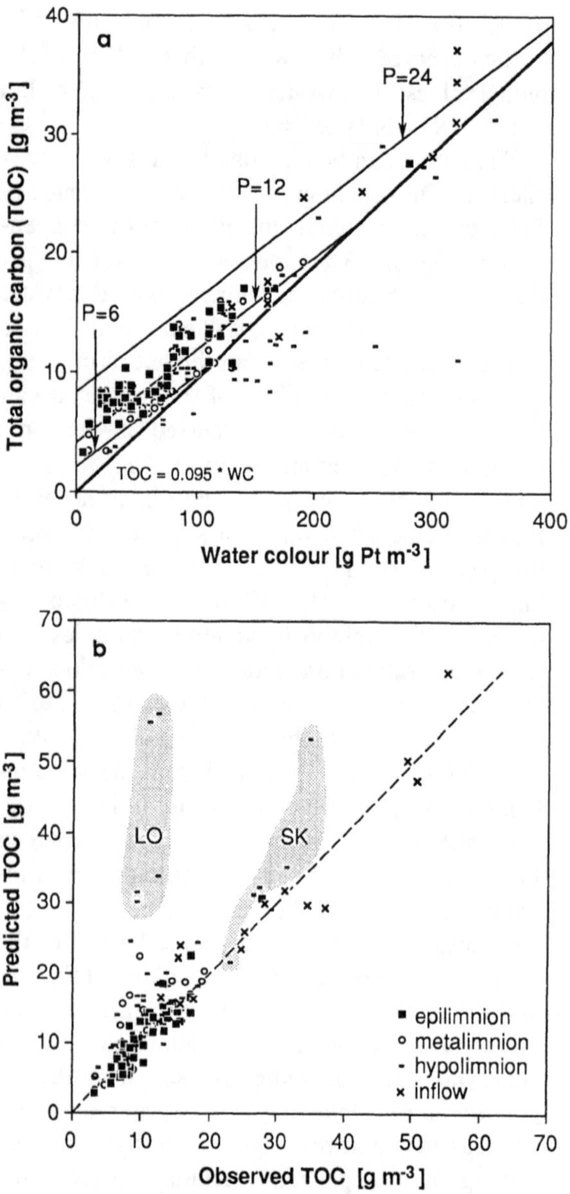


*Fig. 5.* a) Relationship of TOC and water colour in 18 Swedish forest lakes 1985–1986, and a nomogram for predicted TOC in softwater lakes of different water colour and phosphorus concentration (P = 6, 12, 24 mg $m^{-3}$). b) Relationship of predicted ( = 0.095·(WC + 3.7·AP)) and observed concentrations of TOC.

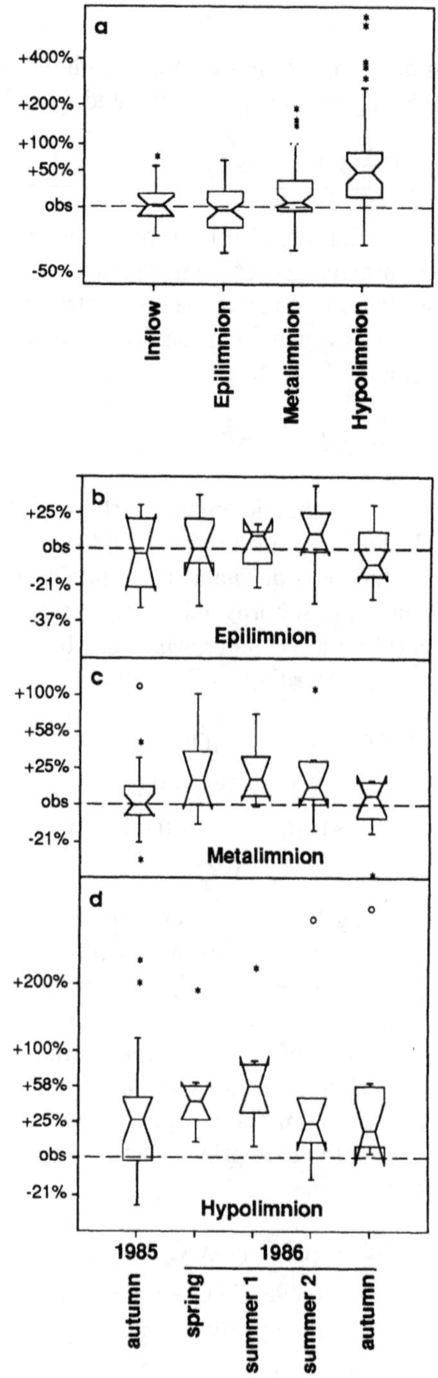


*Fig. 6.* Percentage deviation of predicted concentrations of TOC from observations at different sampling sites and occasions in 18 Swedish forest lakes 1985–1986. The notched box-and-whisker plot shows the median, lower and upper quartiles, min. and max. values (within 1.5 times the interquartile range from the quartiles), and extreme values ('*' = within and 'o' = outside 3 times the interquartile range from the quartiles). The medians of two boxes with notches that do not overlap are different at the 95% confidence level.

samples, although periods of destratification and peak-flow events were included in the study during which TOC, water colour and TP varied dramatically. This robustness of Eq. 11 improves its value as a potential predictive tool for surficial lake waters. On the other hand, the amplitude of seasonal variation increased with water depth, and in the deeper water strata of most lakes, the predictions usually yielded significant overestimates in summer. This confirms the effect of decomposition processes during periods of thermal stratification, during which predictions of TOC from Eq. (11) may only be valid for surface samples, especially in eutrophic lakes.

*Sources, character and composition of organic matter*

Organic matter in natural waters originates from biotic production and occurs in different states of decomposition, which results in a decreasing content of nutrients such as phosphorus. In the following section, an attempt is made to characterize different types of organic matter on the basis of their phosphorus content in order to identify their source and to quantify their quantitative importance in lake waters.

From the two components of Eq. (11), the proportions of allochthonous and autochthonous organic carbon in the water column can be roughly estimated. In Swedish forest lakes, the overall median phosphorus concentration is about 10 to 15 $mg\ m^{-3}$, and the water colour about 70 to 80 $g\ Pt\ m^{-3}$ (Meili *et al.*, 1991; SNV, 1986), which corresponds to a TOC concentration of 9 to 11 $g\ m^{-3}$. According to the influence of AP in Eq. (11), autochthonous TOC (ca. 3 $g\ m^{-3}$) constitutes about 30% of TOC. This indicates that the majority of TOC in a typical forest lake is of allochthonous origin. It can be estimated that in most unpolluted softwater lakes with a water colour of 50 $g\ Pt\ m^{-3}$ or higher, more than half of the organic carbon in the surface water consists of allochthonous humic substances, and in polyhumic lakes (> 200 $g\ Pt\ m^{-3}$) more than 90%.

In oligo-mesotrophic lake waters, the concentration of dissolved inorganic phosphorus is usually very low, and it is the limiting nutrient for algal and bacterial production (Harris, 1986). While according to Harris, the C/P weight ratio in planktonic organisms is between 41 (Redfield ratio, zooplankton) and 64 (phytoplankton), the ratio in particulate organic matter in oligo-mesotrophic lakes is about 60–80 (*op. cit.*). This slight difference is most likely due to the influence of detrital P-depleted particles, which both agrees and disagrees with Harris who states that the C:N:P ratios of living fractions and detritus are indistinguishable, but also that settling detrital particles are nutrient-depleted. The character and amount of detrital particulate organic matter (DPOM) can be assessed in different ways. According to Wetzel (1983) and Harris (1986), roughly half or more of the POM in lakes is detrital, and about 10% are typically algae. Zooplankton accounts for a minor proportion of biotic matter in meso-eutrophic lakes (BPOM), but can exceed the phytoplankton biomass in oligotrophic lakes, especially when the input of allochthonous matter is high (cf. Wetzel, 1983). This is in accordance with data from Swedish oligotrophic lakes, where the biomass of phytoplankton, zooplankton and bacteria often is fairly equal (R. Bell, pers. comm.). For consistency with the C/P ratios given above, the C/P ratio in autochthonous DPOM in the epilimnion of lakes would have to be around 70–100. This ratio is similar to the C/P ratio (TOC/TP) around 100 which was found in the lowest part of the sheltered anoxic hypolimnion of mesotrophic oligomictic Loppesjön (Fig. 1d). This ratio may be representative of more heavily decomposed autochthonous DPOM settling in oligohumic lakes (cf. Boström, 1984) and for terrestrial DPOM occurring in forest streams (cf. Olofsson, 1989). In contrast, the hypolimnetic C/P ratio was up to 1300 in the oligotrophic polyhumic lake Skärhultsjön, which has a similar anoxic hypolimnion. This ratio is approaching the one found for humic matter in forest streams (Fig. 1d), indicating that organic matter settling in polyhumic lakes is dominated by humic colloids.

This element ratio approach is also applicable to assess the characteristics and composition of autochthonous matter. A study of several hundred softwater lakes (Birge & Juday, 1934) showed that in clearwater lakes containing predominantly autochthonous matter, 80–85% of the organic matter in the surface water is typically dissolved, as defined by centrifugation. This dissolved organic matter (EDOM) is a product of excretion and decay of planktonic organisms (Cole *et al.*, 1984). As most phosphorus is associated with particles, the C/P ratio in autochthonous EDOM will be manyfold higher than in POM, probably between 350 which is the lower end of the C/P range of total organic matter (TOM) found in surficial lake waters (Fig. 1d), and infinity which is equivalent to complete P-depletion of EDOM. Given this potential range and a C/P ratio of 70 for POM which accounts for 15–20% of the organic matter, the C/P ratio in autochthonous total organic matter (TOM) including POM and DOM will be within a limited range of 200 to 500, which supports the assumption that the C/P ratio of 350 in fact may be representative of autochthonous matter in lakes.

From the proportion of dissolved carbon and the C/P ratios in TOM, POM, BPOM and DPOM, the partitioning of phosphorus and carbon into different fractions of autochthonous matter can be assessed by solving a system of 9 simple simultaneous equations:

$$C\%(POM) + C\%(EDOM) = 100\%$$
$$P\%\ (POM) + P\%\ (EDOM) = 100\%$$
$$C\%\ (POM) = C\%\ (BPOM) + C\%\ (DPOM)$$
$$P\%\ (POM) = P\%\ (BPOM) + P\%\ (DPOM)$$
$$C\%\ (POM) = 18\%$$

$$C\%\ (TOM)/P\%\ (TOM) = 350$$
$$C\%\ (POM)/P\%\ (POM) = 70$$
$$C\%\ (BPOM)/P\%\ (BPOM) = 50$$
$$C\%\ (DPOM)/P\%\ (DPOM) = 90$$

The results are summarized in Table 3, which also shows that allochthonous and autochthonous matter differ both with respect to physical characteristics and regarding to nutrient content (C/P-ratio) of physically similar fractions. The equations suggest that about 90% of the phosphorus in the surface water of oligotrophic clearwater lakes is particulate, which is in agreement with Wetzel (1983) and Harris (1986). A comparison with the equation in Fig. 3b indicates that

*Table 3.* General character and proportions (weight %) of organic carbon and phosphorus of different type and origin in the surface water of Swedish forest lakes. Values for autochthonous organic matter and estimates derived from oligohumic oligo-mesotrophic lakes. Values for allochthonous organic matter correspond to humic forest stream water after partial settling of particulate matter. See text for abbreviations, analyses, estimates and calculations.

| | C/P | C % | P % |
|---|---|---|---|
| Autochthonous: | | | |
| TOM | 350 | 100 | 100 |
| POM | 70 | 18 | 90 |
| BPOM biotic | 50 | 6 | 45 |
| DPOM detrital | 90 | 12 | 45 |
| EDOM dissolved | 3000 | 82 | ≤10 |
| $PO_4$-P | – | – | ≥0 |
| Allochthonous: | | | |
| TOM | 2000 | 100 | 100 |
| POM detrital | ≥100 | 2 | ≤40 |
| DOM | 3000 | 98 | ≥60 |
| HDOM humic coll. + diss. | 1000 | 20 | 40 |
| FDOM fulvic dissolved | 8000 | 78 | ≤20 |
| $PO_4$-P | – | – | ≥0 |

MRP analyses yield roughly half of the dissolved phosphorus in clearwater lakes (cf. Tranvik, 1988), but most of total phosphorus in polyhumic forest streams. This may be due to the different character of terrestrial and aquatic DOM. The majority of EDOM in lakes consists of high molecular weight substances, whereas the reverse is the case in polyhumic waters where low molecular weight compounds (fulvic acids, FDOM) account for around 70–90% of the organic matter (Cole *et al.*, 1984; Pennanen, 1988; Jackson & Hecky, 1980). Differences with respect to the type of phosphorus binding may also occur due to different contents of iron.

In combination with Eq. (11), the values in Table 3 can be used to estimate the proportion of carbon and phosphorus in organic matter of different type and source. In humic lakes, mean water residence times are typically short, of the order of a few months or less (cf. Meili *et al.*, 1991; Meili, 1991a). This results in a continuous or periodic supply of recalcitrant humic matter, and allows little time for biotic or chemical transformation in the lakes, except a partial settling of particles and heavier colloids. The composition of autochthonous organic matter may be subject to rapid transformations and vary seasonally, but can be regarded constant for the purpose of rough estimates, as spatial and temporal variations with respect to the distribution among biotic and detrital particles and dissolved compounds are not very large (cf. Birge & Juday, 1934), partly because of the slow turnover of most detrital and dissolve matter (Cole *et al.*, 1987). The high MRP/TP-ratios in humic waters, and its similarity in lakes and streams of the same water colour (Fig. 3b) also indicates that little of the phosphorus associated to humic substances is available for planktonic production. Thus assuming constant proportionality between organic fractions of the same origin (allochthonous and autochthonous), the relative contribution of each fraction to the organic matter in lake waters can be calculated for model lakes of different trophic state with respect to dystrophy and eutrophy (Fig. 7). Most of the organic carbon is dissolved, particularly in humic lakes, whereas the majority of phosphorus is associated with particulate or colloidal organic matter of different origin. In polyhumic lakes, humic substances may suppress primary production by competing with biota for phosphorus and by causing light limitation, but at the same time they may serve as an important long-term nutrient storage for primary producers and bacteria (Tranvik, 1988) by preventing phosphorus from settling out from the water column.

The comparison of different analytical methods and calculations also provides a physico-chemical characterization of organic matter. MnOC, which yields 35 to 60% of TOC, seems to include mainly truly dissolved organic matter. Calculations based on data in Table 3 and Figs. 2b and 7 suggest that only about one third of the autochthonous DOC and about half of the allochthonous DOC are susceptible to permanganate oxidation. These proportions may be related to the proportions of low to medium molecular weight (< $10^4$ to $10^5$)

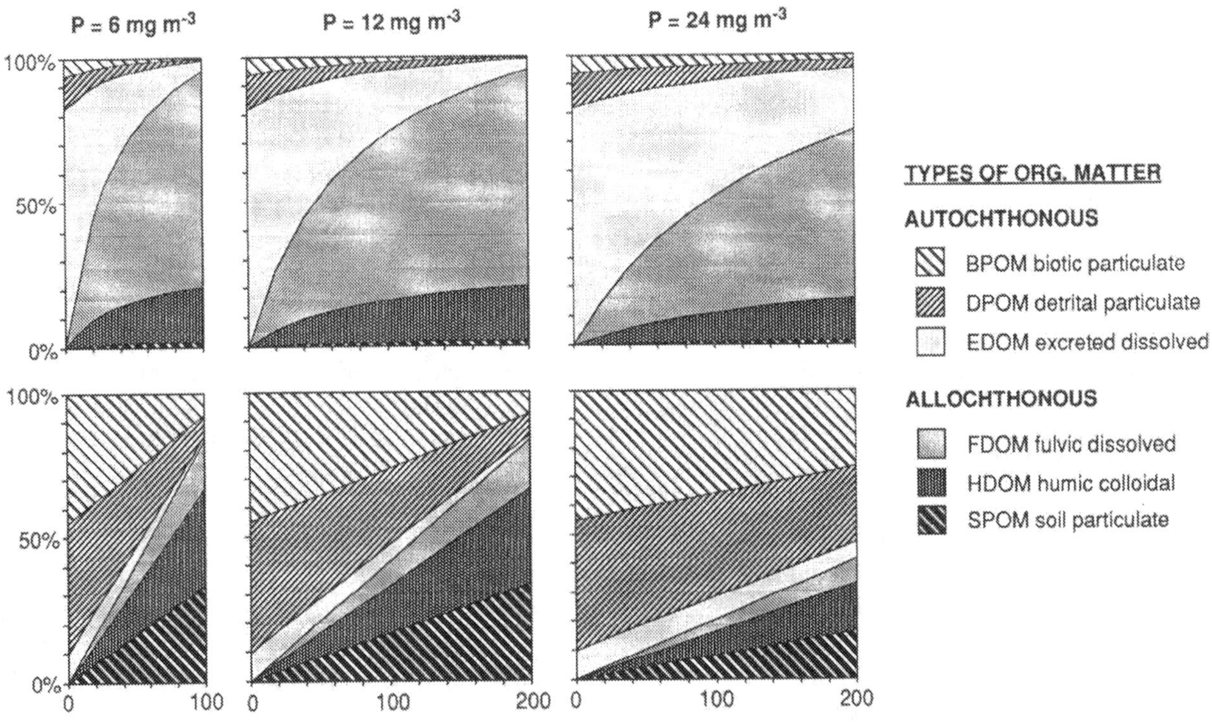


*Fig. 7.* Estimated mean composition of organic matter with respect to carbon (top) and phosphorus (bottom) in boreal forest lakes as a function of water colour (dystrophy) and phosphorus concentration (eutrophy).

compounds in DOM from the two sources (cf. Cole *et al.*, 1984). The higher chemical stability of lacustrine organic matter is also evident from the analysis of hypolimnion water, where the relative yield of MnOC was lower than in stream water at equal TOC (Fig. 1a) or water colour (Fig. 2a,b). These findings indicate that the permanganate method provides information about the molecular weight of the organic substrate, when combined with analysis of TOC.

*Light attenuation by different types of organic matter*

The transparency of lake water is highly dependent on the concentration of organic matter. In the range of Secchi depths (SD) of 1–5 m, reasonable estimates of TOC can be obtained from the inverse proportionality of TOC and SD (Fig. 1c, see above). SD is basically determined by the light absorption by dissolved compounds, the light scattering by particles, and the light absorption by the water itself. In waters with a SD of $<10$ m, the absorption by the water can be neglected. In most softwater lakes $<1\ km^2$ with a mean depth of $>2$ m, the concentration of inorganic particles originating from e.g. sediment resuspension or calcite precipitation will be of little importance. The influence of humic matter on the light attenuation in lakes can be assessed from the linear relationship between 1/SD and water colour (Fig. 2c). Linear regression provided the following relationship (four extreme values were excluded from the calculation):

$$\begin{aligned} 1/SD &= 0.0040 \cdot WC + 1.76 \\ &= 0.0040 \cdot (WC + 44) \end{aligned} \qquad (12)$$

The slope describes the light extinction by humic matter, and the intercept accounts for the interference of autochthonous matter corresponding to 44 Pt colour units on average. The intercept can be replaced by a function of AP (from Eq. 8) which was derived from the relationship of SD with TP in a number of meso-eutrophic oligohumic lakes in the Stockholm area and in central Sweden (Meili, unpubl. reports):

$$1/SD = 0.02 \cdot AP \qquad (13)$$

The coefficient is in good agreement with Carlson (1977) who proposed 0.0208 for summer data using the same formula. Combining the slope of Eq. (12) with Eq. (13) produces:

$$1/SD = 0.0044 \cdot (WC + 5 \cdot AP) \qquad (14)$$

The internal relationship between the allochthonous and the autochthonous term in Eq. (14) is similar to the one in the estimate of TOC (Eq. 11). This indicates that the extinction coefficient per unit carbon is similar for allochthonous and autochthonous matter, although the spectral behaviour may be quite different (Wetzel, 1983), and explains why SD is inversely proportional to TOC (Fig. 1c).

Eq. (14) can be translated into a nomogram for Secchi disk transparency in softwater lakes of different trophic state with respect to dystrophy and eutrophy (Fig. 8a). A comparison of predicted and observed SD (Fig. 8b) shows a scatter of similar magnitude as the empirical relationship between SD and TOC (Fig. 1c).

Secchi disk transparency corresponds to the depth of around 10% of surface light, and is related to the extinction coefficient $[m^{-1}]$ as follows (Idso & Gilbert, 1974):

$$E = 1.7/SD \qquad (15)$$

A combination of Eq. (4),(5),(9),(14) and (15) shows that the carbon-specific extinction coefficient is about 0.07 to 0.1 $m^2\ (gC)^{-1}$ for allochthonous and autochthonous organic matter, respectively.

## Conclusions

Organic matter in surface waters can be classified according to origin (allochthonous or autochthonous), physical state (particulate or dissolved) and nutrient content (carbon/phosphorus ratio). The character as well as the quantity of various organic fractions can be assessed with standard monitoring methods due to differential response

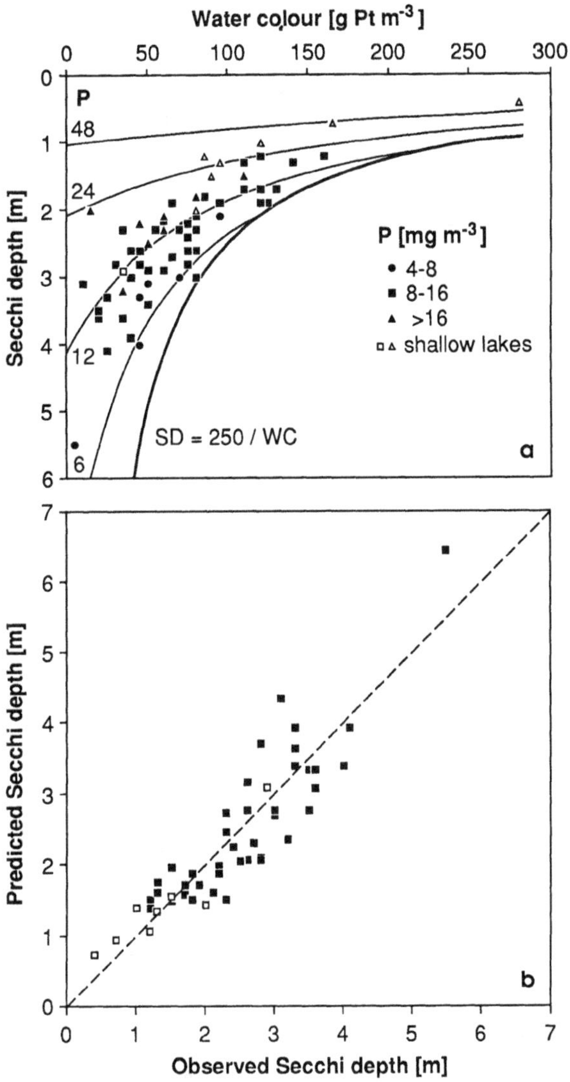


*Fig. 8.* a) Relationship of Secchi disk transparency and water colour in 18 Swedish forest lakes 1985–1987, and a nomogram for predicted SD in softwater lakes of different water colour and phosphorus concentration (P = 6, 12, 24, 48 mg $m^{-3}$). Shallow lakes with a mean depth of <2 m are highlighted. b) Relationship of predicted (= 250/(WC + 5·AP)) and observed Secchi depth.

to different analytical procedures (Table 4). Given constant proportions between organic fractions of the same origin (allochthonous or autochthonous, Table 3), Secchi disk transparency has a similar response to carbon from both sources. As a result, Secchi depths provide good TOC estimates which can be obtained with a single optical measurement (cf. Fig. 1c). Water colour on the other hand is the most selective tool with which allochthonous and autochthonous matter can be clearly distinguished, as most of the brown colour in forest lakes originates from an interaction between iron and humic substances, both of which originate from soils in the drainage area.

A large proportion of phosphorus is associated with biotic particles (Table 3, Fig. 7). Consequently, total phosphorus concentrations provide a rough measure of POM which is little influenced by the relative contribution from different sources. However, the results will vary seasonally

*Table 4.* Apparent relative response of different optical and chemical methods to quantitatively important types of organic matter in lake waters.

| | Secchi depth | Water colour | Permangomate oxidation | Total phosphorus | Molybdate-R. phosphorus |
|---|---|---|---|---|---|
| Autochtohonous | | | | | |
| BPOM biotic particulate | + + | | | + + | |
| DPOM detrital particulate | + + | | | + | |
| EDOM excreted dissolved | + | | + | | + |
| Allochthonous | | | | | |
| DPOM detrital particulate | + + | | | + + | |
| HDOM humic colloidal + dissolved | + + | + + | | + | + + |
| FDOM fulvic dissolved | + | + | + + | | |

and depend on the productivity, the vertical mixing of the water column, and the abundance of iron and other chemical components. MnOC, which yields 35 to 60% of TOC, seems to include mainly truly dissolved organic matter, and when combined with analysis of TOC, the permanganate method provides information about the chemical character of the organic substrate.

The standard methods applied can be used not only to assess lacustrine organic matter with respect to physical and chemical character, but also with respect to the relative contribution from different sources (allochthonous or autochthonous). It is estimated that in most unpolluted softwater lakes of the boreal forest region, the majority of the organic carbon in the surface water consists of allochthonous humic substances, which in polyhumic lakes ($>200\ \mathrm{g\ Pt\ m^{-3}}$) constitute more than 90% of TOC.

## Acknowledgements

I would like to thank Paul Andersson for advice on the selection of lakes, and especially Helena Parkman and Björn Söderbäck for their devoted assistance in collecting and analysing samples as well as for their pleasant company and guidance in the big Swedish forest. Valuable comments on previous drafts of the manuscript were provided by Kalevi Salonen, Ingemar Ahlgren, Anders Wilander and Mike Brett. The study was financially supported by the National Swedish Environmental Protection Agency (SNV) and the Swedish Natural Science Research Council (NFR).

## References

Åberg, B. & W. Rodhe, 1942. Über die Milieufaktoren in einigen südschwedischen Seen. Symbolae Botanicae Upsalienses V:3, 256 pp.

Ahlgren, I. & G. M. Ahlgren, 1976. Analytical methods for water chemistry (revised 1984, in Swedish). Institute of Limnology, Uppsala University, Sweden.

Birge, E. A. & C. Juday, 1934. Particulate and dissolved organic matter in inland lakes. Ecol. Monogr. 4: 440–474.

Boström, B., 1984. Potential mobility of phosphorus in different types of lake sediments. Int. Revue ges. Hydrobiol. 69: 457–474.

Bruenig, E. F., 1987. The forest ecosystem: Tropical and boreal. Ambio 16: 68–79.

Carlson, R. E., 1977. A trophic state index for lakes. Limnol. Oceanogr. 22: 361–369.

Chow-Fraser, P. & H. Duthie, 1987. Response of the phytoplankton community to weekly additions of monoammonium phosphate in a dystrophic lake. Arch. Hydrobiol. 110: 67–82.

Cole, J. J., W. H. McDowell & G. E. Likens, 1984. Sources and molecular weight of 'dissolved' organic carbon in an oligotrophic lake. Oikos 42: 1–9.

Gjessing, E. T., 1976. Physical and Chemical Characteristics of Aquatic Humus. Ann Arbor Science Publ., Ann Arbor MI, 120 pp.

Harris, G. P., 1986. Phytoplankton Ecology. Chapman & Hall, London, 384 pp.

Hutchinson, G. E., 1957. A Treatise on Limnology. Vol. 1: Geography, Physics and Chemistry. Wiley & Sons, New York, 1015 pp.

Idso, S. B. & R. G. Gilbert, 1974. On the universality of the Poole and Atkins Secchi disk – light extinction equation. J. appl. Ecol. 11: 399–401.

Jackson, T. A. & R. E. Hecky, 1980. Depression of primary productivity by humic matter in lake and reservoir waters of the boreal forest zone. Can. J. Fish. aquat. Sci. 37: 2300–2317.

Jones, R. I., K. Salonen & H. De Haan, 1988. Phosphorus transformations in the epilimnion of humic lakes: abiotic interactions between dissolved humic materials and phosphate. Freshwat. Biol. 19: 357–369.

Kämäri, J., M. Forsius & L. Kauppi, 1990. Statistically based lake survey: A representative picture of nutrient status in Finland. Verh. int. Ver. Limnol. 24: 663–666.

Koenings, J. P. & F. F. Hooper, 1976. The influence of colloidal organic matter on iron and iron-phosphorus cycling in an acid bog lake. Limnol. Oceanogr. 21: 684–696.

Meili, M., 1991a. Fluxes, pools and turnover of mercury in Swedish forest lakes. Wat. Air Soil Pollut. 56: 719–727.

Meili, M., 1991b. Mercury in forest lake ecosystems – bioavailability, bioaccumulation and biomagnification. Wat. Air Soil Pollut. 55: 131–157.

Meili, M., Å. Iverfeldt & L. Håkanson, 1991. Mercury in the surface water of Swedish forest lakes: concentrations, speciation and controlling factors. Wat. Air Soil Pollut. 56: 439–453.

Menzel, D. H. & N. Corvin, 1965. The measurement of total phosphorus in seawater based on the liberation of organically bound fractions by persulfate oxidation. Limnol. Oceanogr. 10: 280–282.

Murphy, J. & J. P. Riley, 1962. A modified single solution method for the determination of phosphate in natural waters. Analyt. chim. Acta 27: 31–36.

Olofsson, H., 1989. Torvexploatering på Stormyran – effekter på vattenmiljön. (Peat exploitation at Stormyran – effects on water quality). Report LIU 1989 B: 3, Institute of Limnology, University of Uppsala (in Swedish).

Pennanen, V., 1988. Humic fractions in dimictic lakes in Finland. Dep. of Limnology, Univ. of Helsinki, Finland, Thesis, 28 pp.

Preisendorfer, R. W., 1986. Secchi disk science: Visual optics of natural waters. Limnol. Oceanogr. 31: 909–926.

SNV, 1986. Monitor 1986 (ed. C. Bernes): Sura och försurade vatten (Acidic and acidified waters, in Swedish). National Swedish Environmental Protection Board, Solna, 180 pp.

Stainton, M. P., 1980. Errors in molybdenum blue methods for determining orthophosphate in freshwater. Can. J. Fish. aquat. Sci. 37: 472–478.

Tranvik, L., 1988. Availability of dissolved organic carbon for planktonic bacteria in oligotrophic lakes of differing humic content. Microb. Ecol. 16: 311–322.

Wetzel, R. G., 1983. Limnology. 2nd edn., W.B. Saunders College Publ., Philadelphia, 767 pp.

Wilander, A., 1988. Organic substances in natural water. A comparison of results from different analytical methods (in Swedish, English abstract). Vatten 44: 217–244.

Young, T. C. & W. G. Comstock, 1984. Direct effects and interactions involving iron and humic acid during formation of colloidal phosphorus. In Vernet, J. P. (ed.): Interactions between Sediments and Water, CEP Consultants Ltd., Edinburgh, UK., pp. 95–98.

*Hydrobiologia* **229**: 43–50, 1992.
*K. Salonen, T. Kairesalo & R.I. Jones (eds.),*
*Dissolved Organic Matter in Lacustrine Ecosystems: Energy Source and System Regulator.*



# Leaching from vegetation of soluble polyphenolic compounds, and their abundance in temporary ponds in the Doñana National Park (SW Spain)

Laura Serrano
*Department of Ecology, University of Sevilla, P.O. Box 1095, Sevilla 41080, Spain*

*Key words:* polyphenolic compounds, temporary ponds, leaching, flooding

## Abstract

The origin of allochthonous soluble polyphenols in the water of the temporary ponds of Doñana National Park during flood and heavy rainfall was examined. Soil and litter samples were found to leach large quantities of polyphenolic compounds. The concentration of these compounds showed a significant correlation with the organic content of the soil samples, but indicated no correlation with the litter samples. Rain simulation tests proved that fresh vegetation leached soluble polyphenolic compounds. Mediterranean scrub (*Cystaceae*) from dry poor-nutrient soils yielded the highest concentration (being furthest away from the ponds), whereas wetland (*Ericaceae*) and littoral vegetation (*Juncaceae* and *Cyperaceae*) rated second and third, respectively. The potential of the surrounding vegetation to leach high concentrations of polyphenolic compounds suggests that this is the main source of the rapid polyphenolic enrichment of these ponds during periods of heavy rain.

## Introduction

High concentrations of dissolved natural substances, such as polyphenolic compounds, have been found in the temporary ponds of the Doñana National Park during the 1987–88 winter floods (Toja *et al.*, in press; Serrano & Guisande, 1990). The unusually high polyphenolic concentration has raised the question of the origin of these compounds. Simple phenolic compounds are common microbial degradation products of lignin (Kirk, 1984) and occur in other vegetable substances such as tannins (Hedges, 1988); both are widespread, major classes of ecologically-important secondary products of plant metabolism. Polyphenols are not synonymous with humic compounds, although they are both related: it has been suggested that humic acids are formed by polymerisation of the quinones derived by oxidation of the phenolic constituents of decaying vegetation (Haslam, 1989).

Temporary water bodies are characterized by intense exchanges with the neighbouring terrestrial environments. Woody plants are known to be a quantitatively significant source of allochthonous organic matter in aquatic ecosystems (Henrichs & Doyle, 1986; Fustec *et al.*, 1988), but wetland and littoral flora may also play an important role (Wetzel, 1981; Boon *et al.*, 1982). Allochthonous inputs into the pool of organic matter in the water have been attributed to litter decomposition and plant decay (Day, 1982; Geller, 1986; Bunn, 1988). Decomposition processes include leaching, mechanical fragmentation and microbial activity (Gopal, 1984). Rainfall and temporary flooding take part in the leaching of soluble substances from decomposing litter, but little is known about the effect of standing vegetation on their leaching (Davis & van der Valk, 1978). The aim of this work is to determine the contribution of terrestrial as well as wetland and littoral vegetation to the enrichment of dissolved polyphenolic compounds in the temporary ponds of the Doñana National Park.

## Study area

The Doñana National Park is located on the Atlantic coast of SW Spain. The area has a mediterranean type climate with an average yearly rainfall of 620 mm, concentrated in autumn and spring. Summers are dry, with high temperatures (average for July 26.2 °C; August 27.8 °C) and almost no rainfall (5 mm monthly average).

Four main types of landscape have been described: the coastal dunes, the stabilized sands, the ecotone and the marshes (García-Novo, 1981). Along the NE margin of the dune system, at the boundary with the stabilized sands, the dune water table discharges into a string of ponds called 'lagunas'. These are small ($<0.48$ $Km^2$) shallow ($<3$ m) freshwater bodies that may become desiccated during the driest summers. In the temporary ponds, surface temperature ranged from 25–30 °C during the summer months; it was cooler and more variable during autumn and winter (8–23 °C). Water pH varied between 6 and 8; total alkalinity is usually high and ranged between 0.2 and 2.4 meq $l^{-1}$ (sodium chloride enriched waters); conductivity ranged from 0.35 S $m^{-1}$ during summer to 0.02 S $m^{-1}$ in the flood period.

## Methods

Pond sampling was conducted at 1–8 week intervals between November 1989 and April 1990. Water samples were collected with 2 l Plexiglass bottles from the pond surface and filtered through Whatman GF/C filters. The total concentration of polyphenolic compounds dissolved in the water was measured following Hagerman & Butler's (1978) method. Tannic acid (Merck) was used for the calibration curve. Pond water transparency was estimated by a Secchi-disc of 20 cm diameter.

The amount of rainfall was measured using Hellman's pluviometers of 200 $cm^2$ of collecting surface. PVC piezometers of 4 cm diameter were placed in the pond beds at 0.55–1.15 m depth to record groundwater level during the dry period. Surface water level was measured using 1.5 m long metal rods firmly fixed in the bottom of the ponds.

Soil, litter and vegetation samples were collected along transects (13–88 m) from the center of the ponds to the upper margins (2 m higher) through the littoral, wetland and terrestrial vegetation communities when the ponds were dry (October). Soil samples were sieved through a 2 mm mesh. The fraction larger than 2 mm was labelled as 'litter'. Sieved soil samples (4–78 g dw) were washed in glass beakers with 50 ml of distilled water. 4–20 g dw of litter samples were washed with 100 ml of distilled water. Total polyphenolic concentration was measured in the supernatant water after the suspended materials had settled. Settling time was 7 days for soil and 1 day for litter samples. Dry weight was calculated after drying material for 2 days at 80 °C. Organic matter was estimated by ignition (550 °C, 5 hours) as the Doñana soil is carbonate free.

A rain simulator was used to test the effect of rain leaching on the vegetation using tap water (0.02 S $m^{-1}$ and pH 7.1). The simulator consisted of an atomizer of 1.2 mm diameter held 2 m above the plant, from which water fell with a constant pressure of 2 kg $cm^{-2}$. The average rain for the tests (10 minutes) measured with Hellman's pluviometers was 5 l $m^{-2}$. Fresh vegetation samples were placed in a funnel of 900 $cm^2$ and the rain leachates collected in a glass beaker. The leachate was filtered through Whatman GF/C filters and the dissolved polyphenolic concentration was measured following the method mentioned above.

Aerial photographs (scale 1:10000) were used to estimate pond surfaces by an image analyzer (SKYE SI 700) connected to a 1512 Amstrad PC.

## Results

The rainfall was unusually high during the study (Fig. 1): the sum (658 mm) of autumn and winter rain exceeded the yearly average value (620 mm). 53% of all the rain fell within twenty days (November, 23rd to December, 8th). This brought about the filling of the ponds that had remained dry since late summer (Fig. 2). Until flooding, the groundwater table has been below the surface. The maximum area of the ponds corresponded to the period of higher water level (January–February 1990), whereas the minimum area corresponded to the area limited by the littoral vegetation (*Juncaceae*) and the topography of the bottom during drought (Table 1).

Two important features related to the dissolved organic carbon compounds seemed to be associated with the hydrology of the ponds: water transparency and the concentration of polyphenolic compounds. The amount of total polyphenolic compounds dissolved in the water increased as soon as rainfall began to fill the ponds and the water level rose, flooding the littoral and wetland vegetation around the ponds (Fig. 3). As could be expected, the four ponds showed a markedly decreased transparency as the water level began rising (Table 2). The water colour and polyphenolic concentration in the smallest pond ('Las Verdes') tended to both increase and diminish earlier, whereas both variables were stable or increased later in the rest of the ponds (Table 2 and Fig. 3).

The polyphenolic enrichment of the ponds seemed to be closely related to rainfall evolution. Especially in such very small ponds as 'Las Verdes', the polyphenolic concentration reached its maximum earlier and then started to decrease slowly (Fig. 3). However, the larger ponds exhibited their peaks later, after their highest water level was attained, and tended to increase their polyphenolic concentration through the whole period. These patterns suggest that both rainfall and flooding may have been affecting the release of polyphenolic compounds to the water.

The concentrations of polyphenolic compounds were assayed in the water from the leach-

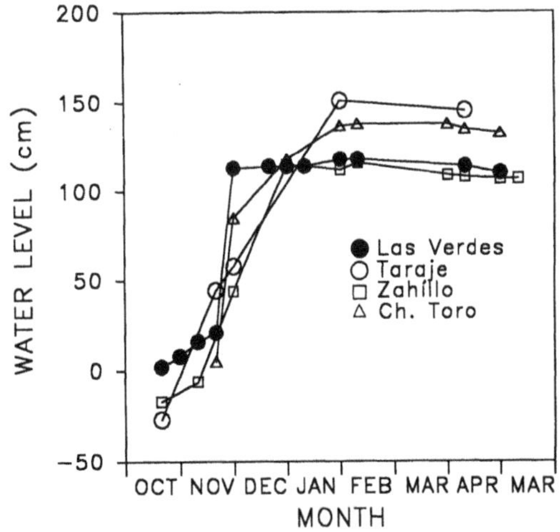


*Fig. 2.* Water level variation with time in each pond. Each month has been divided in three periods of ten days.

*Table 1.* Minimum and maximum area of each pond. The maximum area corresponds to the period of higher water level; the minimum area to the littoral vegetation margin and the topography of the pond bed during drought.

| | Minimum area ($m^2$) | Maximum area ($m^2$) |
|---|---|---|
| Las Verdes | 2200 | 3200 |
| Taraje | 25700 | 61100 |
| Zahíllo | 13800 | 31800 |
| Charco del Toro | 14300 | 19000 |

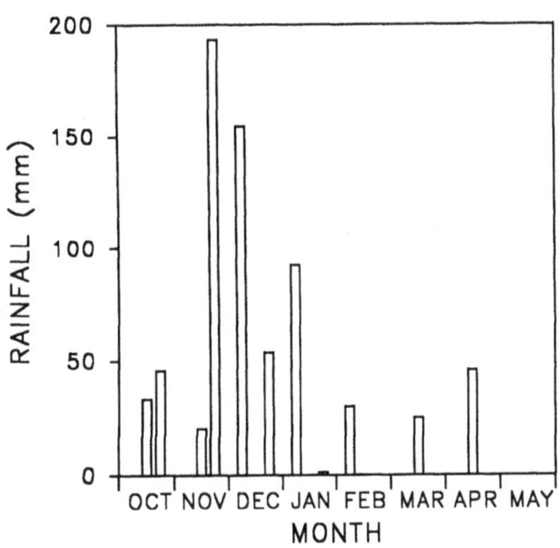


*Fig. 1.* Rainfall distribution through the study period. Each month has been divided in three periods of ten days.

*Table 2.* Variation in Secchi-disc depth in each pond.

| | Secchi-disc dept (cm) | | | |
|---|---|---|---|---|
| | Las Verdes | Taraje | Zahíllo | Charco del Toro |
| Dec | 46 | 53 | 30 | 50 |
| Jan | 27 | 44 | 29 | 47 |
| Feb | 27 | 35 | 26 | 55 |
| Mar | – | – | – | – |
| Apr | 30 | 30 | 27 | 55 |

ing of soil and litter samples from the bottom of the pond to the upper margins (Table 3). There was a high and significant correlation between the concentration of these products in the water from the soil washes and the amount of organic matter of the soil samples ($r = 0.983$, $P < 0.05$ for 'Zahíllo' and $r = 0.973$, $r = 0.948$, $r = 0.99$ $P < 0.001$ for each of 'Las Verdes', 'Taraje' and 'Charco del Toro'). However, the organic matter content of soils was not related to the concentration of polyphenols in the litter washes. The sources of the litter material varied greatly between samples since the vegetation changed much along the transects. The percentage of organic matter was low in the superficial soil samples from the bottom in the first two ponds ('Las Verdes' and 'Taraje'), whereas the other two ('Zahíllo' and 'Charco del Toro') yielded a much higher amount of organic matter. All the ponds exhibited the same decrease of organic matter in the sandy slopes that remained dry for longer.

Rain also caused leaching from fresh vegetation. During the simulated rain assays the amount of polyphenolic compounds leached from vegetation gradually increased from the littoral to the terrestrial communities of the upper margins (Table 4). In these rain leachates the average concentration from *Halimium halimifolium* (*Cystaceae*) was much higher than the rest. Wetland vegetation of hygrophitic species such as *Erica scoparia* (*Ericaceae*) yielded an intermediate concentration, whereas *Juncaceae* and *Cyperaceae* from the littoral vegetation yielded the lowest concentrations.

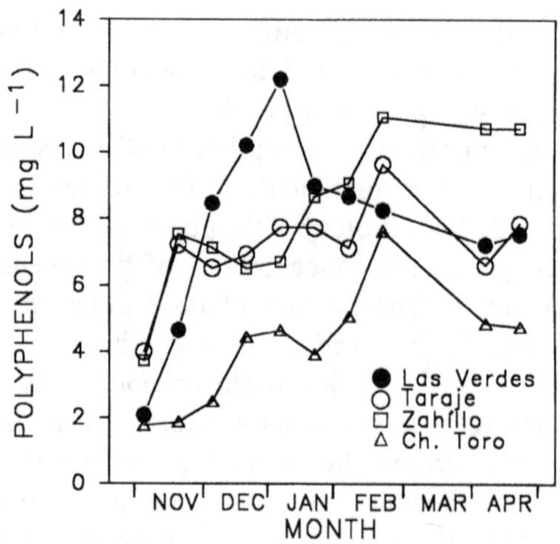


*Fig. 3.* Variation with time of the total dissolved polyphenolic concentration in each pond.

## Discussion

Humic matter concentrations in fresh waters are normally reported to be 1–10 mg $l^{-1}$ (Visser, 1984). Toja *et al.* (in press) found 7 mg $l^{-1}$ of soluble tannin (Folin-Ciocalteu method) in the largest permanent pond of Doñana National Park ('Santa Olalla') after the 1987–88 winter floods. Since dissolved protein and carbohydrate cause interferences in the Folin Ciocalteu method (Serrano & Guisande, unpublished data), we improved the determination of these substances using Hagerman & Butler's (1978) method, which measures total dissolved polyphenols. We then found that after the 1989–90 winter floods, the dissolved polyphenolic concentration rose from 2 to 14 mg $l^{-1}$ in the temporary ponds. This result

*Table 3.* Total dissolved polyphenolic concentration released from the soil and litter samples along a transect from the bottom of the ponds (0 m) to the upper margins. The lack of litter samples corresponds to the bottom and the slopes of the ponds. The first litter sample indicates the presence of littoral vegetation.

| | Distance (m) | Organic matter (%) | Soil polyphenols ($\mu g\ g^{-1}$ dw) | Litter polyphenols ($\mu g\ g^{-1}$ dw) |
|---|---|---|---|---|
| Pond | | | Las Verdes | |
| | 0 | 2.1 | 37 | – |
| | 2 | 2.7 | 32 | – |
| | 3 | 2.2 | 57 | – |
| | 4 | 4.5 | 69 | 538 |
| | 6 | 10.5 | 116 | 280 |
| | 8 | 14.6 | 285 | 480 |
| | 12 | 14.7 | 260 | 747 |
| | 15 | 18.9 | 409 | 327 |
| | 16 | 31.6 | 519 | 299 |
| Pond | | | Taraje | |
| | 0 | 6.7 | 46 | – |
| | 15 | 3.2 | 8 | – |
| | 16 | 3.3 | 9 | – |
| | 25 | 3.8 | 11 | – |
| | 32 | 3.0 | 7 | – |
| | 36 | 12.8 | 114 | 117 |
| | 81 | 5.3 | 28 | 114 |
| | 88 | 7.4 | 91 | 300 |
| Pond | | | Zahíllo | |
| | 0 | 13.4 | 81 | – |
| | 13 | 4.5 | 20 | – |
| | 17 | 9.8 | 46 | 240 |
| | 25 | 4.7 | 22 | 381 |
| Pond | | | Charco del Toro | |
| | 0 | 46.5 | 671 | – |
| | 3 | 1.4 | 3 | 175 |
| | 7 | 2.2 | 2 | 480 |
| | 13 | 1.8 | 8 | 519 |

suggests that high concentrations of these products can be reached very quickly in these ponds. However, we have been unable to compare our figures with other published data that used the same method for the determination of polyphenols in these very same environments.

Several allochthonous sources can contribute to the enrichment of these compounds in the water. Soil and sediments are sources of these products and, more generally, of humic materials which reach the aquatic environments through erosion or underground water seepage (Fustec *et al.*, 1989). Plant litter mineralization can yield polyphenols since they are common end-products of lignin degradation (Kirk, 1984), and they also appear in other plant substances such as tannins.

Soil and litter sample washes yielded high concentrations of polyphenolic compounds and, together with the leaching of flooded vegetation, could account for the later increase of these products in the ponds with larger flood areas ('Taraje', 'Zahíllo' and 'Charco del Toro'). However, rain

*Table 4.* Average dissolved polyphenolic concentration leached from different plant species in the rain simulation tests. SD = standard deviation; n = number of individuals tested.

| | Soluble polyphenolic concentration ($\mu g\ g^{-1}$ dw) | | |
|---|---|---|---|
| | Average | SD | n |
| *Juncus effusus* (*Juncaceae*) | 31 | 13 | 17 |
| *Scirpus holoschoenus* (*Cyperaceae*) | 31 | 12 | 20 |
| *Erica scoparia* (*Ericaceae*) | 43 | 9 | 14 |
| *Halimium halimifolium* (*Cystaceae*) | 89 | 16 | |

can also cause leaching of soluble substances (Gopal, 1984). Runoff and drainage could carry these products down to the water. In fact, it is known that only during heavy rainfall periods do surface drainage and runoff processes appear (Allier *et al.*, 1974), connecting large areas of terrestrial vegetation with the lower aquatic environments.

The rain simulation tests indicated that the polyphenolic leaching from vegetation is considerable even under the low rain intensity of the tests, suggesting that it would be a possible mechanism contributing to the polyphenolic loading of the ponds during rainy periods. Due to their greater lignin content, the amount of polyphenols leached from woody plant communities is expected to be higher than from rushes. However, there was also a different leaching between the xerophytic species (*Halimium halimifolium*) and the hygrophytic species (*Erica scoparia*).

According to several authors, the absence of any general role for most phenolics in the internal economy of plants (Walker, 1975; Zucker, 1983) supports the view that they may owe their wide distribution to selective advantages conferred by their deterrence of herbivores and pathogens (McKey *et al.*, 1978). Janzen cited in McKey *et al.* (1978) predicted that vegetation growing on poor soils would be found to contain higher concentrations of herbivore-deterrent toxic secondary compounds (such as tannins, saponins, and alkaloids) than vegetation growing on more nutrient-rich soils. On this basis, it is precisely the scarcity of water in summer and the nutrient poverty of the soil in the upper areas of the stabilized sands that result in a xerophytic community of lower biomass and lesser productivity as compared to other mediterranean scrub communities under similar rainfall regimes (Merino *et al.*, 1990). Also, other authors have shown that the contribution of biopolymers to soil humic acids is more significant in mediterranean soils supporting *Quercus* or *Cistus* vegetation than in temperate forest soils (Sáiz-Jiménez *et al.*, 1986).

The high potential of the terrestrial vegetation to cause this rapid enrichment of polyphenolic compounds in the ponds suggests that this input of organic matter from terrestrial to aquatic ecosystems follows a dynamic pulse that is triggered off by heavy rainfall. In fact, other authors have shown that this event has already been recorded, although not properly understood. First reports on the freshwater ecosystems of Doñana National Park mentioned that typical ponds were dystrophic (Allier *et al.*, 1974). More recent papers, however, reported that they showed different trophic levels, such as 'Santa Olalla' (eutrophic) and 'Dulce' (mesotrophic) (Montes *et al.*, 1982; López *et al.*, unpublished data). Lately, Toja *et al.* (in press) have recorded a new increase in the dissolved polyphenolic concentration after the 1987–88 floods. All these contradictory findings could be explained by a cyclic alternation of 'dystrophic' and 'non-dystrophic' stages depending on the hydrological conditions. The soil organic content and the concentration of polyphenols washed from the soil decreased on the slopes of the ponds that had remained dry for longer. This suggests that the temporary nature of these water bodies would enhance the degradation processes of these products in the sediment. During drought periods, the sediment would be exposed directly to sunlight and air, when substances can be altered by UV irradiation and sunlight (Geller, 1986). Therefore, little accumulation of these compounds in the pond sediments would be expected from one year to another if the ponds have been dry in between. Sáiz-Jiménez (in press) reported that the pyrolysis pattern of the sedimentary humic acids of 'Santa Olalla' permanent pond was consistent with a mixed cyanobacteria/algal origin in which the more resistant structures, together with some lignin, are more or less preserved from degradation by microbes. He assumed that the larger contribution to the sedimentary humic acids from phytoplankton than from higher plants, was due to the lack of runoff from plant and soil organic matter: both because strong rainfall periods occur every 10–12 years and because plant biomass inputs from the surrounding vegetation were expected to be scarce during dry years in comparison with phytoplanktonic productivity. In fact, the sediment sample he analyzed was taken during a dry period (1986) that was one year before the next flood and almost 14 years since the previous one.

In conclusion, the dissolved polyphenolic concentrations in the pond waters reached very high values after heavy rainfall and floods. Rain leaching of soluble polyphenols from littoral, wetland and terrestrial vegetation can be the main source for this pond enrichment. This event has been cyclical before (as drought and flood are cyclical in the Mediterranean climate), suggesting that the water of the ponds acts as a sink that collects polyphenolic products from the vegetation of their drainage basins.

## Acknowledgements

I am very thankful to Dr. Toja and Dr. García-Novo for their helpful comments. This work was supported by a personal grant from the Ministry of Science and Education and is a contribution from the 'Continental Waters Research Group' of the University of Sevilla.

## References

Allier, C. L., L. Ramírez-Díaz & F. González-Bernáldez, 1974. Doñana: Mapa Ecológico. Publicaciones del CSIC, Sevilla. 12 pp.

Boon, J. J., R. G. Wetzel & G. L. Godshalk, 1982. Pyrolysis mass spectrometry of some *Scirpus* species and their decomposition products. Limnol. Oceanogr. 27: 839–848.

Bunn, S. E., 1988. Processing of leaf litter in two northern jarrah forest streams, Western Australia: II. The role of macroinvertebrates and the influence of soluble polyphenols and inorganic sediment. Hydrobiologia 162: 2111–223.

Davis, C. B. & A. van der Valk, 1978. The decomposition of standing and fallen litter of *Typha glauca* and *Scirpus fluviatilis*. Can. J. Bot. 56: 662–675.

Day, F. P. Jr., 1982. Litter decomposition rates in the seasonally flooded Great Dismal Swamp. Ecology 63: 670–678.

Fustec, E., E. Chauvet, G. Gas, D. Lacaze & M. Escautier, 1988. Degradation of softwood [14C lignin] lignocelluloses and its relation to the formation of humic substances in river and pond environments. Hydrobiologia 159: 169–176.

Fustec, E., E. Chauvet & G. Gas, 1989. Lignin degradation and humus formation in alluvial soils and sediments. Appl. envir. Microbiol. 55: 922–926.

García-Novo, 1981. The ecology of vegetation of the dunes in Doñana National Park (SW Spain). In R. L. Jefferies & A. J. Davy (eds), Ecological Processes in Coastal Environments. Blackwell Scientific Publications: 571–593.

Geller, A., 1986. Comparison of mechanisms enhancing biodegradability of refractory lake constituents. Limnol. Oceanogr. 31: 755–764.

Gopal, B., 1984. Comparison of leaching losses from decomposing litter of the terrestrial and marsh plants on land and in water. Verh. int. Ver. Limnol. 22: 947–950.

Hagerman, A. E. & L. G. Butler, 1978. Protein precipitation method for the quantitative determination of tannins. J. agric. Food Chem. 26: 809–812.

Haslam, E., 1989. Plant polyphenols: vegetable tannins revisited. Cambridge University Press, Cambridge, 230 pp.

Hedges, J. J., 1988. Polymerization of humic substances in natural environments. In F. H. Frimmel & R. F. Christman (eds). J. Wiley & Sons, N.Y.: 45–58.

Henrichs, S. M. & A. P. Doyle, 1986. Decomposition of 14C-labeled organic substances in marine sediments. Limnol. Oceanogr. 31: 765–778.

Kirk, T. K., 1984. Degradation of lignin. Microbiol. Ser. 13: 399–437.

McKey, D., P. G. Waterman, J. S. Gartlan & T. T. Struhsaker, 1978. Phenolic content of vegetation in two African rain forests: ecological implications. Science 202: 61–63.

Merino, O., M. P. Matín, A. Martín & J. Merino, 1990. Successional changes in primary productivity in two Mediterranean scrub ecosystems. Acta Oecologica 1: 81–92.

Montes, C., J. Amat & L. Ramírez-Díaz, 1982. Ecosistemas acuáticos del Bajo Guadalquivir (SW España). Variación estacional de los componentes físico-químicos y biológicos de las aguas. Studia Oecologica 3: 159–180.

Sáiz-Jiménez, C., 1991. Applications of pyrolysis-gas chromatography/mass spectrometry to the study of humic substances: evidence of aliphatic biopolymers in sedimentary and terrestrial humic acids. Sci. Total Envir.: in press.

Sáiz-Jiménez, C., B. L. Hawkins & G. E. Maciel, 1986. Cross polarization, magic angle spinning $^{13}C$ nuclear magnetic resonance spectroscopy of soil humic fractions. Org. Geochem. 9: 277–284.

Serrano, L. & C. Guisande, 1990. Effect of phenolic compounds on phytoplankton. Verh. int. Ver. Limnol. 24: 282–288.

Toja, J., T. López & N. A. Gabellone, in press. Successional changes in two dune ponds (Doñana National Park). Arch. Hydrobiol.

Visser, S. A., 1984. Seasonal changes in the concentration and colour of humic substances in some aquatic environments. Freshwat. Biol. 14: 79–87.

Walker, J. R. I., 1975. The biology of plant phenolics. Edward Arnold Ltd., London. 37 pp.

Wetzel, R. G., 1981. Limnologia. Omega, Barcelona, 679 pp.

Zucker, W. V., 1983. Tannins: does structure determine function? An ecological perspective. Am. Nat. 121: 335–365.

*Hydrobiologia* **229**: 51–58, 1992.
*K. Salonen, T. Kairesalo & R.I. Jones (eds.),*
*Dissolved Organic Matter in Lacustrine Ecosystems: Energy Source and System Regulator.*




# Will an increased greenhouse impact in Fennoscandia give rise to more humic and coloured lakes?

Curt Forsberg
*Institute of Limnology, University of Uppsala, Box 557, 751 22, Uppsala, Sweden*

*Key words:* greenhouse impact, Fennoscandia, humic lakes, water colour

## Abstract

A large-scale increase in the discharge of humic material in rivers, as well as a darkening of Swedish lakes during the last 15 years, are illustrated and discussed. Increased precipitation was responsible for the greatest changes, but a marked part (10–40%) of the increase in river transport of coloured material was independent of discharge. Possible explanations for this increase may be: increased ground water level leading to increased runoff in superficial soil levels, a flow in direct contact with the organic soils; humus fractions in these layers may also be rich in iron resulting in more coloured water; increased primary production giving rise to increased amounts of litter and humic material; reduced capacity of soils to absorb organic matter because of acidification; and effects of modern forest management.

Models of global climate predict a doubling of atmospheric-$CO_2$, an increase in mean annual temperature of up to 1.5 to 4.5 °C, and in Fennoscandia possibly an increase in precipitation. The factors responsible for the observed large-scale change in humus flux may be strengthened by the increasing greenhouse impacts, and may result in more humic and coloured lakes in Fennoscandia during the next century.

## Introduction

Terrestrial vegetation and soil-water systems are the primary sources of humic material to lakes in the boreal landscape. The concentration of humic substances in a lake is regulated by the size and variation of humus loading, and the degradation and sedimentation of this matter within the lake. The size of loading is regulated by the nature of the catchment vegetation and soils, climate and hydrological regime (Engstrom, 1987; Freedman & Clair, 1987; Rasmussen *et al.*, 1989).

Recently, increased interest has been paid to long-term variation of humic content and water colour of Swedish surface waters. An increase in discharge of humic material during the 1970's and 1980's has been reported (Forsberg & Löfgren, 1988; Forsberg & Ahl, unpubl.). Evaluation of time series also demonstrated that Swedish surface waters became more coloured during this time period (Forsberg & Petersen, 1990; Andersson *et al.*, 1990). Finnish lakes also showed increased colour values during the beginning of the 1980's compared with those observed during the 1970's (Laaksonen & Malin, 1984). Andersson *et al.* (1991) found precipitation to be the most important factor for the increase, although high precipitation alone could not explain the entire variation of colour with time. It is also an open question as to what degree the colour increase depended on interference of iron with the coloured organic material (see Pennanen & Frisk, 1984).

The aim of this paper is to further illustrate the change in water colour, and to focus interest on important factors steering the increased humus flux observed in Sweden during the last two decades. As these factors may be strengthened by increasing greenhouse impacts, they may result in more humic and coloured lakes in Fennoscandia.

## Study sites and methods

The following rivers, from north to south, have been arbitrarily chosen to illustrate changes in water colour and transport of humic substances (catchment area in parentheses): 1. Öre Älv (2880 $km^2$, forest dominated), sampling station near its point of discharge into the Bothnian Bay; 2. Ljusnan (365 $km^2$), a mountain-situated sampling station; 3. West Dalälven (8493 $km^2$, forest dominated), with sampling station just before the confluence with East Dalälven; 4. Hedströmmen (1046 $km^2$, forest and agriculture), sampling near its point of discharge into Lake Mälaren; 5. Alsterälven (365 $km^2$, forest and agriculture), sampling near its point of discharge into Lake Vänern; 6. Nissan (2682 $km^2$, forest and agriculture), sampling near its point of discharge into Kattegatt; and 7. Lyckebyån (846 $km^2$, forest and agriculture) sampling near its point of discharge into the Baltic Sea. The first five rivers are situated in very sparsely populated areas, especially the most northern ones. The rivers' geographical positions are shown in Fig. 1, where the position of Lake Öjaren is also indicated. This lake, with a surface area of 20 $km^2$ and maximum depth of 9 m, has a forest dominated catchment area, and serves as a municipal drinking-water supply.

Primary interest is focused on Alsterälven situated in the Värmland County, western Sweden. This river has, 6 km upstream from its discharge into L. Vänern, a reservoir (Lake Alstern, surface area = 5.8 $km^2$, max. depth = 13.2 m), and 6.5 km further upstream a lake, L. Gapern (surface area = 21.9 $km^2$, max. depth = 23 m). The total lake area corresponds to 12.3% of the catchment area of this river. Ground-water level data were available from an observation station situated on an adjacent catchment area, about 30 km from the sampling station of Alsterälven.

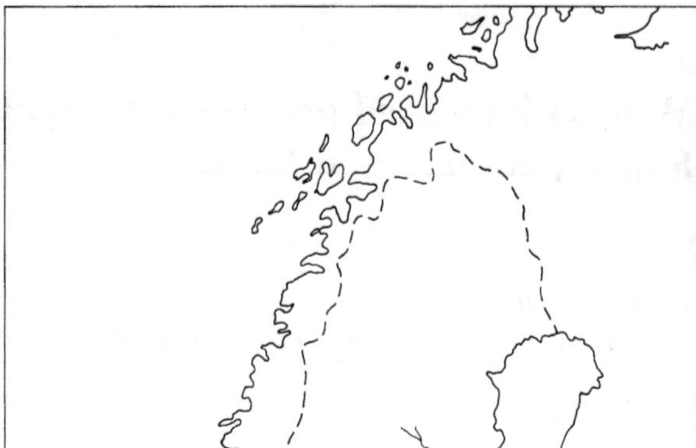


*Fig. 1.* Geographical positions of the river sampling stations: 1. Öre älv; 2. Ljusnan; 3. West Dalälven; 4. Hedströmmen; 5. Alsterälven; 6. Nissan and 7. Lyckebyån. L.Ö. is Lake Öjaren.

The results presented are based on data from the extensive surface-water monitoring program of the Swedish Environmental Protection Board and covering all of Sweden since 1965. River water samples are collected on a monthly basis and are analysed using standard methods (Ahl, 1972, 1985). Thus, for all annual means presented, the number of observations is 12.

The colour of surface water has often been used as an index of dissolved humic matter (see e.g., Rasmussen *et al.*, 1989). The amount of humic material was estimated from filtered water (pore size 0.45 $\mu$m) by measuring the absorbance at 420 nm (50 mm cuvette). The absorbance values are considered to be proportional to concentrations. By multiplying these values by the discharge values, relative figures for the yearly transports of humic material were obtained. Colour values for Lake Öjaren were obtained from the municipality of Sandviken, and annual average values for ground-water level at Brattforsheden from the Geological Survey of Sweden, Uppsala.

## Results and discussion

### *Increase in water colour*

Because concentrations of coloured substances in surface waters are highly regulated by hydrological factors, the variations in discharge in the river Alsterälven and the ground water level at Brattforsheden are given as a background for the discussions on changes in lake water colour (Fig. 2). These hydrological regimes agree with those reported for West Dalälven (Forsberg & Ahl, unpubl.).

The changes in river water colour are illustrated by changes in absorbance (420 nm) in the river Alsterälven, 1968 to 1988 (Fig. 3). The mean annual absorbance declined linearly from the end of the 1960's to the beginning of the 1970's, followed by an almost linear increase up to 1988, during which period the absorbance values rose by about

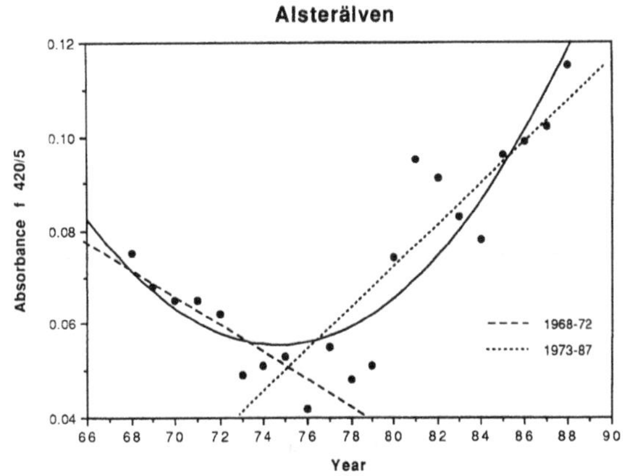


*Fig. 3.* Annual mean absorbance ($n = 12$) at 420 nm for filtered water from Alsterälven.

100%. The year 1976 was extremely dry in Sweden, which may explain the low value for that year (for precipitation and abs., see below). The transports of coloured material in Alsterälven were correlated to transports of organic material and iron as follows (n = 240):

| | |
|---|---|
| Coloured material – $KMnO_4$: | $R^2 = 0.913$ ($p = 0.0001$) |
| Coloured material – Organic-N: | $R^2 = 0.857$ ($p = 0.0001$) |
| Coloured material – Total-Fe: | $R^2 = 0.613$ ($p = 0.0001$) |

Changes in lake water colour are illustrated with data from Lake Öjaren, 1960 to 1988 (Fig. 4).

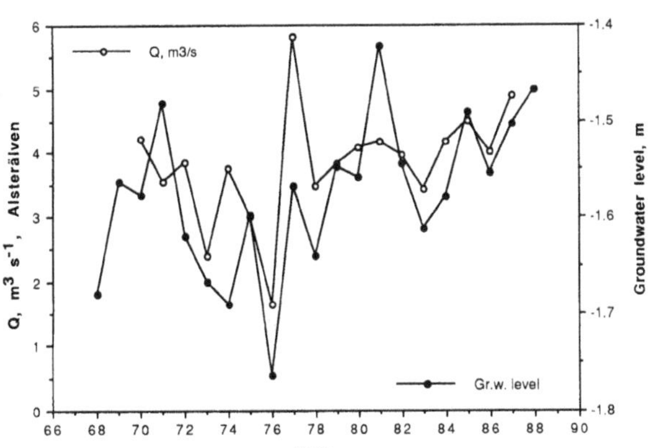


*Fig. 2.* Annual mean discharge in Alsterälven, and ground water level at Brattforsheden.

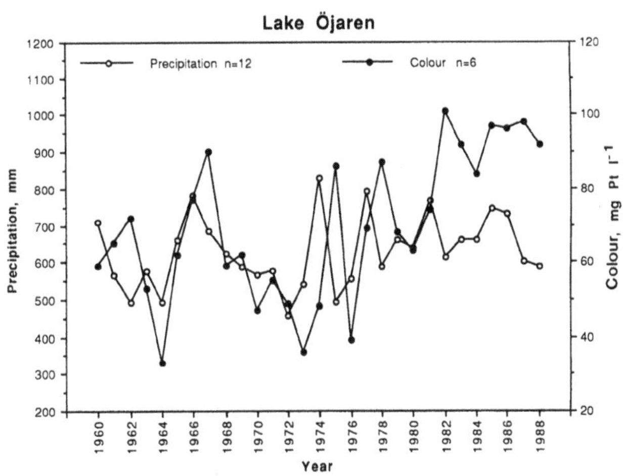


*Fig. 4.* Annual means for precipitation (based on monthly mean values) and for water colour (six observations per year) in Lake Öjaren.

The trends with declining values from the end of the 1960's to the beginning of the 1970's, and then increasing, from a level of 40 mg Pt $l^{-1}$ to about 100 Pt $l^{-1}$, are the same as for the absorbance in Alsterälven (Fig. 3). During the 1980's, most colour values in L. Öjaren were higher than those noted since 1960.

An increase in water colour may depend on an increased flux of coloured organic material, or on an increase in other colour interfering elements, e.g., iron (Pennanen & Frisk, 1984). In Alsterälven, iron appears to play an important role in the colour changes. Thus, concentrations increased from the mid-1970's to the end of the 1980's by a factor of 3, while that for $KMnO_4$-values was 1.5 (Fig. 5). It can be assumed that, due to increased groundwater flow during this time period (Fig. 1), increased runoff from superficial soil levels increased the flux of particulate and colloidal humus fractions. According to Pennanen & Frisk (1984), these fractions contained more iron but less carbon than the dissolved humus fractions. An increasing influence on water colour by substances other than organic matter is also reflected by the ratio $KMnO_4$:absorbance, which during 1974 to 1987 decreased linearly from about 600 to 300 ($R^2 = 0.61$; $p = 0.001$). However, any significant relationships between concentrations of $KMnO_4$ and iron could not be proven.

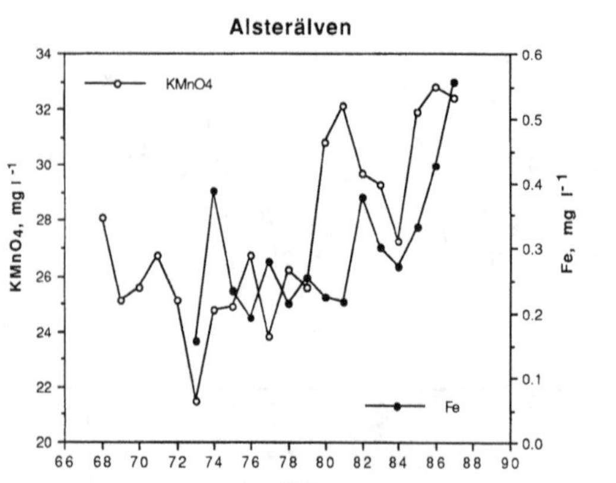


*Fig. 5.* Annual means for $KMnO_4$ and iron in Alsterälven ($n = 12$).

*Precipitation, discharge, ground water level and surface water colour*

For the rivers West Dalälven, Alsterälven, Nissan and Lyckebyån, from which precipitation data were available, precipitation explained 20 to 70% of the variation in water colour (absorbance 420 nm) when annual means were plotted. Correlations between river discharge and water colour may give higher $R^2$-values, as the discharge includes water which has passed through the soil systems on its way to the river. Taking West Dalälven as an example, precipitation explained 49% of the colour increase ($p = 0.0074$) while discharge/runoff explained 69% ($p = 0.0004$). For Alsterälven, the variation in river water colour was best explained by variation in ground water level, demonstrated by the following correlations between annual means, 1968 to 1988:

| | |
|---|---|
| Absorbance – precipitation: | $R^2 = 0.37$ ($p = 0.004$) |
| Absorbance – discharge (Q): | $R^2 = 0.40$ ($p = 0.005$) |
| Absorbance – ground water level: | $R^2 = 0.48$ ($p = 0.0005$) |

The coupling between precipitation and water colour in lakes may be difficult to explain because information on passage of precipitation water through soils towards lakes, temporally and spatially, is lacking. As may be expected, water colour was not correlated to precipitation in L. Öjaren. Plotting precipitation with a time lag of one year gave a $R^2$-value of 0.52 ($p = 0.0001$) for the period 1960 to 1988. This arbitrarily chosen time lag indicated at least the principal importance of precipitation in influencing lake water colour.

*Flow-independent increase in river transport of coloured substances*

By use of correlation analysis, flow-corrected increase in river transport of humic material in Sweden from 1971 to 1985 was found to vary between 0.1 and 3.3% per year (Forsberg & Ahl, unpubl.). Another way to illustrate flow-independent change in flux is to compare river transports based on similar discharge values. This was done by comparing the transports for five sequential yearly years with the corresponding transports for five sequential late years, looking for five-year periods with similar discharges (Table 1). The time periods were not identical for all rivers; early and late respectively for West Dalälven and Hedströmmen was 1965 to 1969 and 1982 to 1986, while the corresponding periods for Alsterälven & Nissan were 1977 to 1982 and 1983 to 1987, respectively. To balance the discharges for the five-year periods in Alsterälven, one year, 1981, was excluded. The examples given show that about 10 to 40% of the increase in river transports of humus/coloured substances was independent of discharge/runoff. The highest flow-independent increase in transport of $KMnO_4$-oxidizable organic material was 28%, noted for Öre Älv (Table 1). The negative values for Nissan and Lyckebyån are probably due to reductions of organic material in industrial and municipal waste water.

The pronounced increase in transport of coloured matter in Alsterälven based on absorbance measurements, can at least in part be explained by an increase in iron concentration (Fig. 5). Thus, from 1983 to 1987, iron transport increased by 56%, compared with the corresponding transport from 1978 to 1982. During the same time periods, Q increased by 8%. The changes in transport (Table 1) demonstrate that large variations may be expected in flux of coloured organic matter, iron and other water colour interfering substances when hydrological conditions are varying.

*Table 1.* Change in percentage between discharge (Q), and river transport of humus (absorbance, 420 nm) and organic matter ($KMnO_4$-oxidizable) during five early and five late years, representing almost identical Q-values, 1965 to 1987. For geographical positions, see Fig. 1.

| River | Changes in % | | |
|---|---|---|---|
| | Q | Humus | Org.-matter |
| 1. Öre älv | 7 | 20 | 28 |
| 2. Ljusnan | 0.3 | 13 | 8 |
| 3. West Dalälven | 7 | 31 | 25 |
| 4. Hedströmmen | 2.1 | 32 | 2 |
| 5. Alsterälven | −0.7 | 39 | 13 |
| 6. Nissan | −3 | 6 | −0.4 |
| 7. Lyckebyån | −3 | 11 | −2 |

## Causal connections behind the increased flux of humus

Precipitation and increased runoff could explain the main increase in water colour and river transports of humic substances, which is in accordance with the data presented by Andersson *et al.* (1989). However, the examples given above demonstrate that 10 to 40% of the increase in river transports of humic substances was independent of the size of discharge/runoff.

One explanation for the flow – independent increase may be an increase in ground water-level leading to increased groundwater flow from superficial soil levels rich in humic material. More superficial runoff may increase the transport of colloidal and particulate humus, resulting in increased water colour due to increase in colour interference by iron. These humus fractions have been found to contain the highest amounts of this element (Pennanen & Frisk, 1984).

Another explanation may be that forest primary production has increased, resulting in higher humus production. An increase in primary production may be explained by increased atmospheric-N deposition, increasing $CO_2$ content in the air, and the effective forest management, which together give rise to increased amounts of litter and humic material. Atmospheric-N deposition has increased markedly in Europe during the 1900's, peaking in Sweden in the early 1970's (Rodhe & Rood, 1986). The increased N deposition may stimulate the basically N-limited forest ecosystem (Abrahamsen, 1980), including the forest ground flora. According to personal information from the Faculty of Forestry at the Swedish University of Agricultural Sciences, forest production in Sweden has increased during the last decades, and larger areas than before are now covered by forests. Therefore, it is reasonable to assume that the primary source for the flux of humus, the vegetation, has increased.

A third explanation may be a reduced capacity of the soils to retain organic material in the humus layer because of increased soil acidification. However, the role of acidification in the regulation of humus output from the vegetation-soil system is not well understood (Degerman, 1987). Andersson *et al.* (1990) found no coupling between increasing water colour and acidification, but concluded that this could not be excluded based merely on correlations. In some soil systems, acidification may bring iron into solution in significant amounts (Reuss & Johnson, 1986), which may increase the colour of the runoff water (see also Pennanen & Frisk, 1984).

As a fourth explanation, modern forest-management methods may cause changes in the decomposition of organic material, in ground water flow, and in humus flux. Considering the large scale of the changes observed, and the comparatively small forest area managed per year (although the declining effects may be long-lasting years), it is assumed that modern forest management alone cannot explain the changes reported.

## Increasing greenhouse impact and lake water colour

Numerical models of the global climate in general agree that a doubling of the atmospheric-$CO_2$ content can be expected during the next century, which will result in an increase in global mean temperature of between 1.5 and 4.5 °C. Based on a three-dimensional, general atmospheric-ocean coupled circulation model (GISS), a climate scenario for Fennoscandia demonstrates in increase in mean annual temperature of 4–5 °C (Boer *et al.*, 1990). Increasing temperature is expected to increase the precipitation due to increased evaporation when the earth surface is warmed (Boer *et al.*, 1990; Swedish Environment Protection Board, 1989). These changes will have a number of impacts, e.g., on geomorphological processes, natural ecosystems, agriculture, forestry and on water resources. As far as is known, no coupling between increased greenhouse impact and lake water colour has yet been discussed.

From the results presented and discussed above, it is evident that precipitation is a key factor in processes regulating the flux of coloured substances from soil systems to surface waters. However, the increase in flux was to varying degrees independent of discharge (Table 1). This flow-independent aspect could be the result of an increase in primary production resulting in more litter and more humus. This may be a reasonable explanation and also very interesting in combination with the increased greenhouse impacts predicted for Fennoscandia (Boer *et al.*, 1990). These impacts include both increased primary production and more rain. Based on limited knowledge of the recently observed changes in humus flux (Forsberg & Petersen, 1990; Andersson *et al.*, 1991; Forsberg & Ahl, unpubl.), and the expected greenhouse impacts (Boer *et al.*, 1990) some principles for the greenhouse impact on lake water color has been outlined in Fig. 6.

According to Van Huis & Ketner (1987) as cited by Boer *et al.* (1990), the most important

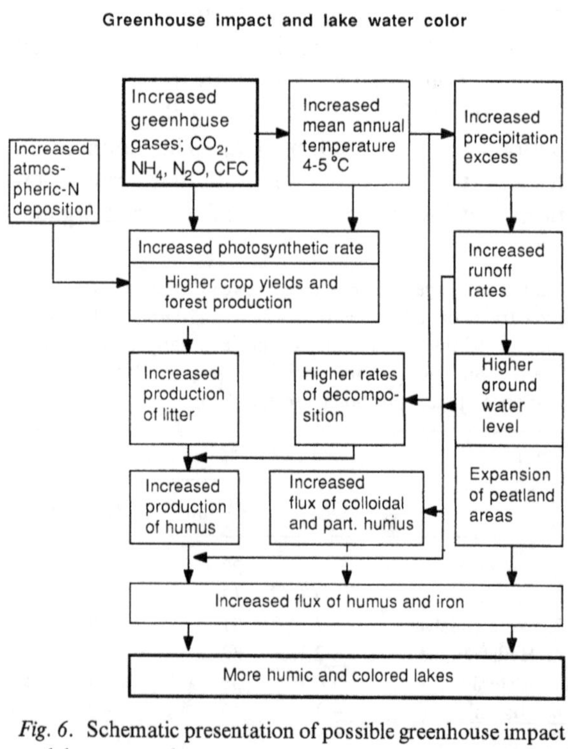


*Fig. 6.* Schematic presentation of possible greenhouse impact on lake water colour.

effects of increasing atmospheric-$CO_2$ content on plants are increasing photosynthetic rates, productivity rates and water use efficiency. However, growth limiting factors may also appear.

Increased greenhouse warming due to increasing $CO_2$ and other greenhouse gases, such as $CH_4$, $N_2O$ and chlorofluorocarbons, will extend the growing season, which will probably result in higher crop yields and forest production. The basically N-limited forest ecosystems (Abrahamsen, 1980) may already be stimulated by the elevated load of atmospheric-N (Rodhe & Rood, 1986; Nihlgård, 1985). It may be expected that this load will continue into the next century as well. Increased temperature will also speed up the decomposition rates and mineralization of the increased organic matter produced, resulting ultimately in increased amounts of humic material.

The increase in precipitation will increase precipitation excess, or the amount of water available for surface water and runoff. According to the GISS-scenario, runoff rates were predicted to rise by 5 to 10% in Norway, 15% in Sweden and 30 to 40% in Finland. Due to increased temperature, larger proportions of this precipitation will fall as rain instead of snow. Such changes will probably alter discharge regimes and lake levels and might induce expansions of peatland areas (Boer *et al.*, 1990), areas which give the strongest primary flux of humus to streams and lakes (see Engstrom, 1987, Fig. 5). Increased precipitation may also result in higher ground water levels, which to a high degree could explain the increase in water colour, as was demonstrated above for Alsterälven.

If the future climate in Fennoscandia will develop according to the scenario outlined, and no unforeseen and overriding plant-growth limiting factors appear, then the production of vegetation/humic substances will increase. As the precipitation excess may increase, then the transport mechanisms of this humus will also be more optimized. Thus, the greenhouse impact during the next century may result in more humic and coloured lakes, in which a large number of physical-chemical and biological processes will be changed (see e.g., Engstrom, 1987; Rasmussen *et al.*, 1989; Forsberg & Petersen, 1990).

In the Experimental Lakes area in Canada, increased greenhouse warming may produce higher evaporation and lower precipitation. These climatic changes will give clearer lakes because of lower imports of dissolved organic matter (DOC), and longer water residence times (Schindler *et al.*, 1990). However, in principle the conclusions on climatic effects on DOC in lakes in Canada and Fennoscandia agree very well.

## Acknowledgement

Many thanks to MSc Robert Titus for his helpful comments on the manuscript, to Mikael Östlund for data treatment and to Allan Petterson for data from Lake Öjaren. This research was supported by grants from the Swedish Environmental Protection Board and the Swedish Council for Forestry and Agricultural Research.

## References

Abrahamsen, G. In D. Drabløs & A. Tollan, (Ed.), 1980. Ecological impact of acid precipitation. – Proc. Int. Conf. Sandefjord, Norway: 58–63.

Ahl, T. (Ed.), 1972. Hydrochemical analyses. A catalogue of procedures used within the IHD in the Nordic countries. National committees for the international hydrological decade in Denmark, Finland, Iceland, Norway and Sweden. Nordic IHD report No. 3. 69 pp. Oslo.

Ahl, T., 1985. Water quality monitoring in Sweden. Wat. Qual. Bull. 10: 82–90.

Andersson, T., Å. Nilsson & M. Jansson, 1991. Coloured substances in Swedish lakes and rivers – temporal variation and regulating factors. In Allard, B., H. Borén & A. Grimvall (eds), 1991. Humic substances in the aquatic and terrestrial environment. Lecture Notes in Earth Sciences 33: 241–253. Springer Verlag, Berlin.

Boer, M. M., E. A. Koster & H. Lundberg, 1990. Greenhouse impact in Fennoscandia – preliminary findings of a European workshop on the effects of climatic changes. AMBIO 19: 2–10.

Degerman, E., 1987. Humösa sjöar. The Swedish Environment Protection Board, Report 3415, 72 pp. In Swedish with Eng. summary.

Engstrom, D. R., 1987. Influence of vegetation and hydrology on the humus budgets of Labrador lakes. Can. J. Fish. aquat. Sci. 44: 1306–1314.

Forsberg, C. & T. Ahl, 1990. Large scale increase in river discharge of colored material in Sweden. In prep.

Forsberg, C. & S. Löfgren, 1988. Dalälvens vatten 1965–1986. Vattenkvalitet och ämnestransport. Länsstyrelsen i Kopparbergs län. N 1988:2, 50 pp.

Forsberg, C. & R. C. Petersen, Jr., 1990. A darkening of Swedish lakes due to increased humus inputs during the last 15 years. Verh. int. Ver. Limnol. 24: 289–292.

Freedman, B. & T. A. Clair, 1987. Ion mass balances and seasonal fluxes from four acid brownwater streams in Nova Scotia. Can. J. Fish. aquat. Sci. 44: 1538–548.

Laaksonen, R. & V. Malin, 1984. Changes in ionic distributions in Finnish lake water during the period 1968–1983. Publ. Wat. Res. Inst., National Board of Waters, Helsinki 57: 59–60.

Nihlgård, B., 1985. The ammonium hypothesis - an additional explanation to the forest dieback in Europe. Ambio 14: 2–8.

Pennanen, V. & T. Frisk, 1984. A statistical model for conversion of absorbance measurements with significant iron interference into organic carbon in a polyhumic lake. Aqua Fennica 14: 171–178.

Rasmussen, J. B., L. Godbout & M. Schallenberg, 1989. The humic content of lake water and its relationship to watershed and lake morphometry. Limnol. Oceanogr. 34: 1336–1343.

Reuss, J. O. & D. W. Johnson, 1986. Acid deposition and the acidification of soils and waters. Springer-Verlag, N.Y., 119 pp.

Rodhe, H. & M. J. Rood, 1986. Temporal evaluation of nitrogen compounds in Swedish precipitation since 1955. Nature 321: 762–764.

Schindler, D. W., K. G. Beaty, E. J. Fee, D. R. Cruikshank, E. R. DeBruyn, D. L. Findlay, G. A. Linsey, J. A. Shearer, M. P. Stainton & M. A. Turner, 1990. Effects of climatic warming on lakes of the central boreal forest. Science 250: 967–970.

Swedish Environment Protection Board, 1989. Växthuseffekten. Orsak, effekter och möjliga åtgärder. 92 pp.

*Hydrobiologia* **229**: 59–71, 1992.
*K. Salonen, T. Kairesalo & R.I. Jones (eds.),*
*Dissolved Organic Matter in Lacustrine Ecosystems: Energy Source and System Regulator.*




# Impacts of environmental changes on the biogeochemistry of aquatic humic substances

H. De Haan
*Limnological Institute, Tjeukemeer Laboratory, De Akkers 47, 8536 VD Oosterzee, The Netherlands*

*Key words:* humic substances, aquatic ecosystems, biogeochemistry, acidification, eutrophication, UV radiation

## Abstract

Humic substances (HS) are the main constituent of the organic carbon pool in stained aquatic ecosystems. HS absorb visible and ultraviolet (UV) light, have acid-base properties and metal and nutrient binding abilities. Based on these characteristics, UV irradiation, pH and the trophic status of aquatic ecosystems will influence the impact of HS on element cycling in surface waters. With climatic change and environmental pollution, UV irradiance, acidification and eutrophication may increase further. In this paper impacts of UV irradiation, pH and eutrophication on the structure, properties and biodegradation of aquatic HS are discussed.

## Introduction

Studies of biotic controls of the chemistry of aquatic humic substances (HS) and of the impact of HS on the biota and element cycling in aquatic ecosystems require that HS can be chemically identified. Despite efforts, the primary chemical structure of a single humic or fulvic acid has not yet been completely described. This failure is connected to the difficulty of extracting homogeneous fractions from the highly complex and polydisperse humic mixtures occurring in nature (Frimmel, 1990). Although the exact chemistry of aquatic HS is still unknown, much work has been done on the impact of HS on the structure and functioning of aquatic ecosystems (Christman & Gjessing, 1983; Aiken *et al.*, 1985; Frimmel & Christman, 1988).

Aquatic HS influence a variety of abiotic and biotic processes in bogs, rivers, lakes and seas. In this context the main chemical properties involved are UV light absorbance, acid-base reactions, redox reactions, ion-exchange, metal complexation and biodegradability. These properties depend on abiotic factors such as light climate, pH and chemical composition of the water, which are currently increasingly influenced by anthropogenic activities. Sustainable management of humic water stressed by more UV irradiation, acidification and eutrophication requires the ability to relate these changes to the functional aspects of aquatic HS. In this paper some recent developments in our knowledge of the interactions between UV radiation, acidification, eutrophication and aquatic HS are discussed.

## Acidification and aquatic HS

The current view is that HS are polycondensates of relatively simple molecules formed during the decay of plant residues. Aromatic rings are evidently an abundant constituent of HS. These rings bear a variety of functional groups in which carboxylic acid and phenolic hydroxyl groups predominate. These groups give HS their acidic properties and also largely determine their metal binding ability. Hence HS are regarded as polyelectrolytes of high molecular weight.

The polyelectrolytic HS can complex metal ions. Thus, depending on the concentration of both components, HS can affect the speciation (and therefore the bioavailability or toxicity) of metals. This speciation applies to essential metals as well as toxic heavy metals. Many humic lakes are characterized by soft water and are oligo- to mesotrophic. Since these systems are relatively sensitive to acidification it is important to understand the acid-base chemistry of HS.

A main property in relation to acidification of humic waters is the pH-dependent impact of HS on the bioavailability of essential elements. Additionally, acidification affects the charge and shape of HS changing their solubility and bioavailability. Although more properties of HS, such as the acid neutralizing capacity and the binding ability of organic micropollutants, are related to acidification, only metal complexing and solubility are discussed here.

### *Acid-base chemistry and metal complexing*

The most direct way to investigate acid-base characteristics of HS is by titration. To predict the acid-base behaviour, all parameters (Table 1), of all different proton binding sites, should be known. However, here we meet the analytical impossibility of determining all these parameters in the complex natural mixture of HS, which may contain several hundreds of different proton binding sites. Nevertheless, various mathematical models have been proposed to describe the acid-base and metal ion complexation chemistry of complex mixtures of organic acids such as HS (Perdue, 1990).

The most common models are polymer models, multi-site models and continuous distribution models; Perdue (1985) and Sposito (1986) have reviewed these. Here special attention is given to the recent introduction of more complex features into the electrostatic polymer model. This increased complexity should account for the effects of conformational changes and Donnan potential effects on proton binding equilibria (Marinsky & Ephraim, 1986; Ephraim *et al.*, 1986).

*Table 1.* Titration parameters of a single proton binding site (After Perdue, 1990).

| | |
|---|---|
| [x] | = concentration of species x |
| $\gamma_x$ | = activity coefficient of species x |
| (x) | = activity of species x, $(x) = \gamma_x[x]$ |
| $C_B$ | = concentration of added base |
| $C_A$ | = concentration of added acid |
| $HL_i$ | = acidic form of i$^{th}$ site |
| $L_i$ | = basic form of i$^{th}$ site |
| $Z_i$ | = charge of $L_i$ |
| $C_i$ | = concentration of i$^{th}$ site, $C_i = [HL_i] + [L_i]$ |
| $K_i$ | = equilibrium constant for ionization of i$^{th}$ site, $K_i = (H)(L_i)/(HL_i)$ |
| $K_i^c$ | = concentration quotient for ionization of i$^{th}$ site, $K_i^c = K_i\, \gamma_{HL}/\gamma_H\, \gamma_{li}$ |

The model proposed by Marinsky & Ephraim (1986) describes polyionic polymers in terms of their rigidity and salt-permeability. They claimed that these properties can be deduced from plots of apparent pK against pH and pH + pX wherein pX = − log[salt anion]. Single unique curves independent of ionic strength would indicate a rigid polymer impermeable to simple salt. Diverging curves would indicate conformational change due to salt ions. Recently, Cäbaniss & Morel (1989) criticized this graphical approach. They demonstrated that the introduction of two acidic sites resulted in ionic strength-dependent pK vs. pH + pX curves. Accidental uniqueness could be achieved in some cases by omitting the titration points at high pH values when the acid groups are dissociated to high extent, i.e., at high $\alpha$ values (where $\alpha$ is the fraction of dissociated groups).

The apparent pK vs. pH plot of fulvic acids extracted from Lake Tjeukemeer (The Netherlands) is shown in Fig. 1. It is clear that this plot is not independent of salt and therefore difficult to explain in terms of rigidity and permeability. However, the divergence of salt dependence is smaller at lower pH and so also at lower $\alpha$. This result indicated the presence of different acidic

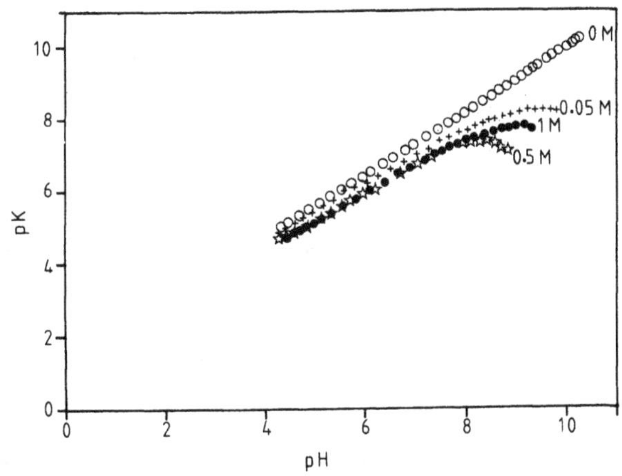


*Fig. 1.* Ephraim-Marinsky analysis of titration data of fulvic acids from Lake Tjeukemeer at different ionic strength.

groups. The convergence of the plots at decreasing pH indicates a hydrophobic polyelectrolyte impermeable to salt at pH ranges characteristic of soft waters suffering from acidification.

The above interpretation would imply decreased reactivity between fulvic acids and metal ions upon acidification. This assumption was confirmed by our recent multicomponent kinetic analysis of iron speciation in Tjeukemeer water (Sojo & De Haan, 1991). Studying the ligand exchange reactions between TPTZ (2.4.6-tri(2′(pyridyl)-s-triazine) and naturally occurring iron ligands, three species were observed (Table 2). The 'immediate' species X is attributed to free ferric iron. The relatively slow species is hydrolyzed ferric iron and the third more rapidly dissociating species is probably iron-fulvate. In accordance with the Marinsky-Ephraim model, the proportion of free ferric iron increased with decreasing pH. The hydrolyzed iron and iron-fulvate fractions decreased with pH.

The multicomponent kinetic study of iron speciation in the humic Tjeukemeer necessitates a reconsideration of our (De Haan *et al.*, 1990) investigation of the abiotic transformation of iron and phosphate in humic lake water in Finland. In this study filtered (0.2 $\mu$m) lake water was double labeled with inorganic $^{55}Fe$ and $^{32}PO_4^{3-}$ and incubated in the dark. After incubation a sample was fractionated by Sephadex G-100 gel filtration at natural pH (ca. 5) and ionic strength (ca. $1.5 \times 10^{-4}$ M). It was found that the simultaneous movement of both isotopes into high molecular weight fractions (10000–20000 daltons) depended on the presence of HS. In the absence of quasi-equilibrium between free and bound phosphate, the high molecular weight complex containing both isotopes readily released $^{32}P$. Compared to the phosphate binding, iron binding was rapid. These results indicate that at pH = 5.0 fulvic acids sequester iron and, to a lesser extent, phosphorus from biotic uptake. The latter statement, and the results of the multicomponent ki-

*Table 2.* Per cent of total iron and dissociation rate of different species X, A1 and A2 according to multicomponent kinetic analysis of solutions of fulvic acids and lake Tjeukemeer water. FA: Peaty polder fulvic adics, TjFA: Lake Tjeukemeer fulvic acids, TjW: Lake Tjeukemeer water, X 'Instantaneous' component ($Fe^{3+}$).

| Sample | Percent of iron | | | | Dissociation rate ($min^{-1}$) | |
|---|---|---|---|---|---|---|
| | pH | X | A1 | A2 | kA1 | kA2 |
| $Fe^{3+}$ | 3.00 | 74.1 | 25.9 | – | 0.02 | – |
| $Fe^{3+}$ | 4.30 | 50.4 | 49.6 | – | 0.02 | – |
| $Fe^{3+}$ | 6.25 | 26.0 | 74.0 | – | 0.01 | – |
| $Fe^{3+}$ – FA | 3.00 | 79.4 | 17.9 | 2.8 | 0.83 | 0.06 |
| $Fe^{3+}$ – FA | 6.02 | 13.2 | 66.5 | 20.2 | 0.22 | 0.04 |
| $Fe^{3+}$ – TjFA | 6.23 | 3.3 | 81.1 | 15.6 | 0.20 | 0.05 |
| TjW | 1.82 | 76.0 | 24.0 | – | 0.01 | – |
| TjW | 3.13 | 44.0 | 56.0 | – | 0.01 | – |
| TjW | 7.21 | – | 62.5 | 37.5 | 0.26 | 0.02 |

netic analysis, may be better reconciled when the results from double isotope labeling experiments at different pH values are available.

The indications that upon acidification smaller proportions of total metal concentrations will occur as humate complexes, raises the question whether higher proportions of these complexed species occur in more alkaline waters. To answer this question we (De Haan *et al.*, 1990a) applied principal component analysis to ultrafiltration data of Dutch lakes with a colour range of 45–130 mg $l^{-1}$ of Pt or a DOC of 15–37 mg $l^{-1}$ and a pH range of 6.5–9.5. The resulting ordination diagrams indicated that organic carbon and metals were intercorrelated mostly in fraction II of the size class 20–35 mm (Table 3). However, in this fraction relatively small proportions of these elements occurred. The highest metal proportions, except of iron, were observed in the fraction containing the smallest (<5 nm) particles, while most organic carbon was found in fraction III (10–20 nm). Most iron and phosphorus appeared to be in the fraction containing the largest particles (35–200 nm) and a minor part of the organic carbon. Apparently, the HS mixture is rather heterogeneous with respect to metal binding properties. Such heterogeneity was also indicated by the Marinsky-Ephraim modelling of the titration results for Lake Tjeukemeer water.

At first sight, the ultrafiltration and above double labeling experiments are no consistent concerning iron, phosphorus and organic carbon (Table 3). The latter technique indicated that iron and phosphorus were bound by high molecular weight fractions representing most of the organic carbon. However, the iron and phosphorus concentrations in the Dutch lake water samples were respectively 4–6 and about 10 times higher, respectively, than in the Finnish samples. Consequently, at the more alkaline conditions in the Dutch lakes, particulate iron species will play a more dominant role as indicated by the high proportions of iron and phosphorus in the largest size class fraction (Table 3). More research is needed to predict the effect of pH on the concentration-dependent interactions between iron, phosphorus and aquatic HS.

Thus future model work might be directed to predict the effects of ionic strength and metal binding on the acid-base chemistry of HS from natural waters. The XAD procedure as recommended by the International Humic Substances Society (IHSS) needs acidification of the sample to pH just below 2 and elution from Amberlite XAD with 0.1 N base. These conditions most certainly influence the chemistry of the HS and will cause artefacts. Already at pH 10, Bowles *et al.* (1989) reported that fulvic acids isolated according to IHSS procedures slowly consumed at least 0.9 meq $l^{-1}$ of base after 18 h. That this uptake was connected to hydrolysis of organic matter was supported by infrared spectroscopy indicating ester hydrolysis. The hydrolysis caused time-dependent titrimetric behaviour of the isolated fulvic acids. It is evident that more sophisticated procedures for isolating HS are needed.

*Table 3.* Mean size distribution (% ± SD) of dissolved (<200 nm) organic carbon, phosphorus, iron and copper and the intercorrelation of these variables, including cobalt, molybdenum and zinc, as % explained variance according to principal component analysis in five size classes after monthly ultrafiltration of four different humic waters during Sept. 1987 and Dec. 1988. The size distributions of cobalt, molybdenum and zinc were similar to that of copper.

| Size (nm) | % ± SD | | | | % explained variance |
|---|---|---|---|---|---|
| | C | P | Fe | Cu | |
| 35–200 | 11 ± 7 | 43 ± 28 | 53 ± 30 | 13 ± 23 | 42 |
| 20–35 | 25 ± 21 | 11 ± 11 | 12 ± 16 | 14 ± 18 | 50 |
| 10–20 | 33 ± 20 | 17 ± 18 | 16 ± 22 | 15 ± 17 | 42 |
| 5–10 | 13 ± 10 | 11 ± 13 | 7 ± 13 | 12 ± 19 | 35 |
| <5 | 17 ± 14 | 18 ± 20 | 12 ± 23 | 46 ± 30 | 43 |

As emphasized by De Haan *et al.* (1983) and Shuman (1990), study of natural untreated water samples high enough in dissolved organic carbon may be an alternative to avoid artefacts during isolation.

*Solubility*

Based on the known acid-base chemistry of HS there should be a reduction in the dissociation rate $\alpha$ with decreasing pH. Consequently, the water solubility should decrease upon acidification. Thus acidification would lead to a decrease in DOC in humic waters which would coincide with a reduced loss of HS from the acidified watershed. However, many other factors such as intensified agriculture, mining and deforestation influence input of organic matter from watersheds and may mask the net flux due to acidification. To date there are few field data to support this hypothesis of decreasing concentrations and inputs of HS with acidification.

Decreased water solubility and increased hydrophobicity of fulvic acids would mean increased lipid solubility of HS. Indeed, Petersen & Kullberg (1985) found a higher octanol soluble proportion of light absorbing ($\lambda = 430$ nm) material with increasing hydrogen activity. Their bog sample exhibited a less pronounced pH-dependent octanol solubility than humic material from a leaf extract and a commercial preparation. The potential octanol solubility of the latter types of HS increased by 500% with a reduction in pH from 6.5 to 3.5. Conversely, the absolute amount of organic carbon extracted by octanol from seven locations in southern Sweden ranged from 23 to 58 $\mu g\,l^{-1}$, representing 0.12 to 0.37% of the organic carbon present (Petersen, 1990).

Because only small amounts of organic carbon seem to be octanol soluble, it is questionable whether octanol extracts fulvic acids or some lipophylic component(s) sorbed to fulvic acids. Organic micropollutants such as pesticides and insecticides belong to the latter category. Perhaps the presence of such materials in preparations of HS can explain the decreasing survival of the crustacean *Daphnia magna* at elevated concentrations of HS (Fig. 2) as found by Petersen & Persson (1987). The initial increase in survival at lower concentrations of HS was explained by masking toxic substances present in the medium. Such different and concentration-dependent impacts of HS on the survival of *Daphnia* would imply careful experimental design to study the effect of HS on the growth of plants and animals. Moreover, the pH-dependent octanol solubility indicates how drastically the potential toxicity of humic samples may change during extraction when the pH is varied. Although the nature of the compounds extracted by octanol from HS is unknown, the available data suggest that the potential toxicity of HS increases with increasing hydrogen activity as caused by acid deposition.

## Eutrophication and aquatic HS

Eutrophication of surface water is caused by inputs of inorganic and organic substances resulting in increased primary productivity. The role of aquatic HS in this process is determined by the interactions between HS and the substances causing eutrophication. On the one hand HS may counteract eutrophication, for example by binding phosphate. On the other hand HS might be mineralized more rapidly in eutrophic waters

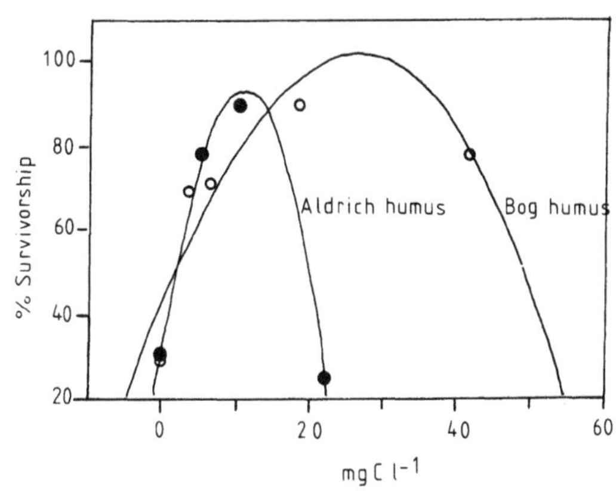


*Fig. 2.* Percent survivorship of *Daphnia magna* at pH 5 with increasing concentrations of HS after Petersen (1990). Curves are the second order polynomial fit to the data.

owing to the elevated production of organic substrates and increased numbers of bacteria utilizing these substrates. However, surface waters do not suffer only from inputs of substances such as phosphorus and nitrogen compounds which regulate the development of algal biomass, but also from other essential substances such as sulphate. Since the interaction between fulvic acids and phosphate has been discussed in a previous section, only the reaction with sulphate is discussed here. Also, the impacts of increased concentrations of relatively simple organic substrates on the biodegradation of HS is treated. Of course, the reaction with sulphate could have been discussed in relation to acidification, but strictly speaking the hydrogen ion and not the sulphate in acid deposition is responsible for the acidification.

*Interaction of HS with sulphate*

Sulphur is a structural component of HS (Schnitzer & Khan, 1972). Most reports give a sulphur content between 0.5 and 1%. Francois (1987) reported evidence that sulphur is incorporated into HS by addition reactions with reduced inorganic sulphur during early diagenesis of sediments. However, under oxygenated and acidic conditions one could also imagine sulphonation of HS to occur in the water column. This reaction is of great importance because of the deposition of sulphuric acid.

Recently Gjessing (1990) presented evidence for the sulphonation of aquatic HS. He found that dialyzed humic water from acidified areas contained more high molecular weight sulphur than did water from less acidified areas. Moreover he found twice as much organic sulphur (0.08 mg $SO_4^{2-}$ $mg^{-1}$ C) in the samples from acidified areas. These results were confirmed by artificial acidification experiments using sulphuric acid.

The sulphonation reaction was observed to decrease the charge density of the HS originating from the acidified area. Lower charge density implies improved transportation across the cell membrane surface and therefore increased bioavailability. This observation is in agreement with the increased toxicity to Atlantic salmon of dialyzed humic water from acidified areas. Similar results were obtained with humic water artificially acidified with sulphuric acid (Gjessing *et al.*, 1988). Acidification has been shown to increase the lipid solubility of HS, and this effect is increased by the sulphonation of HS.

*Biodegradation*

In limnology it has been common practice to regard humic lakes as unproductive. The light absorbing and acidic properties of HS are believed to be the main reasons for the dystrophy of humic waters. Since the humic matter, being mostly of terrestrial origin, is a large potential carbon source for heterotrophic activity, the possible biodegradation of aquatic HS is of great interest. However, HS are considered refractory. Additionally, their complex chemistry has not encouraged investigations into the role of heterotrophic metabolism of allochthonous HS in the carbon cycling of humic lakes. Despite this difficulty a small number of studies has been made on the subject.

From a biochemical point of view, it is difficult to believe that the precursors of HS would resist microbial degradation. If, however, during their polymerization to HS, abiotic conditions in addition to enzymatic reactions control the humification process, non-biodegradable molecular configurations may be formed. On the other hand, soil scientists have demonstrated that abiotic treatments such as ploughing and fertilization enhance the mineralization of humus. Ryhänen (1968) observed a similar effect on the degradation of aquatic HS when he added phosphorus and nitrogen to humic lake water samples.

The above consideration indicates that the degradation of HS may depend on the way HS are formed and on their age. Thus in productive systems with a diverse and active microbial community, HS might be mineralized more rapidly, perhaps by co-metabolism, than in unproductive systems. The term co-metabolism is used to describe the microbial oxidation of substances without the utilization of the derived energy for growth (Horvath, 1972). The possible co-metabolism of aquatic HS was proposed by De Haan (1977) who found that addition of benzoate to humic lake water stimulated bacterial growth and decreased both the fluorescence and the colour of the filtered water. During growth of an *Arthrobacter* sp. from the same lake on benzoate in the presence of fulvic acid, fulvic acid of relatively low molecular weight was formed. These results suggested that aquatic HS are less refractory than previously believed and that the biodegradation of HS in water systems is connected to the trophic level.

Later analogous studies were done with similar results (Rifai & Bertru, 1980; Stabel *et al.*, 1979; Geller, 1986). In addition to benzoate, glutamic acid and glucose stimulated the degradation of macromolecular constituents of the dissolved organic carbon pool. However, no effects of easily degradable substrates have been observed (Steinberg & Hermann, 1981). Strome & Miller (1978) doubted the microbial degradation of fulvic acids and attributed the changed spectral properties to photo-induced splitting of high molecular weight fulvic acids. The products of photolysis would support bacterial growth better than the original fulvic acids. Although it is clear that light influences the microbial degradation of aquatic HS, more work is needed to understand this process in molecular terms. Regardless of our incomplete knowledge of photolysis and co-metabolism of fulvic acids, there is increasing evidence that light and an enlarged dissolved organic carbon pool as caused by eutrophication stimulate the biodegradation of aquatic HS.

Co-metabolism of HS assumes no direct utilization of these substances for bacterial growth. As stated above, biochemical controlled formation might imply that the biodegradation and use of HS for heterotrophic growth is possible. Nevertheless many attempts to demonstrate direct bacterial growth on aquatic HS as the only carbon and energy source failed. However, recently Tranvik (1988; 1989) reported that the potential production of bacterial biomass was positively correlated with the amount of both HS and dissolved organic carbon (Fig. 3). In all waters about the same fraction of the dissolved organic carbon pool was consumed. This observation, and the fact that Tranvik did not measure HS before and after his incubations, do not exclude the possibility that not HS but other compounds served as substrates. But the utilization of equally high organic carbon proportions in both clear and humic lakes does mean that the organic carbon in the latter type of lakes can support more bacteria. In this way allochthonous HS may contribute considerably to the production of the bacterioplankton community.

Eutrophication may also have a negative effect on the biodegradation of aquatic HS. Eutrophication leads to the dominance of cyanobacteria which are known to excrete volatile odorous compounds and toxic substances (reviewed by Steinberg, 1990) which may have antimicrobial activity. In particular some Russian authors (see Carmichael *et al.*, 1985) have reported that cyanobacteria toxins inhibit the growth of various species of bacteria. Thus eutrophication might promote as well as counteract the biodegradation of aquatic HS.

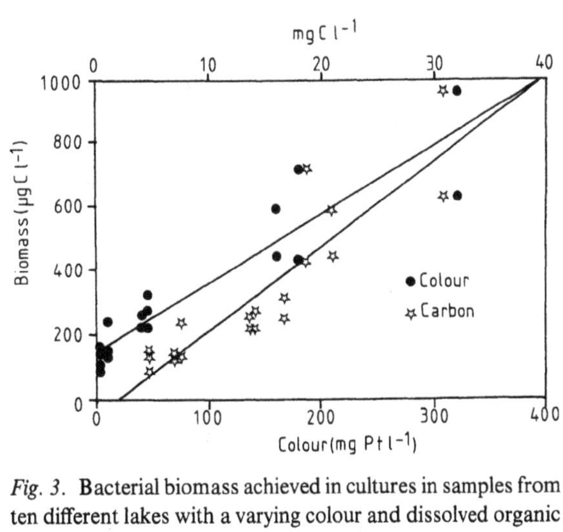


*Fig. 3.* Bacterial biomass achieved in cultures in samples from ten different lakes with a varying colour and dissolved organic carbon concentration after Tranvik (1988). Biomass is plotted versus colour and DOC of the original sterile filtered lake water sample.

## UV radiation and aquatic HS

There are two basic effects of UV radiation on the functioning of aquatic HS. Firstly, the direct degradative effect of UV irradiation may render HS more susceptible to microbial degradation. Moreover, during photolysis, simple and easily degradable molecular fragments may be generated which may stimulate the (co)metabolism of aquatic HS. Secondly, the indirect effect of UV radiation affects the binding and release of biologically important substances from aquatic HS. Both effects will be discussed in this section. The discussion of the indirect effect will focus on the interactions between phosphorus, iron and HS. Additionally, the photochemistry of HS is shortly summarized.

### *Photochemistry*

The non-reflected fraction of sunlight that enters a water body is absorbed by the water and the dissolved and suspended constituents. Part of the entering light will also be back-scattered by the latter components. Humic and fulvic acids strongly absorb the UV part (200–365 nm) of the light spectrum. Thus in humic waters the short wavelengths of sunlight will be absorbed in the top few centimeters of the water column.

The absorbed UV radiation may cause fluorescence, photo-oxidation (photolysis), interactions with binding-release equilibria between aquatic HS and dissolved components and production of several reactive agents such as singlet oxygen, hydrogen peroxide and organic radicals. These topics have been reviewed in Christman & Gjessing (1983), in Aiken *et al.* (1985), by Francko (1990) and by Zika & Cooper (1987).

The reactive products formed on UV irradiation of aquatic HS may influence and react with other dissolved constituents. Thus compounds with a lower redox potential may be photo-oxidized. In this way toxic organics may be inactivated. Such phenomena suggest that aquatic HS act as a buffer, neutralizing agents harmful to biota by photo-induced redox reactions. However, mixing patterns in the water column will govern the extent to which this buffering action of HS will occur within the whole water column (Francko, 1990).

### *Direct effects of UV radiation*

The bleaching of stained waters by UV irradiation is well known. However, little is known of what happens at the molecular level during this process. Steinberg & Münster (1985) reviewed the UV photooxidation of HS and concluded that the bleaching is apparently caused by the oxidation of the macromolecular humic complexes into smaller subunits. Such smaller units may be more susceptible to biodegradation than the originally coloured complexes. Consequently, increased UV radiation intensity may enhance the turnover of aquatic HS.

So far little work has been published on the stimulation of UV radiation of the mineralization of aquatic HS. Geller (1986) reported on the bacterial degradation of macromolecular dissolved organic matter which had been aged under weakly photolytic conditions. Five times concentrated natural lake water samples were exposed to daylight for six weeks. The biodegradability of these light-exposed samples was compared to that of freshly dissolved materials by growth experiments using bacteria isolated from the same water. Selected isolates were able to grow in media with the macromolecular organic matter as the sole organic carbon source. The experiments showed improved accessibility of the samples on exposure to daylight (Fig. 4) since the bacterial growth increased. However, only a *Pseudomonas* sp. grew better indicating that the degree of utilization of the irradiated material depends on the physiological specificity of bacteria. Although the growth data clearly demonstrated enhanced degradability of light-exposed aquatic humic materials, carbon analyses revealed that the degradation increased by only 5–10%. Possibly the small effect of light found by Geller (1986) is connected to exposing the samples to daylight in the laboratory; direct irradiation by UV light might lead to more spectacular results. During the growth experiments, Geller could not demonstrate degra-

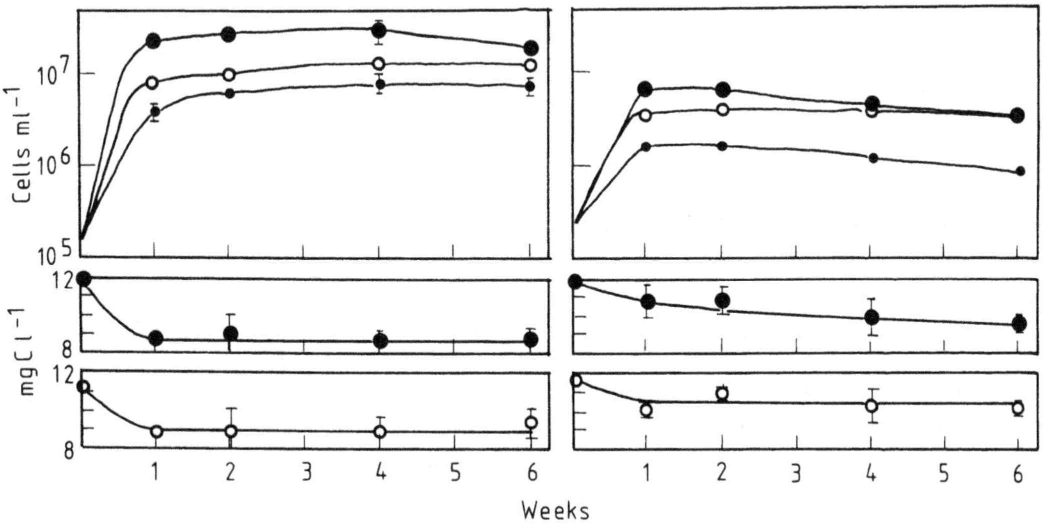


*Fig. 4.* Growth of *Pseudomonas* W1 (left panel) and *Erwinia* S1 (right panel) and degradation of macromolecules in mineral solution without (•) and with aged (●) and original (○) fraction 1 (After Geller, 1986).

dation by change of the spectral properties of the macromolecular substances.

The approach of Geller (1986) did not provide insight into the specific pathway of the degradation of the macromolecular material (i.e. HS). To elucidate what type of substrates are involved in the photooxidation of aquatic HS, more specific techniques are required. Recently, Kieber *et al.* (1989), using HPLC, showed that pyruvate is photochemically formed from dissolved organic carbon having a nominal molecular weight $> 500$. They found that the biological uptake rate and the photoproduction of pyruvate in different sea water samples are linearly correlated (Fig. 5). This is the first evidence that the photooxidation of high molecular weight organic carbon results in the formation of a substrate taken up by heterotrophs. Since this result is based on ultrafiltration data, it does not exclude the possibility that biologically labile substances having nominal molecular weights $> 500$ are the precursors of the pyruvate. More research is needed to prove that the pyruvate originated from refractory substances like HS. On the other hand Kieber *et al.* (1989) pointed out that there is evidence that their samples contained a large pool of refractory dissolved organic matter.

Although Kieber *et al.* (1989) did their measurements with marine samples, there is no reason to believe that the photoproduction of pyru-

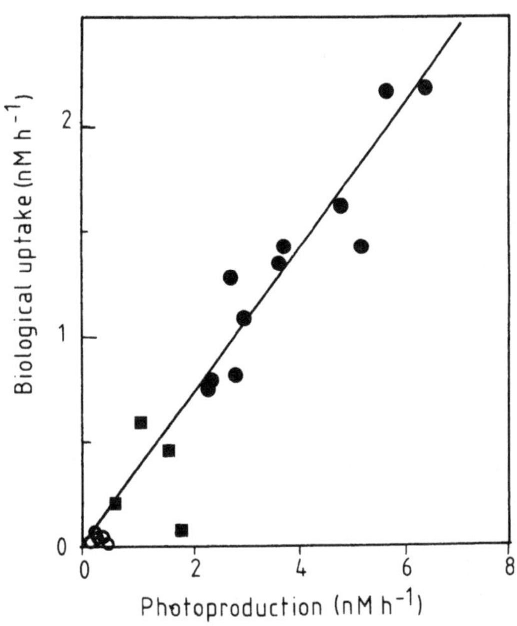


*Fig. 5.* Correlation of biological uptake rate and midday rate of photochemical production of pyruvate at different sampling locations. ○: Sargasso Sea; ■: Gulf of Maine; •: Biscayne Bay (Florida). (After Kieber *et al.*, 1989).

vate should not proceed in humic fresh waters. Thus this study and that of Geller (1986) both provide evidence for the input of labile substrates into the photic zone from high molecular weight and refractory substances in the organic carbon pool. This light driven process implies that not only phytoplankton but allochthonous refractory substances such as HS also serve as a source of dissolved organic carbon for bacteria. If the photochemical production of biologically labile substrates is sufficiently high, then this pathway may contribute more to the productivity of oligotrophic humic lakes than has been believed. Thus not photosynthesis but photoproduction of labile substrates from HS may be the principal source for secondary production in brown water lakes. Kieber *et al.* (1989) estimated that the flux of pyruvate may constitute 2–4% of the total carbon flux to bacteria. This proportion is possibly much higher since a wide variety of labile molecules may be formed during photooxidation of aquatic HS. Thus this light-dependent carbon flux may largely structure the heterotrophic microbial community and their consumers in the upper layers of oligotrophic waters.

From these results an even more important contribution of photochemically produced labile organic carbon from HS to the carbon flux may be expected at increased UV radiation intensity. But this enhanced productivity may be counteracted by increased photoinhibition and photolysis of bacteria. At what intensity these counteracting processes are balanced is unknown.

### *Indirect effects of UV radiation*

The indirect effect of UV radiation on HS involves binding and release of a wide variety of biologically important substances including nutrients, metals and organic micropollutants. Since this topic was recently reviewed by Francko (1990), this section summarizes some results of our work on interactions between aquatic HS, iron, phosphorus and UV light.

Primary production in oligotrophic and moderately eutrophic lakes of the temperate zone is often phosphorus-limited. Under these conditions the binding and release of phosphorus from extracellular sources is of quantitative importance and in this respect HS may influence phosphorus cycling. However, the literature on the effect of HS on different aspects of phosphorus cycling is ambiguous. There are reports of both increased and decreased phosphorus availability to phytoplankton following addition of HS, while opposing effects of HS on phosphatase activity have also been reported (Francko, 1990).

Perhaps the best known effect of HS on phosphorus availability is the binding and sequestering of phosphate by HS in the presence of ferric iron. However, the chemical structure of humic-ferric-phosphate complexes has yet to be identified. Modern techniques such as phosphorus nuclear magnetic resonance spectroscopy ($^{31}$P-NMR) and mass-spectrometry after mild degradation of HS will be helpful for this purpose. But preliminary results of $^{31}$P-NMR for the polyhumic Finnish lake Mekkojärvi, indicated this technique suffers from insufficient sensitivity (Pihlaja, pers. comm.). Despite our lack of knowledge of the structure of humic-ferric-phosphate complexes, there is much evidence that these complexes may serve as sinks for available phosphate. Therefore there is interest in how this bound phosphate may be made available to phytoplankton. Photorelease is one mechanism by which this might happen (see e.g. Francko & Heath, 1979; 1982; 1983; Francko, 1986; Brezonik & Miles, 1981).

To date the most direct evidence for the simultaneous binding of phosphate and ferric iron by aquatic HS comes from our double isotope labeling study of Finnish lake water (De Haan *et al.*, 1990). In this study $^{32}PO_4$-P and $^{55}FeCl_3.6H_2O$ were added to 0.2 $\mu$m filtered humic lake water. After dark incubation the samples were subjected to Sephadex G-100 gel filtration at natural pH and ionic strength. The results showed that both isotopes were transformed into a high molecular weight fraction (10000–20000 daltons). The binding of phosphate appeared to depend on the concentrations of both HS and ferric iron (Jones *et al.*, 1988; De Haan *et al.*, 1990). Preliminary work confirmed that the bound phosphate can be released by exposing the incubated samples to sunlight (Fig. 6). In the lower dark parts of the epilimnion of stained lakes phosphate binding may prevail. This bound phosphorus may be available by UV irradiation if it is transported to the upper few centimeters of the epilimnion. Although the relationships between the wavelength, radiation intensity, mixing and photorelease of phosphate are not yet fully understood, increase in UV radiation intensity may stimulate phosphorus availability. Consequently, this UV light mediated release of phosphorus may increase the productivity of presently oligotrophic coloured lakes.

In addition to UV radiation, acidification may also affect the binding and release of phosphorus in humic lakes. However, so far the effects of pH alone and in combination with UV irradiation on the availability of bound phosphorus have hardly been investigated.

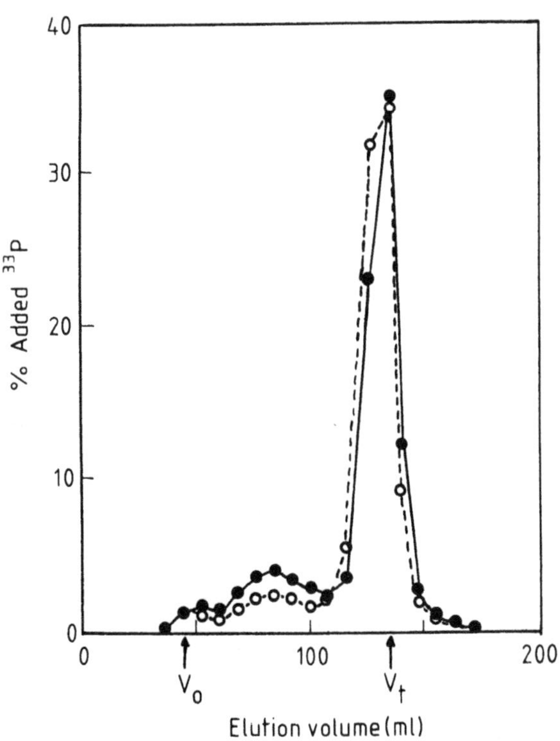


*Fig. 6.* Gel filtration of lake Mekkojärvi water incubated with $^{33}PO_4$-P in the dark after further incubation for 8 h in darkness (●) and for 1 h in bright sunlight (○). (After Jones *et al.*, 1988).

## Conclusions

Anthropogenic activities resulting in acidification, eutrophication and climatic change may have profound impacts on the roles of HS in element cycling in aquatic ecosystems.

Phosphate eutrophication may be counteracted by sequestering of phosphate by HS. However, this influence of HS may be inhibited by acidification and increased UV radiation resulting in release of HS-bound phosphorus. On the other hand acidification will make available toxic metals originally masked by being bound to HS.

Both eutrophication and increased UV radiation can be expected to increase the biodegradation of refractory organic matter by increased cometabolism and photoproduction of biologically labile substrates. These processes will lead to input of non-phytoplankton dissolved organic carbon into the planktonic food webs of humic lakes. This rise in productivity may level off when further increasing acidification and UV radiation intensity start to inhibit heterotrophic organisms.

Thus complex and partly antagonistic, partly synergistic interrelationships exist between environmental change and the functioning of humic aquatic ecosystems.

## Acknowledgements

Drs. H. J. Gons and L. Van Liere (Limnological Institute, NL) and Dr. R. I. Jones (University of Lancaster, UK) critically read an earlier draft of the manuscript.

## References

Aiken, G. R., D. M. McKnight, R. L. Wershaw & P. MacCarthy, 1985. Humic substances in soil, sediment, and water. Geochemistry, isolation, and characterization. J. Wiley & Sons, N.Y., 692 pp.

Bowles, E. C., R. C. Antweiler & P. MacCarthy, 1989. Acid-base titration and hydrolysis of fulvic acid from the Suwannee River. In R. C. Averett, J. A. Leenheer, D. M. McKnight & K. A. Thorn (eds.), Humic substances in the Suwannee River, Georgia: Interactions, properties, and proposed structures. U.S. Geological Survey, Denver, Colorado. Open File Report 87-557: 205–229.

Brezonik, P. L. & C. J. Miles, 1981. Oxygen consumption in humic-colored waters by a photo-chemical ferrous-ferric catalytic cycle. Envir. Sci. Technol. 15: 1089–1095.

Cabaniss, S. E. & F. M. M. Morel, 1989. Comment on 'A unified physicochemical description of the protonation and metal ion complexation equilibria of natural organic acids (humic and fulvic acids)'. Envir. Sci. Technol. 23: 746–747.

Carmichael, W. W., C. L. A. Jones, N. A. Mahmood & W. C. Theiss, 1985. Algal toxins and water-based diseases. CRC Crit. Rev. envir. Cont. 15: 275–313.

Christman, R. F. & E. T. Gjessing, 1983. Aquatic and terrestrial humic materials. Ann Arbor Science, Ann Arbor, Michigan, 538 pp.

De Haan, H., 1977. Effect of benzoate on microbial decomposition of fulvic acids in Tjeukemeer (The Netherlands). Limnol. Oceanogr. 22: 38–44.

De Haan, H., G. Werlemark & T. De Boer, 1983. Effect of pH on molecular weight and size of fulvic acids in drainage water from peaty grassland in NW Netherlands. Plant Soil 75: 63–73.

De Haan, H., R. I. Jones & K. Salonen, 1990. Abiotic transformations of iron and phosphate in humic lake water revealed by double isotope labeling and gel filtration. Limnol. Oceanogr. 35: 491–497.

De Haan, H., T. De Boer, J. Voerman, H. A. Kramer & O. F. R. Van Tongeren, 1990a. Size class distribution of dissolved (200 nm) nutrients and essential metals in shallow, eutrophic and humic lakes. Verh. int. Ver. Limnol. 24: 298–301.

Ephraim, J., S. Alegret, A. Mathuthu, M. Bicking, R. L. Malcolm & J. A. Marinsky, 1986. A unified physicochemical description of the protonation and metal ion complexation equilibria of natural organic acids (humic and fulvic acids). 2. Influence of polyelectrolyte properties and functional group heterogeneity on the protonation equilibria of fulvic acid. Envir. Sci. Technol. 20: 354–366.

Francko, D. A., 1986. Epilimnetic phosphorus cycling: influence of humic materials and iron on coexisting major mechanisms. Can. J. Fish. aquat. Sci. 43: 302–310.

Francko, D. A., 1990. Alteration of bioavailability and toxicity by phototransformation of organic acids. In E. M. Perdue & E. T. Gjessing (eds), Organic acids in aquatic ecosystems. John Wiley & Sons, N.Y.: 167–177.

Francko, D. M. & R. T. Heath, 1979. Functional distinct classes of complex phosphorus compounds in lake water. Limnol. Oceanogr. 24: 463–473.

Francko, D. M. & R. T. Heath, 1982. UV-sensitive complex phosphorus: association with dissolved humic material and iron in a bog lake. Limnol. Oceanogr. 27: 564–569.

Francko, D. M. & R. T. Heath, 1983. Abiotic uptake and photodependent release of phosphate from high-molecular-weight humic-iron complexes in a bog lake. In R. F. Christman & E. T. Gjessing (eds), Aquatic and terrestrial humic materials. Ann Arbor Science, Ann Arbor, Michigan: 467–480.

Francois, R., 1987. A study of sulphur enrichment in the humic fraction of marine sediments during early diagenesis. Geochim. Cosmochim. Acta 51: 17–27.

Frimmel, F. H., 1990. Characterization of organic acids in freshwater: a current status and limitations. In E. M. Perdue & E. T. Gjessing (eds), Organic acids in aquatic ecosystems. John Wiley & Sons, N.Y.: 5–23.

Frimmel, F. H. & R. F. Christman, 1988. Humic substances and their role in the environment. John Wiley & Sons, N.Y.

Geller, A., 1986. Comparison of mechanisms enhancing biodegradability of refractory lake water constituents. Limnol. Oceanogr. 31: 755–764.

Gjessing, E. T., 1990. Mechanisms and effects of reactions of organic acids with anions. In E. M. Perdue & E. T. Gjessing (eds), Organic acids in aquatic ecosystems. John Wiley & Sons, N.Y.: 179–187.

Gjessing, E. T., M. Grande & E. Røgeberg, 1988. Natural organic acids: their role in freshwater acidification and aluminium speciation. Acid rain research report 15/1988. Norwegian Institute for Water Research, Oslo.

Horvath, R. S., 1972. Microbial co-metabolism and the degradation of organic compounds in nature. Bact. Rev. 36: 146–155.

Jones, R. I., K. Salonen & H. De Haan, 1988. Phosphorus transformations in the epilimnion of humic lakes: abiotic interactions between dissolved humic material and phosphate. Freshwat. Biol. 19: 357–369.

Kieber, D. J., J. McDaniel & K. Mopper, 1989. Photochemical source of biological substrates in sea water: implications for carbon cycling. Nature 341: 637–639.

Marinsky, J. A. & J. Ephraim, 1986. A unified physicochemical description of the protonation and metal ion complexation equilibria of natural organic acids (humic and fulvic acids). 1. Analysis of the influence of polyelectrolyte properties on protonation equilibria in ionic media: Fundamental concepts. Envir. Sci. Technol. 20: 349–354.

Perdue, E. M., 1985. Acidic functional groups in humic substances. In G. R. Aiken, D. M. McKnight, R. L. Wershaw & P. MacCarthy (eds), Humic substances in soil, sediment, and water: Geochemistry, isolation, and characterization. John Wiley & Sons, N.Y.: 493–526.

Perdue, E. M., 1990. Modeling the acid-base chemistry of organic acids in laboratory experiments and in freshwaters. In E. M Perdue & E. T. Gjessing (eds), Organic acids in aquatic ecosystems. John Wiley & Sons, N.Y.: 111–126.

Petersen, R. C., Jr., 1990. Effects of ecosystem changes (e.g., acid status) on formation and biotransformation of organic acids. In E. M. Perdue & E. T. Gjessing (eds), Organic acids in aquatic ecosystems. John Wiley & Sons, N.Y.: 151–166.

Petersen, R. C., Jr. & A. Kullberg, 1985. The octanol/

waterpartition coefficient of humic material and its dependence on hydrogen ion activity. Vatten 41: 236–239.

Petersen, R. C., Jr. & U. Persson, 1987. Comparison of the biological effects of humic materials under acidified conditions. Sci. Total. envir. 62: 387–398.

Rifai, N. & G. Bertru, 1980. La biodegradation des acides-fulviques. Hydrobiologia 75: 181–184.

Ryhänen, R., 1968. Die Bedeutung der Humussubstanzen in Stoffhaushalt der Gewässer Finnlands. Mitt. int. Ver. Limnol. 14: 168–178.

Schnitzer, M. & S. U. Khan, 1972. Humic substances in the environment, M. Dekker, N.Y., 327 pp.

Shuman, M. S., 1990. Carboxyl acidity of aquatic organic matter: possible systematic errors introduced by XAD extraction. In E. M. Perdue & E. T. Gjessing (eds), Organic acids in aquatic ecosystems. John Wiley & Sons, N.Y.: 97–109.

Sojo, L. E. & H. De Haan, 1991. Multicomponent kinetic analysis of iron speciation in Lake Tjeukemeer: comparison with iron speciation in fulvic acid solutions extracted from a peaty polder near Lake Tjeukemeer. Envir. Sci. Technol. 25: 935–939.

Sposito, G., 1986. Sorption of trace metals by humic materials in soils and natural waters. CRC Crit. Rev. envir. Cont. 16: 193–229.

Stabel, H. H., K. Moaledj & J. Overbeck, 1979. On the degradation of dissolved organic molecules from Plusssee by oligocarbophilic bacteria. Arch. Hydrobiol. Beih. Ergebn. Limnol. 12: 95–104.

Steinberg, C. E. W., 1990. Alteration of organic substances during eutrophication and effect of the modified organic substances on trophic interactions. In E. M. Perdue & E. T. Gjessing (eds), Organic acids in aquatic ecosystems. John Wiley & Sons, N.Y.: 189–208.

Steinberg, C. E. W. & A. Herrmann, 1981. Utilization of dissolved metal organic compounds by freshwater microorganisms. Verh. int. Ver. Limnol. 21: 231–235.

Steinberg, C. E. W. & U. Muenster, 1985. Geochemistry and ecological role of humic substances in lakewater. In G. R. Aiken, D. M. McKnight, R. I. Wershaw & P. MacCarthy (eds), Humic substances in soil, sediment, and water. Geochemistry, isolation, and characterization. John Wiley & Sons, N.Y.: 105–145.

Strome, D. J. & M. C. Miller, 1978. Photolytic changes in dissolved humic substances. Verh. int. Ver. Limnol. 20: 1248–1254.

Tranvik, L. J., 1988. Availability of dissolved organic carbon for planktonic bacteria in oligotrophic lakes of differing humic content. Microb. Ecol. 16: 311–322.

Tranvik, L. J., 1989. Bacterioplankton in humic lakes. A link between allochthonous organic matter and pelagic food webs. Ph.D. Thesis, Lund University, Lund, 104 pp.

Zika, R. G. & W. J. Cooper, 1987. Photochemistry of environmental aquatic systems. Am. Chemical Soc., Washington, D.C.

*Hydrobiologia* **229**: 73–91, 1992.
*K. Salonen, T. Kairesalo & R.I. Jones (eds.),*
*Dissolved Organic Matter in Lacustrine Ecosystems: Energy Source and System Regulator.*




# The influence of humic substances on lacustrine planktonic food chains

Roger I. Jones
*Division of Biological Sciences, Institute of Environmental and Biological Sciences, University of Lancaster, Lancaster LA1 4YQ, UK*

*Key words:* lakes, humic substances, plankton, food chains

## Abstract

Humic substances (HS) might influence planktonic food chains in lakes in two ways: 1) by altering the physical or chemical environment and thus modifying autotrophic primary production and the dependent food chains; 2) by acting as a direct carbon/energy source for food chains.

HS compete with phytoplankton for available quanta underwater and this effect is seen in the reduced euphotic zone depth in lakes with high concentrations of HS. Thus potential photosynthetic production is lower in the presence of HS. However, this effect can be offset in small lakes in which the depth of mixing is also reduced when HS concentrations are high. Complexation by HS of important nutrients such as iron and phosphorus may also restrict primary production.

Evidence is accumulating that photosynthetic primary production is insufficient to support measured metabolic activity in humic lakes, which implies that metabolism of allochthonous HS underpins much of the observed activity. Studies of bacterial abundance and growth in the presence of HS support the view that bacteria are the most significant utilisers of HS. This use is apparently facilitated by photolysis of HS, particularly by short wavelength radiation. Bacteria are grazed by both micro-zooplankton (heterotrophic and mixotrophic flagellates and ciliates) and macrozooplankton. It is within this microbial community that the food chains derived from autotrophic and allotrophic sources interact. These effects of HS on food chains are discussed in relation to possible implications for the response of different lake types to eutrophication.

## Introduction

According to Ruttner (1963), it was Naumann (1921) who introduced the term 'dystrophic' into his system of lake classification to describe lakes containing water visibly stained brown by the dissolved organic matter washed into these lakes from their peaty catchments. Because the production of such lakes in Sweden is usually very low, Naumann regarded them as an extreme case of oligotrophy. Thienemann (1925) realised that dystrophy is actually a separate concept to that of the production series, and differentiated between clear-water and brown-water lakes. That both groups can in practice range from oligotrophic to eutrophic was explicitly recognised by Järnefelt (1958). Hutchinson (1967) believed the term dystrophic 'suggests a more pathological condition than perhaps exists' and advised the use of Järnefelt's term 'chthoniotrophic' for lakes with pronounced brown staining.

The preoccupation with classification schemes was characteristic of early European limnology, and arguably provided a useful framework for later studies of dynamic processes in lakes, but brown-water lakes always seem to have been viewed as an oddity accommodated at best uncomfortably within mainstream developments in limnology. However, at around the same time in North America Birge & Juday were examining the organic content of numerous small Wisconsin lakes. Birge & Juday (1927) distinguished quite clearly between *autotrophic* lakes, which derive their organic matter wholly from internal sources, and *allotrophic* lakes, which derive organic matter both from within and from drainage. They went on to state that 'in these small bodies of water the external sources may be of great importance and they often completely dominate the situation'. This represented remarkable early insight into the different energy sources which might underpin lacustrine food chains.

The dissolved-organic carbon (DOC) in lakes is dominated by humic substances. Accepted extraction procedures allow these substances to be separated into insoluble humin, humic acids which are not soluble below pH 2 and fulvic acids which are soluble under all pH conditions. This fractionation scheme is somewhat arbitrary and cannot be satisfactorily related to lacustrine processes. Therefore I will simply refer to all these as dissolved humic substances (HS), defined as 'a general category of naturally occurring, biogenic, heterogeneous organic substances that can generally be characterized as being yellow to black in colour, of high molecular weight and refractory' (Aiken *et al.*, 1985). Lacustrine humic substances are generally believed to be of terrestrial origin. However, within the broad definition given above, humic substances can certainly be generated within lakes, particularly from littoral macrophytes (Steinberg & Muenster, 1985). The relative balance between autochthonous and allochthonous inputs of humic substances clearly varies greatly between clear-water and brown-water lakes. However, there is some evidence that, at least within a limited geographical region, the differences in HS between brown-water and clear-water lakes are quantitative rather than qualitative (De Haan *et al.*, 1987). The precise source of humic substances is only likely to be relevant to food chain studies if there are functional differences between humic substances from different sources. Little definite information is available about this, although there is some evidence (see Wetzel, 1983) that autochthonous DOM has a higher nitrogen content (C:N about 12) than does allochthonous DOM (C:N about 50). Such differences may mean that autochthonous DOM is less refractory, at least initially.

Humic substances might influence plankton food chains in lakes in two ways.

1) By altering the physical or chemical environment and thus modifying autotrophic primary production and the dependent food chains.
2) By acting as a direct carbon/energy source for food chains.

**Physical effects**

Dissolved organic compounds, and especially humic substances, have very marked effects on light attenuation in lakes. In clear-water lakes green is usually the most penetrating waveband, but in moderately stained waters red light penetrates as far as green, while in very brown waters red light penetrates furthest (Kirk, 1983). Such changes in spectral quality with depth may have some influence on the phytoplankton community composition (Wall & Briand, 1979), but of far greater importance is the more rapid attenuation of total photosynthetically active radiation (PAR) in brown-water lakes.

The vertical attenuation coefficient for downward irradiance of PAR ($K_{PAR}$) has been shown to be strongly dependent on water colour (Eloranta, 1978; Jones & Arvola, 1984), which is itself strongly dependent on DOC (Jones & Arvola, 1984). As a useful approximation (Kirk, 1983) $K_{PAR}$ may be partitioned into a set of partial attenuation coefficients

$$K_{PAR} = K_w + K_G + K_{TR} + K_{PH} ,$$

where $K_W$, $K_G$, $K_{TR}$ and $K_{PH}$ are the partial attenuation coefficients for PAR due to water, 'gilvin' (humic substances), tripton and phytoplankton respectively. For most lakes the contribution of water and tripton can be assumed small and invariable. Making the further assumption that the contribution of any component to $K_{PAR}$ is linearly related to its concentration, the proportion of total PAR captured by phytoplankton is approximated by

$$K_{PH}/K_{PAR} = B_c \cdot k_c/(B_c \cdot k_c + \mathrm{WC} \cdot k_p)$$

where $B_c$ is the phytoplankton biomass in mg chlorophyll $a$ m$^{-3}$, $k_c$ is the specific vertical attenuation coefficient (m$^2$ mg chl $a^{-1}$) per unit of phytoplankton biomass, WC is the water colour (mg Pt l$^{-1}$) and $k_p$ is the specific vertical attenuation coefficient (m$^2$ g Pt$^{-1}$) per unit of water colour. Although $k_c$ is known to vary between types of algae, a typical mid-range value is 0.015 (see Table 9.1 in Kirk, 1983). A value for $k_p$ of 0.011 can be taken from Fig. 1a in Jones & Arvola (1984). The proportion of total PAR captured by phytoplankton may then be calculated for a range of phytoplankton biomass in lakes of different water colour (Table 1). At low phytoplankton biomass (< 10 mg chl $a$ m$^{-3}$) increases in water colour greatly reduce the proportion of total PAR which can be captured by phytoplankton. At higher phytoplankton biomass, such as might be found in eutrophic lakes, a substantial proportion of the total PAR can still be captured by phytoplankton even with high water colour.

Such calculations provide an indication of how humic substances can compete with phytoplankton for available quanta and hence potentially restrict photosynthetic production. However, this potential effect can be reduced if the photosyn-

*Table 1.* Proportion (%) of total PAR captured by phytoplankton under different conditions of water colour and phytoplankton biomass expressed as chlorophyll $a$.

| Water colour (mgPt l$^{-1}$) | Chlorophyll $a$ (mg m$^{-3}$ | | | |
|---|---|---|---|---|
| | 1 | 10 | 50 | 100 |
| 10 | 13 | 60 | 88 | 94 |
| 50 | 3 | 23 | 60 | 75 |
| 100 | 1 | 13 | 43 | 60 |
| 300 | 0.5 | 5 | 20 | 33 |

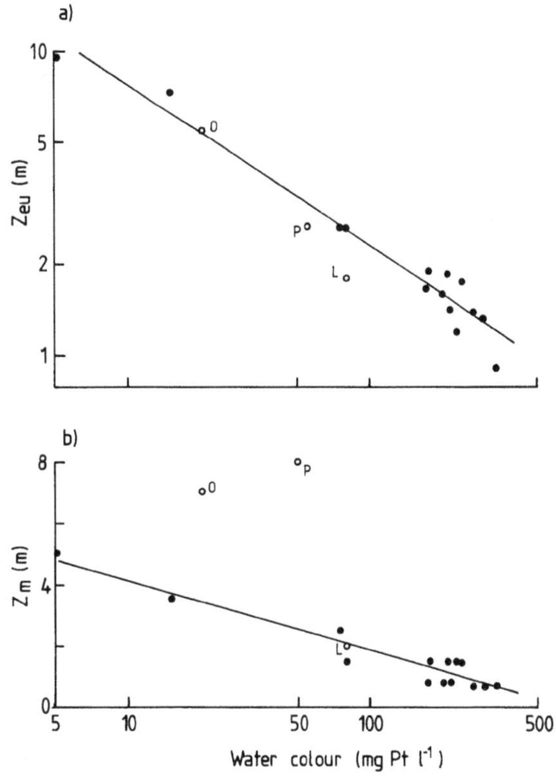


*Fig. 1.* The relation to water colour of a) summer euphotic zone depth and b) summer mixed depth for 14 small forest lakes in southern Finland (data from Jones & Arvola, 1984). The fitted regression lines are a) $Z_{eu} = 26.7\mathrm{WC}^{-0.533}$ ($r = 0.977$) and b) $Z_m = 6.4 - 0.98\ln\mathrm{WC}$ ($r = 0.956$). Also indicated are data from three larger lakes in the same vicinity, Pääjärvi, Ormajärvi and Lovojärvi.

thetic phytoplankton biomass is concentrated in the upper part of the water column. This can be achieved in two ways. Firstly, motile or buoyant algae may be able to regulate their vertical position in order to maximise their photosynthetic efficiency. In fact considerable evidence is available that flagellate algae are particularly prominent in the phytoplankton of brown-water lakes (e.g. Ilmavirta, 1988; Croome & Tyler, 1988). Secondly, if the depth of mixing ($Z_m$) is reduced during the summer stratification period then the phytoplankton biomass will be more concentrated in the upper water column and will also spend less time below the lower limit ($Z_{eu}$) of the euphotic zone.

This second possibility can be realised because the mixed depth depends on the balance between resistance to mixing, created by differential vertical heating of the water column, and the tendency to turbulent mixing from wind action. In clear-water lakes solar radiation penetrates further and the water column tends to be more uniformly heated and hence to offer less resistance to wind mixing. In brown-water lakes the incoming solar radiation is rapidly absorbed by the surface waters and much more abrupt thermal gradients can potentially arise rather rapidly with correspondingly greater thermal stability and resistance to wind mixing (Bowling, 1990). This effect is readily apparent in small, sheltered forest lakes, such as those in the Evo region of southern Finland (see Bowling & Salonen, 1990), where a strong negative correlation exists between $Z_m$ and water colour. Thus increasing water colour will decrease $Z_{eu}$ but, in these small, sheltered lakes, will also lead to decrease $Z_m$ (Fig. 1). The consequence of these opposing effects is that the effective light climate (Ramberg, 1979) experienced by the phytoplankton in these small sheltered lakes varies little (coefficient of variation, C = 23%) in comparison with the water colour (C = 56%). However, in larger brown-water lakes, although the relationship between $Z_{eu}$ and water colour is consistent (Eloranta, 1978), the greater and more variable susceptibility to wind mixing is evident in the lack of correlation between $Z_m$ and water colour (Fig. 1b).

This difference between small, sheltered and larger, more turbulent brown-water lakes is potentially important to the food chains of the lake types. A widely used model for areal photosynthesis for a whole water column is that of Talling (1957). This model can be developed (Jones, 1977b; Kirk, 1983) to provide a useful approximate expression for the critical depth ($Z_c$), the circulation depth beyond which respiratory carbon loss by a phytoplankton population exceeds photosynthetic carbon gain so that net phytoplankton growth cannot occur:

$$Z_c = [N/(24 \cdot K_{PAR} \cdot r)] \cdot \ln (I/0.5\, I_k)\,,$$

where $N$ is the daylength, $r$ is the ratio of respiration rate to light-saturated photosynthetic rate in the phytoplankton, $I$ is the average value of the sub-surface irradiance during daylight hours and $I_k$ is the irradiance value defining the onset of light saturation of photosynthesis. Column photosynthesis is known to be rather independent of the irradiance function (e.g. Jones, 1977a). Therefore the critical depth will be greatly influenced by $K_{PAR}$, which has already been shown to be highly dependent on the concentration of humic substances; higher concentrations of humic substances with the associated higher $K_{PAR}$ will decrease $Z_c$. In large lakes with effective wind mixing independent of water colour, the likelihood of the circulation depth exceeding the critical depth will thus be increased and the maximum potential rate of growth of phytoplankton populations will be decreased. However, in small, sheltered lakes in which both the light attenuation and the depth of mixing are inversely related to water colour (Fig. 1), the critical depth, and hence the potential for phytoplankton growth, will be little affected by variation in water colour.

Evidence concerning these potential effects of humic substances on phytoplankton growth and production in lakes is sparse and unsatisfactory. This is because of a lack of studies comparing processes in otherwise comparable lakes having different concentrations of humic substances. Ramberg (1979) compared two Swedish forest lakes with different effective light climates. The lower value for Botjärn, about two thirds that for Vitalampa, was attributed to the greater water colour combined with the larger size of the lake which led to a greater circulation depth. Annual pelagial primary production in Botjärn was about half that in the clearer and smaller Vitalampa. Ramberg related this difference to the different phytoplankton community composition and associated P/B differences between the two lakes. Arvola (1984) reported phytoplankton production to be only 25% higher in Alinen Mustajärvi (colour 50 mg Pt $l^{-1}$) than in Horkkajärvi (colour 250–300 mg Pt $l^{-1}$), a finding consistent with the low variation in effective light climate in these small, sheltered forest lakes (see above).

Depression of primary productivity by humic matter in northern Manitoba lakes has been reported by Jackson & Hecky (1980), who attributed this to reduced iron availability rather than any effect on the light climate. However, examination of their data shows that none of the lakes they studied had sufficient humic substances (maximum DOC 5 $mg\,l^{-1}$) to markedly affect light attenuation (maximum vertical attenuation coefficient 1.33). Moreover, their samples were collected during only three days in September. In a subsequent limnocorral study in the same region, Guildford *et al.* (1987) reported that additions of moss-peat material lowered primary productivity due to the binding of iron, although they also reported that reduction of light penetration by addition of suspended sediment depressed primary productivity.

## Chemical effects

The biogeochemistry of humic substances is reviewed elsewhere in this volume (De Haan, 1992), but some comments are appropriate on the chemical effects of humic substances as they may influence food chains in lakes.

Complexation of trace metals by humic substances has been widely reported, but the ecological implications are not well understood. Most attention has focused on iron availability to algae. Enhancement of iron uptake by phytoplankton in the presence of humic substances has been demonstrated in laboratory studies (Provasoli, 1963; Prakash *et al.*, 1973), presumably due to improved metal chelation in the medium (Anderson & Morel, 1982). On the other hand, growth inhibition of phytoplankton by high concentrations of humic substances may be caused by overchelation leading to reduced availability of iron (Siegel, 1971; Prakash *et al.*, 1973; Jackson & Hecky, 1980; Guildford *et al.*, 1987). Some bacteria and blue-green algae are known to release siderophores, complexing agents of high affinity for ferric iron, which act as a 'ligand taxi' for transporting iron into the cell (Button, 1985). Microorganisms isolated from a humic water were more efficient at utilizing organic iron compounds than a non-axenic *Chlorella* culture (Steinberg & Hermann, 1981). More information is certainly required on competition between phytoplankton and bacteria for iron in the presence of humic substances.

Phosphorus is more usually cited than iron as a limiting nutrient for phytoplankton growth in lakes, so possible chemical interactions between phosphate and HS may be even more important than iron complexation in regulating plankton growth (Steinberg & Muenster, 1985). Since concentrations of HS in lakes can frequently be several orders of magnitude greater than concentrations of free orthophosphate, even low binding affinities between these two substances could place real constraints upon the availability of phosphorus to plankton (Stewart & Wetzel, 1981). Several studies have demonstrated that under various conditions HS may associate with phosphate in the presence of iron (e.g. Francko & Heath, 1979, 1982; Stevens & Stewart, 1982; De Haan & De Boer, 1986) or manganese (Steinberg & Baltes, 1984). Jones *et al.* (1988) and De Haan *et al.* (1990), using Sephadex gel filtration combined with isotopic labelling of phosphorus and iron, demonstrated abiotic movement of $PO_4$-P to two higher molecular weight fractions (nominal MW > 100000 and 10000–20000) in filtered water from Finnish forest lakes. This movement was most pronounced in waters with high HS concentrations which also had high Fe content. During depletion of ambient free $PO_4$-P these complexes apparently slowly release bound $PO_4$-P and hence may serve as a 'phosphorus buffering system' for plankton. Release of $PO_4$-P from HS-Fe complexes is enhanced by ultra violet light (Cotner & Heath, 1990). Such a system might account for the rather long phosphate turnover times measured in these humic lakes (Jones, 1990).

Naturally occurring HS have been shown to influence plankton metabolism. Stewart & Wetzel (1982) reported that mixed natural assemblages of algae and bacteria exhibited lower rates of $^{14}C$ assimilation and higher rates of dissimilation of recent photosynthate when amended with low concentrations of unfractionated HS. Community alkaline phosphatase activity has been reported to be stimulated by additions of HS

(Stewart & Wetzel, 1982; Auclair *et al.*, 1985; Francko, 1986), as has planktonic orthophosphate uptake rate (Brasard & Auclair, 1984; Francko, 1986). In all cases HS of low apparent molecular weight was more effective in altering metabolic rates than was HS of high apparent molecular weight. Unfortunately the implications of these effects for food chain functioning cannot be stated; we need to know much more about how widespread the effects of HS may be as well knowing more about the general influence on food chain structure of variations in metabolic processes such as alkaline phosphatase activity.

## HS as a carbon/energy source

A number of studies of humic lakes have reported that the annual production of $CO_2$ from respiration of the total plankton community has exceeded the phytoplankton primary production, often several fold (e.g. Sarvala *et al.*, 1981; Salonen *et al.*, 1983; Rask *et al.*, 1986). The consistency and magnitude of this discrepancy suggests that it cannot be explained by methodological error. Abiotic photochemical carbon dioxide production could be a neglected contributory factor in highly humic lakes (Miles & Brezonik, 1981; Salonen & Tulonen, 1990). Miles & Brezonik (1981) reported *in situ* abiotic oxygen consumption up to 0.12 $mg\,l^{-1}\,h^{-1}$ in the surface water of a highly coloured Florida lake. Assuming a molar ratio of $CO_2$ produced to $O_2$ consumed of 2 (based on oxidation of humic carboxyl groups, Miles & Brezonik, 1981), this would represent a $CO_2$ production of 0.3 $mg\,l^{-1}\,h^{-1}$ or around 2160 $mgC\,m^{-3}\,d^{-1}$, which exceeds commonly reported respiration values from humic lakes (e.g. Salonen *et al.*, 1983). Such photochemical carbon dioxide production might lead to underestimation of available $CO_2$ during 14C primary production measurements and hence to underestimation of phytoplankton production in humic lakes, and could contribute to the apparent discrepancy between primary production and respiration. An alternative explanation is that additional carbon/energy sources are available to the plankton, over and above the autotrophic phytoplankton production. In small lakes inputs of organic carbon to the plankton from the littoral region may be important, but in larger lakes, and especially in humic lakes, the most important additional source of carbon is likely to be allochthonous HS.

Dissolved humic substances in lakes might serve as a substrate for bacterioplankton, leading to enhanced microbial production. Any such stimulation of bacterioplankton productivity could influence pelagial food chains in two ways: firstly by providing an alternative base (in addition to autotrophic primary production) for the energetic and nutritional support of consumer organisms, provided that the bacterial production can be effectively grazed; and secondly by increasing bacterial demand for limiting nutrients at the expense of phytoplankton and hence depressing autotrophic primary production. The possibility must also be considered that organisms other than bacteria make direct use of the carbon/energy content of HS.

## Bacterial use of HS

According to Wetzel (1983), numbers of bacteria are markedly lower in acidic dystrophic lakes which contain high concentrations of humic matter. This view is not borne out by recent evidence. Hessen (1985a) and Tranvik (1988) both reported a strong positive correlation between bacterial biomass and humic content in 8 Norwegian lakes (see Fig. 2). Arvola & Kankaala (1989) and Jones (1990), in studies of Finnish forest lakes, found no significant difference in bacterial biomass between lakes with different humic content; however, in both these studies the mean ratio of bacterioplankton biomass to phytoplankton biomass increased significantly with increasing humic content, suggesting a more important contribution from bacterioplankton in humic lakes. Tranvik (1989) reported a bacterial abundance in a humic brown-water lake consistently 2–3 times that in a nearby clear-water lake.

Relative abundance is a poor indicator of the

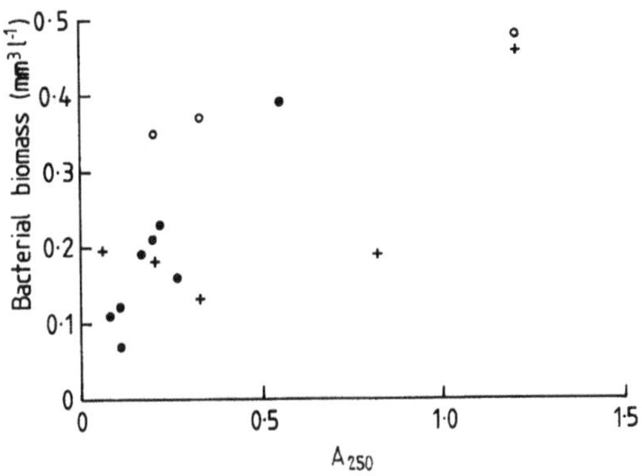


*Fig. 2.* Relation between bacterial biomass and water colour measured as UV absorbance in 8 Norwegian lakes (●, data from Hessen, 1985a), 5 Finnish fores lakes (+, data from Arvola & Kankaala, 1989) and 3 Finnish forest lakes (o, data from Jones, 1990).

likely functional importance of bacterioplankton in lakes; growth rates or measures of production are more desirable. Tranvik & Höfle (1987) set up batch cultures with filter-sterilised water from a humic and a clear-water lake and re-inoculated the cultures with bacteria from each lake. Growth curves and growth rates were similar in all cultures, regardless of inoculum or lakewater, but bacterial biomass produced was double in cultures based on water from the humic lake. Tranvik (1988) grew bacterioplankton from 10 oligotrophic lakes, representing a gradient from clear-water to brown-water, in dilution cultures of sterile filtered lakewater. The bacterial biomass achieved in the stationary phase of the dilution cultures was positively correlated with the HS content of the lakes. About 10% of the total DOC pool was consumed in the dilution cultures with all lakewaters, with approximately the same growth efficiency (26%). Thus the humic lakes could support a higher bacterial biomass because of their larger DOC pools. Tranvik (1989) reported a higher ratio of bacterial production to primary production in a humic lake compared with a clear-water lake, indicating that the bacterioplankton of the humic lake utilized allochthonous substrates in addition to substrates originating from autochthonous primary production. Satoh & Abe (1987) found that 40% of the DOC in coloured water from a mountain bog pool was decomposed by bacteria in four days. The remaining part appeared highly refractory over 90 days, with a decomposition rate constant only 5% that for the readily decomposable fraction.

Studies such as those cited provide strong evidence that HS in lakes do act as a substrate for bacteria, despite the general view of HS as refractory substances. Nevertheless, there are still few specific studies of the biodegradation of HS. De Haan (1974) showed that a *Pseudomonas* sp. isolated from the eutrophic, humic lake Tjeukemeer could not utilise fulvic acid. However, in the presence of added lactate, fulvic acid was partially degraded, leading to an increase in the cell yield, which can be explained in terms of co-metabolism. Similar results were reported by Stabel *et al.* (1979). De Haan (1977) further demonstrated that an *Arthrobacter* sp. from Tjeukemeer was able to utilise fulvic acid, which was only partially degraded, giving a small cell yield compared with that on benzoate. In media containing both benzoate and fulvic acid, the *Arthrobacter* showed a higher growth rate as well as higher cell yield than in media with only benzoate or fulvic acid. De Haan interpreted these results as a priming effect of benzoate. Correlation between fluctuations in the content of fulvic acids and the amount of benzoate-oxidizing bacteria suggested that the priming effect might be more important than co-metabolism during the decomposition of fulvic acids in Tjeukemeer.

Photodegradation of HS in lakewater is known to occur (e.g. Strome & Miller, 1978; Geller, 1985a) and there is accumulating evidence that photochemical breakdown of biologically refractory HS into lower molecular weight, biologically labile organic products may be an important initial step in the bacterial decomposition of HS (Strome & Miller, 1978; Geller, 1985b; Amador *et al.*, 1989; Kieber *et al.*, 1989). This evidence derives from laboratory experiments, often with exposure to unrealistically intense short wavelength radiation. Since penetration of shorter wavelengths into humic lakes is very restricted, more information is needed on the likely *in situ* importance of these photochemical processes.

Francko & Heath (1982) did observe a diurnal fluctuation in an acid bog lake of both the concentration of HS and in the amount of UV-sensitive complex P. Slow photolysis of HS may also be induced by non-UV radiation (Geller, 1985a). Some authors have argued (e.g. Currie & Kalff, 1984; Bratbak & Thingstad, 1985) that bacteria in aquatic systems should be carbon limited rather than mineral nutrient limited. In oligotrophic, clear-water Mirror Lake, enrichment of lakewater with N and P caused a two- to three-fold increase in chlorophyll concentration; in bottles to which 4 mg $l^{-1}$ organic C (glucose, peptone, yeast extract) was also added, the chlorophyll concentration increased only 36% and when 20 mg $l^{-1}$ organic C was added the chlorophyll concentration decreased by 19% (Makarewicz *et al.*, 1985). If HS in lakes can serve as a carbon substrate, this would be expected to release bacteria from dependence on carbon supply and to increase their dependence on mineral nutrient supplies in humic lakes. Evidence for this is conflicting. Some authors have reported that bacterial growth in humic waters, or bacterial degradation of HS, is enhanced by additions of phosphorus or nitrogen (Ryhänen, 1968; Sederholm *et al.*, 1973; Satoh & Abe, 1987; Salonen *et al.*, 1987), but Strome & Miller (1978) found no stimulation of bacterial degradation of HS upon addition of N and P. Watanabe & Goldman (1984) studied the heterotrophic bacterial community in the clear-water, ultraoligotrophic Lake Tahoe. They found that growth of bacteria was not stimulated by addition of glucose, N, P or N + P, but that addition of glucose and mineral nutrients produced a remarkable stimulation of bacterial growth. Jones (1990) found that phosphate uptake by both bacteria and algae from a highly humic lake was stimulated after storage of water for several days. The nutritional status of bacterioplankton is evidently delicately balanced, but there is insufficient evidence to make distinctions between lakes with different concentrations of HS. There are few comparative studies of phosphate uptake partitioning between algae and bacteria in brown-water and clear-water lakes. In a study of three lakes of widely differing water colour, Jones (1990) found no significant difference between the lakes in the orthophosphate uptake partitioning between algae and bacteria. At present there seems no reason to believe that phytoplankton production in brown-water lakes is suppressed by diversion of mineral nutrients to bacteria which are metabolising HS.

Given that bacteria are indeed able to make use of HS as a substrate, the question then arises of the relative importance to bacterial production of allochthonous HS and DOM generated within a lake, principally through release of photosynthetically fixed dissolved organic carbon (PDOC) from phytoplankton. (I am deliberately ignoring influx of DOM from the littoral zone, which is reviewed elsewhere in this volume by Wetzel.) Again, a lack of concensus over the operation of these processes in clear-water lakes makes it difficult to identify a distinctive pattern in humic lakes.

Reliance on PDOC by bacterioplankton and a tight coupling between production and decomposition of organic matter in the pelagial zone of lakes have been emphasised in many recent publications (e.g. Bell & Kuparinen, 1984; Jones & Cannon, 1986; Chrost *et al.*, 1989; Vadstein *et al.*, 1989). In some circumstances 60% or more of phytoplankton production may cycle through bacterioplankton (Chrost *et al.*, 1989; Vadstein *et al.*, 1989). Bacteria evidently may assimilate and use for growth a large fraction of the PDOC, especially in eutrophic lakes where algal PDOC may support most of the heterotrophic bacterial production Coveney, 1982; Bell & Kuparinen, 1984; Vadstein *et al.*, 1989). Even in oligotrophic lakes PDOC may account for about one third of bacterial production (Cole *et al.*, 1982).

Little information is available concerning the likely importance of PDOC to bacterial production in humic lakes. During the period of ice melt and spring phytoplankton growth in small forest lakes of various water colour, the biomass of bacterioplankton also increases (Arvola & Kankaala, 1989; Jones, 1990). A similar broad coincidence between bacterial production and phytoplankton biomass was noted during the summer in the large humic lake Pääjärvi (Salonen, 1981). It is not clear whether this kind of coincidental occurrence reflects a real dependence of bacterial production upon phytoplankton. At this time of year, inflowing water may provide a high loading of both inorganic nutrients and more readily utilizable organic material. It is possible that both groups of organisms are responding to changing environmental conditions, particularly if the increased irradiance around ice melt stimulates both phytoplankton production and photochemical facilitation of bacterial usage of HS (see above).

Jones & Salonen (1985) found that in humic forest lakes up to 40% of PDOC was utilised by bacteria within four hours but the remainder was hardly used. Although the magnitude of PDOC release reported by Jones & Salonen (1985) for humic forest lakes was comparable with values reported from other lakes, the absolute daily PDOC release was only about 0.1% of the total DOC pool in these lakes. Their rough calculations suggested a bacterial use of PDOC of 2.8 mg C $m^{-3}$ $d^{-1}$ in the epilimnion compared with an estimated bacterial respiration of 60 mg C $m^{-3}$ $d^{-1}$. A rather low dependence of bacterial production of PDOC in a moderately humic lake was also reported by Sundh (1989). More studies are needed, but these results indicate that PDOC does not have an important quantitative role for bacterial production in humic lakes. However, PDOC may be qualitatively important if it contains substances that are essential for optimal growth of bacteria.

## Grazing on bacteria

If bacterial utilization of HS is to be important in the food chains of humic lakes, the resulting bacterial production must be effectively utilised by grazers. There is no doubt that this is generally the case, but there is conflicting evidence about the extent to which bacteria are grazed directly by macrozooplankton or grazed mainly by microzooplankton, including flagellates, which themselves are grazed by macrozooplankton.

The ability to feed on bacteria has been demonstrated for many species of macrozooplankton (Peterson *et al.*, 1978; Geller & Müller, 1981; Pedros-Alio & Brock, 1983; Forsyth & James, 1984; Hessen, 1985b; Kankaala, 1988). Such ability may be particularly important in humic lakes with a high bacterial production relative to that of phytoplankton. Hessen *et al.* (1989) studied crustacean zooplankton grazing and resource utilization in a humic lake. They found species-specific differences in the selection ratio of large to small particles, and large species-specific differences in the assimilation efficiency with algae and bacteria. In general, assimilation efficiencies for algae were considerably higher than those for bacteria, but the high biomass of bacteria compared to algae (3:1) made bacteria an important food source for the zooplankton (Hessen *et al.*, 1990). Kankaala (1988) used $^{14}C$ labelled *Chlamydomonas* and $^{3}H$ labelled natural bacteria to study grazing by *Daphnia longispina*, the dominant macrozooplankter in the highly humic lake Mekkojärvi. She also found that filtering rates on bacteria were lower (26–92%) than those on algae. Assuming that the *Daphnia* nonselectively ingested the natural phytoplankton (mainly small flagellates) at the same rate as the $^{14}C$ labelled *Chlamydomonas*, from June to August algae comprised 56–93% of the food ingested by *Daphnia*. In September and October, when the *Daphnia* population was declining and the algal biomass was low, bacteria comprised 73% and 55% respectively of the food of *Daphnia*. However, it should be recognised that these results are highly sensitive to the value selected to convert bacterial biovolume to carbon.

In many humic lakes in Fenno-Scandia, cladocerans appear to predominate in the zooplankton (Rask *et al.*, 1986; Kankaala, 1988; Salonen & Arvola, 1988), although in some cases rotifers are a substantial component (Latja, 1974; Hessen & Schartau, 1988). Fish predation on large crustacean zooplankton might be less effective in dark, humic water, although in fishless, humic Triangle Lake, Ohio, rotifers were dominant (Havens, 1989).

In recent years fundamental changes have taken place in the accepted view of pelagial food chains in aquatic systems with the realization that microheterotrophic processes can account for a substantial part of the primary production (e.g. Williams, 1981; Azam *et al.*, 1983). Initial work by marine scientists focused on non-pigmented, heterotrophic nanoflagellates. More recently it has become apparent that pigmented, phagotrophic phytoflagellates are also important in these processes (e.g. Estep *et al.*, 1986; Sanders & Porter, 1988; Boraas *et al.*, 1988), and in freshwaters these mixotrophic flagellates may be the most significant microheterotrophic grazers on bacteria.

Many incidences of algal phagotrophy have been reported in the literature (see Sanders & Porter, 1988), but most of these derive from casual observations and almost all from laboratory studies. Quantifying the importance of phagotrophy in the ecology of phytoflagellates and the role of phagotrophic phytoflagellates in food chains in lakes is still in the very early stages. In a rare field study, Bird & Kalff (1987) reported that a metalimnetic peak of *Dinobryon* in a small humic lake obtained the greater part of its carbon through phagocytosis of bacteria. Several other chrysomonad genera were found to ingest particles and, in another oligotrophic lake, 30% of the phytoplankton were actively ingesting small particles so that these phytoplankton were responsible for most bacterial grazing in the lake. Subsequently Bird & Kalff (1989) reported that about 80% of algal community carbon assimilation by a concentrated layer of chrysomonads (*Dinobryon* and *Ochromonas*) in the metalimnion of Lac Gilbert was accounted for by phagotrophy during an overcast day. Vadstein *et al.* (1989) reported that heterotrophic flagellates were the main grazers on bacteria in the photic zone of a eutrophic lake. However, not all reports support such an important role for phagotrophy by flagellates. Pratt & Chappell (1989) suggested that numbers of planktonic bacteria in a eutrophic lake might not be controlled by microflagellate predation; moreover, clearance rates of heterotrophic microflagellates were insufficient to maintain flagellate growth, suggesting that other energy sources were needed. Veen (1990) found that bacteria were of only minor importance as a carbon source for growth of *Dinobryon divergens* in a Dutch lake and suggested that phagotrophy by the flagellate might be more important as a source of trace nutrients.

With this sort of conflicting evidence, it is difficult to say whether the relative magnitude of bacterial grazing by heterotrophic flagellates is any different in humic lakes. The particular importance of flagellates in the phytoplankton of humic lakes (Ilmavirta, 1984, 1988; Croome & Tyler, 1988), coupled with the widespread occurrence of phagotrophy amongst flagellates, could indicate a particularly important role for microheterotrophy in humic lakes. Tranvik (1989), in a comparison of a humic and a clearwater lake, reported that most bacterial loss in both lakes could be attributed to flagellates. However, since the ratio of bacterial to primary production was higher in the humic lake, the relative carbon/energy flow through flagellate heterotrophy must also have been greater in the humic lake. In the small, highly humic lake, Mekkojärvi, where the phytoplankton during June to August was dominated by flagellates, a large proportion were found to ingest bacteria-sized particles (Salonen & Jokinen, 1988). Some species, notably *Ochromonas* and *Chromulina*, seemed to acquire a high proportion of their carbon from ingestion of bacteria. Other species apparently acquired little carbon from bacteria, while some *Mallomonas* and *Cryptomonas* species were not observed to ingest particles. Tranvik *et al.* (1989) also reported only very low rates of bacterivory by *Cryptomonas* from a humic lake. Salonen & Jokinen (1988) calculated that heterotrophic production by flagellate grazing on bacteria was only about 6% of phytoplankton primary production in the euphotic zone of Mekkojärvi. Nevertheless, since during the June to August period the dominant macrozooplankton in the lake were feeding on algae rather than bacteria (Kankaala, 1988), and if $<5\%$ of the carbon of zooplankton from such a highly humic lake was of photosynthetic origin (Salonen & Hammar, 1986), it follows that an energy pathway from HS to macrozooplankton via heterotrophic flagellates must be of considerable importance. Again, this conclusion is sensitive to the conversion factor from bacterial biovolume to carbon.

## Direct use of HS by other organisms

The capacity for uptake of dissolved organic matter is known to be widespread amongst invertebrate phyla, but appears to proceed at only a very slow rate, or to be completely absent in freshwater invertebrates, possibly because the uptake mechanism is incompatible with the processes of osmoregulation (Sepers, 1977). Many protozoa, both freshwater and marine, are known to be capable of utilizing dissolved organic compounds for growth, either alone or as a supplement to other energy sources (e.g. Sleigh, 1989), and these include photosynthetic types. Many common phytoplankton have been shown to be facultatively heterotrophic, capable of uptake of organic solutes (e.g. Sepers, 1977; Sandgren, 1988). However, in virtually all cases the organic compounds involved have been simple sugars, organic acids and amino acids, which are typically present at very low concentrations in lakes (Wetzel, 1983), although they may be at higher concentrations in eutrophic lakes (e.g. De Haan & De Boer, 1979). Algae and protozoa are unlikely to successfully compete with bacteria for these compounds because of surface to volume considerations.

Evidence for direct use of HS by organisms other than bacteria is slender. Sherr (1988) has demonstrated direct use of high molecular weight polysaccharide by heterotrophic flagellates from a salt marsh estuary and a small pond. The pond flagellates were capable of ingesting molecules of the polysaccharide dextran of relative molecular mass > 500 000, and showed enhanced growth of in the presence of 2 000 000, but not 40 000, dextran. DOC generated within a lake may contain a greater proportion of high molecular weight (> 100 000) carbon than DOC in inflow streams (Cole *et al.*, 1984) and might be quantitatively important in clear-water lakes. However, Sephadex gel filtration of HS from Finnish humic lakes suggested only 5% of the HS carbon had a nominal molecular weight > 100 000 while the bulk of the HS carbon (55–75%) had a nominal molecular weight around 30 000 (De Haan *et al.*, 1987). Although such figures do not suggest that the kind of high molecular weight polysaccharides identified by Sherr (1988) are likely to be of importance in the direct nutrition of flagellates in humic lakes, it should be remembered that even 5% of HS carbon can still represent a considerable quantity in humic lakes.

Salonen & Hammar (1986) demonstrated that crustacean zooplankton from several humic lakes obtained a very high proportion of their carbon from allochthonous HS. They did not rule out the possibility of direct utilization of HS by zooplankton, but their results could be explained by a food chain from HS to zooplankton via an intermediate trophic level. Recently Salonen *et al.* (1990) have reported enhanced growth of axenic cultures of *Chlamydomonas* sp. and *Cryptomonas* sp. from a humic lake in inorganic medium supplemented with HS, but only in the light. The possibility of significant direct utilization of HS by algae and even zooplankton certainly needs more study, but pending more conclusive evidence to the contrary, it is reasonable to consider that bacteria are the only important direct consumers of HS in lakes.

## Food web interactions

Until relatively recently planktonic food webs were regarded as examples of simple systems in which phytoplankton primary production passed up the food chain through rather discrete trophic levels of metazoans to fish. Energy was lost at each stage by respiration and excretion. Bacteria were also present as 'mineralizers' of dead and excreted organic matter. Protozoans were noticed in passing by those studying phytoplankton and by those studying zooplankton, but were conveniently ignored by both. Perceptions of planktonic food webs have now changed greatly, with a focus of attention on the role of bacteria and protozoa. Much of the impetus for this change has come from marine scientists and their recognition of the 'microbial loop' (Pomeroy, 1974; Williams, 1981;

Pomeroy & Wiebe, 1988). However, in marine plankton systems the DOM which drives the microbial loop can be considered to derive entirely from phytoplankton production via excretion, death or organisms and messy feeding of grazers. Hence the argument has concentrated on the extent to which the microbial loop can provide an energy link to metazoa and thus recycle the 'lost' primary production. Although in principle such a link may exist, in practice its significance must depend upon the number of transfer steps and the efficiency of transfer at each step. Similar arguments apply to lake plankton, and probably particularly so in eutrophic lakes where most DOM derives from phytoplankton production (e.g. Vadstein *et al.*, 1989). In clear-water oligotrophic lakes a substantial part of the DOM is also autochthonous in origin (e.g. Cole *et al.*, 1984). However, in humic lakes the overwhelming majority of the DOM is allochthonous, so that in these lakes the most important function of a microbial sub-system may be as a link for transferring energy from allochthonous HS to metazoans, rather than as a loop for recycling lost primary production.

The quantitative role of these trophic interactions, only recently identified in aquatic systems, is not yet understood in any kind of lake, so it is hardly possible to do more than speculate on the likely significance of these processes in humic lakes. Recent conceptual advances in understanding of plankton interactions have relied heavily on laboratory measurements of rates of processes in model systems (e.g. Güde, 1985; Caron *et al.*, 1988). However, there is growing evidence that in aquatic systems interactions within the microbial components of plankton food webs are particularly finely balanced, possibly as a result of internal, biotic feedback controls (Sherr *et al.*, 1988). Disruption of these controls by removal of a critical component may then lead to measurement of processes at rates which bear little relation to the magnitude of those rates in nature. Extrapolation from such simplified model systems to *in situ* planktonic food webs must be approached with the greatest care.

The use of microcosms (or mesocosms, depending on one's view of scale) has become popular as a possible means of bridging the gap between oversimplified laboratory systems and the open lake, although mesocosms themselves inevitably introduce various degrees of deviation from the latter. Nevertheless, this approach is one which might be used more to study the influence of HS on planktonic food webs.

Salonen *et al.* (1992a, 1992b) have used mesocosms to investigate food web interactions in the plankton of a steeply stratified and highly humic lake, Mekkojärvi. The results from their two mesocosm experiments suggest a shifting pattern of food web interactions through the year in Mekkojärvi. In spring, when macrozooplankton are rare, phytoplankton biomass develops to a peak and production exceeds bacterial production. Grazing by developing populations of macrozooplankton, and by some micrograzers such as *Ochromonas* which are able to graze the smaller phytoplankton, reduces the phytoplankton biomass and production to low levels. The zooplankton populations are then mainly sustained by the bacterial production from HS, in part by direct grazing on bacteria and in part via heterotrophic nanoflagellates which graze the bacteria.

The consistency of this pattern between lakes and in successive years needs to be verified, and more seasonal resolution of the changing relative importance of autotrophic and heterotrophic production is also needed. In particular there is a need to know to what extent such a pattern may be applicable to larger humic lakes with different morphometric and physical conditions, and in which more efficient mixing will generate a less favourable light climate for autotrophic phytoplankton. Consequently, at this point it is only possible to speculate about some of the ways in which humic substances might affect food web structure.

Hessen *et al.* (1990) also used a microcosm to investigate summer carbon metabolism in a humic lake. They found that DOC was 80–85% of total carbon, while 75% of POC was detritus. Bacterial biomass and production exceeded those of phytoplankton, and bacteria contributed more to zooplankton body carbon. However, most zooplankton body carbon (46–82%) was apparently derived from direct ingestion of the large detrital carbon pool. The loop of ingestion and defecation was important, giving a detritus particle turnover rate of $0.39\ d^{-1}$. Although it was not clear what proportion of this detrital carbon had originated autochthonously and what proportion allochthonously, Hessen *et al.* concluded that carbon cycling in humic lakes is essentially different from that in clear-water lakes.

In lakes without a significant allochthonous loading of HS, autochthonous primary production by phytoplankton must underpin the food chain (Fig. 3a), and a variable proportion of this is channelled through the microbial loop. Allochthonous inputs of DOM, which represent a quite separate carbon/energy source, are also channelled through the microbial loop (Fig. 3b). So far as is known, the organisms which make up the microbial loop can be the same in both cases. Thus two conceptually separate food chains, powered by distinct energy sources, come together in the microbial loop (Fig. 4). This gives us good reason to believe that the key to understanding the relative contribution to plankton dynamics of these separate energy sources lies with better understanding of these microbial processes.

In terms of food chain structure, phytoplankton and bacteria can therefore be viewed as occupying the same trophic level, what I have termed 'energy mobilisers' (Fig. 4). That is, they are organisms which serve to mobilise, and to make available to higher trophic levels, energy which is otherwise unavailable, whether in a physical form (light) or a chemical form (DOM). The 'energy dissipators' are then those organisms which take advantage of this mobilization of energy within the food chain but which dissipate it through their respiration. Some of the carbon/energy passing up the food chain will be returned to the DOM pool by excretion of PDOC from phytoplankton and by excretion and messy feeding of the energy dissipators. Of course, in practice such returned DOM will be identical irrespective of whether its carbon originated autochthonously or allochthonously. However, in order to allow a complete assessment of the relative importance of autochthonous and allochthonous inputs of energy, it will be necessary to know the extent to which

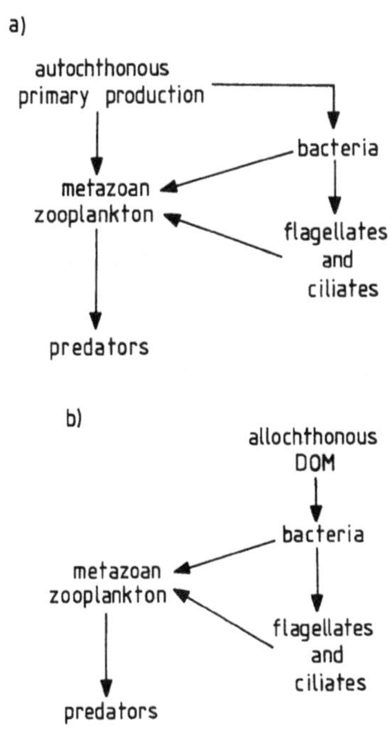


*Fig. 3.* Plankton food chains based on a) autochthonous primary production or b) allochthonous inputs of dissolved organic matter.

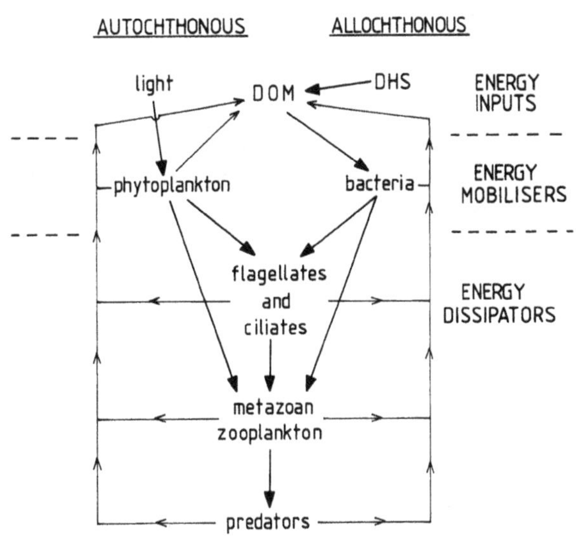


*Fig. 4.* Plankton food chain structure illustrating the separate autochthonous and allochthonous energy sources becoming intertwined within the 'microbial loop'.

carbon from each source is recycled through the DOM pool, or through the pool of particulate detritus. It is possible that some insight into the involvement in lacustrine food chains of allochthonous carbon of terrestrial origin will be obtained from studies of carbon isotope ratios (e.g. Schell, 1983). However, in humic lakes this approach may prove to be fraught with difficulty if much of the inorganic carbon available for phytoplankton photosynthesis originates not from atmospheric invasion but from release from allochthonous HS, either through biological mineralization or through photochemical oxidation.

In temperate zone lakes plankton dynamics is strongly linked to the seasonal variability in solar energy input. An important question is whether in humic lakes the continual presence of a large energy reservoir of DOM of allochthonous origin serves to stabilise seasonal fluctuations of populations within food chains. For this to be the case bacterial production would have to show less seasonal variability in humic lakes than in clear-water lakes. Insufficient evidence is available to evaluate this. However, bacterial biomass appears to show similar seasonal trends in lakes with different water colour (Arvola & Kankaala, 1989; Jones, 1990), with lower values during the winter. Control of bacterioplankton density and production is generally assumed to be effected by substrate supply and grazing (Wright, 1988). Grazing rate may be assumed to be lower during the winter (although more information is needed about winter populations of heterotrophic flagellates) so lower winter bacterial densities presumably reflect reduced substrate supply, although lower winter temperatures are also likely to reduce bacterial growth rates. Since the total DOM pool in humic lakes, like that in oligotrophic lakes (Cole *et al.*, 1984), fluctuates very little on a seasonal basis, a winter reduction in substrate supply to bacteria is likely to reflect a reduction in the labile fraction of the total DOM. This could either mean a reduced supply of important autochthonous material or a change in the availability of HS. Inputs of total HS are linked to hydrology and can show marked seasonal fluctuations in catchments with prolonged ice and snow cover (e.g. Hakala, 1974). Also the quality of HS in lakes can vary seasonally (e.g. Visser, 1984). Such variations, coupled with the reduced photochemical breakdown of DOM at this time, raise the possibility that supply of readily available substrate to bacteria might be seasonally highly variable even in lakes with a relatively constant DOM concentration.

During the winter, then, in humic lakes the bacterioplankton may be primarily a dormant, substrate-limited community (Wright, 1984) able to sustain only a small community of microzooplankton grazers. In spring fresh inputs of labile allochthonous HS and autochthonous PDOC, perhaps with enhanced photochemical activity, stimulates an increase in bacterial production. This in turn promotes a rapid development of grazing flagellates until a quasi steady state is reached with an active, grazer-controlled bacterioplankton (Wright, 1984). Subsequent and slower development of macrozooplankton populations utilizing the nanoflagellates reduces the direct grazing pressure on bacteria, which may then oscillate between substrate-limitation and grazer-control. Identification of valid differences in any such seasonal trends between humic and clear-water lakes will require carefully designed comparative studies.

In these speculations about food chain structure in humic lakes I have made little reference to the possible role of limiting nutrients. Phosphorus is most often the limiting nutrient in lakes, and there is as yet no reason to suppose the situation is fundamentally different in humic lakes. For example, chemical budgets for a small humic lake in southern Finland (Arvola *et al.*, 1990) show an increase in both C:N and C:P ratios between the inflows and the outflow, but most markedly in the C:P ratio (Table 2). This indicates that nitrogen and especially phosphorus are being retained in the lake to a greater degree than carbon. However, fluctuations in the concentration of phosphorus available to plankton may be less marked in humic lakes because of phosphate association with and disassociation from DHM-Fe-$PO_4$ complexes (Jones *et al.*, 1988; De Haan *et al.*,

*Table 2.* C:N and C:P ratios for organic carbon, total nitrogen and total phosphorus inputs (streamwater and groundwater) and outputs (outflow) for the small humic lake, Nimetön, in southern Finland. Recalculated from Arvola *et al.* (1990).

| | C:N | | | C:P | | |
|---|---|---|---|---|---|---|
| | in | out | out/in | in | out | out/in |
| 1981 | 27 | 33 | 1.2 | 268 | 353 | 1.3 |
| 1982 | 20 | 39 | 2.0 | 190 | 459 | 2.4 |
| 1983 | 27 | 28 | 1.0 | 229 | 382 | 1.7 |
| Mean | – | – | 1.4 | – | -- | 1.8 |

1990; Jones, 1990). Thus nutrient-induced population oscillations might be dampened in humic lakes.

## Humic substances and trophic state

Returning to the concept of Birge & Juday (1927) of autotrophic and allotrophic lakes, we can expect that humic lakes should sit at one end of a continuum between the two conditions. However, the balance between autotrophy and allotrophy should shift with changing eutrophy (Fig. 5). Oligotrophic clear-water lakes must be predominantly autotrophic, but will have a small allotrophic element due to allochthonous inputs of DOM. With increased nutrient loading and a corresponding move to greater eutrophy, the increase in autochthonous primary production while allochthonous inputs remain unchanged will result in an asymptotic approach to absolute autotrophy. At the other extreme, an oligotrophic but highly humic lake may be predominantly allotrophic in character, especially larger lakes with more effective mixing and hence less favourable light climates for autotrophic primary production. With increased nutrient loading such a lake may pass through several stages (Fig. 5). 1) The increased nutrient input stimulates bacterial metabolism of allochthonous DHM; since bacteria are more efficient nutrient scavengers than algae, there is a disproportionately small increase in autochthonous primary production. 2) With further increases in nutrient loading, bacteria are no longer nutrient limited and an increasing proportion of the inorganic nutrients is utilised in autochthonous primary production. 3) As nutrient saturation is approached the upper limit of phytoplankton production is determined by the effective light climate, so that physical parameters set the ratio of autotrophy to allotrophy. These postulated different responses of clear-water and brown-water lakes to eutrophication should be testable using experimental microcosms with controlled additions of nutrients.

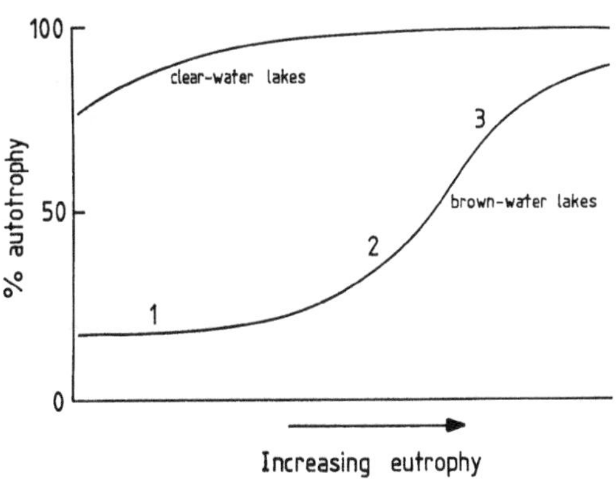


*Fig. 5.* Representation of the effect of increasing eutrophy on the proportion of total energy inputs originating autotrophically for clear-water and brown-water lakes.

Much of the discussion in this paper has been more speculation than review of confirmed knowledge about food chains in humic lakes. This is inevitable when perceptions of food chain structure in all aquatic ecosystems are currently in a state of flux. But it also reflects a paucity of experimental evidence on the influence of humic substances on food chains in lakes. Until more work is undertaken it will not be possible to distinguish between the particular and the general. I have expressed the view that the 'microbial loop' can be the key stage at which differences between clear-water and brown-water lakes manifest themselves. I also believe that most rapid progress in understanding such differences can be brought about by well designed comparative studies between lakes of different humic content and between humic lakes of different size and morphometry.

## References

Aiken, G. R., D. M. McKnight, R. L. Wershaw & P. MacCarthy (eds), 1985. Humic substances in soil, sediment and water. J. Wiley & Sons, N.Y., 692 pp.

Amador, J. A., M. Alexander & R. G. Zika, 1989. Sequential photochemical and microbial degradation of organic molecules bound to humic acid. Appl. envir. Microbiol. 55: 2843–2849.

Anderson, M. A. & F. M. M. Morel, 1982. The influence of aqueous iron chemistry on the uptake of iron by the coastal diatom *Thalassiosira weissflogii*. Limnol. Oceanogr. 27: 789–813.

Arvola, L., 1984. Vertical distribution of primary production and phytoplankton in two small lakes with different humus concentration in southern Finland. Holarct. Ecol. 7: 390–398.

Arvola, L. & P. Kankaala, 1989. Winter and spring variability in phyto- and bacterioplankton in lakes with different water colour. Aqua fenn. 19: 29–39.

Arvola, L., K. Salonen & M. Rask, 1990. Chemical budgets for a small dystrophic lake in southern Finland. Limnologica (Berlin) In Press.

Auclair, J. C., P. Brassard & P. Couture, 1985. Effects of two molecular weight fractions on phosphorus cycling in natural phytoplankton communities. Wat. Res. 19: 1447–1453.

Azam, F., T. Fenchel, J. G. Field, J. S. Gray, L.-A. Meyer-Reil & F. Thingstad, 1983. The ecological role of water-column microbes in the sea. Mar. Ecol. Prog. Ser. 10: 257–263.

Bell, R. T. & J. Kuparinen, 1984. Assessing phytoplankton and bacterioplankton production during early spring in Lake Erken, Sweden. Appl. envir. Microbiol. 45: 1709–1721.

Bird, D. F. & J. Kalff, 1987. Algal phagotrophy: Regulating factors and importance relative to photosynthesis in *Dinobryon* (Chrysophyceae). Limnol. Oceanogr. 32: 277–284.

Bird, D. F. & J. Kalff, 1989. Phagotrophic sustenance of a metalimnetic phytoplankton peak. Limnol. Oceanogr. 34: 155–162.

Birge, E. A. & C. Juday, 1927. The organic content of the water of small lakes. Proc. Amer. Phil. Soc. 66: 357–372.

Boraas, M. E., K. W. Estep, P. W. Johnson & J. McN. Sieburth, 1988. Phagotrophic phototrophs: The ecological significance of mixotrophy. J. Protozool. 35: 249–252.

Bowling, L. C., 1990. Heat contents, thermal stabilities and Birgean wind work in dystrophic Tasmanian lakes and reservoirs. Aust. J. Mar. Freshwat. Res. 41: 429–441.

Bowling, L. C. & K. Salonen, 1990. Heat uptake and resistance to mixing in small humic forest lakes in southern Finland. Aust. J. Mar. Freshwat. Res. 41: 747–759.

Brassard, P. & J. C. Auclair, 1984. Orthophosphate uptake rate constants are mediated by the $10^3$–$10^4$ molecular weight fraction in Shield lakewater. Can. J. Fish. aquat. Sci. 41: 166–173.

Bratbak, G. & T. F. Thingstad, 1985. Phytoplankton-bacteria interactions: an apparent paradox? Analysis of a model system with both competition and commensalism. Mar. Ecol. Prog. Ser. 25: 23–30.

Button, D. K., 1985. Kinetics of nutrient-limited transport and microbial growth. Microb. Rev. 49: 270–297.

Caron, D. A., J. C. Goldman & M. R. Dennett, 1988. Experimental demonstration of the roles of bacteria and bacterivorous protozoa in plankton nutrient cycles. Hydrobiologia 159: 27–40.

Chrost, R. J., U. Münster, H. Rai, D. Albrecht, P. K. Witzel & J. Overbeck, 1989. Photosynthetic production and exoenzymatic degradation of organic matter in the euphotic zone of a eutrophic lake. J. Plankton Res. 11: 223–242.

Cole, J. J., G. E. Likens & D. L. Strayer, 1982. Photosynthetically produced dissolved organic carbon: an important carbon source for planktonic bacteria. Limnol. Oceanogr. 27: 1080–1090.

Cole, J. J., W. H. McDowell & G. E. Likens, 1984. Sources and molecular weight of 'dissolved' organic carbon in an oligotrophic lake. Oikos 42: 1–9.

Cotner, J. B. & R. T. Heath, 1990. Iron redox effects on photosensitive phosphorus release from dissolved humic materials. Limnol. Oceanogr. 35: 1175–1181.

Coveney, M. F., 1982. Bacterial uptake of photosynthetic carbon from freshwater phytoplankton. Oikos, 38: 8–20.

Croome, R. L. & P. A. Tyler, 1988. Phytoflagellates and their ecology in Tasmanian polyhumic lakes. Hydrobiologia 161: 245–253.

Currie, D. J. & J. Kalff, 1984. A comparison of the abilities of freshwater algae and bacteria to acquire and retain phosphorus. Limnol. Oceanogr. 29: 298–310.

De Haan, H., 1974. Effect of a fulvic acid fraction on the growth of a *Pseudomonas* from Tjeukemeer (The Netherlands). Freshwat. Biol. 4: 301–309.

De Haan, H., 1977. Effect of benzoate on microbial decomposition of fulvic acid in Tjeukemeer (The Netherlands). Limnol. Oceanogr. 22: 38–44.

De Haan, H., 1992. Impacts of environmental changes on the biogeochemistry of aquatic humic substances. Hydrobiologia 229: 59–71.

De Haan, H. & T. De Boer, 1979. Seasonal variations of fulvic acids, amino acids, and sugars in Tjeukemeer, The Netherlands. Arch. Hydrobiol. 85: 30–40.

De Haan, H. & T. De Boer, 1986. Geochemical aspects of aqueoua iron, phosphorus and dissolved organic carbon in the humic Lake Tjeukemeer, The Netherlands. Freshwat. Biol. 16: 661–672.

De Haan, H., R. I. Jones & K. Salonen, 1987. Does ionic strength affect the configuration of aquatic humic substances, as indicated by gel filtration? Freshwat. Biol. 17: 453–459.

De Haan, H., R. I. Jones & K. Salonen, 1990. Abiotic transformations of iron and phosphate in humic lake water, revealed by double isotope labelling and gel filtration. Limnol. Oceanogr. 35: 35: 491–497.

Eloranta, P., 1978. Light penetration in different types of lakes in Central Finland. Holarct. Ecol. 1: 362–366.

Estep, K. W., P. G. Davis, M. D. Keller & J. McN. Sieburth, 1986. How important are oceanic algal nanoflagellates in bacterivory? Limnol. Oceanogr. 31: 646–650.

Forsyth, D. J. & M. R. James, 1984. Zooplankton grazing on lake bacterioplankton and phytoplankton. J. Plankton Res. 6: 803–810.

Francko, D. A., 1986. Epilimnetic phosphorus cycling: Influence of humic materials and iron coexisting major mechanisms. Can. J. Fish. aquat. Sci. 43: 302–310.

Francko, D. A. & R. T. Heath, 1979. Functionally distinct classes of complex phosphorus compounds in lake water. Limnol. Oceanogr. 24: 463–473.

Francko, D. A. & R. T. Heath, 1982. UV-sensitive complex phosphorus: association with dissolved humic material and iron in a bog lake. Limnol. Oceanogr. 27: 564–569.

Geller, A., 1985a. Light-induced conversion of refractory, high molecular weight lake water constituents. Schweiz. Z. Hydrol. 47: 21–26.

Geller, A., 1985b. Degradation and formation of refractory DOM by bacteria during simultaneous growth on labile substrates and persistent lake water constituents. Schweiz. Z. Hydrol. 47: 27–44.

Geller, W. & H. Müller, 1981. The filtration apparatus of Cladocera: filter mesh-sizes and their implications on food selectivity. Oecologia 49: 316–321.

Guildford, S. J., F. P. Healey & R. E. Hecky, 1987. Depression of primary production by humic matter and suspended sediment in limnocorral experiments at Southern Indian Lake, Northern Manitoba. Can. J. Fish. aquat. Sci. 45: 1408–1417.

Güde, H., 1985. Influence of phagotrophic processes on the regeneration of nutrients in two-stage continuous culture systems. Microb. Ecol. 11: 193–204.

Hakala, I., 1974. Sedimentaatio Pääjärvessä. Luonnon Tutkija 78: 108–110.

Havens, K. E. III., 1989. Seasonal succession in the plankton of a naturally acidic, highly humic lake in Northeastern Ohio, USA. J. Plankton Res. 11: 1321–1327.

Hessen, D. O., 1985a. The relation between bacterial carbon and dissolved humic compounds in oligotrophic lakes. FEMS Microbiol. Ecol. 31: 215–223.

Hessen, D. O., 1985b. Filtering structures and particle size selection in coexisting cladocerans. Oecologia 66: 368–372.

Hessen, D. O. & A. K. Schartau, 1988. Seasonal and spatial overlap between cladocerans in humic lakes. Int. Revue ges. Hydrobiol. 73: 379–405.

Hessen, D. O., T. Andersen & A. Lyche, 1989. Differential grazing and resource utilization of zooplankton in a humic lake. Arch. Hydrobiol. 114: 321–347.

Hessen, D. O., T. Andersen & A. Lyche, 1990. Carbon metabolism in a humic lake: pool sizes and cycling through zooplankton. Limnol. Oceanogr. 35: 84–99.

Ilmavirta, V., 1984. The ecology of flagellated phytoplankton in brown-water lakes. Verh. int. Ver. Limnol. 22: 817–821.

Ilmavirta, V., 1988. Phytoflagellates and their ecology in Finnish brown-water lakes. Hydrobiologia 161: 255–270.

Jackson, T. A. & R. E. Hecky, 1980. Depression of primary productivity by humic matter in lake and reservoir waters of the boreal forest zone. Can. J. Fish. aquat. Sci. 37: 2300–2317.

Järnefelt, H., 1958. On the typology of the northern lakes. Verh. int. Ver. Limnol. 13: 228–235.

Jones, A. K. & R. C. Cannon, 1986. The release of microalgal photosynthate and associated bacterial uptake and heterotrophic growth. Br. phycol. J. 21: 341–358.

Jones, R. I., 1977a. Factors controlling phytoplankton production and succession in a highly eutrophic lake (Kinnego Bay, Lough Neagh). II. Phytoplankton production and its chief determinants. J. Ecol. 65: 561–577.

Jones, R. I., 1977a. Factors controlling phytoplankton production and succession in a highly eutrophic lake (Kinnego Bay, Lough Neagh). III. Interspecific competition in relation to irradiance and temperature. J. Ecol. 65: 579–586.

Jones, R. I., 1990. Phosphorus transformations in the epilimnion of humic lakes: biological uptake of phosphate. Freshwat. Biol. 23: 323–337.

Jones, R. I. & L. Arvola, 1984. Light penetration and some related characteristics in small forest lakes in southern Finland. Verh. int. Ver. Limnol. 22: 811–816.

Jones, R. I. & K. Salonen, 1985. The importance of bacterial utilization of released phytoplankton photosynthate in two humic forest lakes in southern Finland. Holarct. Ecol. 8: 133–140.

Jones, R. I., K. Salonen & H. De Haan, 1988. Phosphorus transformations in the epilimnion of humic lakes: abiotic interactions between dissolved humic materials and phosphate. Freshwat. Biol. 19: 357–369.

Kankaala, P., 1988. The relative importance of algae and bacteria as food for *Daphnia longispina* (Cladocera) in a polyhumic lake. Freshwat. Biol. 19: 285–296.

Kieber, D. J., J. McDaniel & K. Mopper, 1989. Photochemical source of biological substrates in seawater: implications for carbon cycling. Nature 341: 637–639.

Kirk, J. T. O., 1983. Light and photosynthesis in aquatic ecosystems. Cambridge University Press, Cambridge, 401 pp.

Latja, R., 1974. Pääjärven eläinplankton. Luonnon Tutkija 78: 153–156.

Makarewicz, J. C., G. E. Likens & M. J. Jordan, 1985. Interactions between bacteria and phytoplankton. In G. E. Likens (ed.), An Ecosystem Approach to Aquatic Ecology. Springer-Verlag, New York: 323–324.

Miles, C. J. & P. L. Brezonik, 1981. Oxygen consumption in humic-colored waters by a photochemical ferrous-ferric catalytic cycle. Envir. Sci. Technol. 15: 1089–1095.

Pedros-Alio, C. & T. D. Brock, 1983. The impact of zooplankton feeding on the epilimnetic bacteria of a eutrophic lake. Freshwat. Biol. 13: 227–239.

Peterson, B. J., J. E. Hobbie & J. F. Haney, 1978. *Daphnia* grazing on natural bacteria. Limnol. Oceanogr. 23: 1039–1045.

Pomeroy, L. R., 1974. The ocean's food web, a changing paradigm. BioScience 24: 499–504.

Pomeroy, L. R. & W. J. Wiebe, 1988. Energetics of microbial food webs. Hydrobiologia 159: 7–18.

Prakash, A., M. A. Rashid, A. Jensen & D. V. Subba Rao, 1973. Influence of humic substances on the growth of marine phytoplankton: diatoms. Limnol. Oceanogr. 18: 516–524.

Pratt, J. R. & J. D. Chappell, 1989. Abundance and feeding of microheterotrophic flagellates from a eutrophic lake. Hydrobiologia 182: 165–169.

Provasoli, L., 1963. Organic regulation of phytoplankton fertility. In The Sea, Vol. 2. Wiley-Interscience, New York: 165–219.

Ramberg, L., 1979. Relations between phytoplankton and light climate in two Swedish forest lakes. Int. Revue ges. Hydrobiol. 64: 749–782.

Rask, M., A. Heinänen, K. Salonen, L. Arvola, I. Bergström, M. Liukkonen & A. Ojala, 1986. The limnology of a small, naturally acidic, highly humic lake. Arch. Hydrobiol. 106: 351–371.

Ruttner, F., 1963. Fundamentals of limnology. 3rd edn. University of Toronto Press, Toronto, 307 pp.

Ryhänen, R., 1968. Die Bedeutung der Humussubstanzen im Stoffhaushalt der Gewässer Finnlands. Mitt. int. Ver. Limnol. 14: 168–178.

Salonen, K., 1981. The ecosystem of the oligotrophic Lake Pääjärvi. 2. Bacterioplankton. Verh. int. Ver. Limnol. 21: 448–453.

Salonen, K. & L. Arvola, 1988. A radiotracer study of zooplankton grazing in two small humic lakes. Verh. int. Ver. Limnol. 23: 462–469.

Salonen, K. & T. Hammar, 1986. On the importance of dissolved organic matter in the nutrition of zooplankton in some lake waters. Oecologia 68: 246–253.

Salonen, K. & S. Jokinen, 1988. Flagellate grazing on bacteria in a small dystrophic lake. Hydrobiologia 161: 203–209.

Salonen, K. & T. Tulonen, 1990. Photochemical and biological transformations of dissolved humic substances. (Abstract). Verh. int. Ver. Limnol. 24: 294.

Salonen, K., K. Kononen & L. Arvola, 1983. Respiration of plankton in two small, polyhumic lakes. Hydrobiologia 101: 65–70.

Salonen, K., L. Arvola, H. De Haan, T. Hammar, S. Jokinen, R. Jones, P. Kankaala, A. Lehtovaara, A. Ojala & U. Smolander, 1987. Progress reports: Research on humic lakes. Lammi Notes 14: 6–7.

Salonen, K., T. Kairesalo, L. Arvola, T. Hammar, P. Kankaala, A. Lehtovaara, A. Ojala & T. Tulonen, 1990. Progress reports: Food chains of humic lakes. Lammi Notes 17: 1.

Salonen, K., L. Arvola, T. Tulonen, T. Hammar, T.-R. Metsälä, P. Kankaala & U. Münster, 1992a. Planktonic food chains of a highly humic lake. I. A mesocosm experiment during the spring primary production maximum. Hydrobiologia 229: 125–142.

Salonen, K., P. Kankaala, T. Tulonen, T. Hammar, M. James, T.-R. Metsälä & L. Arvola, 1992b. Planktonic food chains of a highly humic lake. II. A mesocosm experiment in summer during dominance of heterotrophic processes. Hydrobiologia 229: 143–157.

Sarvala, J., V. Ilmavirta, L. Paasivirta & K. Salonen, 1981. The ecosystem of the oligotrophic Lake Pääjärvi 3. Secondary production and an ecological energy budget of the lake. Verh. int. Ver. Limnol. 21: 422–427.

Sanders, R. W. & K. G. Porter, 1988. Phagotrophic phytoflagellates. Adv. microb. Ecol. 10: 167–192.

Sandgren, C. D., 1988. The ecology of chrysophyte flagellates: their growth and perennation strategies as freshwater phytoplankton. In C. D. Sandgren (ed.), Growth and Reproductive Strategies of Freshwater Phytoplankton. Cambridge University Press, Cambridge: 9–104.

Satoh, Y. & H. Abe, 1987. Dissolved organic matter in colored water from mountain bog pools in Japan. II. Biological decomposability. Arch. Hydrobiol. 111: 25–35.

Schell, D. M., 1983. Carbon-13 and carbon-14 abundances in Alaskan aquatic organisms: delayed production from peat in Arctic food webs. Science, 219: 1068–1071.

Sederholm, H., A. Mauranen & L. Montonen, 1973. Some observations on the microbial degradation of humus substances in water. Verh. int. Ver. Limnol. 18: 1301–1305.

Sepers, A. B. J., 1977. The utilization of dissolved organic compounds in aquatic environments. Hydrobiologia 52: 39–54.

Sherr, E. B., 1988. Direct use of high molecular weight polysaccharide by heterotrophic flagellates. Nature 335: 348–351.

Sherr, B. F., E. B. Sherr & C. S. Hopkinson, 1988. Trophic interactions within pelagic microbial communities: Indications of feedback regulation of carbon flow. Hydrobiologia 159: 19–26.

Siegel, A., 1971. Metal-organic interactions in the marine environment. In S. D. Faust & J. V. Hunder (eds), Organic Compounds in Aquatic Environment. Marcel Dekker: 265–295.

Sleigh, M. A., 1989. Protozoa and other protists. Edward Arnold, London, 342 pp.

Stabel, H.-H., K. Moaledj & J. Overbeck, 1979. On the degradation of dissolved organic molecules from Plussee by oligocarbophilic bacteria. Arch. Hydrobiol. Beih. Ergebn. Limnol. 12: 95–104.

Steinberg, C. & G. F. Baltes, 1984. Influence of metal compounds on fulvic acid/molybdenum blue reactive phosphate associations. Arch. Hydrobiol. 100: 61–71.

Steinberg, C. & A. Herrmann, 1981. Utilization of dissolved metal organic compounds by freshwater microorganisms. Verh. int. Ver. Limnol. 21: 231–235.

Steinberg, C. & U. Muenster, 1985. Geochemistry and ecological role of humic substances in lake water. In G. R. Aiken *et al.* (eds), Humic Substances in Soil, Sediment and Water. J. Wiley & Sons, N.Y.: 104–145.

Stevens, R. J. & B. M. Stewart, 1982. Concentration, fractionation and characterization of soluble organic phosphorus in river water entering Lough Neagh. Wat. Res. 16: 1507–1519.

Stewart, A. J. & R. G. Wetzel, 1981. Dissolved humic materials: Photodegradation, sediment effects, and reactivity with phosphate and calcium carbonate precipitation. Arch. Hydrobiol. 92: 265–286.

Stewart, A. J. & R. G. Wetzel, 1982. Influence of dissolved humic materials on carbon assimilation and alkaline phosphatase activity in natural algal-bacterial assemblages. Freshwat. Biol. 12: 369–380.

Strome, D. J. & M. C. Miller, 1978. Photolytic changes in dissolved humic substances. Verh. int. Ver. Limnol. 20: 1248–1254.

Sundh, I., 1989. Characterization of phytoplankton extracellular products (PDOC) and their subsequent uptake by heterotrophic organisms in a mesotrophic fores lake. J. Plankton Res. 11: 463–486.

Talling, J. F., 1957. The phytoplankton population as a compound photosynthetic system. New Phytol. 56: 133–149.

Thienemann, A., 1925. Die Binnengewässer Mitteleuropas. Die Binnengewässer, 1, 255 pp.

Tranvik, L. J., 1988. Availability of dissolved organic carbon for planktonic bacteria in oligotrophic lakes of differing humic content. Microb. Ecol. 16: 311–322.

Tranvik, L. J., 1989. Bacterioplankton growth, grazing mortality and quantitative relationship to primary production in a humic and a clearwater lake. J. Plankton Res. 11: 985–1000.

Tranvik, L. J. & M. G. Höfle, 1987. Bacterial growth in mixed cultures on dissolved organic carbon from humic and clear waters. Appl. envir. Microbiol. 53: 482–488.

Tranvik, L. J., K. G. Porter & J. McN. Sieburth, 1989. Occurrence of bacterivory in *Cryptomonas*, a common freshwater phytoplankter. Oecologia 78: 473–476.

Vadstein, O., B. O. Harkjerr, A. Jensen, Y. Olsen & H. Reinertsen, 1989. Cycling of organic carbon in the photic zone of a eutrophic lake with special reference to the heterotrophic bacteria. Limnol. Oceanogr. 34: 840–855.

Veen, A., 1990. Phagotrophy by *Dynobryon*: a survival strategy in a low-nutrient environment? (Abstract). Br. phycol. J. 25: 98–99.

Visser, S. A., 1984. Seasonal changes in the concentration and colour of humic substances in some aquatic environments. Freshwat. Biol. 14: 79–87.

Wall, D. & F. Briand, 1979. Response of lake phytoplankton communities to *in situ* manipulations of light intensity and colour. J. Plankton Res. 1: 103–112.

Watanabe, Y. & C. R. Goldman, 1984. Heterotrophic bacterial community in oligotrophic Lake Tahoe. Verh. int. Ver. Limnol. 22: 584–590.

Wetzel, R. G., 1983. Limnology, 2nd edn. W.B. Saunders Co., Philadelphia.

Williams, P. J. leB., 1981. Incorporation of microheterotrophic processes into the classical paradigm of the planktonic food web. Kieler Meeresforsch. Sonderh. 5: 1–28.

Wright, R. T., 1984. Dynamics of pools of dissolved organic carbon. In J. E. Hobbie & P. J. leB. Williams (eds), Heterotrophic Activity in the Sea. Proc. NATO ARI, Cascais, Portugal, 1981. Plenum, NY: 121–154.

Wright, R. T., 1988. A model for short-term control of the bacterioplankton by substrate and grazing. Hydrobiologia 159: 111–117.

*Hydrobiologia* **229**: 93–106, 1992.
*K. Salonen, T. Kairesalo & R.I. Jones (eds.),*
*Dissolved Organic Matter in Lacustrine Ecosystems: Energy Source and System Regulator.*




# Extracellular dissolved organic carbon released from phytoplankton as a source of carbon for heterotrophic bacteria in lakes of different humic content

Ingvar Sundh[1] & Russell T. Bell
*Uppsala University, Institute of Limnology, Box 557, S-751 22 Uppsala, Sweden; [1]Present address: Department of Microbiology, SLU, Box 7025, S-750 07 Uppsala, Sweden*

*Key words:* dissolved organic carbon, PDOC, bacterial production, humic lakes

## Abstract

The quantitative importance of photosynthetically produced dissolved organic carbon (PDOC) released from phytoplankton as a source of carbon for pelagic, heterotrophic bacteria was investigated in four temperate Swedish lakes, of which two had low ($\sim 20$ mg Pt $l^{-1}$), and two moderately high (60–80 mg Pt $l^{-1}$) humic content. The bacterial assimilation of PDOC was estimated with the $^{14}C$ method, and the total production of the heterotrophic bacteria was estimated with the [$^{3}H$]thymidine incorporation method. The release of PDOC from natural communities of phytoplankton was not restricted to periods of photosynthesis, but often continued during periods of darkness. Heterotrophic bacteria often assimilated the labile components of the PDOC at high rates (up to 73% of the released PDOC was assimilated during the incubation in our experiments). The contribution of PDOC to bacterial production exhibited large within-lake seasonal variations, but PDOC was at certain times, both in humic and non-humic lakes, a quantitatively very important carbon source for the heterotrophic bacteria. Under periods of comparatively low primary production, heterotrophic bacteria in humic lakes appear to utilize allochthonous, humic substances as a substrate.

## Introduction

The dissolved organic carbon (DOC) in the pelagic zone of lakes consists of both allochthonous substances produced in the watershed and transported to the lake in its inflows, and autochthonous substances produced by phytoplankton and macrophytes within the lake, and entering the DOC pool by direct release by living organisms or degradation of dead material. The DOC pool is normally dominated by the allochthonous compounds (Wetzel, 1983). In theory, the whole pool of DOC is available to the heterotrophic bacteria, but while the large pool of allochthonous substances are relatively refractory to bacterial utilization (Søndergaard & Schierup, 1982; Cole *et al.*, 1984), the photosynthetically produced dissolved organic carbon released from phytoplankton (PDOC) may be readily available to bacterial utilization and assimilated at high rates (Coveney, 1982; Chróst, 1983; Jensen, 1983; Feuillade *et al.*, 1988). Bacterial dependence on PDOC released from phytoplankton has been suggested to be a major reason for the positive relationship between phytoplankton biomass (often measured as concentration of chlorophyll *a*) and bacterial numbers in pelagic environments (reviews by Bird & Kalff, 1984; Cole *et al.*, 1988). This relationship is not always clear-cut however, and Scavia & Laird (1987) found an almost inverse relationship between chlorophyll concentration and bacterial numbers in Lake Michigan. In their recent review of simultaneous measurements of phyto- and bacterioplankton production rates from both marine and freshwater environments, Cole *et al.* (1988) found a significant positive relationship between phytoplankton primary production and bacterial secondary production. The relationship contained considerable scatter, however, which could be due to fluctuations both in phytoplankton release rates and bacterial uptake rates of PDOC, as well as bacterial utilization of allochthonous DOC.

To assess the quantitative importance of PDOC as a carbon source for the bacteria, the rates of bacterial uptake of PDOC should be directly compared with the total bacterial carbon production. These kinds of comparisons have demonstrated that PDOC often contributes $> 50\%$ to the total bacterial carbon demand (Cole *et al.*, 1982; Bell *et al.*, 1983; Brock & Clyne, 1984; Søndergaard *et al.*, 1985; Chrzanowski & Hubbard, 1989).

These general trends have been established in non-humic systems, and may not hold for humic brown-water lakes, where allochthonous humic substances constitute an even more dominant portion of the total DOC than in clear-water lakes. For example, bacterial isolates from humic lakes are able to degrade dissolved humic substances (Sederholm *et al.*, 1973; De Haan, 1977), and Hessen (1985) reported that the bacterial biomass in brown-water lakes was not correlated with phytoplankton biomass but rather was positively correlated with water colour. Also, Jones & Salonen (1985) found that PDOC was not a significant source of substrate for total bacterial respiration in two polyhumic lakes. Additionally, Tranvik (1989) compared the primary and bacterial secondary production in two lakes of similar abiotic and biotic characteristics but different humic content, and found that the ratio of bacterial production to primary production was significantly higher in the humic lake. Taken together, these results strongly imply that in humic lakes, allochthonous dissolved organic carbon is a significant carbon source for bacterioplankton.

In this paper we present estimates of the epilimnetic bacterial assimilation of PDOC (measured with the $^{14}C$-bicarbonate tracer technique involving fractionated filtration) and the total bacterial secondary production (measured as [$^{3}H$]thymidine incorporation) in two non-humic and two moderately humic temperate lakes in central Sweden. If the heterotrophic bacteria of the humic lakes utilize allochthonous, humic material as substrate, we expect the uptake of PDOC to make a smaller contribution to the total bacterial production in these lakes.

## Materials and methods

### *Description of lakes*

Some basic abiotic characteristics of the lakes are presented in Table 1. Of the two non-humic lakes, Lake Vallentunasjön is small, shallow and highly eutrophic whereas Lake Erken is larger, and gen-

*Table 1.* Some general characteristics of the investigation lakes. Values given for total phosphorus and water colour are approximate mean values for the ice-free season[1].

| | Lake area km$^2$ | Maximum depth m | Mean depth m | Total phosphorus $\mu g$ P $l^{-1}$ | Water colour mg Pt $l^{-1}$ |
|---|---|---|---|---|---|
| Lake Vallentunasjön | 6.2 | 5.0 | 2.7 | 100 | 20 |
| Lake Erken | 24 | 20 | 9.0 | 20 | 20 |
| Lake Siggeforasjön | 0.76 | 11 | 4.2 | 15 | 80 |
| Lake Sörmogen | 3.7 | 10 | – | 15 | 80 |

[1] For more details on the abiotic and biotic characteristics of the lakes consult Boström *et al.* (1989) (Lake Vallentunasjön), Petterson (1990) (Lake Erken), Heyman & Blomqvist (1984) (Lake Siggeforasjön) and Sundh (1989) (Lake Sörmogen).

erally of a mesotrophic character. Both the humic lakes (Lakes Sörmogen and Siggeforasjön) are comparatively small forest lakes and have mesotrophic to oligotrophic character.

*Sampling*

All details on the sampling, incubation and filtration procedures refer to the diel experiments in Lakes Vallentunasjön, Erken and Siggeforasjön. No diel incubations, and slightly different filtration methods (Sundh, 1989) were used in Lake Sörmogen.

Composite samples, from the euphotic zone during circulation and from the epilimnion during stratification, were taken with a 2 m long tube sampler. A subsample was preserved with formaldehyde (final concentration 4%) for enumeration and biomass determination of bacteria. In five experiments (Lake Vallentunasjön April 29, 1988 and September 27, 1989 and Lake Erken May 5, and August 30, 1988 and August 1, 1989) sampling was done in the afternoon of the day before start of the experiments and the water samples stored in darkness at *in situ* temperature until experiments were started the following morning. On the other occasions, sampling was performed in the morning (07.00–09.00) and incubations with $^{14}C$ were started within three hours of sample collection.

*Bacterial abundance and biomass*

Bacteria were enumerated and cell volumes estimated by acridine orange staining and epifluorescent microscopy according to Bell *et al.* (1983). For biomass determinations the dimensions of at least 125 randomly chosen cells were measured and biovolumes calculated using geometric formulas.

*Heterotrophic bacterial production*

Heterotrophic bacterial production was assessed via [*methyl*-$^3H$]thymidine (40–60 Ci mmol$^{-1}$; Amersham) incorporation into macromolecules (Bell *et al.*, 1983). Triplicate 10–20 ml samples (100-$\mu$m prefiltered) were incubated with [$^3H$]thymidine (final concentration 25–30 nM in Lakes Vallentunasjön and Erken and 40–50 nM in Lakes Sörmogen and Siggeforasjön) at *in situ* temperature for $\approx 1$ hour. The incubations were stopped by adding formaldehyde to a final concentration of 2%. A formaldehyde-killed blank (2% final conc.) was always included. In these lakes, 70–85% of the [$^3H$]thymidine incorporation into macromolecules (material insoluble in cold 5% trichloroacetic acid) is in DNA (Bell & Kuparinen, 1984; Bell, 1986; Bell, 1990). Bacterial production in Lake Siggeforasjön on August 8, 1988 was assessed via [$^3H$]adenine incorporation into DNA (Bell & Riemann, 1989). Triplicate 15 ml subsamples were prefiltered (2 $\mu$m) to separate algae from bacteria, and incubated with 100 nM [$^3H$]adenine (25 Ci mmol$^{-1}$; Amersham) for $\approx 4$ hours. The samples were subsequently extracted and filtered according to Bell & Riemann (1989). Thymidine and adenine incorporation were converted to carbon units using a conversion factor of $2 \times 10^{18}$ cells mol$^{-1}$ (Bell *et al.*, 1983; Smits & Riemann, 1988; Riemann & Bell, 1990) and assuming a bacterial carbon content of 0.22 pg $\mu m^{-3}$ (Bratbak & Dundas, 1984). Bacterial production was assessed twice during each diel cycle.

The phytoplankton community during autumn in Lake Vallentunasjön was dominated by colonial cyanobacteria (primarily *Microcystis* spp.). These contain numerous bacteria in their gelatinous sheaths (Pearl, 1982; Bern, 1985). To assess the respective contribution of attached and non-attached bacteria to total bacterial production the [$^3H$]thymidine uptake during these periods was size-fractionated by post-filtration through a 2 $\mu$m filter prior to TCA extraction.

*Primary production, percentage extracellular release and bacterial assimilation of PDO$^{14}$C*

Before the start of the $^{14}C$ tracer incubations, the mixed water samples were gently sieved through a 100 $\mu$m net to remove large zooplankton. Three transparent glass flasks with plastic screw caps were each filled with 500 ml of the 100 μm filtered water, and $^{14}C$ labelled bicarbonate (0.23–1.6 μCi ml$^{-1}$ sample) was added to two of them. The unlabelled flask was subsequently used to take subsamples for measurements of the inorganic carbon uptake in the bacterial size fraction (0.2–2 μm) and for [$^{3}H$]thymidine incorporation assays. All three flasks were incubated *in situ* at ≈0.7 m depth (Lake Erken May 25, 1987 and July 20, 1988) or under *in situ* temperature in the laboratory at 110–250 μE m$^{-2}$ s$^{-1}$ during the light periods and in darkness during the night. The incubations lasted at least 24 hours, and always included a dark period between two light periods. At intervals, subsamples were taken for determination of $^{14}C$ in the phytoplankton (>2 μm), the bacterial (0.2–2 μm) and dissolved fractions (<0.2 μm). Aliquots of 5–15 ml were filtered sequentially through a 2 μm (Nuclepore polycarbonate) and 0.2 μm (Schleicher & Schüll membrane) filter. Aliquots for zero time radioactivity of the different size fractions were taken immediately after addition of the $^{14}C$. To avoid breakage of phytoplankton cells retained by the 2 μm filter, the filtration was allowed to proceed under gravity, but a slight vacuum (<100 mm Hg) was necessary for the 0.2 μm filtration. The filters were placed in scintillation vials and a few drops of 0.5 M HCl were added to remove radioactive inorganic carbon. The filtrates were collected in glass scintillation vials. The pH was lowered (to ≈2) by addition of 0.5 M HCl and excess radioactive inorganic carbon was removed by air bubbling for 45 minutes. The radioactivity of the filters and filtrates was subsequently measured by liquid scintillation counting, with quench correction according to the external standard channels ratio method. In the calculations of the $^{14}C$ activity in the particulate and dissolved fractions, the zero time values were subtracted.

To correct for possible autotrophic uptake of $^{14}C$ labelled bicarbonate in the bacterial size fraction (0.2–2 μm), we prefiltered aliquots from the unlabelled flasks through a 2 μm (Nuclepore polycarbonate) filter. After addition of $_{14}C$ labelled bicarbonate to the filtrates (0.4–3.9 μCi ml$^{-1}$) and 3.3–12 hours incubation at *in situ* temperature, the incubation was terminated by filtration through a 0.2 μm filter (Schleicher & Schüll membrane). The filters and filtrates were subsequently treated as described above. The inorganic carbon uptake in the bacterial size fraction was determined during both the first day and the night.

The total concentration of inorganic carbon was calculated from pH (measured on a Radiometer pHM 62), alkalinity (titration with HCl under $N_2$ flow and with a mixed indicator) and conductivity (measured on a Radiometer CDM 2d). For transformation of $^{14}C$ in the different size fractions to units of carbon, the same specific $^{14}C$ activity was assumed in these fractions as the specific activity of inorganic carbon at the start of the experiment. The total carbon assimilation was calculated as the sum of radioactivity of the two filters and the filtrate, and the percentage extracellular release (PER) as the sum of the filtrate and the 0.2 μm filter (corrected for bicarbonate uptake in the 0.2–2 μm size fraction) in percent of the total assimilation. The bacterial incorporation of PDO$^{14}$C was obtained from the radioactivity of the 0.2 μm filter (corrected for bicarbonate uptake in this size fraction), and transformed to total carbon uptake assuming 50% respiration of PDO$^{14}$C (Cole *et al.*, 1982; Chrost & Faust, 1983). The total $^{14}C$ assimilation, PER and bacterial incorporation of PDO$^{14}$C was calculated both for subsamples taken after the first light period and after the first light + the dark period.

## Results

### *Bacterial production and abundance*

Bacterial abundance and biomass was highest and the within-lake variation greatest in eutrophic Lake Vallentunasjön (Table 2). Bacterial abundance and biomass was comparable in Lake Erken and the humic Lake Siggeforasjön, whereas abundance and biomass was ≈3 times greater in the equally humic Lake Sörmogen.

Except for Lake Erken on August 1 1989, when the bacterial production was four times higher

*Table 2.* Bacterial abundance and biomass and total bacterial production. Columns denoted 'light' shows results from subsamples taken after the first light period of incubation, and 'light + dark' from subsamples taken after the dark period.

| | Bacterial abundance $10^9\ l^{-1}$ | Bacterial biomass $\mu g\ C\ l^{-1}$ | Total bacterial production[1] $\mu g\ C\ l^{-1}\ h^{-1}$ | Specific bacterial production $day^{-1}$ | Bacterial production in % of prim. prod. | |
|---|---|---|---|---|---|---|
| | | | | | Light | Light + dark |
| Lake Vallentunasjön | | | | | | |
| April 29 1988 | 4.3 | 140 | 0.44 | 0.07 | 1.0 | 1.7 |
| Sep 5 1988 | 7.1 | 323 | 6.2 | 0.46 | 3.4 | 8.4 |
| Sep 27 1989 | 15.6 | 717 | 6.1 | 0.20 | 1.6 | 4.0 |
| Lake Erken | | | | | | |
| May 25 1987 | – | – | 0.09 | – | 5.5 | 8.0 |
| May 5 1988 | 1.1 | 25 | 0.12 | 0.12 | 0.6 | 1.2 |
| July 20 1988 | 1.3 | 28 | 1.19 | 1.02 | 12 | 19 |
| Aug 30 1988 | 1.6 | 30 | 0.30 | 0.24 | 5.4 | 12 |
| August 1 1989 | 1.8 | 42 | 0.75 | 0.43 | 6.3 | 15 |
| Lake Siggeforasjön | | | | | | |
| Aug 8 1988 | 1.5 | 34 | 0.34[2] | 0.24 | 5.0 | 8.3 |
| Sep 19 1988 | 1.4 | 29 | 0.20 | 0.17 | 19 | 40 |
| Aug 1 1989 | 1.2 | 26 | 0.15 | 0.14 | 1.5 | 3.0 |
| Lake Sörmogen | | | | | | |
| June 25 1986 | 2.6 | 85 | 0.95[3] | 0.27 | 23 | – |
| July 9 1986 | 3.5 | 117 | 0.95[3] | 0.19 | 17 | – |
| Sep 3 1986 | 4.5 | 149 | 0.45[3] | 0.07 | 14 | – |

[1] Mean of the afternoon and morning determinations.
[2] Bacterial production measured as adenine incorporation into DNA.
[3] Mean of two determinations a few days before and after the dates of $^{14}C$ incorporation experiments.

after the night, and Lake Vallentunasjön on September 5 1988 (21% higher after the night), the bacterial production was somewhat lower (4–33%) at the end of the dark period than during the first light period. Because the differences between the two determinations were moderate, the mean was used for all calculations involving total bacterial production. Considering the non-humic lakes, the total bacterial production was higher in eutrophic Lake Vallentunasjön (0.44–6.2 $\mu g\ C\ l^{-1}\ h^{-1}$) than in mesotrophic Lake Erken (0.092–1.19 $\mu g\ C\ l^{-1}\ h^{-1}$, Table 2). The bacterial production rates in the humic lakes exhibited the same pattern as the bacterial biomass, i.e., a higher production rate in Lake Sörmogen (0.45–0.95 $\mu g\ C\ l^{-1}\ h^{-1}$) than in Lake Siggeforasjön (0.15–0.34 $\mu g\ C\ l^{-1}\ h^{-1}$). The specific bacterial production ($day^{-1}$) did not exhibit any significant difference between the lakes, and except for Lake Erken on July 20, 1988 (1.02 $day^{-1}$), it varied between 0.072 and 0.46 $day^{-1}$ (Table 2).

### *Primary production, PER and bacterial uptake of PDO$^{14}$C*

Three examples of the diel time course of $^{14}C$ labelling of the phytoplankton, bacterioplankton and dissolved fractions are shown in Fig. 1. The most common pattern was that the $^{14}C$ of the phytoplankton size fraction increased during the first day, decreased slightly during the night and increased again during the second light period, while the labelling of the bacterial size fraction, and particularly the filtrate, continued to increase during the night, implying that phytoplankton release of PDO$^{14}$C did not cease when photosynthesis stopped.

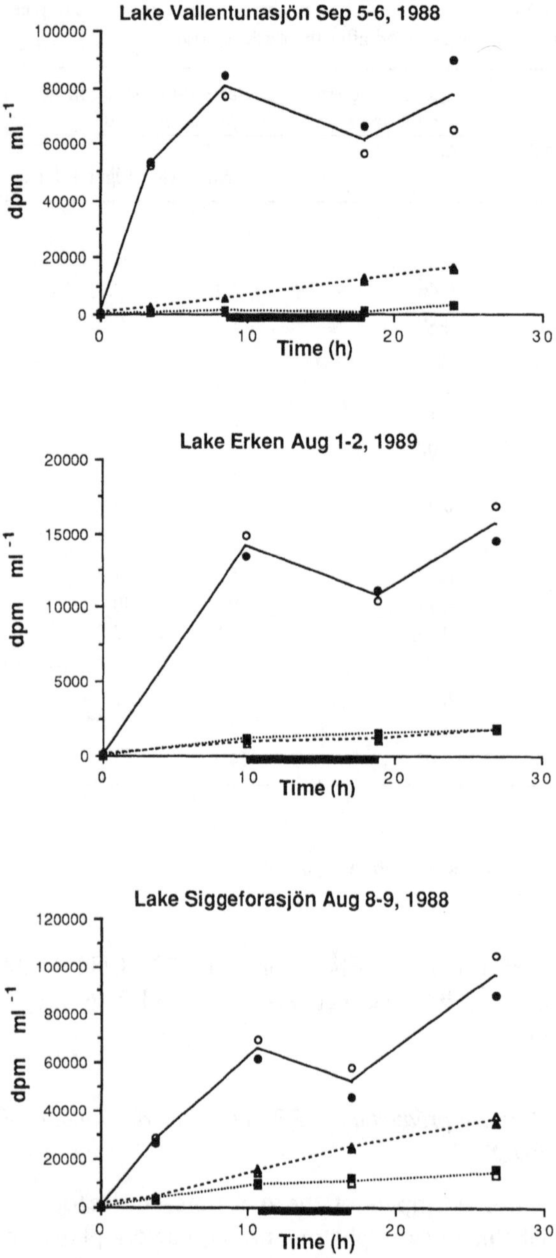


*Fig. 1.* Three examples of the diel time-course of $^{14}C$ labelling of phytoplankton (circles), bacterioplankton (squares) and dissolved (triangles) fractions. Light and dark symbols represent replicate flasks, and the lines connect the mean values. The thickened section of the x-axis shows the duration of the dark period.

The total $^{14}C$ assimilation in Lakes Erken, Siggeforasjön and Sörmogen was comparable (diel assimilation ranging between 1.15–10.4 $\mu g$ $C\,l^{-1}\,h^{-1}$ in Lake Erken and 0.50–4.97 $\mu g$ $C\,l^{-1}\,h^{-1}$ in Lake Siggeforasjön), but the total assimilation in eutrophic Lake Vallentunasjön was an order of magnitude higher (range of 25.6–151 $\mu g\,C\,l^{-1}\,h^{-1}$, Table 3). The short incubations in Lake Sörmogen yielded PER values between 36% and 55% of the total assimilation. In Lakes Vallentunasjön, Erken and Siggeforasjön, the continued release of $PDO^{14}C$ during the night consistently resulted in significantly higher release rates (in several experiments about twice as high) if the whole period (light + dark) is considered, and the diel rates of release ranged between 6–29%, 9–35% and 21–40% in Lakes Vallentunasjön, Erken and Siggeforasjön, respectively (Table 3).

The percentage of the released $PDO^{14}C$ recovered in the bacterial size-fraction after the incubation varied between 5% and 57% (Table 3). The $PDO^{14}C$ appears to be more efficiently utilized in Lake Erken (23–57% of the total released $PDO^{14}C$ incorporated on a diel basis) compared to the other lakes (5–17% in Lake Vallentunasjön and 6–31% in Lake Siggeforasjön). The same trend is apparent when the bacterial incorporation of $PDO^{14}C$ is expressed as a percentage of the total primary production (Table 3).

### *Uptake of $PDO^{14}C$ in relation to total bacterial carbon demand*

To assess the total bacterial uptake of carbon and the contribution of PDOC, the bacterial incorporation of $PDO^{14}C$ was transformed to gross carbon uptake assuming 50% respiration of PDOC, and the bacterial production measured as thymidine incorporation transformed to total bacterial carbon uptake by assuming a growth yield of 50% (Table 4). Frequently, the uptake of $PDO^{14}C$ exceeded the estimated total uptake of carbon measured with the [$^{3}H$]thymidine technique, thus resulting in higher than 100% contribution of PDOC. Except for Lake Sörmogen, the contribution of PDOC exhibited very large variations within the lakes (Table 4), and consequently there is a large overlap between the lakes.

*Table 3.* Total $^{14}C$ assimilation, percentage extracellular release and bacterial assimilation of $PDO^{14}C$. Columns denoted 'light' shows results from subsamples taken after the first light period of incubation, and 'light + dark' from subsamples taken after the dark period.

| | Total $^{14}C$ ass. $\mu g\ C\ l^{-1}\ h^{-1}$ | | PER % | | Bacterial ass. of $PDO^{14}C$ % of release | | Bacterial ass. of $PDO^{14}C$ % of prim. prod. | |
|---|---|---|---|---|---|---|---|---|
| | Light | Light + dark | Light | Light + dark | Light | Light + dark | Light | Light + dark |
| Lake Vallentunasjön | | | | | | | | |
| April 29 1988 | 42.6 | 25.6 | 14 | 29 | 10 | 10 | 1.5 | 2.9 |
| Sep 5 1988 | 183 | 74 | 7 | 17 | 13 | 5 | 0.9 | 0.9 |
| Sep 27 1989 | 383 | 151 | 4[1] | 6[1] | 16[1] | 17[1] | 0.7[1] | 0.9[1] |
| Lake Erken | | | | | | | | |
| May 25 1987 | 1.68 | 1.15 | 34 | 35 | 49 | 42 | 16 | 14 |
| May 5 1988 | 19.9 | 10.4 | 5[1] | 9[1] | 39[1] | 28[1] | 1.9[1] | 2.7[1] |
| July 20 1988 | 10.3 | 6.12 | 21 | 33 | 16 | 23 | 3.2 | 7.4 |
| Aug 30 1988 | 5.55 | 2.45 | 15 | 22 | 21 | 38 | 3.1 | 8.6 |
| Aug 1 1989 | 12.0 | 5.11 | 11 | 17 | 57 | 57 | 6.3 | 9.6 |
| Lake Siggeforasjön | | | | | | | | |
| Aug 8 1988 | 6.78 | 4.11 | 26 | 40 | 40 | 31 | 10 | 9.6 |
| Sep 19 1988 | 1.04 | 0.50 | 25 | 37 | 12 | 6 | 2.8 | 2.4 |
| Aug 1 1989 | 10.1 | 4.97 | 10 | 21 | 28 | 13 | 2.7 | 2.6 |
| Lake Sörmogen | | | | | | | | |
| June 25 1986 | 4.13 | – | 55 | – | 31 | – | 17 | – |
| July 9 1986 | 5.62 | – | 36 | – | 14 | – | 5.0 | – |
| Sep 3 1986 | 3.25 | – | 45 | – | 13 | – | 5.8 | – |

[1] No correction for bicarbonate uptake in the bacterial (0.2–2 $\mu m$) size fraction

During September in Lake Vallentunasjön, when colonial cyanobacteria dominated the phytoplankton, more than half the total [$^3H$]thymidine incorporation was by the $>2\ \mu m$ fraction (Table 4). Since the $PDO^{14}C$ uptake is measured for the 0.2–2 $\mu m$ fraction, it is more accurate to compare the measured $PDO^{14}C$ uptake with the [$^3H$]thymidine uptake based estimate of carbon uptake for only the 0.2–2 $\mu m$ size fraction. Based on this later comparison PDOC constitutes $\approx 120\%$ of total bacterial production at this time (Table 4). Thus, when considering only the $<2\ \mu m$ bacterial size fraction, the contribution of PDOC (short light incubations) is higher in the non-humic (147%) than the humic (90%) lakes. This difference, however, is not significant at the 0.05 level (student's t-test). The contribution of PDOC to total bacterial production in Lake Erken was large during the spring experiments but smaller in summer.

## Discussion

### *Phytoplankton PDOC release*

In most experiments, the radioactivity of the phytoplankton size fraction decreased during the night while the activity in the bacterial size fraction and filtrate continued to increase (Fig. 1). A continued night-time release, although at lower rates, has been noted earlier (Saunders, 1972; Berman & Kaplan, 1984). Clearly, the decrease in phytoplankton radioactivity during darkness is not only due to respiration losses, but also to continued release of $PDO^{14}C$. In consequence,

*Table 4.* Bacterial uptake of PDO$^{14}$C and its contribution to total bacterial carbon uptake, assuming growth yields of 50% both for PDO$^{14}$C uptake and total carbon uptake. Columns denoted 'light' shows results from subsamples taken after the first light period of incubation, and 'light + dark' from subsamples taken after the dark period.

| | Bacterial uptake of PDO$^{14}$C $\mu g\ C\ l^{-1}\ h^{-1}$ | | Total bacterial carbon uptake $\mu g\ C\ l^{-1}\ h^{-1}$ | Bacterial carbon uptake, 0.2–2 $\mu m$ fraction $\mu g\ C\ l^{-1}\ h^{-1}$ | Uptake PDO$^{14}$C, % of total bacterial carbon uptake % | | Uptake PDO$^{14}$C, % of bacterial carbon uptake in 0.2–2 $\mu m$ fraction % | |
|---|---|---|---|---|---|---|---|---|
| | Light | Light + dark | | | Light | Light + dark | Light | Light + dark |
| Lake Vallentunasjön | | | | | | | | |
| April 29 1988 | 1.28 | 1.50 | 0.88 | – | 145 | 170 | – | – |
| Sep 5 1988 | 3.24 | 1.28 | 12.4 | 2.96 | 26 | 10 | 109 | 43 |
| Sep 27 1989 | 5.26 | 2.80 | 12.2 | 3.96 | 43 | 23 | 133 | 71 |
| Lake Erken | | | | | | | | |
| May 25 1987 | 0.54 | 0.32 | 0.18 | – | 293 | 174 | – | – |
| May 5 1988 | 0.74 | 0.56 | 0.24 | – | 308 | 233 | – | – |
| July 20 1988 | 0.66 | 0.90 | 2.4 | – | 28 | 38 | – | – |
| Aug 30 1988 | 0.34 | 0.42 | 0.60 | – | 57 | 70 | – | – |
| Aug 1 1989 | 1.52 | 0.98 | 1.51 | – | 101 | 65 | – | – |
| Lake Siggeforasjön | | | | | | | | |
| Aug 8 1988 | 1.36 | 0.98 | 0.68 | – | 200 | 144 | – | – |
| Sep 19 1988 | 0.058 | 0.024 | 0.40 | – | 15 | 6 | – | – |
| Aug 1 1989 | 0.54 | 0.26 | 0.30 | – | 180 | 87 | – | – |
| Lake Sörmogen | | | | | | | | |
| June 25 1986 | 1.16 | – | 1.90 | – | 74 | – | – | – |
| July 9 1986 | 0.38 | – | 1.90 | – | 29 | – | – | – |
| Sep 3 1986 | 0.40 | – | 0.90 | – | 42 | – | – | – |

the PER from the long incubations (light + dark period) are consistently higher than when only the first light period is considered (Table 3). This indicates, as suggested by Søndergaard *et al.* (1985) that diel incubations are necessary to obtain reliable PER estimates. A continued night-time release is not unexpected, when the possible mechanisms of release are considered. Death and subsequent autolysis may be an important loss factor for phytoplankton (Jassby & Goldmann, 1974; Pollinger & Serruya, 1976; Heyman & Blomqvist, 1984). The activity of zooplankton ('sloppy feeding', excretion or defecation) may contribute to the continued release of PDO$^{14}$C at night (Lampert, 1978; Olsen *et al.*, 1986; Jumars *et al.*, 1989). In our experiments, large zooplankton were removed by filtering the sample through 100 $\mu$m, but small crustacean zooplankton, rotifers, ciliates and phagotrophic flagellates were not removed by this procedure. Another possible explanation for continued release in darkness has been suggested by Björnsen (1988). He hypothesized that phytoplankton release of PDOC can be explained by diffusive leakage of organic molecules over the plasma membrane, driven by the steep concentration gradient, and that the release should be interpreted in terms of biomass as opposed to production of biomass. This release mechanism is only possible for low molecular weight metabolic intermediates, however, and it has been shown repeatedly (Sundh, 1989 and references therein) that high molecular weight substances often make large contributions to *in situ* PDOC. Clearly, the release of PDOC from *in situ* phytoplankton communities is a result of several mechanisms acting simultaneously, and the PDOC is a mixture, containing compounds released by direct excretion from intact cells as well as autolysis products and substances released as a result of the activities of grazing zooplankton and protozoans. Diffusive leakage may be an important mechanism for the direct excretion of low molecular weight compounds from intact cells, while the grazing component of the release is obviously dependent on the prevailing grazing pressure. In addition, it is theoretically possible that the $PDO^{14}C$ contains $DO^{14}C$ released by bacteria. It has been suggested, however, that $DO^{14}C$ released by bacteria is not important when incubations are shorter than 24 hours (Dunstall & Nalewajko, 1975; Iturriaga & Zsolnay, 1981).

Overall, the large seasonal variations of PER within the lakes probably result from the taxonomic succession of phytoplankton. The release rates differ widely between different species in laboratory cultures (Hellebust, 1965). We observed that the PER was frequently higher when small forms of flagellates and diatoms (the humic lakes, and spring conditions in Lakes Vallentunasjön and Erken) dominated the phytoplankton biomass (Sundh, in prep.), perhaps partly due to microzooplankton grazing on those edible forms of phytoplankton, leading to higher contribution to $PDO^{14}C$ release from microzooplankton activities.

### *Bacterial assimilation of $PDO^{14}C$*

Although the bacterial uptake rates of $PDO^{14}C$ (expressed as percent of released $PDO^{14}C$ processed during the incubation) did not vary significantly between the lakes, there was a tendency for higher utilization rates in Lake Erken. Assuming 50% respiration of PDOC, the bacteria removed 18–28% (10–29% on a diel basis) in Lake Vallentunasjön, 28–73% in Lake Erken (37–73% on a diel basis), 21–57% in Lake Siggeforasjön (11–47% on a diel basis) and 23–47% in Lake Sörmogen. Our results thus corroborate earlier findings of high uptake rates of the labile components of PDOC.

When large colonial cyanobacteria dominated the phytoplankton biomass (September experiments in Lake Vallentunasjön), more than half the [$^3$H]thymidine incorporation was by the $>2\ \mu m$ size fraction (Table 4). If the bacterial incorporation of $PDO^{14}C$ is likewise predominately by the large and attached bacteria not passing the $2\ \mu m$ filter, it follows that the incorporation of $PDO^{14}C$ (and to some extent the PER) in these experiments are seriously underestimated. Thus, it is possible that the actual uptake rates in the September experiments in Lake Vallentunasjön were about twice as high as indicated in Table 3. We can not be sure whether attached bacteria made a significant contribution to the heterotrophic activity in the other experiments. We know, however, from the microscopic direct counts of bacteria, that aggregates of bacteria or bacteria attached to large particles made a negligible contribution to total bacterial biomass. Thus, in these experiments it is unlikely that the bacterial assimilation of $PDO^{14}C$ is underestimated due to uptake by attached bacteria.

### *$PDO^{14}C$ contribution to total carbon uptake*

The goal of this study was to assess the quantitative importance of newly released PDOC to total bacterial carbon demand. This kind of comparison places demands on the methodology. Both the measurements of release and bacterial assimilation of PDOC, as well as thymidine incorporation, have uncertainties that are difficult to assess for every sample. We discuss these problems below, but are convinced that our results are internally consistent and the general trends are real. For example, accumulating evidence shows that thymidine uptake and other methods give equivalent estimates of bacterial production (Bell & Riemann, 1989; Simon & Azam, 1989; Chin-Leo & Kirchman, 1988; Riemann *et al.*, 1990).

The bacterial uptake of PDOC exceeded the estimated total bacterial carbon uptake in several of our experiments. Likewise, Søndergaard *et al.* (1985) and Chrzanowski & Hubbard (1989) (also using the $^{14}C$ and [$^{3}H$]thymidine methods) frequently obtained PDOC contribution to total carbon production exceeding 100%. The contribution of PDOC to total bacterial production of biomass estimated from [$^{3}H$]thymidine incorporation, however, must not necessarily be lower than 100%. Large bacterial cells contain more carbon per unit of DNA than smaller cells (Simon & Azam, 1989; Riemann *et al.*, 1990). Thus, when cells are increasing in size, bacterial carbon uptake would exceed the carbon uptake estimated from DNA synthesis. Temporal rhythms of carbon uptake and DNA synthesis would also uncouple PDOC uptake and thymidine incorporation. Riemann & Bell (1990) found that the ratio of leucine incorporation ( = estimate of protein synthesis) to thymidine incorporation varied by a factor of $\approx 5$ over periods of weeks to months in enclosures in both an oligotrophic and a eutrophic lake. Diurnal rhythms in carbon incorporation and DNA synthesis could also result in actual carbon uptake exceeding the uptake calculated from the thymidine incorporation. Diurnal variations in DNA synthesis are comparatively small, however (Riemann & Söndergaard, 1984), compared to diurnal variation in leucine incorporation (Bell *et al.*, in prep).

We used a constant factor for converting moles of thymidine incorporated into bacterial cell production, and in the literature this factor varies by more than an order of magnitude (Bell, 1990). However, empirically determined conversion factors are very dependent on the concentration of thymidine used for calibration, and it is difficult to compare different studies. Bell (1990) argues that using a high ($\geq 20$ nM) concentration of [$^{3}H$]thymidine will minimize *de novo* synthesis, giving conversion factors of $\approx 2 \times 10^{18}$ cells $mol^{-1}$. This factor has been derived for several Swedish (Bell *et al.*, 1983; Bell & Kuparinen, 1984; Bell, 1990) and Danish lakes (Smits & Riemann, 1988) and approaches the factor expected on theoretical grounds (Bell, 1990; Riemann & Bell, 1990). The weakest links in converting rates of thymidine incorporation into carbon production are the estimation of bacterial cell volumes and the factor for carbon per unit bacterial biovolume. At present, most workers use the factor of 0.22 pg $\mu m^{-3}$ of Bratbak & Dundas (1984) which lies in the middle of the range of literature values (Riemann & Bell, 1990).

To convert thymidine incorporation to gross bacterial carbon uptake, we assumed a bacterial growth yield of 50%. It has been suggested that 30% is a more realistic estimate for natural bacterial communities (Bjørnsen & Riemann, 1988; Schwaerter *et al.*, 1988). The most realistic estimate of bacterial respiration of PDOC is 50%, however (Cole *et al.*, 1982; Chrost & Faust, 1983; Coveney & Wetzel, 1989), and our calculations reveal that on several occasions, PDOC is the dominant (or the sole) bacterial substrate. Therefore, we do not think that use of different respiration quotas for PDOC uptake and total bacterial carbon uptake can be justified for our experiments.

Another reason for the fact that the $PDO^{14}C$ uptake at times exceeds total bacterial production concerns the methodology of $PDO^{14}C$ uptake measurements. Several studies have shown that inorganic carbon uptake in the 'bacterial' size fraction may lead to serious overestimations of the bacterial assimilation of $PDO^{14}C$ measured with the fractionated filtration method (Larsson & Hagström, 1982; Sundh, 1989). With the exception of Lake Erken on May 5, 1988 (when the correction resulted in negative uptake rate) and Lake Vallentunasjön on September 27, 1989 (the prefiltered samples were lost), the diel assimilation of inorganic carbon in the bacterial size fraction was 5–10%, 14–39% and 7–58% of the total radioactivity of the bacterial filter after fractionated filtration, in Lakes Vallentunasjön, Erken and Siggeforasjön, respectively, stressing the importance of this type of correction. In addition, it is possible that fragile phytoplankton cells break when they are trapped on the 2 $\mu$m filter (Kuosa, 1988) and, if so, fragments of the cells may pass the filter and be trapped on the 0.2 $\mu$m filter instead, resulting in overestimations of the $PDO^{14}C$ uptake. The same effect may result from postfiltration rinsing of filters (Goldman & Dennet, 1985). In the diel experiments in Lakes Vallentunasjön, Erken and Siggeforasjön the effect of such errors was minimized by not using vacuum for the 2 μm filtration, and excluding the rinsing of filters, and should thus be moderate. The bacterial $PDO^{14}C$ uptake measurements, however, may be underestimated in two ways. Besides underestimation because isotopic equilibrium in the $PDO^{14}C$ released from the cells was not attained (Smith, 1982; Jensen *et al.*, 1985), presence in the water of unlabelled substances identical to the released PDOC, leads to dilution of the released substances and underestimation of uptake. Evidence presented elsewhere (Sundh, 1989) for experiments in Lake Sörmogen, however, suggests that isotopic equilibrium may not have been a major problem in our experiments.

The mean contribution of PDOC to total bacterial carbon uptake was lower in the humic than in the non-humic lakes (Table 4) and thus our results, in agreement with other studies (Hessen, 1985; Jones & Salonen, 1985; Tranvik, 1989), suggest that pelagic bacteria in humic lakes utilize allochthonous, humic substances as a significant, additional carbon source. The lower contribution of PDOC to bacterial production in the humic lakes was not statistically significant on the 0.05 level, however (student's t-test). 'Humic lakes' is an arbitrary distinction, however, and compared to the lakes studied by Jones & Salonen (1985, water colour up to 300 mg Pt $l^{-1}$) and Tranvik (1989, water colour up to 180 mg Pt $l^{-1}$), Lakes Sörmogen and Siggeforasjön must be considered moderately humic. It should be emphasized, therefore, that the small difference between non-humic and humic lakes in our data may be an effect of the comparatively low humic content of Lakes Siggeforasjön and Sörmogen. Another factor that would tend to render the difference in PDOC contributions to bacterial production in the humic and clear-water lakes an underestimate, is that we only used one light level (probably close to the level of maximum photosynthesis) for the $^{14}C$ incubation, and thus our data are volume- and not area-based. Assuming that in the epilimnion, bacterial production is less variable with depth than primary production (Bell *et al.*, in prep.), and that the depth of the euphotic zone relative to the depth of the epilimnion is smaller in the humic lakes, it is probable that on an areal basis, the bacterial production in relation to primary production and bacterial assimilation of PDOC, differs more between the humic and the non-humic lakes than is suggested from our data.

*Seasonal variations in activity of heterotrophic bacteria*

The contribution of PDOC to total bacterial production (Table 4), exhibited large within-lake variation. Especially in Lake Erken, the tendency is that PDOC contributes more to bacterial production during spring conditions than later in summer. A possible explanation for this situation is that since quite a large fraction ($\approx 50\%$ in Lake Erken) of the released PDOC is not immediately assimilated by bacteria, these substances will accumulate in the water. In spring, after the ice-covered period with very low primary production, the water may be almost depleted of DOC utilizable to the bacteria. When primary production (and thus production of PDOC) increases during the spring bloom and with the passing of summer, the PDOC not immediately assimilated by bacteria accumulates, and may subsequently serve as an additional bacterial substrate during summer stratification. Similarly, Scavia & Laird (1987) found that during summer stratification in Lake Michigan, not even the total primary production could meet the total bacterial carbon demand, and their explanation was that in summer the bacteria utilized PDOC that had accumulated in the water during the spring chlorophyll peak.

In Lake Siggeforasjön, the contribution of PDOC to bacterial production was much higher (87% and 144%, which is well within the range of the non-humic lakes) during summer stratification (August experiments) than in the September experiment (6%), indicating larger utilization of allochthonous, humic substances during autumn conditions. Interestingly, the low contribution of PDOC in the September experiment occurs simultaneously with the lowest rates of $PDO^{14}C$ assimilation (6% of released $PDO^{14}C$ recovered in the bacteria). Tranvik (1989) obtained the highest rates of bacterial production to primary production in a humic lake in autumn, compared to earlier in the summer. Thus, the bacterial assemblage of humic lakes appears to be able to utilize other substrates than recently released PDOC when necessary to keep the production on a high level, but nevertheless seems to prefer PDOC under conditions of high primary production and release of PDOC.

Bacterial growth rates (production/biomass) also showed seasonal trends. The bacterial growth rates tended to be lowest during spring and autumn when the water temperature is lower than during summer stratification. Likewise, Hobbie & Cole (1984) found a close relationship between [$^3$H]thymidine incorporation and water temperature in marine enclosures. Also, Scavia & Laird (1987) found that when water temperatures were lower than 10 °C, bacterial production was positively correlated to the water temperature, but not to primary production. These studies strongly imply that water temperature exerts a major influence on the activity of pelagic communities of heterotrophic bacteria.

It is interesting that, overall, the bacterioplankton in Lake Erken utilized the highest percentage of released PDOC (Table 3). The role of bacteria in pelagic carbon and phosphorus cycling has attracted considerable attention (Vadstein *et al.*, 1988, 1989; Jürgens & Güde, 1990). The common paradigm has been that bacteria, while outcompeting algae for phosphorus, are nonetheless dependent on the algae for carbon, creating a type of mutualism (Bratbak & Thingstad, 1985). This view is being modified since recent evidence suggests that bacteria may at times also be limited by phosphorus (Vadstein *et al.*, 1988). Two lines of evidence suggest that bacterioplankton in Erken may be more limited by carbon than phosphorus. The molar C:P ratio of bacteria in Erken is ≈30 (Pettersson & Bell, unpubl.) which, based on chemostat studies, suggests C-limitation (Vadstein & Olsen, 1989). Likewise, from May through August 1989, ≈15% of the alkaline phosphatase activity was by the bacterial size-fraction (Pettersson *et al.*, unpubl.) suggesting that it is the phytoplankton that are P-limited. In this context it seems logical that the bacteria in Erken would utilize a higher percentage of PDOC. Perhaps the percentage of released PDOC that is utilized by the bacteria can be an index of the extent to which bacteria are 'C-limited' and of the availability of other carbon sources.

In short, the most striking conclusion in terms of seasonal variations of bacterial dependence of PDOC and total bacterial activity is that the variations within the lakes appear to be at least as large as the variations between the non-humic and the humic lakes. The most important governing factors for these within lakes seasonal variations can probably be found in the seasonal succession of phytoplankton, both in terms of biomass (influencing the quantity of PDOC) and taxonomic composition (influencing both the quantity and quality of PDOC), as well as the seasonal fluctuations of water temperature.

## Acknowledgements

This work was supported in part by grants from Swedish Natural Science Research Council (NFR) to Bengt Boström and Russell T. Bell, and from Malmen's Foundation of Uppsala University to Ingvar Sundh.

## References

Bell, R. T., 1986. Further verification of the isotope dilution approach for estimating the degree of participation of [$^3$H]thymidine in studies of aquatic bacterial production. Appl. envir. Microbiol. 52: 1212–1214.

Bell, R. T., 1990. An explanation for the variability in the conversion factor deriving bacterial cell production from incorporation of [$^3$H]thymidine. Limnol. Oceanogr. 35: 910–915.

Bell, R. T. & J. Kuparinen, 1984. Assessing phytoplankton and bacterioplankton production during early spring in Lake Erken, Sweden. Appl. envir. Microbiol. 48: 1221–1230.

Bell, R. T. & B. Riemann, 1989. Adenine incorporation into DNA as a measure of microbial production in freshwaters. Limnol. Oceanogr. 34: 435–444.

Bell, R. T., G. M. Ahlgren & I. Ahlgren, 1983. Estimating bacterioplankton production by measuring [$^3$H]thymidine incorporation in a eutrophic Swedish lake. Appl. envir. Microbiol. 45: 1709–1721.

Berman, T. & B. Kaplan, 1984. Diffusion chamber studies of carbon flux from living algae to heterotrophic bacteria. Hydrobiologia 108: 127–134.

Bern, L., 1985. Autoradiographic studies of [*methyl*-$^3$H]thymidine incorporation in a cyanobacterium (*Microcystis wesenbergii*)-bacterium association and in selected algae and bacteria. Appl. envir. Microbiol. 49: 232–233.

Bird, D. F. & J. Kalff, 1984. Empirical relationships between bacterial abundance and chlorophyll concentration in fresh and marine waters. Can. J. Fish. aquat. Sci. 41: 1015–1023.

Björnsen, P. K., 1988. Phytoplankton exudation of organic matter: *Why* do healthy cells do it? Limnol. Oceanogr. 33: 151–154.

Björnsen, P. K. & B. Riemann, 1988. Towards a quantitative stage in the study of microbial processes in pelagic carbon flows. Arch. Hydrobiol. Beih. Ergebn. Limnol. 31: 185–193.

Boström, B., A.-K. Pettersson & I. Ahlgren, 1989. Seasonal dynamics of a cyanobacteria dominated microbial community in surface sediments of a shallow, eutrophic lake. Aquat. Sci. 51: 153–178.

Bratbak, G. & I. Dundas, 1984. Bacterial dry matter content and biomass estimations. Appl. envir. Microbiol. 48: 755–757.

Bratbak, G. & T. F. Thingstad, 1985. Phytoplankton-bacteria interactions: an apparent paradox? Analysis of a model system with both competition and comensalism. Mar. Ecol. Prog. Ser. 25: 23–30.

Brock, T. D. & J. Clyne, 1984. Significance of algal excretory products for growth of epilimnetic bacteria. Appl. envir. Microbiol. 47: 731–734.

Chin-Leo, G. & D. L. Kirchman, 1988. Estimating bacterial production in marine waters from the simultaneous incorporation of thymidine and leucine. Appl. envir. Microbiol. 54: 1394–1399.

Chróst, R. J., 1983. Plankton photosynthesis, extracellular release, and bacterial utilization of released dissolved organic carbon (RDOC) in lakes of different trophy. Acta Microbiol. Polon. 32: 275–287.

Chróst, R. J. & M. A. Faust, 1983. Organic carbon release by phytoplankton: its composition and utilization by bacterioplankton. J. Plankton Res. 5: 477–493.

Chrzanowski, T. H. & J. G. Hubbard, 1989. Bacterial utilization of algal extracellular products in a southwestern reservoir. Hydrobiologia 179: 61–71.

Cole, J. J., S. Findlay & M. L. Pace, 1988. Bacterial production in fresh and saltwater ecosystems: A cross-system overview. Mar. Ecol. Prog. Ser. 42: 1–10.

Cole, J. J., G. E. Likens & D. L. Strayer, 1982. Photosynthetically produced dissolved organic carbon: an important carbon source for planktonic bacteria. Limnol. Oceanogr. 27: 1080–1090.

Cole, J. J., W. H. McDowell & G. E. Likens, 1984. Sources and molecular weight of 'dissolved' organic carbon in an oligotrophic lake. Oikos 42: 1–9.

Coveney, M. F., 1982. Bacterial uptake of photosynthetic carbon from freshwater phytoplankton. Oikos 38: 8–20.

Coveney, M. F. & R. G. Wetzel, 1989. Bacterial metabolism of algal extracellular carbon. Hydrobiologia 173: 141–149.

De Haan, H., 1977. Effect of benzoate on microbial decomposition of fulvic acids in Tjeukemeer (the Netherlands). Limnol. Oceanogr. 22: 38–44.

Dunstall, T. G. & C. Nalewajko, 1975. Extracellular release in planktonic bacteria. Verh. Int. Ver. Limnol. 19: 2643–2649.

Feuillade, M., P. Dufour & J. Feuillade, 1988. Organic carbon release by phytoplankton and bacterial reassimilation. Schweiz. Z. Hydrol. 50: 115–135.

Goldman, J. C. & M. R. Dennet, 1985. Susceptibility of some marine phytoplankton species to cell breakage during filtration and post-filtration rinsing. J. exp. mar. Biol. Ecol. 86: 47–58.

Hellebust, J. A., 1965. Excretion of some organic compounds by marine phytoplankton. Limnol. Oceanogr. 10: 192–206.

Hessen, D. O., 1985. The relation between bacterial carbon and dissolved humic compounds in oligotrophic lakes. FEMS Microbiol. Ecol. 31: 215–223.

Heyman, U. & P. Blomqvist, 1984. Diurnal variations in phytoplankton cell numbers and primary productivity in Siggeforasjön. Arch. Hydrobiol. 100: 219–233.

Hobbie, J. E. & J. J. Cole, 1984. Response of a detrital foodweb to eutrophication. Bull. mar. Sci. 35: 357–363.

Iturriaga, R. & A. Zsolnay, 1981. Transformation of some dissolved organic compounds by a natural heterotrophic population. Mar. Biol. 62: 125–129.

Jassby, A. D. & C. R. Goldmann, 1974. Loss rates from a lake phytoplankton community. Limnol. Oceanogr. 19: 618–627.

Jensen, L. M., 1983. Phytoplankton release of extracellular organic carbon, molecular weight composition, and bacterial assimilation. Mar. Ecol. Prog. Ser. 11: 39–48.

Jensen, L. M., N. O. G. Jörgensen & M. Söndergaard, 1985. Specific activity. Significance in estimating release rates of extracellular dissolved organic carbon (EOC) by algae. Verh. int. Ver. Limnol. 22: 2893–2897.

Jones, R. I. & K. Salonen, 1985. The importance of bacterial utilization of released phytoplankton photosynthate in two humic forest lakes in southern Finland. Holarct. Ecol. 8: 133–140.

Jumars, P. A., D. L. Penry, J. A. Baross, M. J. Perry & B. W. Frost, 1989. Closing the microbial loop: dissolved carbon pathway to heterotrophic bacteria from incomplete ingestion, digestion and absorption in animals. Deep Sea Res. 36: 483–495.

Jürgens, K. & H. Güde, 1990. Incorporation and release of phosphorus by planktonic bacteria and phagotrophic flagellates. Mar. Ecol. Prog. Ser. 59: 271–284.

Kuosa, H., 1988. Enumeration of autotrophic and heterotrophic flagellates in Baltic Sea samples – a comparison of microscopical methods. Arch. Hydrobiol. Beih. Ergebn. Limnol. 31: 301–306.

Lampert, W., 1978. Release of dissolved organic carbon by grazing zooplankton. Limnol. Oceanogr. 23: 831–834.

Larsson, U. & Å. Hagström, 1982. Fractionated phytoplankton primary production, exudate release and bacterial production in a Baltic eutrophication gradient. Mar. Biol. 67: 57–70.

Olsen, Y., K. M. Vårum & A. Jensen, 1986. Some characteristics of the carbon compounds released by *Daphnia*. J. Plankton Res. 8: 505–517.

Pearl, H. W., 1982. Interactions with bacteria. In N. G. Carr & B. A. Whitton (eds), The Biology of Cyanobacteria. University of California Press, Berkley & Los Angeles: 441–461.

Pettersson, K., 1990. The spring development of phytoplankton in Lake Erken: species composition, biomass, primary production and nutrient conditions – a review. Hydrobiologia 191: 9–14.

Pollinger, U. & C. Serruya, 1976. Phased division of *Peridinium cinctum* f. Westii (Dinophyceae) and development of the Lake Kinneret (Israel) bloom. J. Phycol. 12: 162–170.

Riemann, B. & R. T. Bell, 1990. Advances in estimating bacterial biomass and growth in aquatic systems. Arch. Hydrobiol. 118: 385–402.

Riemann, B. & M. Söndergaard, 1984. Measurements of diel rates of bacterial secondary production in aquatic environments. Appl. envir. Microbiol. 47: 632–638.

Riemann, B., R. T. Bell & N. O. G. Jørgensen, 1990. Incorporation of thymidine, adenine, and leucine into natural bacterial communities. Mar. Ecol. Prog. Ser. 65: 87–94.

Saunders, G. W., 1972. The kinetics of extracellular release of soluble organic matter by plankton. Verh. int. Ver. Limnol. 18: 140–146.

Scavia, D. & G. A. Laird, 1987. Bacterioplankton in Lake Michigan: Dynamics, controls and significance to carbon flux. Limnol. Oceanogr. 32: 1017–1033.

Schwaerter, S., M. Søndergaard, B. Riemann & L. M. Jensen, 1988. Respiration in eutrophic lakes: the contribution of bacterioplankton and bacterial growth yield. J. Plankton Res. 10: 515–531.

Sederholm, H., A. Mauranen & L. Montonen, 1973. Some observations on the microbial degradation of humous substances in water. Verh. int. Ver. Limnol. 18: 1301–1305.

Simon, M. & F. Azam, 1989. Protein content and protein synthesis rates of planktonic marine bacteria. Mar. Ecol. Prog. Ser. 51: 201–213.

Smith, R. E. H., 1982. The estimation of phytoplankton production and excretion by carbon-14. Mar. Biol. Lett. 3: 325–334.

Smits, J. & B. Riemann, 1988. Cell production derived from [$^3$H]thymidine incorporation using freshwater bacteria. Appl. envir. Microbiol. 54: 2213–2219.

Sundh, I., 1989. Characterization of phytoplankton extracellular products (PDOC) and their subsequent uptake by heterotrophic organisms in a mesotrophic forest lake. J. Plankton Res. 11: 463–486.

Søndergaard, M. & H.-H. Shierup, 1982. Dissolved organic carbon during a spring diatom bloom in lake Mossö, Denmark. Wat. Res. 16: 815–821.

Søndergaard, M., B. Riemann & N. O. G. Jörgensen, 1985. Extracellular organic carbon (EOC) released by phytoplankton and bacterial production. Oikos 45: 323–332.

Tranvik, L. J., 1989. Bacterioplankton growth, grazing mortality and quantitative relationship to primary production in a humic and a clearwater lake. J. Plankton Res. 11: 985–1000.

Vadstein, O. & Y. Olsen, 1989. Chemical composition and phosphate uptake kinetics of limnetic bacterial communities cultured in chemostats under phosphorus limitation. Limnol. Oceanogr. 34: 939–946.

Vadstein, O., B. O. Harkjerr, A. Jensen, Y. Olsen & H. Reinerstein, 1989. Cycling of organic carbon in the photic zone of a eutrophic lake with special reference to the heterotrophic bacteria. Limnol. Oceanogr. 34: 840–855.

Vadstein, O., A. Jensen, Y. Olsen & H. Reinerstein, 1988. Growth and phosphorus status of limnetic phytoplankton and bacteria. Limnol. Oceanogr. 33: 489–503.

Wetzel, R. G., 1983. Limnology (Second edition), Saunders College Publishing, 858 pp.

*Hydrobiologia* **229**: 107–114, 1992.
*K. Salonen, T. Kairesalo & R.I. Jones (eds.),*
*Dissolved Organic Matter in Lacustrine Ecosystems: Energy Source and System Regulator.*




# Allochthonous dissolved organic matter as an energy source for pelagic bacteria and the concept of the microbial loop

Lars J. Tranvik
*Institute of Ecology, Limnology, University of Lund, Box 65, S-221 00 Lund, Sweden*

*Key words:* allochthonous DOM, recalcitrant DOM, bacterioplankton, microbial loop


## Abstract

Substantial evidence exists that allochthonous dissolved organic matter (DOM) can provide an important carbon source for pelagic bacteria. On the other hand, it is implicit in the concept of the 'microbial loop' that the degradation of recalcitrant, allochthonous DOM should be retarded in the pelagic environment, as bacteria able to utilize recalcitrant DOM compounds for slow growth would be outcompeted by faster-growing bacteria utilizing more labile DOM compounds. Several possible solutions of this apparent paradox are suggested in this paper, including formation of labile DOM from recalcitrant DOM by e.g. photochemical reactions, and mechanisms enabling the maintenance of a metabolically diverse bacterioplankton. These mechanisms include an explanation analogous to Hutchinson's classical solution to the 'paradox of plankton', and differential mortality of different populations within the bacterioplankton enabled by selective grazing, infections by bacteriophages and predatory bacteria, and spatial micropatchiness.


## Introduction

An important change in the understanding of pelagic ecosystems took place with the recognition of the 'microbial loop' (the term coined by Azam *et al.*, 1983, although the concept was presented earlier, e.g. by Pomeroy, 1974; Williams, 1981; Sieburth & Davis, 1982). According to several observations during the 1970's and 80's, a substantial fraction of primary production is transformed into dissolved organic matter (DOM) through extracellular release from algal and cyanobacterial cells, and this DOM is subsequently utilized by heterotrophic bacteria (e.g. Williams & Yentsch, 1976; Larsson & Hagström, 1979). Heterotrophic, planktonic bacteria make up a large biomass with a high turnover rate (Hagström *et al.*, 1979; Fuhrman & Azam, 1980; 1982). A major sink for heterotrophic bacteria, as well as the smallest autotrophic cells, is through grazing by protozoa, earlier not considered important (Fenchel, 1982; Sieburth & Davis, 1982; Sherr & Sherr, 1984). Thus, to the conventional linear food chain a microbial loop can be added, transferring much of the energy of pelagic systems through a DOM-bacterial-protozoan pathway (Azam *et al.*, 1983). The consequences of this concept are discussed by Williams (1981), and Azam *et al.* (1983). Major effects of the microbial loop are that more trophic levels exist than are accounted for in the conventional food chain, and that a smaller fraction of the primarily produced organic carbon reaches larger organisms than previously believed, as more of it is respired by microorganisms. Bacteria play a key role, reincorporating DOM lost from phagotrophic food chains into particulate matter.

The research that generated the concept of the microbial loop was almost exclusively carried out in marine environments. There are only minor differences between marine and freshwater habitats, and 'a scientist working in aquatic ecology may quite safely ignore the differences and use the literature from freshwater research to understand marine problems and vice versa' (Hobbie, 1988). However, some differences still exist. Macrozooplankton (Cladocera) that may be efficient grazers of bacterioplankton are ubiquitous in fresh waters, but virtually absent in the marine pelagial (Hobbie, 1988). These may sometimes be the major bacterivores, short-circuiting the microbial loop by a direct bacterial-macrozooplankton transfer (Bosselmann & Riemann, 1986). However, bacterial-protozoan food chains in fresh water similar to those of marine systems have also been described (Güde, 1986; Scavia & Laird, 1987; Tranvik, 1989). Another important difference is that the bacteria of inland waters are to a much higher degree than the bacteria at oceanic sites exposed to dissolved organic matter from soil and stream runoff (allochthonous matter). These compounds have undergone some degradation and transformation before entering the pelagic environment, and thus are less labile than autochthonous substrates originating from indigenous algal production (Hobbie, 1988). As the modern concepts of aquatic microbial ecology mainly arise from marine science focused on the open ocean – which is influenced by allochthonous DOM to much less extent than inland waters – such substances may largely have been overlooked by microbial ecologists applying these concepts to inland waters.

## Evidence for allochthonous DOM as an energy source in pelagic ecosystems

In the light of 'the microbial loop', much recent research has been conducted concerning the coupling of autochthonous primary production and microheterotrophic activity (for reviews of the functioning of pelagic ecosystems without input of allochthonous DOM, see Fenchel, 1988; Porter *et al.*, 1985). This is parallelled by a growing literature suggesting a microbial link through which allochthonous DOM is imported to pelagic phagotrophic food chains via heterotrophic bacterioplankton, even in pristine freshwater environments without significant anthropogenic input of degradable DOM. High bacterial biomass in the water of humic lakes has been recorded by several investigators (Rodina, 1967; Salonen, 1981; Hessen, 1985), and more DOM has been demonstrated to be available for bacterial growth in lakes with a high load of allochthonous humic matter, than in clearwater lakes less influenced by such compounds (Tranvik, 1988). High ratios of respiration to primary production in the water of humic lakes suggest the presence of an important carbon source in addition to that supplied by primary producers (Salonen, 1981; Salonen *et al.*, 1983). Accordingly, dissolved photosynthates released from algae may be a minor bacterial substrate in humic lakes (Jones & Salonen, 1985; Tranvik, 1989). Hessen *et al.* (1990) demonstrated that most of the carbon biomass of zooplankton in a humic lake was derived from detritus and bacteria, while phytoplankton accounted for a minor fraction of the zooplankton biomass. It has been suggested that allochthonous DOM is a major food source for zooplankton in humic lakes, either directly or via bacteria and microzooplankton (Salonen & Hammar, 1986).

Thus, there is reason to believe that allochthonous organic matter is an important energy source for the pelagic microbiota of humic lakes. Accordingly, it has been suggested that bacterial growth on allochthonous DOM provides stability to freshwater bacterioplankton, whereby a large pool of allochthonous recalcitrant DOM supports continuous slow growth, independent of the intermittent growth associated with the less constant presence of autochthonous labile DOM (Wetzel, 1984; Hobbie, 1988). In Fig. 1 is presented a graphic model of a pelagic food web including the microbial loop and allochthonous DOM.

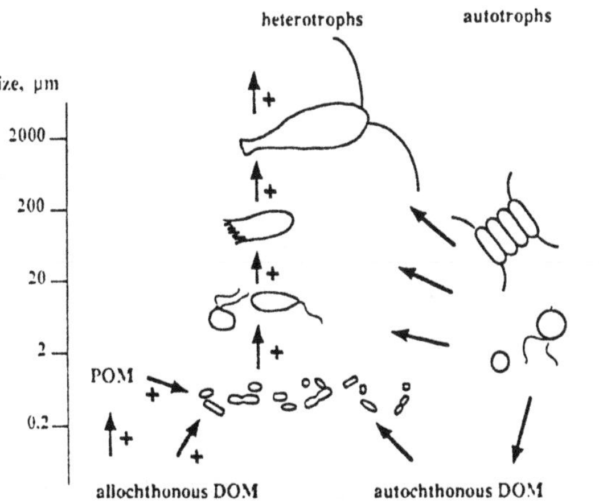


*Fig. 1.* Outline of a pelagic food web including the microbial loop, reincorporating dissolved organic mater (DOM) lost from autotrophs into the phagotrophic food chain via bacteria and protozoan grazers, and a similar microbial link incorporating allochthonous DOM into the food chain, directly or via the formation of particulate organic matter (POM). Processes added or potentially enhanced through the influence of allochthonous DOM are indicated by '+'. Modified from Fenchel (1986).

## Utilization of recalcitrant DOM by microheterotrophs of the microbial loop – is it possible?

In most current research on the fate of pelagic bacterial production, the bacterioplankton is viewed as one homogeneous population with a uniform mortality rate, and thus by implication also a uniform growth rate. This has been a fruitful concept, generating understanding of trophic interactions and carbon flow within pelagic food webs. It has also been a necessary operational concept, as the methods available for the measurement of bacterioplankton dynamics have not allowed the separate study of different 'subpopulations' of pelagic bacteria. Accordingly, Wright & Coffin (1984) stated that 'for practical and theoretical reasons, then, the assemblage of bacteria in the plankton can be treated essentially as a population, recognizing that when our methods and understanding improve we may be able to refine this approach'. As will be shown below, the concept of a homogeneous bacterioplankton population is not compatible with the possible existence of 'subpopulations' growing at different rates, at the expense of substrates of different recalcitrance. To understand the bacterial degradation of recalcitrant DOM in the pelagial we now need to refine this approach. Some of the methods usually employed for the measurement of protozoan bacterivory, which is currently believed to be the major mechanism of mortality for pelagic bacteria (Fenchel, 1988), provide a clear example of the perception of the bacterioplankton as a homogeneous population. According to these methods, grazing mortality is measured from the incorporation into bacterivores of tracer particles such as fluorescent labeled cells of a bacterial isolate (Sherr *et al.*, 1987), radioactive labelled bacteria (Gast, 1985), genetically labelled *E. coli* minicells (Wikner *et al.*, 1986), or inert particles, such as artificial fluorescent microparticles (Börsheim, 1984; McManus & Fuhrman, 1986). Although there is some concern about bacterivores preferentially grazing indigeneous bacteria (e.g. Sherr *et al.*, 1987; Pace & Bailiff, 1987; Bloem *et al.*, 1988; Nygaard *et al.*, 1988) when offered tracer particles, these methods provide at least in some cases, bacterial grazing mortality rates similar within a factor of two to the growth rates (Bloem *et al.*, 1989; Sherr *et al.*, 1989; Wikner *et al.*, 1986). Some investigators of bacterivory have also pointed out that factors of bacterial mortality other than bacterivory may be important, but generally within the concept of a homogeneous population, i.e. in order to assess loss factors that would help balance 'the' growth rate (e.g. Pace, 1988; Sherr *et al.*, 1989).

With these reservations, bacterivory is often believed to be largely non-selective and to occur at rates that balance the rates of growth of the total bacterial assemblage (Fenchel, 1982; Azam *et al.*, 1983). Consequently, the bacterioplankton is concluded to be subject to one homogeneous loss rate. Thus, 'subpopulations' that grow slowly (or not at all, i.e. dormant bacteria: Stevenson, 1978) at the expense of recalcitrant substrates are not possible, as they would be outcompeted by 'subpopulations' that grow at higher rates, utilizing labile compounds. Considering the intense grazing pressure acting on bacteria in the sea,

Azam & Cho (1987), declared that 'it is difficult to see how a large bacterial assemblage with a 1–2 day turnover time could maintain a large dormant component'. According to these premises, and assuming that 1) growth is slower on recalcitrant substrates than on labile ones, and 2) labile DOM is utilized by bacteria prior to more recalcitrant DOM, the continuous excretion of labile compounds from various plankton organisms should inhibit the utilization of recalcitrant DOM. Further more, if allochthonous DOM (that has been subject to microbial degradation before entering the pelagic environment) is assumed to be more recalcitrant than autochthonous DOM (that is available immediately after being produced by e.g. phytoplankton), the continuous regeneration of labile DOM within the microbial loop prevents microbial utilization of allochthonous substrates. The (implicitly non-selective) grazing mortality has been suggested to nurture the bacterioplankton 'by keeping the bacteria in a state of logarithmic growth, preventing senescent cells from accumulating in the population, by keeping the bacterial abundance low enough to minimize the possibility of substrate limitation, and by supplying a certain amount of dissolved organic carbon through their grazing activity for re-uptake by the bacteria' (Porter *et al.*, 1985). Moreover, grazing mortality 'might be analogous to a high flow chemostat effect with the rapid selection for less energetically efficient, fast growth species (i.e. r-selected) over more efficient slow growing forms (i.e. K-selected)' (Sherr *et al.*, 1988).

## Mechanisms potentially enabling simultaneous degradation of labile and recalcitrant substrates

As shown above, substantial evidence exists that allochthonous DOM may provide an important carbon source for pelagic bacteria. On the other hand, it is implicit in the concept of the microbial loop that degradation of recalcitrant, allochthonous DOM should be retarded in the pelagic environment. Below I will suggest some possible solutions of this apparent paradox.

### *Simultaneous degradation of labile and recalcitrant DOM within the same population*

Assuming that bacteria preferentially utilize labile substrates, this is difficult to conceive. Among several species with similar specific mortality rates, any one species utilizing more recalcitrant substrates than the others (thereby spending relatively more energy in the assimilation of the substrate) should have a lower growth efficiency (less biomass produced per unit of substrate utilized), and consequently be less competitive. Accordingly, a freshwater bacterial assemblage in mixed batch culture utilized a labile (glucose) and a recalcitrant (phenol) substrate in a sequential pattern (Tranvik & Höfle, 1987). Independent of these theoretical arguments, simultaneous substrate utilization is known to occur. This can take place as cometabolism (Horvath, 1972), whereby recalcitrant substrates are cometabolized with labile compounds. De Haan (1977) demonstrated that an *Arthrobacter* strain, isolated from a humic lake and able to grow at the expense of fulvic acids, exhibited increased growth on fulvic acids during simultaneous degradation of a labile cosubstrate (benzoate). This was suggested to be due to a priming effect of benzoate on the otherwise slow degradation of fulvic acids. Thus, at least in pure cultures, bacterial growth at the expense of recalcitrant compounds may occur during the simultaneous utilization of labile substrates.

### *The available fraction of the allochthonous DOM is not recalcitrant*

The assumption that allochthonous DOM is recalcitrant derives from the hypothesis that labile moieties are degraded before the DOM reaches the pelagic zone. Contradictory to this assumption, the available fraction of the total DOM facilitates bacterial growth at similar rates in batch cultures from lakes that are to different extents influenced by allochthonous DOM (Tranvik, 1988). Similarly, specific growth rates of bacteria were similar in a humic lake, where allochthonous

DOM was the dominating carbon source of the bacteria, and a clearwater lake, where autochthonous primary production was the main carbon source (Tranvik, 1989).

Mechanisms that may be speculated to allow continuous production of labile substrates from the recalcitrant bulk of DOM include photochemical reactions (Mopper *et al.*, 1991), and exoenzymes, diffusing into the pelagial from littoral and sediment or from allochthonous sources, that enables the depolymerization of recalcitrant macromolecules.

*Bacterioplankton is a heterogeneous mixture of slow-growing degraders of recalcitrant substrates and fast-growing degraders of labile substrates*

Coexistence of different bacterial species within the bacterioplankton is well known from taxonomic studies, either using conventional culture techniques for the isolation and identification of bacterial strains (e.g. Moaledj, 1978) or recently developed molecular biological tools for the direct assessment of the diversity of nucleic acids in water samples (Giovannoni *et al.*, 1990; Höfle, 1990). Assuming that the factors of bacterial mortality act non-selectively on all these strains (see above) the taxonomic diversity can be viewed as a paradox analogous to 'the paradox of plankton' identified by Hutchinson (1961), who demonstrated an apparent contradiction between theoretical population ecology (the principle of competitive exclusion) and empirical data (phytoplankton generally comprise a large number of species, apparently occupying the same ecological niche). The paradox can be resolved in a way identical to a major solution that Hutchinson imposed upon to solve his classical paradox of phytoplankton: For competitive exclusion to occur, the competing populations are required to operate within constant environmental conditions. However, the pelagic environment is perturbed in different ways (light, temperature, fluctuations in organic and inorganic chemistry of importance to bacterial metabolism) on a time scale short enough to alter the arena of competing bacterial populations before competitive exclusion occurs. For example, reasonably frequent periods of shortage of labile substrates due to decreases in primary production (e.g. on the basis of day-night cycles) provide a temporal refuge for degraders of recalcitrant compounds.

Additional potential mechanisms enabling coexistence of slow- and fast-growing bacterial populations include differential mortality rates enabled by selective grazing by bacterivores, selective infections by bacteriophages and predatory bacteria, and the occurrence of microzones in the water with different turnover rates of DOM and microorganisms.

Selective grazing by bacterivores has been considered as an artefact in measurements of bacterivory utilizing tracer particles (see above). Accordingly, bacterivorous flagellates are known to be size-selective in their prey ingestion (Andersson *et al.*, 1986; Gonzales *et al.*, 1990) and some species of bacterivorous protozoa isolated from aquatic environments also exhibit chemosensory preference of specific bacterial strains (Sibbald *et al.*, 1987; Mitchell *et al.*, 1988).

Viruses and bacteriophage-infected cells can be observed in marine water by electron microscopy (Sieburth, 1979) and their ubiquity in pelagic environments has been demonstrated (Bergh *et al.*, 1989; Proctor & Fuhrman, 1990). They are speculated to be a quantitatively important factor for bacterioplankton mortality. Host-specific bacteriophages have an obvious potential for providing a mechanism of differential bacterial mortality, as only some bacterial species would be infected and lysed.

Spatial structuring of planktonic communities into microzones of different turnover rates would allow the parallel existence of slow- and fast growing bacteria. Suspended detrital particles, such as marine snow (Caron *et al.*, 1982) and humic flocs (Tranvik & Sieburth, 1989), might provide niches for bacterial and protozoan communities with growth and loss characteristics potentially different from those in the surrounding water. It has also been speculated that chemosensory mechanisms allow plankton organisms to gather up into loosely connected microaggregates with enhanced population turnover, e.g. with bacteria being attracted to the surface of phytoplankton cells (Goldman, 1984; Mitchell *et al.*, 1985).

*Utilization of recalcitrant DOM by organisms other than bacteria*

Utilization of DOM by free-living protozoa has been considered improbable, as 'no protozoan groups have evolved free-living (non-photosynthetic) forms which do not ingest particulate materials' (Fenchel, 1986), and because bacteria are competitively superior in the uptake of dissolved nutrients (Fenchel, 1986). Nevertheless, it has been indicated that planktonic, heterotrophic flagellates, hitherto believed to be exclusively phagotrophic, can feed on DOM compounds with molecular weights of 500 000 or more (Sherr, 1988). Bacteria must degrade such macromolecules into monomers before ingesting them into the cell, while protozoa are able to directly incorporate them suggesting that they have an advantage over bacteria in the utilization of DOM of high molecular weight (Sherr, 1988). Moreover, recalcitrant DOM might be indirectly available to phagotrophic organisms after flocculation to particulate organic matter, by e.g. bubble coagulation (Kepkay & Johnson, 1989).

In conclusion, a multitude of mechanisms exist, that might explain the utilization of allochthonous recalcitrant DOM by pelagic microbes. Future research should aim to study pelagic DOM turnover in the light of these mechanisms.

## Acknowledgements

Participants of the symposium at Lammi Biological Station, Finland, May 1990, 'Dissolved Organic Matter in Lacustrine Ecosystems: Energy Source and System Regulator', are acknowledged for stimulating discussions and encouragement in the preparation of this manuscript. This work was funded by the Swedish Environment Protection Board and the Swedish Natural Science Research Council.

## References

Andersson, A., U. Larsson & Å. Hagström, 1986. Size-selective grazing by a microflagellate on pelagic bacteria. Mar. Ecol. Prog. Ser. 33: 51–57.

Azam, F. & B. C. Cho, 1987. Bacterial utilization of organic matter in the sea. In M. Fletcher, T. R. G. Gray & J. G. Jones (eds), Ecology of microbial Communities. Cambridge University Press, Cambridge: 261–281.

Azam, F., T. Fenchel, F. G. Field, J. S. Gray, L.-A. Meyer-Reil & F. Thingstad, 1983. The ecological role of water-column microbes in the sea. Mar. Ecol. Prog. Ser. 10: 257–263.

Bergh, Ö., K. Y. Börsheim, G. Bratbak & M. Heldal, 1989. High abundance of viruses found in aquatic environments. Nature 340: 467–468.

Bloem, J., M. Starink, M.-J. B. Bär-Gilissen & T. E. Cappenberg, 1988. Protozoan grazing, bacterial activity, and mineralization in two-stage continuous cultures. Appl. envir. Microbiol. 54: 3113–3121.

Bloem, J., F. M. Ellenbroek, M.-J. B. Bär-Gilissen & T. E. Cappenberg, 1989. Protozoan grazing and bacterial production in stratified lake Vechten estimated with fluorescently labeled bacteria and by thymidine incorporation. Appl. envir. Microbiol. 55: 1787–1795.

Bosselmann, S. & B. Riemann, 1986. Zooplankton. In B. Riemann & M. Søndergaard (eds), Carbon Dynamics in eutrophic, temperate Lakes. Elsevier: 199–236.

Börsheim, K. Y., 1984. Clearance rates of bacteria-sized particles by freshwater ciliates, measured with monodisperse fluorescent latex beads. Oecologia 63: 286–288.

Caron, D. A., P. G. Davis, L. P. Madin & J. McN. Sieburth, 1982. Heterotrophic bacteria and bacterivorous protozoa in oceanic macroaggregates. Science 218: 795–797.

De Haan, H., 1977. Effect of benzoate on microbial decomposition of fulvic acids in Tjeukemeer (The Netherlands). Limnol. Oceanogr. 22: 38–44.

Fenchel, T., 1982. Ecology of heterotrophic microflagellates. IV. Quantitative occurrence and importance as bacterial consumers. Mar. Ecol. Prog. Ser. 9: 35–42.

Fenchel, T., 1986. Ecology of protozoa. The biology of free-living phagotrophic protists. Brock/Springer Series in Contemporary Bioscience. Springer-Verlag, Berlin.

Fenchel, T., 1988. Marine plankton food chains. Ann. Rev. Ecol. Syst. 19: 19–38.

Fuhrman, J. A. & F. Azam, 1980. Bacterioplankton secondary production estimates for coastal waters of British Columbia, Antarctica, and California. Appl. envir. Microbiol. 39: 1085–1095.

Fuhrman, J. A. & F. Azam, 1982. Thymidine incorporation as a measure of heterotrophic bacterioplankton production in marine surface waters: Evaluation and field results. Mar. Biol. 66: 109–120.

Gast, V., 1985. Bacteria as a food source for microzooplankton in the Schlei Fjord and Baltic Sea with special reference to ciliates. Mar. Ecol. Prog. Ser. 22: 107–120.

Giovannoni, S. J., T. B. Britschgi, C. L. Moyer & K. G. Field, 1990. Genetic diversity of Sargasso Sea bacterioplankton. Nature 345: 60–63.

Goldman, J. C., 1984. Conceptual role for microaggregates in pelagic waters. Bull. mar. Sci. 35: 462–476.

Gonzalez, J. M., E. B. Sherr & B. F. Sherr, 1990. Size-selective grazing on bacteria by natural assemblages of estuarine flagellates and ciliates. Appl. envir. Microbiol. 56: 583–589.

Güde, H., 1986. Loss processes influencing growth of planktonic bacterial populations in Lake Constance. J. Plankton Res. 8: 795–810.

Hagström, Å., U. Larsson, P. Hörstedt & S. Normark, 1979. Frequency of dividing cells, a new approach to the determination of bacterial growth rates in aquatic environments. Appl. envir. Microbiol. 37: 805–812.

Hessen, D. O., 1985. The relation between bacterial carbon and dissolved humic compounds in oligotrophic lakes. FEMS Microbial. Ecol. 31: 215–223.

Hessen, D. O., T. Andersen & A. Lyche, 1990. Carbon metabolism in a humic lake; pool sizes and cycling through zooplankton. Limnol. Oceanogr. 35: 84–99.

Hobbie, J. E., 1988. A comparison of the ecology of planktonic bacteria in fresh and salt water. Limnol. Oceanogr. 33: 750–764.

Horvath, R. S., 1972. Microbial co-metabolism and the degradation of organic compounds in nature. Bact. Rev. 36: 146–155.

Hutchinson, G. E., 1961. The paradox of plankton. Am. Nat. 95: 137–145.

Höfle, M. G., 1990. RNA chemotaxonomy of bacterial isolates and natural microbial communities. In J. Overbeck & R. J. Chróst (eds), Aquatic microbial Ecology – biochemical and molecular Approaches. Brock/Springer Series in Contemporary Bioscience. Springer-Verlag, Berlin: 129–159.

Jones, R. I. & K. Salonen, 1985. The importance of bacterial utilization of released phytoplankton photosynthate in two humic forest lakes in southern Finland. Holarct. Ecol. 8: 133–140.

Kepkay, P. E. & B. D. Johnson, 1989. Coagulation on bubbles allows microbial respiration of oceanic dissolved organic carbon. Nature 338: 63–65.

Larsson, U. & Å. Hagström, 1979. Phytoplankton exudate release as an energy source for the growth of pelagic bacteria. Mar. Biol. 52: 199–206.

McManus, G. B. & J. A. Fuhrman, 1986. Bacterivory in seawater studied with the use of inert fluorescent particles. Limnol. Oceanogr. 31: 420–426.

Mitchell, G. C., J. H. Baker & M. A. Sleigh, 1988. Feeding of a freshwater flagellate, *Bodo saltans*, on diverse bacteria. J. Protozool. 35: 219–222.

Mitchell, J. G., A. Okubo & J. A. Fuhrman, 1985. Microzones surrounding phytoplankton form the basis for a stratified marine microbial ecosystem. Nature 316: 58–59.

Moaledj, K., 1978. Qualitative analysis of an oligocarbophilic aquatic microflora in the Plussee. Arch. Hydrobiol. 82: 98–113.

Mopper, K., X. Zhou, R. J. Kieber, D. J. Kieber, R. J. Sikorski & R. D. Jones, 1991. Photochemical degradation of dissolved organic carbon and its impact on the oceanic carbon cycle. Nature 353: 60–62.

Nygaard, K., K. Y. Börsheim & T. F. Thingstad, 1988. Grazing rates on bacteria by marine heterotrophic microflagellates compared to uptake rates of bacterial-sized monodisperse fluorescent latex beads. Mar. Ecol. Prog. Ser. 44: 159–165.

Pace, M. L., 1988. Bacterial mortality and the fate of bacterial production. Hydrobiologia 159: 41–49.

Pace, M. L. & M. D. Bailiff, 1987. Evaluation of a fluorescent microsphere technique for measuring grazing rates of phagotrophic microorganisms. Mar. Ecol. Prog. Ser. 40: 185–193.

Pomeroy, L. R., 1974. The oceans food web, a changing paradigm. BioScience 24: 499–504.

Porter, K. G., E. B. Sherr, B. F. Sherr, M. Pace & R. W. Sanders, 1985. Protozoa in planktonic food webs. J. Protozool. 32: 409–415.

Proctor, L. M. & J. A. Fuhrman, 1990. Viral mortality of marine bacteria and cyanobacteria. Nature 343: 60–62.

Rodina, A. G., 1967. Bacterial populations of humified lakes. Microbiologiya 38: 531–537.

Salonen, K., 1981. The ecosystem of the oligotrophic lake Pääjärvi. 2. Bacterioplankton. Int. Verein Theor. Angew. Limnol. Verh. 21: 448–553.

Salonen, K. & T. Hammar, 1986. On the importance of dissolved organic matter in the nutrition of zooplankton in some lake waters. Oecologia 68: 246–253.

Salonen, K., K. Kolonen & L. Arvola, 1983. Respiration of plankton in two small, polyhumic lakes. Hydrobiologia 101: 65–70.

Scavia, D. & G. A. Laird, 1987. Bacterioplankton in Lake Michigan: dynamics, control, and significance to carbon flux. Limnol. Oceanogr. 32: 1017–1033.

Sherr, E. B., 1988. Direct use of high molecular weight polysaccharide by heterotrophic flagellates. Nature 335: 348–351.

Sherr, B. F. & E. B. Sherr, 1984. Role of heterotrophic protozoa in carbon and energy flow in aquatic ecosystems. In M. J. Klug & C. A. Reddy (eds), Current Perspectives in microbial Ecology. American Society for Microbiology, Washington, D.C.: 412–423.

Sherr, B. F., E. B. Sherr & R. D. Fallon, 1987. Use of monodispersed, fluorescently labeled bacteria to estimate *in situ* protozoan bacterivory. Appl. envir. Microbiol. 53: 958–965.

Sherr, B. F., E. B. Sherr & C. S. Hopkinson, 1988. Trophic interactions within pelagic microbial communities: Indications of feedback regulation of carbon flow. Hydrobiologia 159: 19–26.

Sherr, B. F., E. B. Sherr & C. Pedrós-Alió, 1989. Simultaneous measurement of bacterioplankton production and protozoan bacterivory in estuarine water. Mar. Ecol. Prog. Ser. 54: 209–219.

Sibbald, M. J., L. J. Albright & P. R. Sibbald, 1987. Chemosensory responses of a heterotrophic microflagellate to bacteria and several nitrogen compounds. Mar. Ecol. Prog. Ser. 36: 201–204.

Sieburth, J. McN., 1979. Sea Microbes. Oxford University Press, New York.

Sieburth, J. McN. & P. G. Davis, 1982. The role of heterotrophic nanoplankton in the grazing and nurturing of planktonic bacteria in the Sargasso and Caribbean seas. Ann. Inst. Oceanogr. 58: 285–296.

Stevenson, L. H., 1978. A case for bacterial dormancy in aquatic systems. Microb. Ecol. 4: 127–133.

Tranvik, L. J., 1988. Availability of dissolved organic carbon for planktonic bacteria in oligotrophic lakes of differing humic content. Microb. Ecol. 16: 311–322.

Tranvik, L. J., 1989. Bacterioplankton growth, grazing mortality and quantitative relationship to primary production in a humic and a clearwater lake. J. Plankton Res. 11: 985–1000.

Tranvik, L. J. & M. G. Höfle, 1987. Bacterial growth in mixed cultures on dissolved organic carbon from humic and clear waters. Appl. envir. Microbiol. 53: 482–488.

Tranvik, L. J. & J. McN. Sieburth, 1989. Effects of flocculated humic matter on free and attached pelagic microorganisms. Limnol. Oceanogr. 34: 688–699.

Wetzel, R. G., 1984. Detrital dissolved and particulate organic carbon functions in aquatic ecosystems. Bull. mar. Sci. 35: 503–509.

Wikner, J., A. Andersson, S. Normark & Å Hagström, 1986. Use of genetically marked minicells as a probe in measurement of predation on bacteria in aquatic environments. Appl. envir. Microbiol. 52: 4–8.

Williams, P. J. LeB., 1981. Incorporation of microheterotrophic processes into the classical paradigm of the planktonic food web. Kieler Meeresforsch. Sonderh. 5: 1–28.

Williams, P. J. LeB. & C. S. Yentsch, 1976. An examination of photosynthetic production, excretion of photosynthetic products, and heterotrophic utilization of dissolved organic compounds with reference to results from a coastal subtropical sea. Mar. Biol. 35: 31–40.

Wright, R. T. & R. B. Coffin, 1984. Factors affecting bacterioplankton density and productivity in salt marsh estuaries. In M. J. Klug and C. A. Reddy (eds), Current perspectives in microbial ecology. American Society for Microbiology, Washington, D.C.: 485–494.

*Hydrobiologia* **229**: 115–123, 1992.
*K. Salonen, T. Kairesalo & R.I. Jones (eds.),*
*Dissolved Organic Matter in Lacustrine Ecosystems: Energy Source and System Regulator.*




# Dissolved organic carbon in a humic lake: effects on bacterial production and respiration

Dag O. Hessen
*Norwegian Institute for Water Research, P.O. Box 69 Korsvoll, N-0808 Oslo 8, Norway*

*Key words:* DOC, humus, bacteria, production, respiration

## Abstract

Allochthonous matter was the main source of carbon for pelagic bacteria in a humic lake, accounting for almost 90% of the carbon required to support observed bacterial growth. The estimated contribution from zooplankton excretion was of the same magnitude as direct phytoplankton release, both accounting for 5–7% of bacterial demands for dissolved carbon. Bacteria were an important source of carbon both for heterotrophic phytoplankton and for filter feeding zooplankton species, further stressing the role of humus DOC in overall lake productivity.

The high contribution of allochthonous DOC implies a stoichiometry of dissolved nutrients with a surplus of C relative to P. The high P cell quota of bacteria suggest that under such conditions they are P-limited and act like net consumers of P. Excess C will be disposed of, and bacterial respiration rate will increase following a transition from carbon-limited bacterial growth towards mineral-nutrient-limited growth. Thus the high community respiration and frequent $CO_2$-supersaturation in humic lakes may be caused not only by the absolute supply of organic C, but also by the stoichiometry of the dissolved nutrient pool.

## Introduction

Growth and production of pelagic bacteria are governed by at least four different sources of dissolved organic substratum: 1) release from living phytoplankton cells; 2) release from lysis or degradation of (recently) dead planktonic organisms; 3) release from zooplankton (and fish); and 4) dissolved organic matter from littoral and terrestrial sources (eg. humic compounds). The quantitative and qualitative importance of these sources undergo strong seasonal variations and one may also expect differences between localities depending on their trophy and food chain structure.

In most clearwater lakes, bacterial secondary production seems to correlate well with phytoplankton biomass and primary production. This may in part be an autocorrelation caused by temperature changes or nutrient loadings, but is more likely attributed to release of dissolved organic carbon from phytoplankton (PDOC) (Cole *et al.*, 1982; Bell & Kuparinen, 1984; Brock & Clyne, 1984; Jensen & Schwærter, 1988). Although some estimates indicate a rather restricted importance of PDOC (Cf. Lovell & Konopka, 1985), most authors report a contribution from PDOC to bacterial gross production of more than 50% in mean. However, the contribution from algal exudates undergoes strong seasonal (Coveney, 1982; Brock

& Clyne, 1984; Vadstein *et al.*, 1989) and depth variations (Søndergaard *et al.*, 1985; Vadstein *et al.*, 1989).

Zooplankton release may also be a substantial source of dissolved matter (Lampert, 1978), although one should keep in mind that most of zooplankton releases are of algal origin. Riemann & Søndergaard (1986) demonstrated that macrozooplankton may become the major contributor of POC for bacterial growth during low fish predation pressure and a subsequent high grazing pressure due to high biomass of large grazers like *Daphnia*.

That part of bacterial production not accounted for by algal- or zooplankton release must be of terrestrial origin. This may be large even in clearwater localities (Rieman *et al.*, 1982). Humic lakes are often characterized by high inputs of nonphotosynthetic DOC, and have typically a high bacteria/phytoplankton production ratio (Salonen, 1981; Johansson, 1983; Jones & Salonen, 1985; Hessen *et al.*, 1990). One might thus suspect that the relatively low phytoplankton production can support only a modest share of bacterial production, and the major source of energy in such ecosystems is probably terrestrial inputs as humic carbon. Although present in high quantities, humic substances must be regarded as a qualitatively deficient substratum for bacterial growth, not only due to its recalcitrant nature, but also due to its low content of phosphorus and nitrogen.

The present study focuses upon the role of humic DOC. I present data on carbon fluxes through pelagic bacteria in a humic lake, and further discuss the overall importance of humus DOC relative to other sources of DOC for aquatic secondary production

## Material and methods

This study was performed in the small, humic Lake Kjelsåsputten, S.E. Norway. The lake has a surface area of 8000 m$^2$ and a max. depth of 8 m. pH is stable at 4.50–4.65, and water color ranges 60 (mid summer) to 140 (autumn) mg Pt l$^{-1}$. During most of the summer the lake is anoxic below 3.5 m, corresponding to mean depth of the thermocline, and almost the entire biomass and production of phyto- and zooplankton are restricted to the upper 3 m. Hence the considerations presented in this paper, based on integrated data from the upper 2 m, should be fairly representative for the lake epilimnion. The lake has been described in a series of papers (Hessen, 1989; Hessen *et al.*, 1989, 1990). During summer 1985, an extensive programme was performed, in order to characterize biotic and abiotic pools and fluxes of carbon (see references above).

The experiments were performed in two bags (1 m diam., 2.7 m depth) anchored in the central part of the lake. One bag received $^{14}$C-protein hydrolysate (Amersham CFB.25), 22 kBq l$^{-1}$ for specific labeling of bacteria, the other received $^{14}C$-$CO_2$, 33 kBq l$^{-1}$, for labeling of phytoplankton. Specific activity was followed in all dissolved and particulate fractions for 12 days in bag 1 to estimate tracer flux through the bacteria. As the lower ends of the bags were sealed, the flux of inorganic carbon during the incubation period was assumed to represent epilimnetic processes, and not to be influenced by hypolimnetic respiration. The development of biotic and abiotic parameters were closely parallel in the two bags. With the exception of grazing estimates on bacteria, biomasses and rates presented below are based on measurements and estimates from the $CO_2$-labeled bag.

All organic carbon samples were analyzed on a Carlo-Erba CHN 1106 Elemental Analyzer. DOC-samples were screened through a Whatman GF/F-filter, acidified to pH = 2 and purged with air. A 2 ml subsample was evaporated on a strip of pre-ignited GF/F-filter and total DOC determined by CHN-analysis. POC was collected on washed, pre-ignited GF/F-filters. Dissolved inorganic carbon was analyzed on an infrared $CO_2$-analyzer. A more detailed description of analytical methods is given in Hessen *et al.* (1990).

Phytoplankton were counted using an inverted microscope, and volumes estimated from geometrical models. Carbon was estimated as 10% of wet weight (Vollenweider, 1969). Bacteria were counted after staining with acridine orange (Hobbie *et al.*, 1977). Cell volumes were determined on 100 randomly chosen cells on each filter. Volumes were estimated by use of a micrometer ocular, calibrated with monodisperse, fluorescent beads. Mean volumes were converted to carbon assuming 12% C in wet weight (Watson *et al.*, 1977). Bacterial production was measured in two series of 250 ml Pyrex flasks, filled and incubated at 0, 1, 2 and 3 m depth. The first series was screened through a 45 $\mu$m net to remove macrozooplankton. The second series was filtered through 3 $\mu$m Nucleopore membranes to remove microzooplankton. Bacterial growth was monitored by subsampling the bottles at 5 times during the 18 h incubation period. Net production was calculated from increase in numbers and biomass for each sampling interval for both incubation series. Phytoplankton production was measured by a standard $^{14}$C-method (Vollenweider, 1969) by incubating light and dark bottles at 6 different depths. Filters (0.45 $\mu$m membrane) were kept in open scintillation vials before counting. Zooplankton grazing on bacteria and algae was estimated by measuring grazing on labelled seston in bags 1 and 2 respectively. Further details are given in (Hessen *et al.*, 1989, 1990).

Total respiration from epilimnetic biota was estimated from total efflux of radioactive DIC, assuming that other loss rates (sedimentation and wall growth) were negligible (Hessen *et al.*, 1990). The isotopic loss rate over time gave a second order polynomial fit:

$$\mathrm{DIC(DPM\ ml^{-1})} = 1976.0 - 208.4\mathrm{D} + 9.99D^2 \; (R = 0.98)$$

where $D$ is number of days.

Zooplankton respiration was directly calculated from release of $^{14}$C from labelled animals placed in non-radioactive medium. Specific activity was determined every days, and these experiments were performed after 10 days, when even the slowest growing species (*Acanthodiaptomus*) had reached isotopic equilibrium and thus was homogenously labelled. Zooplankton gross production was calculated from biomass increase and egg production, with respiration losses added (Hessen *et al.*, 1990). As the phytoplankton biomass remained stable, respiration from this compartment was calculated from mass balance considerations as the difference between fixed carbon and grazing losses. Then bacterial respiration would be the total efflux minus net phytoplankton respiration and zooplankton respiration. Although the labeled DOC-pool in the bacteria-bag was strongly depleted within 2 days, there was a lower, but continuous uptake of label also during the following days. Hence a direct estimate of bacterial respiration from losses of labeled DIC from day 2 onwards would underestimate the sum of true respiratory losses and grazing losses.

## Results and discussion

### *DOC and production*

In Lake Kjelsåsputten the mid summer carbon pool was totally dominated by dissolved organic carbon (DOC) which accounted for approximately 7.5 mg C l$^{-1}$, or near 80% of total C. Dissolved inorganic carbon (DIC) accounted for 0.57 mg C l$^{-1}$ in mean, while the total particulate pool (POC) ranges 0.3–0.5, with a mid-summer average near 0.4 mg C l$^{-1}$. In general, detrital particles (total POC minus bacteria, phyto- and zooplankton carbon) made up more than 75% of the particulate pool, with a mean biomass of zooplankton and bacteria close to 60 $\mu$g l$^{-1}$. Quantitative phytoplankton data are few, but mid summer counts in general yielded less than 20 $\mu$g C l$^{-1}$, and Chl. *a* rarely exceeded 2 $\mu$g l$^{-1}$.

Mid-summer production estimates for the 0–2 m layer were 24 and 32 $\mu$g C l$^{-1}$ d$^{-1}$ for phytoplankton and bacteria, respectively (Table 1). Bacterial production was estimated from biomass increase, and is thus a net production estimate, whereas the $^{14}$C-fixation approaches gross primary production. Mean total atmospheric efflux of DIC was near 7% d$^{-1}$, or close to 45 $\mu$g C l$^{-1}$ d$^{-1}$. Bacterial respiration was 24 $\mu$g C l$^{-1}$ d$^{-1}$, i.e. 43% of biomass respired d$^{-1}$, and thus bacterial gross production

*Table 1.* Mean biomass ($\mu g\ C\ l^{-1}$), gross production and respiration rates ($\mu g\ C\ l^{-1}\ d^{-1}$) for phytoplankton, bacteria and zooplankton in the $CO_2$-labeled bag. The high respiration/production ratio for phytoplankton is caused by a large fraction of heterotrophs.

| | Biomass (SD) | Gross production | Respiration |
|---|---|---|---|
| Phytoplankton | 16 (±2) | 24 | 26 |
| Bacteria | 45 (±2) | 56 | 24 |
| Zooplankton | 61 (±9) | 23 | 18 |

would amount to 56 (32 + 24) $\mu g\ C\ l^{-1}\ d^{-1}$. Release of algal exudates (PDOC) was not measured directly. This fraction may amount to <10–>50% of fixed carbon (Jordan & Likens, 1980; Søndergaard *et al.*, 1985; Rieman & Søndergaard, 1986), although most estimates are within the range 15–30%. If we adopt a DOC-release according to 20% of primary production estimated from humic lakes by Jones & Salonen (1985), this amounts to less than 5 $\mu g\ C\ l^{-1}\ d^{-1}$, or less than 10% of bacterial gross production (Fig. 1). Assuming a maximum assimilation efficiency of 80% of algal DOC (i.e. 35% growth efficiency), this source would at most account for 6% of bacterial gross production. Even if 40% of fixed algal carbon is released as DOC, this contribution would only amount to a maximum of 12% of bacterial gross production.

The other source of biologically released DOC for bacterial growth is through zooplankton release. Total excretory release of organic carbon was estimated from the equation:

$$R_t = (I_a * A_a + I_b * A_b + I_d * A_d) - P_z - r \quad (1)$$

where $R_t$ is total release; $I_a$, $I_b$, $I_d$ are gross ingestion rates of bacteria, algae and detritus respectively; $A_a$, $A_b$ and $A_d$ are corresponding as-

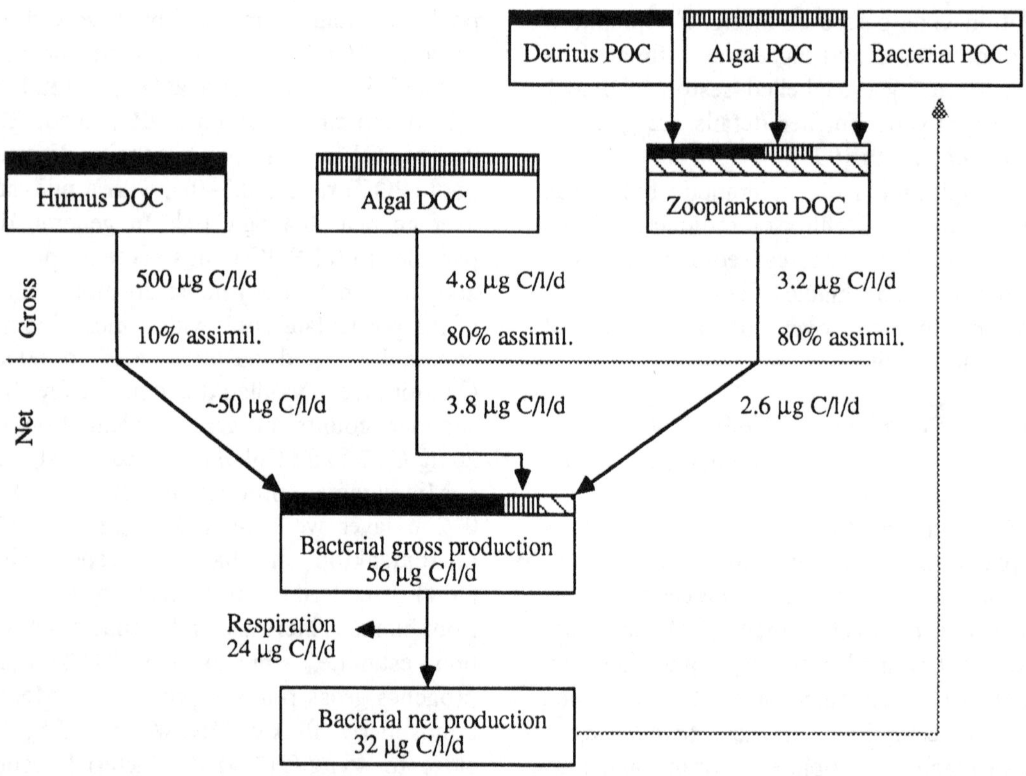


*Fig. 1.* Idealized flow-scheme for the major fluxes of DOC in Lake Kjelsåsputten. Gross (above vertical line) denotes gross intake of carbon, net (below line) denote net intake (gross intake * assimilation efficiency).

similation efficiencies; $P_z$ is net zooplankton production and $r$ is zooplankton community respiration. Release of DOC from zooplankton through 'sloppy feeding' was considered negligible in this locality, where the algal community was entirely composed of small species (cf. Lampert, 1978). Total daily net ingestion of POC was estimated as 22 $\mu$g C $l^{-1}$ $d^{-1}$ based on data from Hessen *et al.* (1989). Net ingestion was calculated from gross ingestion using assimilation efficiencies for algal, bacterial and detrital carbon of 80, 50 and 10% respectively (cf. Hessen *et al.*, 1989, 1990). Based on net zooplankton production (which was low during mid summer) and a 30% zooplankton community respiratory loss, total release of organic C from zooplankton was estimated to be 15 $\mu$g C $l^{-1}$ $d^{-1}$. As much as 80% of *Daphnia* excretion may belong to the particulate fraction (Olsen *et al.*, 1986). For copepods the particulate fraction is probably even higher. Assuming a mean of 20% DOC of total non-respiratory release, zooplankton release of DOC would be close to 3 $\mu$g C $l^{-1}$ $d^{-1}$. If a maximum 80% assimilation efficiency is applied also to this fraction, zooplankton release would contribute only 4% to bacterial gross production. Even if 50% of total zooplankton release is DOC, this contribution would still be less than 8% of bacterial gross production. However, although phytoplankton and zooplankton release are almost negligible in the context of carbon, these releases may still be qualitatively important as a source of vitamins, aminoacids or dissolved P, especially in P-deficient systems.

The DOC released from zooplankton originates from detrital, bacterial and algal POC. Although species dependent, detritus in general accounts for more than 50% of the zooplankton body carbon, with the remaining 50% almost equally divided between algae and bacteria (Hessen *et al.*, 1990). If the ratio of released compounds reflects that of the ingested, the main bulk of DOC released through zooplankton would originate from particulate detritus.

These approximations suggest that the remaining 80–95% of DOC required to support the observed bacterial production must be obtained from humus/detritus DOC (Fig. 1). The recalcitrant humic compounds are probably poorly assimilated. An approximate assimilation efficiency of 10% (a growth efficiency of 4.3%) would require ~500 $\mu$g C $l^{-1}$ $d^{-1}$ to support bacterial production, implying a turnover-time of 15 d for the humus DOC-pool. Based on a mean pool of 5 $\mu$g dissolved P, a bacterial gross production of 56 $\mu$g C $d^{-1}$, and a bacterial P:C ratio of 40 $\mu$g P (mg C)$^{-1}$, bacteria would give a corresponding turnover of dissolved P of 0.45 $d^{-1}$.

DOC from algal release is considered a major source of energy for bacteria in most clearwater lakes. A literature review (Table 2) shows that on average algal release supports 50% or more of bacterial production in most clearwater localities. Although macrophyte release of DOC may give an important contribution to bacterial growth in shallow humic systems (Murray & Hodson, 1985), this does not hold for the pelagial of humic lakes, where in general the influence of PDOC on bacterial production is negligible (Jones & Salonen, 1985). Repeated experiments have verified that humic substances, although recalcitrant, may provide a significant source of energy for bacteria (De Haan, 1974; Tranvik & Hofle, 1987; Tranvik, 1988). Hessen (1985) found a good correlation between humic content and bacterial biomass in nutrient-poor lakes, and even strongly acidic lakes may have a considerable bacterial biomass and production if supplied with humic influx. As DOC increases to surplus concentrations, P may become the limiting element and one may expect a saturation level where additional concentrations of humus DOC will no longer be accompanied by increased bacterial production.

At least 30% of phytoplankton production in Lake Kjelsåsputten was supported from organic carbon by heterotrophic activity (Hessen *et al.*, 1990). Moreover, the zooplankton community was totally dominated by filter feeding crustaceans, and it was previously estimated that in this locality bacteria may account for almost 30% of body carbon in common filterfeeding species like *Daphnia* and *Diaphanosoma* (Hessen *et al.*, 1990). Thus, humus DOC converted into bacterial biomass does not represent a 'sink' of carbon in this

*Table 2.* Literature review on the percent of bacterial gross production (in most cases assuming 50% respiration) sustained by phytoplankton release of DOC.

| Lake | Type | % of bact. gross prod; Mean (range) | Reference |
|---|---|---|---|
| Almind | Oligotrophic | (20–27) | Jensen & Søndergaard (1985) |
| Mirror | Oligotrophic | ~30 | Cole *et al.* (1982) |
| Nimeton | Humic | <5 | Jones & Salonen (1985) |
| Kjelsåspt. | Humic | (3–10) | Present study |
| Crooked | Eutrophic | 3.4 (0.5–10.8 | Lovell & Konopka (1985) |
| Little Crooked | Eutrophic | 5.1 (0.7–9.7) | Lovell & Konopka (1985) |
| Mendota | Eutrophic | 13.6 (0.01–100) | Brock & Clyne (1984) |
| Esrom | Eutrophic | (15–97) | Jensen & Søndergaard (1985) |
| Mossø | Eutrophic | (26–77) | Jensen & Søndergaard (1985) |
| Ørn | Eutrophic | (40–74) | Jensen & Søndergaard (1985) |
| Fr.B. Slottsø[1] | Eutrophic | (36–55) | Riemann & Søndergaard (1986) |
| Norrviken | Eutrophic | 39 (9–83) | Bell *et al.* (1983) |
| Fr.B. Slottsø | Eutrophic | 44 | Jenesen & Schwærter (1988) |
| Nesjøvatn | Eutrophic | 45 (2–180) | Vadstein *et al.* (1989) |
| Bysjøn | Eutrophic | 52 (32–95) | Coveney (1982) |
| Erken[2] | Eutrophic | >75 | Bell & Kuparinen (1984) |
| Danish coast | Marine | (38–80) | Riemann & Søndergaard (1986) |
| Baltic Sea | Marine | ~50 | Larsson & Hagstrøm (1982) |

[1] Maximum estimates from enclosures.
[2] Estimated during spring bloom.

system, but certainly influences productivity of higher trophic levels.

*DOC and respiration*

The dominance of dissolved humus as a source of energy for the bacteria may also have implications for their respiratory rate and relative turnover of elements. While humus is poor in nutrient elements like phosphorus (P) and nitrogen (N), aquatic bacteria have high relative contents of these elements. The P:C ratio of the dissolved fraction in Lake Kjelsåsputten never exceeds 0.8 $\mu$g P (mg C)$^{-1}$, whereas even P-starved bacteria have a P:C ratio of more than 40 (42, Vadstein *et al.*, 1989; 50, Olsen *et al.*, 1986; 65; Fuhs *et al.*, 1972). An approximation from Lake Kjelsåsputten gave a bacterial P:C ratio of 40 (Hessen & Andersen, 1990). As ambient P:C-ratios affect cell quotas of nutrients, it has obvious consequences also for growth rate following from the Droop model (Droop, 1974), where growth rate $\mu$ is related to cell quota $Q$ through

$$\mu = \mu_{\max}(Q - Q_o)/Q\,, \tag{2}$$

where $Q_o$ is the threshold value of Q at which growth ceases and $\mu$ max is maximum cell growth. $\mu$ has the dimension growth h$^{-1}$, and $Q$ is $\mu$g P (mg C)$^{-1}$. Thingstad (1987) used a multidimensional approach of the Droop model to estimate areas of C, N and P-limitation respectively, and also modelled an increased respiration following increased P-deficiency. Based on the assumption that respiration depends on P-starvation, an increase in the respiration coefficient will occur when going from a carbon-limited to a mineral-limited situation as indicated in Fig. 2 (cf. Tempest *et al.*, 1985; Thingstad, 1987), and thus bacteria become net consumers of P (Vadstein *et al.*, 1989). Lack of sufficient data on $Q$ and $\mu_{\max}$ does not allow this approach in Lake Kjelsåsputten. Assuming that C-assimilation is independent of

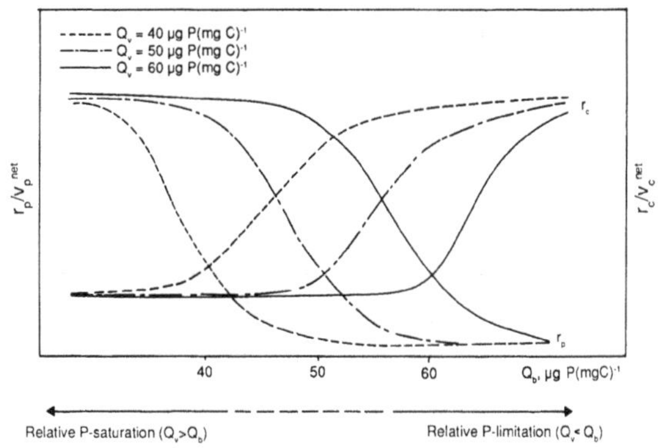


*Fig. 2.* Generalized graphical presentation of the dependency of cell releases of P ($r_p$) and C ($r_c$ – mainly respiration) on cell stoichiometry ($Q_b$) and ratio of net assimilated P:C ($Q_v$). Axes represents the ratio of release of net incorporation. The situation to the left represents P-saturated conditions (C-limitation) with a high release of P and minimum respiratory losses. Towards right the transition towards P-limitation with P-release approaching zero and increased respiratory losses. The situation is idealized as it neglects variable $Q_b$ (see text).

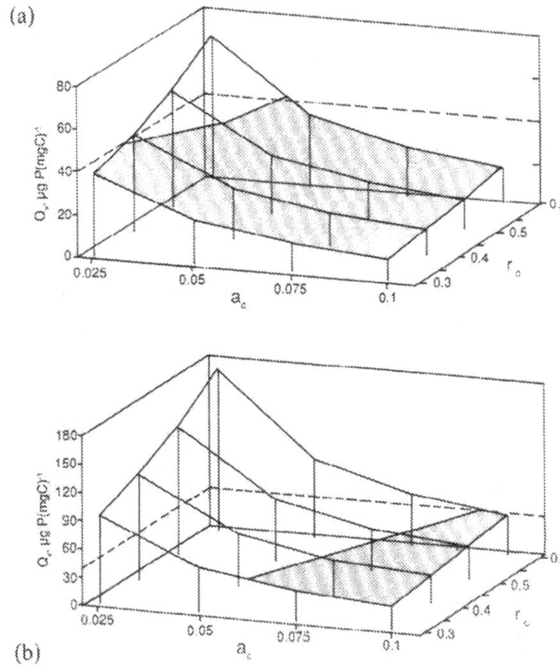


*Fig. 3a.* (Upper) The interrelation between net assimilated P:C ($Q_v$), assimilation efficiency of carbon ($a_c$) and respiratory losses ($r_c$), assuming ambient concentrations of 7500 $\mu$g DOC and 5 $\mu$g dissolved P $l^{-1}$, an assimilation efficiency of P = 1 (100%) and no release of P. Setting bacterial P:C-ratio = 40 $\mu$g P (mg C)$^{-1}$ (dotted line), the shaded area represents theoretical P-limitation. This approximation does not account for changes in respiration following from changes in $\mu$ (see text).

*Fig. 3b.* Same as 3a, but reducing the pool of available DOC to 3000 $\mu$g $l^{-1}$.

ambient C-concentrations, surplus incorporated C, relative to bacterial cell quota, must be respired in order to maintain the cell stoichiometry. In most cases this is an oversimplification as the P:C-ratio of bacteria to some extent adjusts to medium concentrations of nutrients. One may, however, assume a rather constant stoichiometry of bacteria with a constant minimum P:C-ratio in the constant nutrient-poor Lake Kjelsåsputten. Thus the approach below may be used to describe how low P:C-ratios affect the relative release of minerals, being aware that respiration also depends on growth rate, which in turn varies with temperature and ambient concentrations of mineral nutrients. The dependence between respiration and the ratio of net incorporated P:C ($Q_v$) may be described as:

$$Q_v = \frac{V_p^{\text{gross}} a_p - V_p^{\text{net}} r_p}{V_c^{\text{gross}} a_c - V_c^{\text{net}} r_c}, \qquad (3)$$

where $V_c^{\text{gross}}$ and $V_p^{\text{gross}}$are gross uptake of C and P; $a_c$ and $a_p$ are the assimilation efficiencies (range 0–1) of C and P; $r_c$ and $r_p$ are total cell release of C and P (range 0–1). $V_c^{\text{net}}$ and $V_p^{\text{net}}$ are net incorporation of C and P, i.e. $V_i^{\text{net}} = V_i^{\text{gross}} * a_j$. If cell quota of nutrients reflects the demands for cell growth, P deficiency occur when $Q_v$ (reported as $\mu$g P (mg C)$^{-1}$) $< Q_b$ (bacterial P:C ratio). For simplicity, it may be assumed that cell release of P is negligible when approaching P deficiency, and thus $r_p = 0$.

Although $R_c$ here is named 'respiration', it is actually the total cell release of C and while the major bulk of this is probably lost as $CO_2$, it is reasonable that some release also occurs as accumulation of polysaccarides (which is a common observation under P-deficiency). From an energetic point of view, it is also likely that not all C is totally oxidized, or that fermentation accounts for some of the metabolic activity in deeper layers. In Fig. 3a may be seen how $Q_v$ varies with changing assimilation efficiencies of C ($a_c$) and varying assimilation rates ($r_c$) assuming a fixed assimilation efficiency of P = 1 (100%), and zero cell release of P. The shaded area represents $Q_v < Q_b$, where $Q_b$ is set = 40. External P:C ratio is here set to 0.67 μg P (mg C)$^{-1}$ (7500 μg DOC l$^{-1}$ and 5 μg DIP l$^{-1}$). However, the entire pool of DOC is unlikely to be available for bacterial growth, although a two-phase uptake kinetics (PDOC and humus DOC) was not considered (cf. Tranvik, 1989). Assuming that only 3000 μg of the total DOC is available for bacterial growth (P:C = 1.67 μg P (mg C)$^{-1}$), the area of P-limitation is strongly reduced (Fig. 3b). Variations in available P, within the maximum range 3–10 μg P l$^{-1}$, have only a minor impact on $Q_v$. It should be emphasized here that the selected range of $a_c$ in Fig. 3 is no more than tentative. Calculated $r_c$ from the Lake Kjelsåsputten bacteria was close to 0.5 (50%) (Hessen *et al.*, 1990). Assuming a medium P:C ratio of 0.67, $a_c$ should not exceed 0.03 (3%) in order to maintain a cell quota of 40 μg P (mg C)$^{-1}$, which probably is a minimum. If the medium P:C ratio is elevated to 1.67, $a_c$ should not exceed 0.08. If $a_c$ increases, a corresponding increase in $r_c$ should occur. These estimates indicate an assimilation of humus DOC of < 10%, or a growth efficiency of less than 5% which is low compared with the near 20% estimated by Tranvik & Høfle (1987) and Tranvik (1988). These authors used a biomass to carbon conversion factor for bacteria of 0.56 pg C per μm$^3$, but point out that a 0.121 pg C conversion factor (which was applied in this study) would give a growth efficiency of only 4%, corresponding well to the above estimates.

Lake Kjelsåsputten has a substantial net production of $CO_2$, with a more than 400% supersaturation (Hessen *et al.*, 1990). The bacterial respiration is almost twice that of algal C-fixation. This vigorous respiration in this lake, with a net release of DIC, supports other studies in humic lakes (Salonen, 1981; Salonen *et al.*, 1983). This is in part of natural consequence of low primary production, high inputs of allochthonous matter and high bacterial biomass, but may be reinforced by the superfluous supply of DOC, relative to the bacterial P:C cell ratio. Estimates of bacterial respiration rates are often based on a 30% rate (Bell & Sakshaug, 1980), but respiration rates close to 50% are also reported from eutrophic lakes (Riemann & Søndergaard, 1986). High respiration rates may indicate that under most situations bacteria are P-limited (Cf. Vadstein & Olsen, 1989), and thus dispose of excess carbon through respiration.

## Acknowledgements

Thanks to M. Laake and an anonymous referee for valuable comments and suggestions. Financial support was received from NIVA (Norwegian Institute for Water Research) fundings.

## Literature

Bell, W. & E. Sakshaug, 1980. Bacterial utilization of algal extracellular products. 2. A kinetic study of natural populations. Limnol. Oceanogr. 25: 1021–1033.

Bell, R. T., G. M. Ahlgren & I. Ahlgren, 1983. Estimating bacterioplankton production by measuring [$^3$H] thymidine incorporation in a eutrophic swedish lake. Appl. envir. Microbiol. 45: 1709–1721.

Bell, R. T. & J. Kuparinen, 1984. Assessing phytoplankton and bacterioplankton production during early spring in Lake Erken, Sweden. Appl. envir. Microbiol. 48: 1221–1230.

Brock, T. D. & J. Clyne, 1984. Significance of algal excretory products for growth of epilimnic bacteria. Appl. envir. Microbiol. 47: 731–734.

Cole, J. J., G. E. Likens & D. L. Strayer, 1982. Photosynthetically produced dissolved organic carbon: An important carbon source for planktonic bacteria. Limnol. Oceanogr. 27: 1080–1090.

Coveney, M. F., 1982. Bacterial uptake of photosynthetic carbon from freshwater phytoplankton. Oikos 38: 8–20.

Droop, M. R., 1974. The nutrient status of algal cells in continuous culture. J. mar. biol. Assoc. U.K. 54: 825–855.

Fuhs, G. W., S. D. Demerle, E. Canelli & M. Chen, 1972. Characterization of phosphorus-limited algae. Am. Soc. Limnol. Oceanogr. Spec. Symp. 1: 113–132.

De Haan, H., 1974. Effect of a fulvic acid fraction on the growth of a *Pseudomonas* from Tjeukemeer (the Netherlands). Freshwat. Biol. 4: 301–310.

Hessen, D. O., 1985. The relation between bacterial carbon and dissolved humic compounds in oligotrophic lakes. FEMS Microbiol. Ecol. 31: 215–223.

Hessen, D. O., 1989. Factors determining the nutritive status and production of zooplankton in a humic lake. J. Plankton Res. 11: 649–664.

Hessen, D. O., T. Andersen & A. Lyche, 1989. Differential grazing and resource utilization of zooplankton in a humic lake. Arch. Hydrobiol. 114: 321–347.

Hessen, D. O., T. Andersen & A. Lyche, 1990. Carbon metabolism in a humic lake; pool sizes and cycling through zooplankton. Limnol. Oceanogr. 35: 84–99.

Hessen, D. O. & T. Andersen, 1990. Bacteria as a source of phosphorus for zooplankton. Hydrobiologia 206: 217–223.

Hobbie, J. E., R. J. Daley & S. Jasper, 1977. Use of nucleopore filters for counting bacteria by fluorescent microscopy. Appl. Envir. Microbiol. 33: 1225–1228.

Jensen, L. M. & S. Schwærter, 1988. Major pathways involved in the utilization of primary production in a temperature eutrophic lake. Verh. int. Ver. Limnol. 23: 445–450.

Jensen, L. M. & M. Søndergaard, 1985. Comparison of two methods to measure algal release of dissolved organic carbon and the subsequent uptake by bacteria. J. Plankton Res. 7: 41–56.

Johansson, J. Å., 1983. Seasonal development of bacterioplankton in two forest lakes in central Sweden. Hydrobiologia 101: 71–88.

Jones, R. I. & K. Salonen, 1985. The importance of bacterial utilization of released phytoplankton photosynthate in two humic forest lakes in southern Finland. Holarct. Ecol. 8: 133–140.

Jordan, M. & G. E. Likens, 1980. Measurement of planktonic bacterial production in an oligotrophic lake. Limnol. Oceanogr. 25: 719–732.

Lampert, W., 1978. Release of dissolved organic carbon by grazing zooplankton. Limnol. Oceanogr. 23: 831–834.

Larsson, U. & Å. Hagstrøm, 1982. Fractionated phytoplankton primary production, exudate release, and bacterial production in a baltic eutrophication gradient. Mar. Biol. 67: 57–70.

Lovell, C. R. & A. Konopka, 1985. Primary and bacterial production in two dimictic Indiana lakes. Appl. envir. Microbiol. 49: 485–491.

Murray, R. E. & R. E. Hodson, 1985. Annual cycle of bacterial production in five aquatic habitats of the Okefenokee Swamp ecosystem. Appl. envir. Microbiol. 49: 650–655.

Olsen, Y., A. Jensen, H. Reinertsen, Y. Børsheim, M. Heldal & A. Langeland, 1986. Dependence on the rate of release of phosphorus by zooplankton on the P:C-ratio in the food supply as calculated by the recycling model. Limnol. Oceanogr. 31: 34–44.

Riemann, B., M. Søndergaard, H.-H. Schierup, S. Bosselmann, G. Christensen, J. Hansen & B. Nilsen, 1982. Carbon metabolism during a spring peak in the eutrophic Lake Mossø. Int. Revue ges. Hydrobiol. 67: 145–185.

Riemann, B. & M. Søndergaard, 1986. Regulations of bacterial secondary production in two eutrophic lakes and in experimental enclosures. J. Plankton Res. 8: 519–536.

Salonen, K., 1981. The ecosystem of the oligotrophic Lake Paajarvi. 2. Bacterioplankton. Verh. int. Ver. Limnol. 21: 448–453.

Salonen, K., K. Kolonen & L. Arvola, 1983. Respiration of plankton in two small, polyhumic lakes. Hydrobiologia 101: 65–70.

Søndergaard, M. B., B. Riemann & N. O. G. Jørgensen, 1985. Extracellular organic carbon (EOC) released by phytoplankton and bacterial production. Oikos 45: 323–332.

Tempest, D. W., O. M. Neijssel & M. J. Teixeira de Mattos, 1985. Regulation of carbon substrate metabolism in bacteria growing in chemostat culture. In I. S. Kulaev, E. W. Dawes & D. W. Tempest (eds) Environmental regulation of bacterial metabolism. Acad. Press., New York, p. 53–69.

Thingstad, T. F., 1987. Utilization of N, P, and organic C by heterotrophic bacteria. I. Outline of a chemostate theory with a consistent concept of 'maintenance' metabolism. Mar. Ecol. Progr. Ser. 35: 99–109.

Tranvik, L., 1988. Availability of dissolved organic carbon for planktonic bacteria in oligotrophic lakes of differing humic content. Microb. Ecol. 16: 311–322.

Tranvik, L. & M. Hofle, 1987. Bacterial growth in mixed cultures on dissolved organic carbon from humic and clear waters. Appl. envir. Microbiol. 53: 482–488.

Vadstein, O. & Y. Olsen, 1989. Chemical composition and phosphate uptake kinetics of limnetic bacterial communities cultured in chemostats under phosphorus limitation. Limnol. Oceanogr. 34: 939–946.

Vadstein, O., B. O. Harkjerr, A. Jensen, Y. Olsen & H. Reinertsen, 1989. Cycling of organic carbon in the photic zone of a eutrophic lake with special reference to the heterotrophic bacteria. Limnol. Oceanogr. 34: 840–855.

Vollenweider, R. A. (Ed.), 1969. A manual on methods for measuring primary production in aquatic environments. IBP-handbook no. 12, Blackwell Sci. Publ. Oxford. 214 pp.

Watson, S. W., T. J. Novitsky, H. L. Quinby & F. M. Velois, 1977. Determination on bacterial number and biomass in the marine environment. Appl. envir. Microbiol. 33: 940–946.

*Hydrobiologia* **229**: 125–142, 1992.
*K. Salonen, T. Kairesalo & R.I. Jones (eds.),*
*Dissolved Organic Matter in Lacustrine Ecosystems: Energy Source and System Regulator.*



# Planktonic food chains of a highly humic lake

## *I. A mesocosm experiment during the spring primary production maximum*

K. Salonen, L. Arvola, T. Tulonen, T. Hammar, T.-R. Metsälä, P. Kankaala & U. Münster
*University of Helsinki, Lammi Biological Station, SF-16900 Lammi, Finland*

*Key words:* algae, bacteria, Protozoa, food chains, growth rate, humic lakes, succession

**Abstract**

The development and metabolism of the plankton of a highly humic lake were followed over the vernal primary production maximum. The study was made in a mesocosm in which large filter feeders, typical of this lake in summer, were absent. During the rising phase of phytoplankton, the community was predominantly autotrophic. The most important constituents in the algal biomass were a dinoflagellate, *Gymnodinium* sp. (40–50%), and a prasinophycean, *Scourfieldia cordiformis* (7%). The biomasses of *Chlamydomonas* spp. and *Chrysococcus* spp. reached their maxima a few days later and *Cryptomonas* sp. became most abundant at the end of the experiment. After the phytoplankton maximum, about one week from the beginning of the experiment, grazing of algae by phagotrophic protozoans and phosphate depletion led to a rapid decrease of algal biomass and the community became predominantly heterotrophic. In spite of a large variation in algal biomass and primary production, the biomass of bacteria remained of the same order of magnitude as in algae both before and after the algal maximum. Bacteria were mostly responsible for the plankton respiration, which also showed no dependence on primary production. Since exudation by phytoplankton was also low, the nutrition of bacterioplankton was probably mainly based on allochthonous dissolved organic matter rather than on primary production. Thus the production of bacteria was an additional food source for higher trophic levels along with phytoplankton. Because filter feeding zooplankton was absent in the experiment, protozoans were the only grazers utilizing algae and bacteria. Essentially all growth of bacteria was used by bacterivores.

## Introduction

In humic lakes dissolved organic matter (DOM) is the dominant organic matter pool (cf. Sarvala *et al.*, 1981; Wetzel, 1983; Arvola *et al.*, 1990; Hessen *et al.*, 1990). Particularly in small headwater lakes, most DOM originates from the catchment area. It is generally considered to consist of poorly defined refractory humic substances. However, there are indications that microorganisms are able partially to utilize allochthonous DOM (De Haan, 1974; Wetzel, 1983; Johansson, 1983; Hessen, 1985; Tranvik & Höfle, 1987; Tranvik, 1988; 1990). Thus in highly humic lakes DOM may represent a carbon source for higher trophic levels comparable with that from primary producers (cf. Salonen *et al.*, 1983; Salonen & Hammar, 1986). Consequently trophic interactions in humic lakes (Hessen, 1985; Salonen & Hammar, 1986; Salonen & Jokinen, 1988) may be different from those in clear lakes, in which autotrophic production is responsible for the bulk of the energy flow through planktonic food chains.

On the assumption that with higher concentrations of DOM these features specific to the food chains of humic lakes will be more striking, we studied the structure and functioning of the food chains in a highly humic headwater lake. In spring the relatively high concentration of dissolved inorganic phosphorus and nitrogen, together with decreasing hydraulic washout and low grazing pressure by macrozooplankton, favour the growth of algae in such lakes (cf. Similä, 1988). Marked changes in the composition of plankton may occur within short time intervals (Arvola, 1983; Similä, 1988). Such a situation offers a possibility to study the effects of phytoplankton primary production on the utilization of allochthonous DOM. In this paper we describe the metabolic processes and the trophic relationships in the plankton from a highly humic lake, Mekkojärvi, during the wax and wane of the spring algal maximum.

## Material and methods

### *Study lake*

The experiment was made with water from a small, highly humic lake, Mekkojärvi. Various characteristics of this lake have been described by Kankaala (1988), Kuuppo-Leinikki & Salonen (1992), Münster (1992), Salonen & Lehtovaara (1992).

### *Experimental design*

Due to the difficulties in controlling limnological variables in nature, the study was performed in a mesocosm. With this arrangement we could avoid problems associated with a time delay between sampling and determinations, and could increase sampling frequency and the number of replicate determinations. The experiment was performed in May immediately after the melting of ice. Epilimnetic water from Mekkojärvi was taken with buckets to fill two 560 l polyethylene tanks. The zooplankton of Mekkojärvi is dominated by *Daphnia longispina* (Cladocera), but at the time of sampling it had not yet hatched from ephippia and was almost absent. To be sure that no macrozooplankton was present in the experiment, water was poured through a 50 $\mu$m plankton net. After filling, the tanks were sealed with polyethylene sheets and transported the 25 km to Lammi Biological Station. One tank was used as an experimental mesocosm and the other was kept as an incubator. In this way we could minimize the risk of disturbance and contamination in the experimental tank. The walls of the tanks were insulated by glass fibre wool and wrapped with black paper to reduce temperature changes caused by the fluctuations of air temperature and to prevent light penetration through the walls. The tanks were covered by a 3 mm thick plexiglass plate lying on three aluminium rods so that air could enter the tank. Both tanks were kept side by side outside the laboratory building. One of the major differences from the situation in the lake was that the tanks contained only epilimnetic water which was continuously mixed with a large (150 mm long), slowly rotating (*ca.* 100 rpm) propeller. In the lake thermal convection mixes the epilimnion during the night. The efficacy of water mixing was monitored by 2 temperature recorders, one at the bottom and one at the top of the water column. These measurements showed complete mixing of the water column. Samples for chemical and biological determinations were siphoned from the 0.15–0.2 m depth in the middle of the tank with silicone tubing. Phosphate ($KH_2PO_4$) diluted into a small volume of tank water (corresponding to an increase of 6.5 $\mu$g P $l^{-1}$ in the concentration of phosphorus) was added to the experimental tanks 3 days before the end of the study. All samples were taken daily, but only phytoplankton and some chemical sampling were continued after the addition of phosphate.

### *Physico-chemical determinations*

Irradiance was measured beside the tanks by a Kipp & Zonen solarimeter. Penetration of photosynthetically active radiation (PAR, $\mu E m^{-2} s^{-1}$) into the water column was measured *in situ* with a LiCor Li 1000 quantum sensor. Temperature and oxygen concentration were measured daily with a combined probe (Yellow Springs Instruments).

All glassware, filtration equipment etc. were acid washed and rinsed with deionized (MQ, Millipore Corp.) water and then dried at 60–100 °C. Dissolved nutrients were determined immediately after sampling (09.00) by filtering the samples through Whatman GF/C filters rinsed with deionized water and sample water. Inorganic phosphorus (Murphy & Riley, 1963, nitrate + nitrite (Wood *et al.*, 1967) and ammonium nitrogen (Solorzano, 1969) were determined with an Akea autoanalyser. Subsamples of unfiltered and filtered water were frozen for later determination of total and dissolved organic phosphorus and nitrogen with the persulphate digestion method of Koroleff (1979).

Water pH, alkalinity (Gran titration), electrical conductivity and colour were measured daily. Cation samples (Ca, Mg, Fe, Na, K) taken at the beginning and at the end of the experiment were frozen and determined later by AAS. Chloride and sulphate were measured daily by ion chromatography (Dionex).

### *Bacteria*

Samples for bacteria were taken daily at 08.00–08.30 and fixed with 0.2 $\mu$m filtered formaldehyde (final concentration 3.8%). Samples were filtered onto black Nuclepore membranes (0.2 $\mu$m) and stained with acriflavine (Bergström *et al.*, 1986). Bacteria were counted with a Nikon Optiphot epifluorescent microscope at 1200 × magnification in 20 randomly taken fields (576 $\mu m^2$ each). The volume of bacteria was calculated from the linear dimensions (length and width; assuming spherical ends) of 100 bacteria measured in each sample with an eyepiece graticule (Patterson Globe and Circle G1). Volumes were converted to carbon using a factor of 0.38 pg C $\mu m^{-3}$ (Lee & Fuhrman, 1987; Salonen *et al.*, unpubl.).

For the estimation of bacterial production, growth rate and respiration sample water was siphoned through a 1.0 $\mu$m filter cartridge (Nuclepore No 720100, pressure < 5 kPa) to remove phytoplankton and protozoans. After use the cartridge was immediately backwashed with MQ water, sonicated and rinsed and finally dried at 60 °C for at least 12 h. Filtrate < 1 $\mu$m was incubated for 24 h in duplicate 250 ml glass bottles for 24 h at 5 and 50 cm depths. The growth of bacteria was determined microscopically as an increase in bacterial biomass during the incubation. The growth rate (divisions $d^{-1}$) of bacteria was calculated from the following equation (Guillard, 1973):

$$k = \ln (N_t/N_0) * 1/(t * \ln 2) \quad (1)$$

where $N_0$ = biomass or cell number in filtered sample at the beginning of incubation, $N_t$ = biomass or cell number in filtered sample at end of incubation and $t$ = incubation time in days.

Grazing on bacteria by heterotrophic flagellates and ciliates was estimated from the change in cell number or biomass of bacteria in unfiltered and filtered samples after 24 h. Bacterial grazing ($\mu$g C or cells $l^{-1} d^{-1}$) was calculated from the following equation:

$$G = [(N_t - N_0) - (n_t - n_0)]/t \quad (2)$$

where $n_0$ = cell number of bacteria in unfiltered sample at the beginning of incubation and $n_t$ = cell number of bacteria in unfiltered samples at the end of incubation.

Grazing on bacteria was also estimated using fluorescent carboxylated latex beads (diameter 0.57 $\mu$m; Polyscience Inc.). Beads of 2.0 $\mu$m and 3.9 $\mu$m diameter were used in the estimation of grazing on algae. The beads were added at a concentration of *ca.* 20–30% bacterial density and in excess of algal density. Before use the bead solutions were diluted with 0.2 $\mu$m filtered sample water and homogenized with a sonifier. The grazing experiments were made in time series lasting 20 or 40 min in 22 ml scintillation vials in light or darkness at the same temperature as in the tank.

The incubation was terminated by the addition of glutaraldehyde (final concentration 0.3%) and the cells with ingested beads were counted with the settling chamber technique using a Nikon Optiphot-TMD inverted microscope under phase contrast and epifluorescence illumination. The counting was continued until the minimum number of cells (25 for *Ochromonas* spp. and 50 for *Desmarella*) with ingested beads was found. Each determination was made in duplicate and each time series included 2 blanks.

*Respiration and community net production*

Respiration was measured in triplicate as an increase of DIC concentration in unfiltered and 1 $\mu$m filtered (approximate bacterial fraction; Jones & Salonen, 1985) water during 24 h (8.00–8.00) in darkness at the depths of 5, 25 and 50 cm. The bottles were filled through a divider so that water for the determination of the initial concentration of DIC as well as the cell and particle concentrations were identical with water in the bottle to be incubated in the tank. The community net production of plankton was determined in a similar way except that bottles were incubated in light instead of darkness and incubation time was 10 h (8.00–18.00). DIC was determined according to Salonen (1981).

*Primary production and chlorophyll* a

Primary production of phytoplankton and exudation was measured with the radiocarbon method in parallel with community net primary production. The samples were incubated in 1 l glass bottles with $Na_2{}^{14}CO_3$ solution (final radioactivity 925 kBq $l^{-1}$). After incubation, 8 ml of water was transferred to a glass vial and biological activity stopped with 2 drops of 10% phosphorus acid and inorganic $^{14}C$ was allowed to outgas in a vacuum hood (Niemi *et al.*, 1983) for 48 h. After that 10 ml Lumagel was added into 8 ml sample and radioactivity measured with a Wallac Ultrobeta 1210 liquid scintillation counter. Exudation by phytoplankton was measured in 0.2 $\mu$m Nuclepore filtrates treated as described above.

Chlorophyll *a* was determined in duplicate or triplicate 20 ml subsamples filtered onto Whatman GF/C filters and stored in 10 ml plastic syringes wrapped in aluminium foil at −20 °C. Pigments were extracted in syringes with 5 ml of 100% ethanol. To increase the extracting efficiency samples were homogenized for 3 min with an ultrasonic cell disruptor. This treatment raised the temperature to *ca.* 65–70 °C and denatured chlorophyllase enzymes which otherwise may decompose chlorophyll (Jeffrey & Hallegraeff, 1987). The extract was filtered through Whatman GF/F and 0.2 $\mu$m silver filter (Selas, Flotronics) directly into the microflow cell (volume 18 $\mu$l) of a Hitachi F-4000 Fluorescence spectrophotometer. The wavelengths for excitation and emission were 430 nm and 670 nm, respectively.

*Algae and protozoa*

Duplicate samples of algae and protozoa were taken daily at 9.00 am., preserved in glutaraldehyde (0.3%) and sedimented in 10 ml settling chambers for 24 h. Cells were counted under phase contrast illumination in 80 fields with a Nikon Diaphot-TMD inverted microscope at magnification of 400 and 200 ×. The mean volumes for algal species were estimated by counting cells in generally five size classes or by measuring the linear dimensions and calculating the mean volume of 20 cells in each sample. To differentiate between autotrophic and heterotrophic species by autofluorescence of photosynthetic pigments we used fluorescence microscopy (excitation at 450–490 nm or 535–550 nm) alternatively with phase contrast illumination. Volumes of preserved algae and protozoans were converted to carbon assuming a ratio of 0.2 pg C $\mu m^{-3}$ (Rocha & Duncan, 1985; Børsheim & Bratbak, 1987; Arvola *et al.*, unpubl.). The growth rates of algae and zooflagellates were estimated for species exhibiting clear development of maximum density. The growth rate was calculated by linear least square fit of logarithmic transformed numbers during the exponential phase. Because of cyclic variation in the cell size, biovolume was used instead of numbers for *Gymnodinium* sp.. Particulate organic carbon (POC) was determined on preignited Whatman GF/C filters by high temperature combustion (Salonen, 1979).

## Results

### *Environmental factors during the experiment*

Before the experiment water temperature in the lake was *ca.* 7.5 °C, but sunny days led to a rapid warming of the water (Fig. 1). The diurnal variation in temperature was usually 2–4 °C. At the surface of a full tank, irradiance was 89% of the measured above the plexiglass cover. Due to the dark colour of water (Table 1) only 4% of light was left at a depth of 0.5 m (15 cm above the bottom of the tank).

Phosphate concentration declined during the first couple of days and stayed thereafter at a very low level, 1–2 $\mu g\ P\ l^{-1}$ (Fig. 2), until the 9th day of the experiment. Then, prior to the addition of phosphate (6.5 $\mu g\ l^{-1}$) 3 days before the end of the experiment, the concentration started to increase. Phosphate was still so depleted that after only a few hours, the added phosphate had mostly been consumed and the concentration was about the same as before the addition (Fig. 3). Later the concentration increased slightly, probably because of the changes in the community (see later). Nitrate concentration decreased more than 70 $\mu g\ l^{-1}$ in 8 days. Similarly during the first 5 days ammonium nitrogen declined from 23 to 12 $\mu g\ l^{-1}$. No changes were observed in the concentration of dissolved organic nitrogen and phosphorus (cf. Table 1). However, total nitrogen and phosphorus decreased slightly during the first 10 days, which indicates losses through sedimentation or absorption on the walls. At the end of the experiment a brown layer with plenty of bacteria (practically no algae) developed on the walls and some sediment was found in the bottom corners of the tank in spite of continuous mixing. The atomic ratio between dissolved inorganic nitrogen

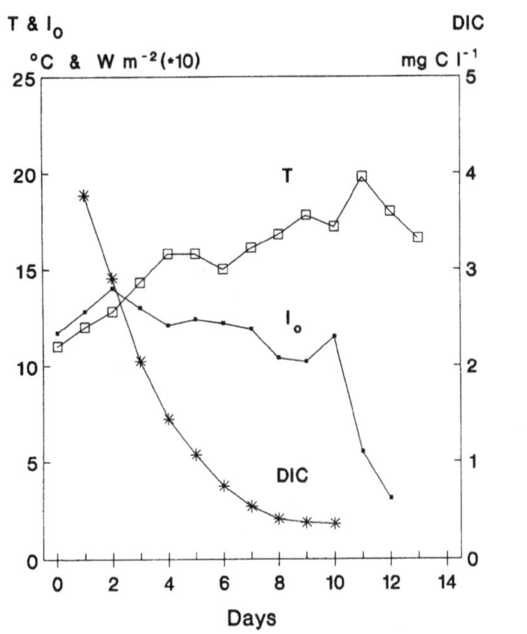


*Fig. 1.* Water temperature (T), dissolved inorganic carbon (DIC) and light intensity (above the tank) during the experiment.

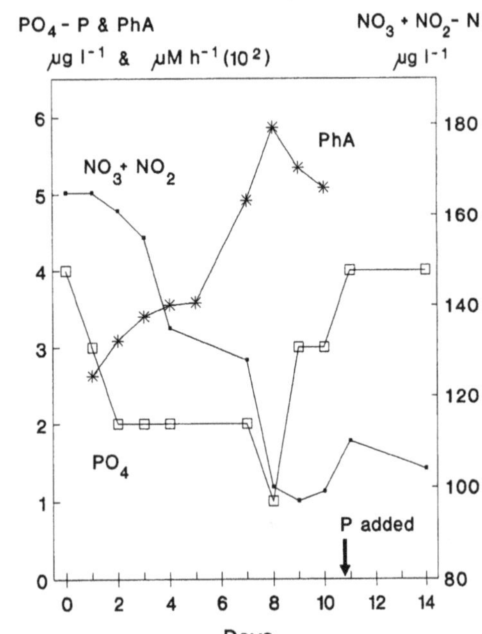


*Fig. 2.* Phosphatase activity (PhA) and the concentrations of phosphate and nitrate + nitrite in the tank.

*Table 1*. Some physical and chemical properties of the water in the tank during the experiment. All concentrtations are given as mg $l^{-1}$.

| Variable | Mean | SD | Range | N |
|---|---|---|---|---|
| pH | 5.2 | 0.2 | 5.0–5.4 | 11 |
| Conductivity (mS $m^{-1}$) | 3.4 | 0.2 | 3.6–3.1 | 11 |
| Alkalinity (mmol $l^{-1}$) | 0.02 | 0.01 | 0.01–0.03 | 4 |
| Absorbance at 260 nm | 0.93 | - | 0.92–0.93 | 2 |
| Total P | 0.015 | 0.003 | 0.010–0.019 | 10 |
| Phosphate P | 0.003 | 0.001 | 0.001–0.004 | 11 |
| Dissolved organic P | 0.007 | 0.001 | 0.004–0.008 | 10 |
| Total N | 0.74 | 0.04 | 0.69–0.79 | 9 |
| Ammonium N | 0.019 | 0.007 | 0.012–0.023 | 7 |
| Nitrate + nitrite N | 0.13 | 0.03 | 0.09–0.17 | 11 |
| Dissolved organic N | 0.58 | 0.02 | 0.53–0.63 | 10 |
| Calcium | 4.9 | 0.2 | 4.7–5.0 | 3 |
| Magnesium | 0.94 | 0.02 | 0.93–0.96 | 3 |
| Iron | 0.52 | 0.02 | 0.50–0.55 | 3 |
| Chloride | 0.78 | 0.03 | 0.75–0.80 | 3 |
| sulphate | 7.5 | 0.1 | 7.4–7.6 | 3 |
| Dissolved organic C | 24.1 | 0.3 | 24.7–23.6 | 11 |
| Particulate organic C | 0.34 | 0.08 | 0.20–0.46 | 10 |

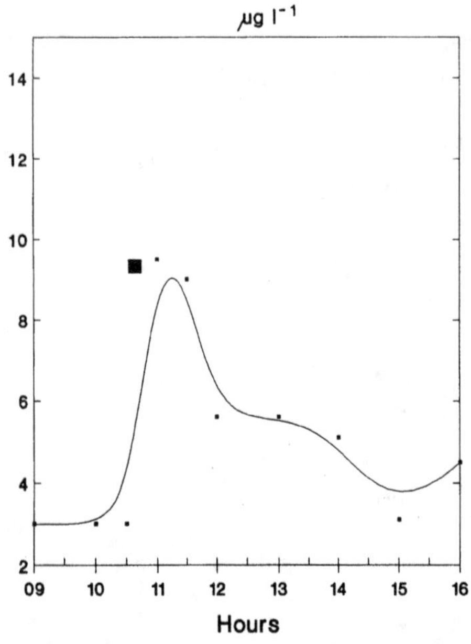


*Fig. 3*. The development of the phosphate-P concentration after the addition of $KH_2PO_4$ on the 11th day of the experiment. Large square indicates the calculated concentration of phosphate-P at the time of addition.

($NH_4 + NO_2 + NO_3$-N) and phosphate phosphorus varied from 95–264. Phosphatase activity showed an opposite trend to that of phosphate concentration and reached a high value approaching 600 $\mu M\ h^{-1}$ at the time of phosphate minimum (Fig. 2). This strongly suggests that the availability of phosphorus was limiting microorganisms.

The concentrations of cations, chloride and sulphate were stable during the experiment. The concentration of dissolved organic carbon also stayed stable, decreasing only from 24.7 to *ca.* 24.0 mg C $l^{-1}$ which is not a significant difference in terms of the precision of the method. Dissolved inorganic carbon (DIC) decreased along with the increase in temperature (Fig. 1). Although primary production contributed to the decline, it explained only less than 25% of the total decline during the first half of the experiment. As a result of primary production and escape of DIC from the tank, pH and alkalinity slightly increased during the experiment.

*The development of populations*

During the experiment a pronounced phytoplankton maximum developed about six days after the beginning of the experiment (Fig. 4). Biomass then declined rapidly due to increasing phosphorus depletion (Fig. 2) and grazing on algae. At its maximum density 7 days after the beginning of the experiment, a dinoflagellate, *Gymnodinium* sp., contributed 40–48% of the total algal biomass (Fig. 5). At the same time a small prasinophycean, *Scourfieldia cordiformis* Takeda, was abundant (max. *ca.* 4500 cells $ml^{-1}$, 7% of total biomass). A few days later *Chlamydomonas* spp. (Chlorophyceae) and *Chrysococcus* spp. (Chrysophyceae) reached their maxima and *Cryptomonas* sp. (Cryptophyceae) increased at the end of the experiment. *Pedinella* spp. (Chrysophyceae) were the only taxa which declined at the beginning of the experiment. Total phytoplankton biomass explained 77% ($p<0.01$, $n=8$) of the variation of chlorophyll *a* (Fig. 6). In spite of continuous mixing, which was able to maintain homogenous, water temperature in the tank chlorophyll *a* was 17 and 19% lower at the surface than at 0.25 and 0.5 m depths, respectively. This difference was statistically significant (sign test, $p<0.05$ and $p<0.01$). Although POC was not determined during the two days when the maximum algal biomass occurred, the results do suggest a maximum in POC as well. It seems that the sum of the estimated carbon biomasses of organisms (Fig. 4) was of the same order as POC except at the algal maximum, when the sum of estimated biomasses exceeded POC.

The cell volume of *Gymnodinium* sp. showed regular fluctuations (Fig. 7) with a cycle of two-three days which is in agreement with the generation time of the species (cf. Table 2). The generation times for the other algal species during exponential growth varied between 11 and 42 hours (Table 2). These estimates are based on the assumption that there were no losses (particularly from grazing) of algae before the increase in their density began to go down. That certainly could not be true for all species, because the population

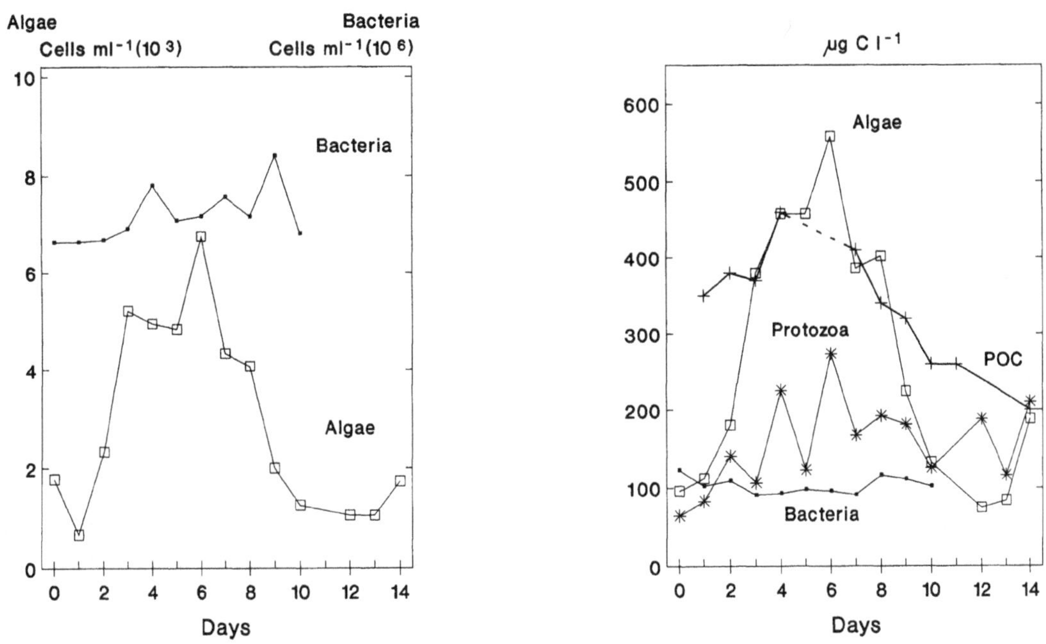


*Fig. 4.* The time course of algal, bacterial and protozoan densities and biomasses in the tank,. *Ochromonas* sp. and other mixotrophs are included in algae. Particulate organic carbon is shown as a reference.

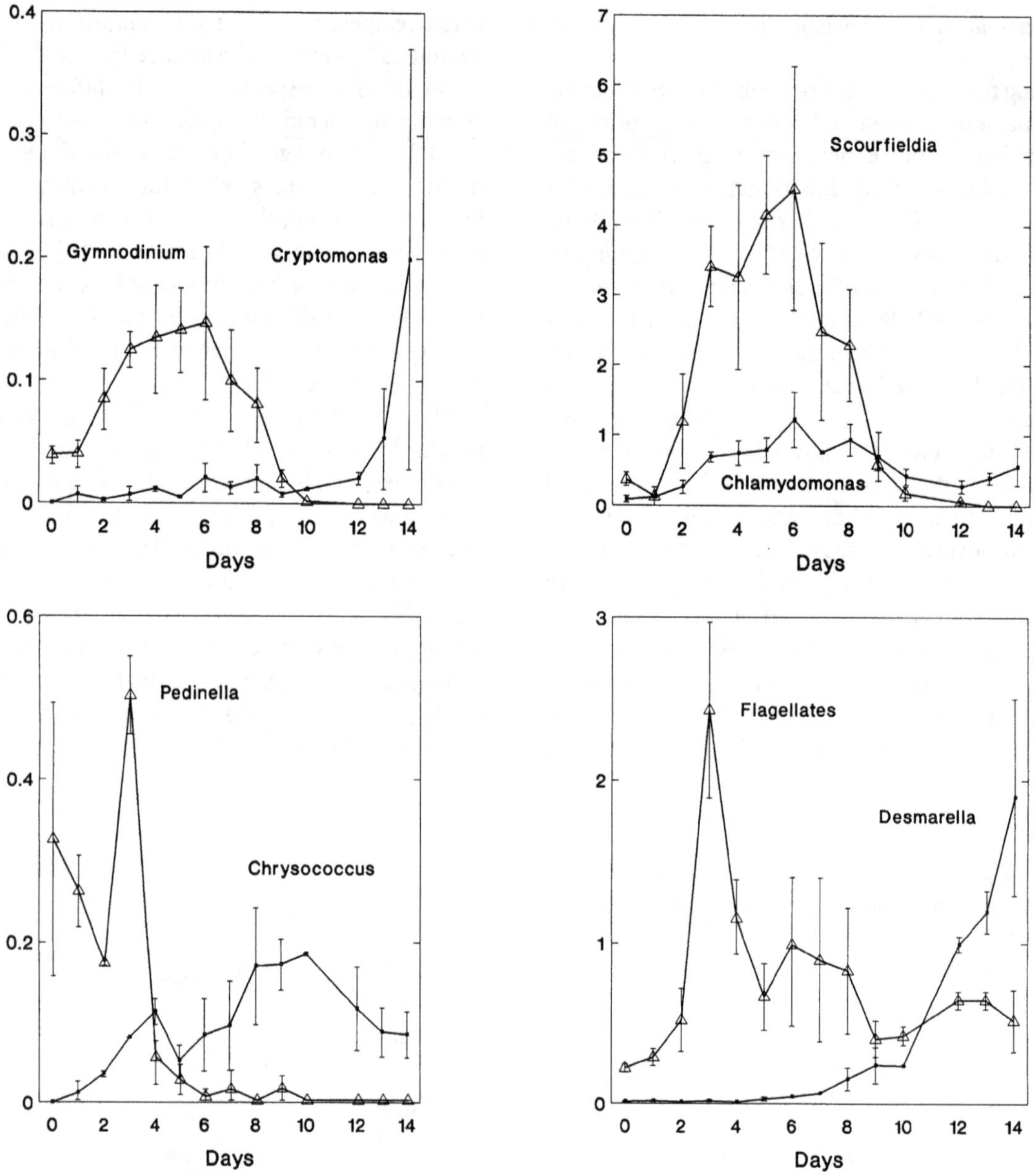


*Fig. 5.* The development of the densities ($10^3$ cells $ml^{-1}$; means with the range of two replicates) of the most abundant algal and protozoan taxa during the experiment.

of larger protozoans was more or less stable (Fig. 8). Therefore the growth rates are minimum estimates.

The abundance and biomass of bacteria were more or less stable (Fig. 4). Bacteria were generally small with an average cell volume of 0.038 $\mu m^3$. The share of smallest cells in the biomass of bacteria was rather stable while the largest bacteria clearly decreased towards the end of the experiment (Fig. 9). Although bacteria smaller than 0.016 $\mu m^3$ made up on average 56% of total numbers, their proportion was not more than 15% of total bacterial biomass. During the phytoplankton maximum, bacterial biomass was *ca.*

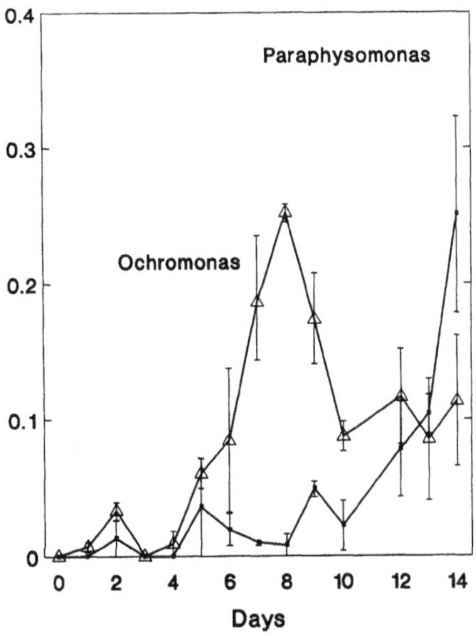


*Fig. 5.* Continued

20% of algal biomass, but at the beginning and the end of the experiment both were roughly equal. Compared to the growth rates of other micro-organisms, the growth rate of bacteria was low (Table 2). In terms of biomass the production of bacteria was considerably higher than in terms of numbers (Table 3). This indicates that some bacteria grew faster than the others.

The number and biomass of heterotrophic nanoflagellates increased strongly after the first week of the experiment (Fig. 8) and probably contributed to the observed decrease in the size of bacteria (Fig. 9). *Desmarella moniliformis* Kent (Fig. 5) and *Salpingoeca* sp. were the most common species. Choanoflagellates, *Monosiga* sp. and *Sphaerocoeca* sp., as well as the heterotrophic flagellates, *Bodo* sp. and *Rhynchomonas* sp., were present only in few samples and their numbers were low. The biomass of large (> 10 μm) protozoans was rather stable. In most cases *Strombidium* spp. and *Paramaecium* spp. contributed 40–60% of the biomass of protozoans. At the end of the experiment *Actinosphaerium(?)* sp. was rather abundant (up to 260 cells $ml^{-1}$; 19% in protozoan biomass). The growth rates of choanoflagellates were similar to those found for most algal species with photosynthetic pigments (Table 2). With the exception of the last 3–4 days the biomass of heterotrophic eukaryotic species was lower than that of the autotrophic ones.

*Table 2.* Growth rates of bacteria, algae and protozoans. The estimates for algae and protozoans were derived from the peaks in the numbers observed for the species during the experiment

| Organism | Growth rate (div. $d^{-1}$) | Generation time (h) |
|---|---|---|
| Bacteria | 0.48 | 50 |
| Algae | | |
| *Gymnodinium* sp. | 0.63 | 38 |
| *Cryptomonas* sp. (small) | 0.96 | 25 |
| *Chilomonas ?bacillans* | 1.30 | 19 |
| *Ochromonas* sp. | 0.58 | 42 |
| *Paraphysomonas* spp. | 1.07 | 23 |
| *Chrysococcus* sp. | 1.00 | 24 |
| *Scourfieldia cordiformis* | 2.18 | 11 |
| *Chlamydomonas* spp. | 0.75 | 32 |
| *Koliella spiculiformis* | 0.97 | 25 |
| Protozoa | | |
| *Desmarella moniliformis* | 1.23 | 31 |
| *Salpingoeca* sp. | 0.33 | 50 |
| Unidentified flagellates | 1.62 | 15 |

## *Production and respiration*

Phytoplankton primary production increased during the first 4 days and then started to decline (Fig. 6). In the rising phase of primary production, community net production was higher and after that lower than primary production determined with the $^{14}C$ method. This suggests weak-

*Table 3.* Mean production and grazing (± SD) of bacteria estimated from numbers and biomasses.

| Depth (m) | $10^9$ cells $^{-1}$ $d^{-1}$ | %N | $\mu g\ C\ l^{-1}\ d^{-1}$ | %B |
|---|---|---|---|---|
| | | Production | | |
| 0.05 | 1.5 ± 0.7 | 20 | 46 ± 35 | 43 |
| 0.50 | 0.7 ± 0.6 | 9 | 33 ± 20 | 31 |
| | | Grazing | | |
| 0.05 | 1.3 ± 1.0 | 17 | 48 ± 29 | 45 |
| 0.50 | 1.3 ± 1.1 | 17 | 31 ± 17 | 29 |

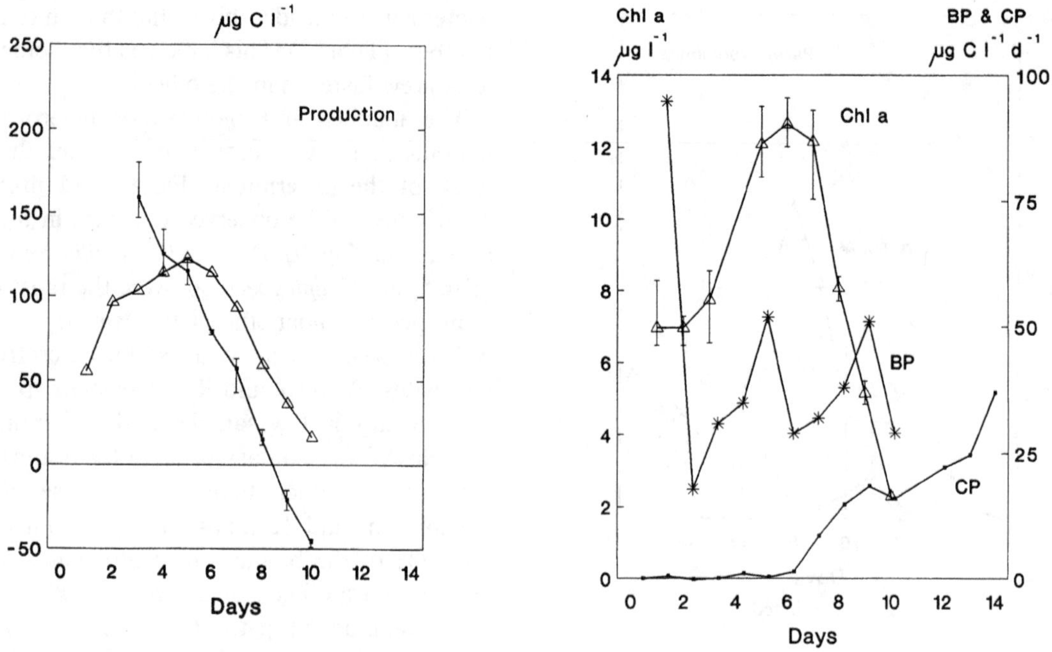


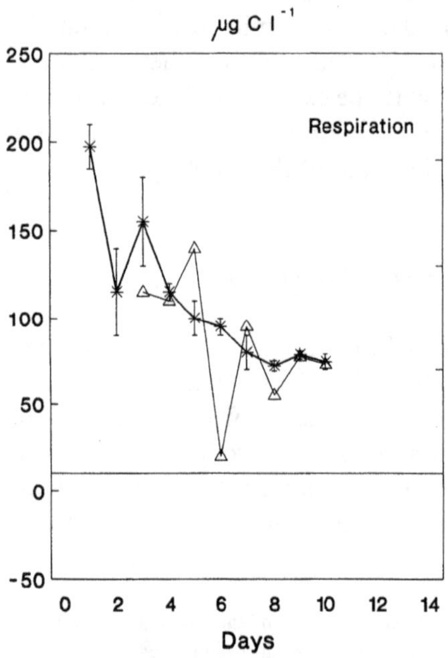


*Fig. 6.* Primary production (in 10 h) determined with the radiocarbon method (triangles), community net production (in 24 h, with range for three replicates, dots), total (asterisks) and bacterial (< 1 μm fraction, triangles) respiration, chlorophyll *a*, and the growth of bacteria (BP) and choanoflagellates (CP) during the experiment.

ening conditions for primary producers well before the maximum primary production or/and a phase difference in the development of producer and consumer populations. Primary production decreased strongly as a function of water depth (Fig. 10) which demonstrates the poor light penetration. However, the decrease of production (*ca.* 60%) from 0.05 to 0.5 m was not as distinct as the reduction of light (95%). Average primary production for the first 10 days in the 0.5 m water column was 81 $\mu g\ C\ l^{-1}\ 10\ h^{-1}$. Exudates represented only a minor proportion (1.0–9.9%) of primary production, indicating low excretion or rapid uptake by bacteria. Although there was no

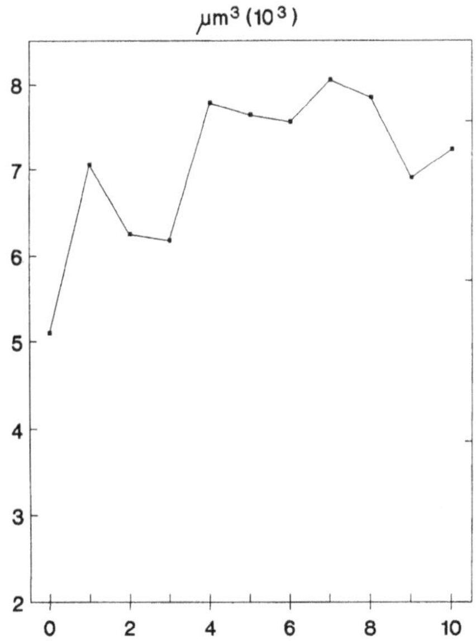


*Fig. 7.* The variation of the cell volume of *Gymnodinium* sp. during the experiment.

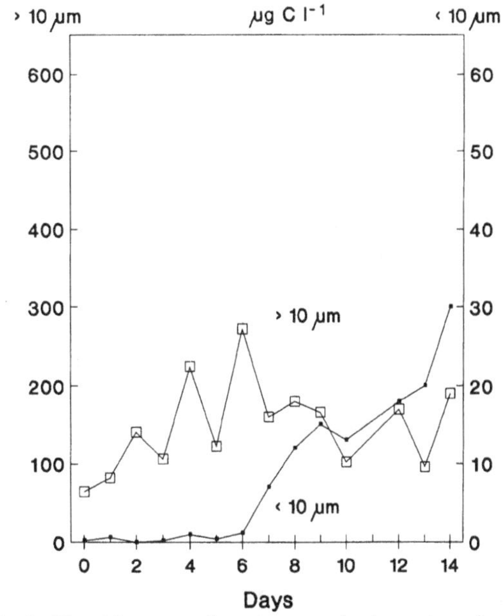


*Fig. 8.* The biomass of protozoans in two size classes, $>10\ \mu m$ and $<10\ \mu m$.

clear difference in the absolute exudation between the surface and 0.5 m depths the percentage of extracellular production in total primary production increased from 3.4% to 6.3%, respectively.

Total respiration of plankton and respiration in $<1\ \mu m$ size fraction declined steadily during the experiment (Fig. 6). Due to the satisfactory precision of DIC determinations ($\pm 0.6\%$ of mean) the mean standard deviation for respiration was reasonable ($\pm 20 \pm 8\%$ of mean (with SD); $n = 36$; range (6–38%). Respiration in $<1\ \mu m$ fraction was not significantly different (ANOVA) from total respiration, which indicates that bacteria were mainly responsible for the decomposition of organic matter. In $<1\ \mu m$ fraction respiration was higher near the surface than at the bottom ($F = 2.16$, $P = 0.17$), but in total respiration there was no significant difference ($F = 1.68$, $P = 0.54$). During the first 10 days cumulative $^{14}C$ production was 80% of total respiration. However, at the end of the experiment respiration was already 3 times higher than primary production which is in good agreement with the shift in the community structure from predominantly autotrophic organisms to heterotrophic ones.

The highest growth (95 $\mu g\ C\ l^{-1}\ d^{-1}$) of bacteria was found at the beginning of the experiment. Although this result represents only a single data point, it is in agreement with clearly the highest respiration at the same time. Due to the

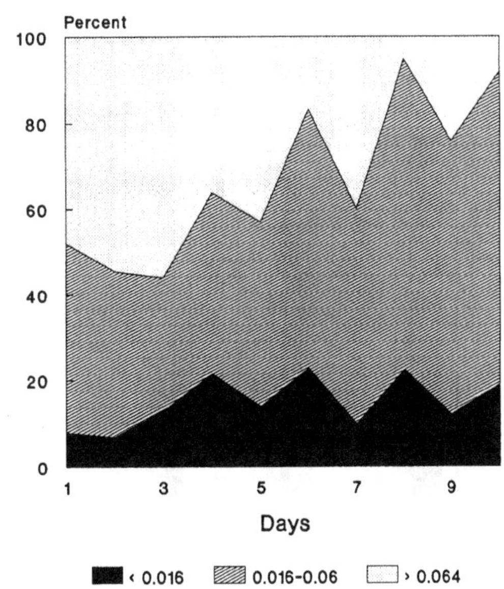


*Fig. 9.* The size distribution of bacterial biomass during the experiment. The size classes are expressed as volume ($\mu m^3$).

rather low precision of the microscopic method for the determination of the growth of bacteria, the other maxima appearing at the time of maximum primary production and at the collapse of phytoplankton biomass cannot be considered significant. The mean growth of bacteria (39 $\mu$g C $l^{-1}$ $d^{-1}$) during 10 days was 38% (range 18–160%) of phytoplankton primary production measured using the inorganic carbon exchange modification of the radiocarbon method. Thus even during the spring phytoplankton maximum bacteria are a quantitatively important constituent in the plankton.

*Trophic relationships*

In the experiment the growth and grazing on bacteria were equal (Table 3), implying that protozoans control the density and biomass of the bacterial populations. Because 29–45% of the biomass, but only 17% of the cell numbers were consumed daily, larger bacteria were evidently preferred by grazers. Also the increase in the size of bacteria during the incubation of < 1 $\mu$m fraction (Table 4), and the disappearance of larger cells at the end of the experiment together with the increase in the numbers of bacterivores, suggest size-selective grazing.

*Table 4.* Mean cell volume (± SE) of bacteria before and after the 24 h incubation with and without the grazers.

| Depth (m) | | Volume ($\mu m^3$) | | Volume change (%) |
|---|---|---|---|---|
| | | $T_0$ | $T_{24}$ | |
| 0.05 | With grazers | 0.031 + 0.003 | 0.041 + 0.010 | 33 |
| | No grazers | 0.036 + 0.007 | 0.036 + 0.006 | 0 |
| 0.50 | With grazers | 0.031 + 0.003 | 0.041 + 0.007 | 33 |
| | No grazers | 0.038 + 0.006 | 0.041 + 0.005 | 9.5 |

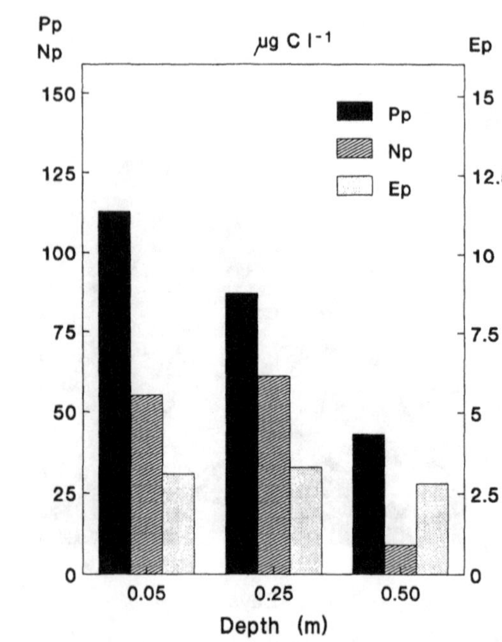


*Fig. 10.* Average primary production (PP), community net production (NP), and extracellular production (EP) at different depths in the tank during the experiment.

Experiments with spherical 0.57 $\mu$m fluorescent beads showed that heterotrophic 300 flagellates (*Desmarella* sp., *Salpingoeca* sp., *Bodo* sp., *Sphaerocoeca* sp. and *Monosiga* sp.) and some algae, (*Paraphysomonas takahashii* Cronberg & Kristiansen and *P. vestita* (Stokes) De Saedleer and *Chilomonas* sp.) engulfed bacteria. No ciliates were observed with ingested fluorescent beads. The grazing by *Desmarella moniliformis* was similar in light and darkness (Fig. 11). At the end of the experiment, when *D. moniliformis* was most abundant (1800 cells $ml^{-1}$) it cleared 2.5–3 × $10^{-6}$ ml $ind.^{-1}$ $h^{-1}$ (9 bacteria $ind.^{-1}$ $h^{-1}$). Then the population of *Desmarella* consumed 0.4 × $10^6$ bacteria $ml^{-1}$ $d^{-1}$ representing 32% of total grazing on bacteria. The estimated growth of *Desmarella* was 14 pg C $ind.^{-1}$ $d^{-1}$ which is *ca.* 78% of its consumption of bacteria (18 pg C $ind.^{-1}$ $d^{-1}$).

Grazing of bacteria by protozoans as estimated by fluorescent beads was < 1% of the grazing estimated with the microscopic method, by which the development of bacteria was followed in the presence and absence of grazers. Thus although large ciliates and heliozoans did not ingest fluorescent beads, they may still have been responsible for much grazing on bacteria. Assuming an ingestion rate of 400 bacteria $ind.^{-1}$ $h^{-1}$ (Sherr *et al.*, 1988), 85 large protozoans (the observed maximum density in the experiment) could engulf

$>0.8 \times 10^6$ bacteria $ml^{-1}$ $d^{-1}$. Thus the potential consumption of bacteria by nanoflagellates and large protozoans were in close agreement with the results obtained in the microscopic determination of total grazing (mean $1.3 \times 10^6$ bacteria $ml^{-1}$ $d^{-1}$).

Grazing experiments made with 2.0 and 3.9 μm fluorescent beads (Fig. 11), indicated that only a non-pigmented flagellate (tentatively identified as *Ochromonas* sp.) ingested large particles. This small (8–10 μm in diameter) species, variable in shape, with two long flagella and generally surrounded by a mucilage layer, did not ingest 0.57 μm particles. In an experiment in which large beads occurred in excess of algal cells grazed by *Ochromonas*, there was no difference between the uptake of two bead sizes. Together with fluorescent beads, autofluorescent *Chlamydomonas* cells were observed inside *Ochromonas* sp., proving that *Ochromonas* sp. ingested small algae. Assuming constant grazing, the *Ochromonas* population ingested more than $1.3 \times 10^4$ cells $ml^{-1}$ $d^{-1}$. As the maximum density of this species occurred at the time of highest phosphatase activity, the grazing by *Ochromonas* markedly amplified the decline of algal populations. Experiments with 0.57 μm beads provided no evidence for grazing on bacteria by this species. No autofluorescent cells or plastids were found inside the larger protozoans, indicating that ciliates and other large protozoans grazed mainly on bacteria rather than on nanoplanktonic algae and smaller protozoa.

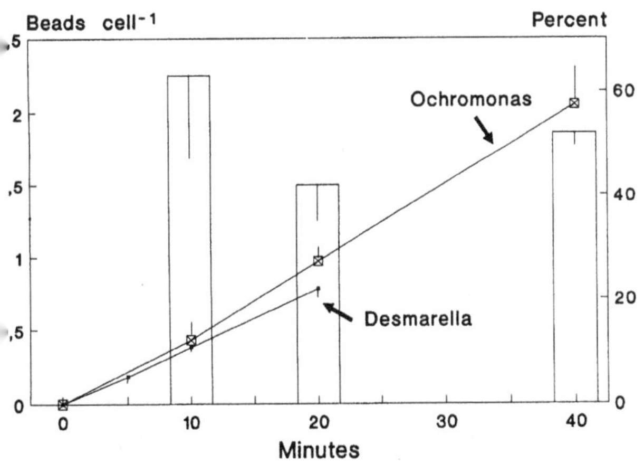


*Fig. 11.* The ingestion rate of fluorescent latex beads by *Desmarella moniliformis* (0.57 μm beads) and *Ochromonas* sp. (2 μm beads) during the day and night (with SD for three measurements of *Ochromonas* and two of *Desmarella*). The bars show the proportion of 2 μm diameter fluorescent labelled beads ingested by *Ochromonas* at three different times of the day (thin bars indicate SD), when both 2 and 3.9 μm beads were in excess of algal density in the tank water.

## Discussion

Algal assemblages found during the experiment were typical of Mekkojärvi (cf. Salonen & Jokinen, 1988; Arvola & Kankaala, 1989). The dominant algae, *Gymnodinium* sp., *Chlamydomonas* spp. and *Scourfieldia cordiformis*, are generally common already under the ice (Arvola & Kankaala, 1989) which indicates that these species are adapted to low temperature (cf. Schiller, 1954; Arvola, 1986; Similä, 1988). Indeed, in summer *Scourfieldia cordiformis* is restricted to the upper hypolimnion of Mekkojärvi. Thus rising temperature is a potential reason for the decline of these species during the present experiment. However, our results suggest more that these species might have declined due to nutrient limitation (Fig. 2) and grazing by *Ochromonas* sp. (Fig. 5). In general, it is difficult to be sure of the reasons behind the succession of species in the tank. For example, the nearly similar occurrence of *Gymnodinium* and *Scourfieldia* populations could have arisen from different reasons: *Gymnodinium* is not as heavily grazed as *Scourfieldia*, and the latter is probably a more efficient nutrient competitor (e.g. Reynolds, 1988) than *Gymnodinium*.

In the first phase of the experiment autotrophic processes dominated. Later, along with the depletion of inorganic nutrients, the plankton became predominantly heterotrophic at the time of maximum phosphatase activity. In spite of a remarkable maximum in the biomass and primary production of phytoplankton the growth and biomass of bacterioplankton showed little variation. Hessen & Schartau (1988) found similarly stable bacterial biomass in three humic lakes irrespective of any clear algal maximum. Perhaps the most definitive sign of the independence of bacterioplankton from primary producers is seen in respiration. Bacteria were apparently responsible for most of the respiration (Fig. 6), although the rather long 24 h incubation time may have allowed the density of bacteria to increase so much that the difference between respiration in the total and $< 1\ \mu m$ size fractions is meaningless. In spite of this reservation there is little doubt that bacteria had a major role in the respiration of plankton in Mekkojärvi, but showed a quite different time course from that of primary production. Low exudation by algae is also in accordance with other observations, although it might also be interpreted as a rapid consumption of exudates by bacteria. Further, during the decline of phytoplankton biomass, algae were mostly consumed by flagellates, so that bacteria may have utilized allochthonous DOM instead of primary production. Thus for higher trophic levels bacterial production was likely a supplementary food resource in addition to primary production rather than a sink of autotrophic production.

Filter feeding macrozooplankton is negligible in Mekkojärvi in early spring, so protozoans were the main consumers of bacteria. Bacterial production was practically completely harvested by bacterivores (Table 3) as has been found elsewhere on many occasions (Güde, 1986; Bloem *et al.*, 1989; Nagata, 1988). Since the feeding of *Desmarella* was found to be diurnally constant, which is consistent with the results of Aaronson (1980), our grazing estimates may not be biased by diurnal fluctuations. In Mekkojärvi the grazing of bacteria by this species was similar to that found for heterotrophic nanoflagellates by Fenchel (1987) and Bloem *et al.* (1988), but lower rates are also often reported (e.g. Sherr *et al.*, 1983; 1988; McManus & Fuhrman, 1986). Different results, even if not biased by methodological aspects, are to be expected given the variable abundance, size and physiological state of the grazer and food organisms as well as various environmental conditions and available variety of nutritional alternatives.

Grazing by bacterivores can generate large fluctuations in the density of bacteria within some hours or days (Andersen & Fenchel, 1985; Davis *et al.*, 1985; Fuhrman *et al.*, 1985; Güde, 1986; Bloem *et al.*, 1988; Hagström *et al.*, 1988). In spite of dense populations of heterotrophic protozoans, such fluctuations were not found in our experiment. This could probably be explained by the structure of bacterial and grazer communities. The bacterial population in the epilimnion of Mekkojärvi was mainly composed of very small cells, but less abundant large bacteria contribute a significant part of the biomass (Fig. 9). In the absence of grazers the size of bacteria increased so much that the growth rate of bacteria calculated from the increase in biomass was four times higher than when calculated from that of numbers. This, and the similarity of grazing and growth of bacteria, suggests that large cells were consumed almost as soon as they were produced. Size selective grazing (cf. Gonzales *et al.*, 1990) might have important consequences for the bacterial community. If there is a continuous production of small cells, the latter could keep the growth potential of bacteria high in spite of intensive grazing. This hypothesis would also allow a higher variety of species to coexist even in stable and homogeneous environments, since slowly growing species of bacteria are not excluded under heavy grazing (cf. Tranvik, 1992).

In part microscopic methods may also influence observed patterns of cycling in the biomass of bacteria. Particularly when bacteria preparations show background fluorescence, small bacteria, often with rapidly fading fluorescence, are easily overlooked. Then the proportion of tiny cells will become small and this might lead to more pronounced fluctuations in the bacterial counts. With acriflavine staining the background fluorescence was low and stable, thus enabling the counting of small cells. However, since large bacteria, although less numerous, are probably the most important food source for bacterivorous protozoans, precise information about the size distributions of bacteria would greatly aid interpretation of bacteria results. This goal may be achieved with improved image analysis which is becoming able to resolve the smallest bacteria seen by light microscopy.

In addition to the carbon flow, high grazing on bacteria may have quantitatively significant consequences for the cycling of nutrients in Mekkojärvi. Protozoans have been found to be important mineralizers of nutrients otherwise bound in bacterial biomass (cf. Caron *et al.*, 1985; Goldman *et al.*, 1985; Fenchel, 1987; Bloem *et al.*, 1988; Hagström *et al.*, 1988; Sherr *et al.*, 1988). In Mekkojärvi, where allochthonous organic matter is important for the nutrition of plankton (Salonen & Hammar, 1986), bacteria can compete with algae for nutrients (Salonen *et al.*, unpubl.) and regeneration through grazing may be particularly important. Further, most algae present in the experiment are potentially mixotrophic (Sanders & Porter, 1988) and may have been able to enhance their autotrophic production directly by ingesting other micro-organisms or by using dissolved organic carbon. This is particularly relevant in the context of the present study where phosphate concentration was low and phosphatase activity and N:P ratio were high.

In the absence of filter feeding macrozooplankton, part of the algal population was grazed by *Ochromonas* sp.. Its maximum grazing on *Chlamydomonas* and *Scourfieldia* was more than 100 cells $d^{-1}$ and thus *Ochromonas* would have markedly contributed to the decline of their populations during the second half of the experiment. The observed algal ingestion rate for *Ochromonas* sp. is in agreement with the results of Cole & Wynne (1974) for *O. danica* (> 2 *Microcystis* cells $ind.^{-1} h^{-1}$), but was much lower than Fenchel (1982) estimated for *Ochromonas* grazing on bacteria. Being mixotrophic, *Ochromonas* may also have contributed to primary production. *Ochromonas* is known to engulf bacteria (Aaronson, 1973), but it may prefer small cyanobacteria (Daley *et al.*, 1973; Cole & Wynne, 1974) and eukaryotes (Andersson *et al.*, 1986; Boraas *et al.*, 1988). However, even the small additional amount of phosphorus obtained from bacteria, which are assumed to have higher relative proportion of P in their cells than other organisms (e.g. Olsen *et al.*, 1986), may have been extremely important at the time of coinciding maxima of phosphatase activity and density of *Ochromonas*. Another potentially important aspect in the mixotrophy of *Ochromonas* is that carbon dioxide from the heterotrophic utilization of ingested algae might be recycled internally so that primary production of this species is underestimated.

The growth rates of the heterotrophic nanoflagellates (Table 2) were comparable with those of the photosynthetic algae, but lower than the maximum growth rates reported in the literature (cf. Fenchel, 1987; Andersson *et al.*, 1989). Although the experiment was made in spring, temperature was not particularly low (11–20 °C) and cannot probably explain the difference. The absence of larger filter feeding zooplankton may have been more important, because without their grazing the populations of heterotrophic protozoans may have approached the carrying capacity of the environment and thus the latter may have become food limited. The growth rate of bacteria (on average 0.5 divisions $d^{-1}$) was slower than that of protozoans. In spite of evident close coupling between the growth of bacteria and grazing by protozoans this is not unexpected, since, as discussed above, fast growing bacteria may have represented only a minor proportion in the bacterial assemblage.

A common difficulty in any enclosure experiment is the time dependent divergence of the community from the structure and functioning of the natural ecosystem. In our experiment some clear changes following the isolation of epilimnetic water were observed and some others probably occurred. Because of the shallow and steep thermal and chemical stratification of Mekkojärvi, the hypolimnion may have a particularly pronounced effect on the chemistry of the epilimnion. Although the phytoplankton maximum and the rise in water temperature both decreased the concentration of DIC, they cannot explain why DIC reached *ca.* 0.2 mg C $l^{-1}$ instead of the typical summer value of *ca.* 2–3 mg C $l^{-1}$. This suggests that the diffusion from hypolimnion might be a significant factor keeping DIC high. The concentration of DIC is known to affect the rate of photosynthesis (e.g. Jaworski *et al.*, 1981). The decrease of DIC may have contributed to the collapse of algal biomass and primary production, but, as suggested by the very high phosphatase activity, it is unlikely to have been the primary reason. Other compounds, which may affect the metabolism of organisms even more than DIC, may also diffuse vertically. The decrease of respiration, although less dramatic than for DIC, might support an idea that compounds diffusing from the anoxic hypolimnion are significant for the heterotrophic metabolism of epilimnetic plankton. Further, the absence of vertical migrations of phyto- and zooplankton between epi- and hypolimnion might have similar effect, if their possible uptake of vital substances from hypolimnion is excluded. In spite of these obvious limitations the isolation of epilimnetic water is useful in studying the functioning of plankton, because it reduces the complexity of the system to be studied. The effects of vertical migrations and vertical diffusion should be studied in other experiments to be able to get a more general view of the functioning of plankton in steeply stratified humic lakes.

In conclusion, in the tank with water from the epilimnion of highly humic Mekkojärvi, the biomass, production and respiration of bacterioplankton were rather independent of the spring maximum of phytoplankton primary production. Although there are no direct observations about the utilization of allochthonous organic matter by plankton organisms, the latter probably explained much of the growth of bacterioplankton. Some nutritional relationships between different species were demonstrated, but still the changes in the species composition of plankton cannot be adequately explained. The unambiguous understanding of species succession would require physiological approaches and the scope of research should be directed more to the species and individual level rather than the community one. This would be particularly important in studying the food chains of humic lakes, because in those one should be able to trace, in addition to the fate of autochthonous production, also that of the allochthonous input of organic matter. Another potentially important aspect for further studies is to evaluate the importance of the movements of inorganic and organic compounds between anoxic hypolimnion and oxic epilimnion by diffusion and migrations of organisms.

## Acknowledgements

This study was supported by the Academy of Finland and the Maj and Tor Nessling Foundation. Our thanks are due to R. Ilola and J. Vainionpää for making most of chemical determinations, S. Hällfors for the identification of *Paraphysomonas* species and Roger Jones for the revision of English of the manuscript.

## References

Aaronson, S., 1973. Particle aggregation and phagotrophy by *Ochromonas*. Arch. Mikrobiol. 92: 39–44.

Aaronson, S., 1980. Descriptive biochemistry and physiology of the Chrysophyceae (with some comparisons to the Prymnesiophyceae), pp. 118–169. In Levandowsky, M. & S. H. Hunter (eds.), Biochemistry and physiology of the Protozoa. Academic.

Andersen, P. & T. Fenchel, 1985. Bacterivory by microheterotrophic flagellates in seawater samples. Limnol. Oceanogr. 30: 198–202.

Andersson, A., C. Lee, F. Azam & Å. Hagström, 1985. Release of amino acids and inorganic nutrients by heterotrophic marine microflagellates. Mar. Ecol. Progr. Ser. 23: 99–106.

Andersson, A., V. Larsson & Å. Hagström, 1986. Size-selective grazing by a microflagellate on pelagic bacteria. Mar. Ecol. Progr. Ser. 33: 51–57.

Andersson, A., S. Falk, G. Samuelsson & Å. Hagström, 1989. Nutritional characteristics of a mixotrophic nanoflagellate, *Ochromonas* sp.. Microb. Ecol. 17: 251–262.

Arvola, L., 1983. Primary production and phytoplankton in two small, polyhumic forest lakes in S. Finland. Hydrobiologia 101: 105–110.

Arvola, L., 1986. Spring phytoplankton of 54 small lakes in southern Finland. Hydrobiologia 137: 125–134.

Arvola, L. & P. Kankaala, 1989. Winter and spring variability in phyto- and bacterioplankton in lakes with different water colour. Aqua fenn. 19: 29–39.

Arvola, L., K. Salonen & M. Rask, 1990. Chemical budgets for a small dystrophic lake in southern Finland. Limnologica 20: 243–251.

Bergström, I., A. Heinänen & K. Salonen, 1986. Comparison of acridine orange, acriflavine and bisbenzimide stains for enumeration of bacteria in clear and humic waters. Appl. envir. Microbiol. 51: 664–667.

Bloem, J., F. M. Ellenbroek, M.-J. B. Gilissen & T. E. Cappenberg, 1989. Protozoan grazing and bacterial production in stratified lake Vechten estimated with fluorescently labeled bacteria and thymidine incorporation. Appl. envir. Microbiol. 55: 1787–1795.

Bloem, J., M. Starink, M.-J. B. Gilissen & T. E. Cappenberg, 1988. Protozoan grazing, bacterial activity and mineralization in two-stage continuous cultures. Appl. envir. Microbiol. 54: 3113–3121.

Boraas, M., K. W. Estep, P. W. Johnson & J. M. Sieburth, 1988. Phagotrophic phototrophs: The ecological significance of mixotrophy. J. Protozool. 25: 249–252.

Børsheim, K. Y. & G. Bratbak, 1987. Cell volume to cell carbon conversion factors for a bacteriovorous *Monas* sp. enriched from sea water. Mar. Ecol. Progr. Ser. 36: 171–175.

Caron, D. A., J. C. Goldman, O. K. Andersen & M. R. Dennett, 1985. Nutrient cycling in a microflagellate food chain: II. Population dynamics and carbon cycling. Mar. Ecol. Progr. Ser. 24: 243–254.

Cole, G. T. & M. J. Wynne, 1974. Endocytosis of *Microcystis aeruginosa* by *Ochromonas danica*. J. Phycol. 10: 397–410.

Daley, R. J., G. P. Morris & S. R. Brown, 1973. Phagotrophic ingestion of a blue-green alga by *Ochromonas*. J. Protozool. 20: 58–61.

Davis, G. P., D. A. Caron, P. W. Johnson & J. McN. Sieburth, 1985. Phototrophic and apochlorotic components of picoplankton and nanoplankton in the North Atlantic: geographic, vertical, seasonal and diel distributions. Mar. Ecol. Progr. Ser. 21: 15–26.

De Haan, H., 1974. Effect of a fulvic acid fraction on the growth of a *Pseudomonas* from Tjeukemeer (the Netherlands). Freshwat. Biol. 4: 301–310.

Fenchel, T., 1982. Ecology of heterotrophic microflagellates II. Bioenergetics and growth. Mar. Ecol. Progr. Ser. 8: 225–231.

Fenchel, T., 1987. Ecology of protozoa. The biology of free-living phagotrophic protists. Brock/Springer Series In Contemporary Bioscience, Ann Arbor. 197 pp.

Fuhrman, J. A., R. W. Eppley, Å. Hagström & F. Azam, 1985. Diel variations in bacterioplankton, phytoplankton, and related parameters in the Southern California Bight. Mar. Ecol. Progr. Ser. 27: 9–20.

Goldman, J. C., D. A. Caron, O. K. Andersen & M. R. Dennett, 1985. Nutrient cycling in a microflagellate food chain: I. Nitrogen dynamics. Mar. Ecol. Progr. Ser. 24: 231–242.

Gonzales, J. M., E. B. Sherr & B. F. Sherr, 1990. Size-selective grazing on bacteria by natural assemblages of estuarine flagellates and ciliates. Appl. envir. Microbiol. 56: 583–589.

Guillard, R. R. L., 1973. Division rates. In J. R. Stein (ed.), Handbook of phycological methods, culture methods and growth measurements. Cambridge Univ. Press, Cambridge, pp. 289–311.

Güde, H., 1985. Influence of phagotrophic processes on the regeneration of nutrients in two-stage continuous culture systems. Microb. Ecol. 11: 193–204.

Güde, H., 1986. Loss processes influencing growth of planktonic bacterial populations in Lake Constance. J. Plankton Res. 8: 795–810.

Hagström, Å., F. Azam, A. Andersson, J. Wikner & F. Rassoulzadegan, 1988. Microbial loop in an oligotrophic pelagic marine ecosystem: possible roles of cyanobacteria and nanoflagellates in the pelagic fluxes. Mar. Ecol. Progr. Ser. 49: 171–178.

Hessen, D. O., 1985. The relation between bacterial carbon and dissolved organic humic compounds in oligotrophic lakes. FEMS Microbiol. Ecol. 31: 215–223.

Hessen, D. O. & A. K. Schartau, 1988. Seasonal and spatial overlap between cladocerans in humic lakes. Int. Revue ges. Hydrobiol. 73: 379–405.

Hessen, D. O., T. Andersen & A. Lyche, 1990. Carbon metabolism in a humic lake: Pool sizes and cycling through zooplankton. Limnol. Oceanogr. 35: 84–99.

Ishida, Y. & B. Kimura, 1986. Photosynthetic phagotrophy of Chrysophyceae: evolutionary aspects. Microbiol. Sciences 3: 132–135.

Jaworski, G. H. M., J. F. Talling & S. I. Heaney, 1981. The influence of carbon dioxide-depletion on growth and sinking rate of two planktonic diatoms in culture. Br. phycol. J. 16: 395–410.

Jeffrey, S. W. & G. M. Hallegraeff, 1987. Chlorophyllase distribution in ten classes of phytoplankton: a problem for chlorophyll analysis. Mar. Ecol. Progr. Ser. 35: 293–304.

Johansson, J.-Å., 1983. Seasonal development of bacterioplankton in two forest lakes in central Sweden. Hydrobiologia 101: 71–88.

Jones, R. I. & K. Salonen, 1985. The importance of bacterial utilization of released phytoplankton photosynthate in two humic forest lakes in southern Finland. Holarct. Ecol. 8: 133–140.

Kankaala, P., 1988. The relative importance of algae and bacteria as food for *Daphnia longispina* (Cladocera) in a polyhumic lake. Freshwat. Biol. 19: 285–296.

Koroleff, F., 1979. Methods for the chemical analysis of seawater. Meri 7: 1–60 (in Finnish).

Kuuppo-Leinikki, P. & K. Salonen, 1992. Bacterioplankton in a small polyhumic lake with an anoxic hypolimnion. Hydrobiologia 229: 159–168.

Lee, S. & J. A. Fuhrman, 1987. Relationships between biovolume and biomass of naturally derived marine bacterioplankton. Appl. envir. Microbiol. 53: 1298–1303.

McManus, G. B. & J. A. Fuhrman, 1986. Bacterivory in seawater studied with the use of inert fluorescent particles. Limnol. Oceanogr. 31: 420–426.

Murphy, J. & J. P. Riley, 1963. A modified singlesolution method for the determination of phosphate in natural waters. Analyt. chim. Acta 27: 31–36.

Münster, U., 1992. Extracellular enzymes in a polyhumic lake: important regulators in detritus processing. Hydrobiologia 229: 225–238.

Nagata, T., 1988. The microflagellate-picoplankton food linkage in the water column of lake Biwa. Limnol. Oceanogr. 33: 504–517.

Niemi, M., J. Kuparinen, A. Uusi-Rauva & K. Korhonen, 1983. Preparation of $^{14}$C labelled algal samples for liquid scintillation counting. Hydrobiologia 106: 149–156.

Olsen, Y., A. Jensen, H. Reinertsen, K. Y. Borsheim, M. Heldal & A. Langeland, 1986. Dependence of the rate of release of phosphorus by zooplankton on the P:C ratio in the food supply, as calculated by a recycling model. Limnol. Oceanogr. 31: 34–44.

Reynolds, C. S., 1988. Functional morphology and adaptive strategies of freshwater phytoplankton. In Growth and reproductive strategies of freshwater phytoplankton, Sandgren, C. D. (ed.), Cambridge Univ. Press, Cambridge, pp. 388–433.

Rocha, O. & A. Duncan, 1985. The relationship between cell carbon and cell volume in freshwater algae species used in zooplankton studies. J. Plankton Res. 7: 279–294.

Sanders, R. W. & P. G. Porter, 1988. Phagotrophic phytoflagellates. In Marshall, K. G. (ed.), Advances in microbial ecology. Plenum Press, New York, Vol. 10, pp. 167–192.

Salonen, K., 1979. A versatile method for the rapid and accurate determination of carbon by high temperature combustion. Limnol. Oceanogr. 24: 177–183.

Salonen, K., 1981. Rapid and precise determination of total inorganic carbon and some gases in aqueous solutions. Wat. Res. 15: 403–406.

Salonen, K. & T. Hammar, 1986. On the importance of dissolved organic matter in the nutrition of zooplankton in some lake waters. Oecologia (Berl.) 8: 246–253.

Salonen, K. & S. Jokinen, 1988. Flagellate grazing on bacteria in a small dystrophic lake. Hydrobiologia 161: 203–309.

Salonen, K. & A. Lehtovaara, 1992. Migration of haemoglobin-rich *Daphnia longispina* in a small, steeply stratified, humic lake with an anoxic hypolimnion. Hydrobiologia 229: 271–288.

Salonen, K., K. Kononen & L. Arvola, 1983. Respiration of plankton in two small, polyhumic lakes. Hydrobiologia 101: 65–70.

Sarvala, J., V. Ilmavirta, L. Paasivirta & K. Salonen, 1981. The ecosystem of the oligotrophic Lake Pääjärvi. 3. Secondary production and an ecological energy budget of the lake. Verh. int. Ver. Limnol. 21: 454–459.

Schiller, 1954. Uber winterliche pflanzliche Bewohner des Wassers, Eises und des daraufliegenden Schneebreies. I. sterr. Bot. Z. 101 (3).

Sherr, B. F., E. B. Sherr & T. Berman, 1983. Grazing, growth, and ammonium excretion rates of a heterotrophic microflagellate fed with four species of bacteria. Appl. envir. Microbiol. 45: 1196–1201.

Sherr, B. F., E. B. Sherr & R. D. Fallon, 1987. Use of monodispersed, fluorescently labelled bacteria to estimate *in situ* protozoan bacterivory. Appl. envir. Microbiol. 53: 958–965.

Sherr, B. F., E. B. Sherr & F. Rassoulzadegan, 1988. Rates of digestion of bacteria by marine phagotrophic protozoa: temperature dependence. Appl. envir. Microbiol. 54: 1091–1095.

Similä, A., 1988. Spring development of a *Chlamydomonas* population in Lake Nimetön, a small humic forest lake in southern Finland. Hydrobiologia 161: 149–157.

Solorzano, L., 1969. Determination of ammonia in natural waters by the phenolhypochlorite method. Limnol. Oceanogr. 14: 799–801.

Tranvik, L. J., 1988. Availability of dissolved organic carbon for planktonic bacteria in oligotrophic lakes of differing humic content. Microb. Ecol. 16: 311–322.

Tranvik, L. J., 1990. Bacterioplankton growth on fractions of dissolved organic carbon of different molecular weights from humic and clear waters. Appl. envir. Microbiol. 56: 1672–1677.

Tranvik, L. J., 1992. Allochthonous dissolved organic matter as an energy source for pelagic bacteria and the concept of microbial the loop. Hydrobiologia 229: 107–114.

Tranvik, L. & M. G. Höfle, 1987. Bacterial growth in mixed cultures on dissolved organic carbon from humic and clear waters. Appl. envir. Microbiol. 53: 482–488.

Wetzel, R. G., 1983. Limnology, 2nd edition, W. B. Saunders Co., Philadelphia. 767 pp.

Wood, E. D., F. A. J. Armstrong & F. A. Richards, 1967. Determination of nitrate in sea water by cadmium-copper reduction to nitrite. J. mar. biol. Ass. U.K. 47: 23–31.

*Hydrobiologia* **229**: 143–157, 1992.
*K. Salonen, T. Kairesalo & R.I. Jones (eds.),*
*Dissolved Organic Matter in Lacustrine Ecosystems: Energy Source and System Regulator.*




# Planktonic food chains of a highly humic lake

## *II. A mesocosm experiment in summer during dominance of heterotrophic processes*

K. Salonen, P. Kankaala, T. Tulonen, T. Hammar, M. James, T.-R. Metsälä & L. Arvola
*University of Helsinki, Lammi Biological Station, SF-16900 Lammi, Finland*

*Key words: Daphnia longispina*, plankton, heterotrophy, mesocosm, food chains, humic lake


## Abstract

The development and metabolism of epilimnetic plankton from a highly humic lake was followed in late summer, when the predominant zooplankton species, *Daphnia longispina*, was very abundant (*ca.* 200 ind. $l^{-1}$). The experiment was made in two tanks: one with an unaltered plankton assemblage and one with larger zooplankton removed. The scarce phytoplankton community was also simple, consisting mainly of one *Cryptomonas* and two *Mallomonas* species. The abundance and species composition of smaller plankton was heavily influenced by grazing of *Daphnia*. In particular, the biomass of heterotrophic flagellates increased after the removal of *Daphnia*. The biomass and production of bacterioplankton were not affected, and remained several times higher than that of phytoplankton. Bacterial production and grazing on bacteria were balanced, and when *Daphnia* was removed its grazing activity was compensated by flagellates. The removal of *Daphnia* did not affect the respiration or community net production of plankton. Among organisms smaller than zooplankton, bacteria seemed to be responsible for most of the respiration. The community net production was consistently negative even at the water surface, indicating an allochthonous carbon source. The results suggest that phytoplankton primary production was insufficient for the secondary production in the epilimnetic water of the study lake. The food requirements of bacteria and zooplankton, as well as of flagellates, each exceeded that supplied by phytoplankton primary production. The simple food chains in this experiment made it possible to reveal the functioning of the community so completely that dissolved organic matter is certainly comparable to or exceeds the importance of phytoplankton primary production as an energy and carbon source for food webs in this humic lake.


## Introduction

In summer the high water colour of small humic lakes leads to steep thermal and chemical stratification (e.g. Jones & Arvola, 1984). Inorganic nutrients are often depleted in the shallow epilimnion and primary production of phytoplankton is low. Under such conditions allochthonous dissolved organic matter (DOM) seems to be a major energy and carbon source for the food chains (Salonen & Hammar, 1986; Tranvik, 1989; Moran & Hodson, 1990). Bacteria are particularly abundant in humic waters (Johansson, 1983; Hessen, 1985) and are probably an important link converting allochthonous DOM to biomass available for higher trophic levels. A similar DOM-bacteria-protozoa food chain as described in marine waters by Azam *et al.* (1983) and termed 'microbial loop' seems to operate in humic lakes (Salonen & Hammar, 1986; Salonen & Jokinen, 1988). However, in contrast to the case in the ocean, in humic lakes DOM exuded by phytoplankton seems to be quantitatively unimportant (Jones & Salonen, 1985).

The aim of this study was to describe the microbial communities which serve as food for zooplankton and to quantify the relative importance of allochthonous and autochthonous sources of energy. As high grazing pressure by crustacean zooplankton may greatly influence phytoplankton and other microbial communities (Kankaala, 1988; Hessen *et al.*, 1989), particular emphasis was placed on the role of zooplankton as a system regulator. To simplify the approach the study was made in large tanks where the vertical gradients and migrations of organisms in the water column (Arvola *et al.*, 1992; Salonen & Lehtovaara, 1992) were eliminated and epilimnetic water was isolated from the underlying hypolimnion. A similar experiment was also made with water from the same lake (Salonen *et al.*, 1992) during the short spring phytoplankton maximum when zooplankton was not yet developed.

## Material and methods

### *Experimental design*

The study was performed between 25 July and 2 August 1988 in three tanks filled with water from the epilimnion of Lake Mekkojärvi. Two of the tanks were used to follow physical, chemical and biological processes. The third was used as an incubation tank to minimize disturbances in the experimental tanks. The sampling started the next day after filling the tanks and was carried out daily between 8–9 am. In the evening of the second day the dominant zooplankton species, *Daphnia longispina*, was removed from one tank by sieving the water through a 200 $\mu$m net. Otherwise the experimental design was similar to that used in the earlier mesocosm experiment during the spring phytoplankton maximum (Salonen *et al.*, 1992). Unless stated otherwise, chemical and other methods were the same as in the spring experiment.

Samples for cations, chloride and sulphate were taken immediately after filling the tanks. Alkalinity and absorbance at 260 nm (describing water colour) were measured on the first and last day of the experiment. All other chemical determinations and pH and conductivity were made daily (Table 1).

The growth rates of algae and zooflagellates were estimated from the increase in cell numbers in 1 liter acid-washed glass bottles filled with 200 $\mu$m filtered tank water and incubated in the tank for 24 h. Exuded primary production was determined using 0.2 $\mu$m Nuclepore filters. To exclude small bacteria and viruses we also used 0.02 $\mu$m Anopore filters which, except for the pore size, have rather similar particle retention characteristics to Nuclepore filters. Flagellate grazing on bacteria was estimated as a the difference between bacterial biomass in unfiltered and 1 $\mu$m-filtered 250 ml samples after 24 h incubation (see Salonen *et al.*, 1992). As some *Daphnia longispina* individuals were present in unfiltered samples, their number was counted and their share in the grazing (based on grazing experiments with $^3$H-(methyl)thymidine labelled bacteria as will be described later) was subtracted. Samplings and incubations were carried out at the depths of 0.05 and 0.5 m, but most results are shown as means of these two depths. Zooplankton samples were taken daily with an acrylic plastic tube (diameter 0.05 m, length 0.5 m) submerged vertically in the middle of the tank. Formaldehyde (4%) killed animals were counted for 0.5 mm interval size classes under a dissecting microscope. Zooplankton grazing experiments were carried out only with *Daphnia longispina*, which formed >90% of the macrozooplankton numbers and biomass. Algal grazing was measured as uptake of $^{32}$P-labelled natural algae (size fractions 1–10 $\mu$m and >10 $\mu$m) until the fifth day of the experiment (see James & Salonen, 1992). For bacterial grazing experiments, epilimnetic bacteria from the same lake were labelled with $^3$H-(methyl)thymidine. Water taken from the experimental tank was daily filtered through a 2 $\mu$m

*Table 1.* Physical and chemical properties of water in the experimental tanks. N-Number of observations. All concentrations are expressed as mg $l^{-1}$.

| | *Daphnia* removed | | | With *Daphnia* | | |
|---|---|---|---|---|---|---|
| | Mean | SD | N | Mean | SD | N |
| pH | 5.3 | 0.1 | 8 | 5.2 | 0.2 | 8 |
| Conductivity (mS $m^{-1}$) | 3.1 | 0.07 | 8 | 3.2 | 0.07 | 8 |
| Alkalinity (mmol $l^{-1}$) | 0.04 | - | 2 | 0.04 | - | 2 |
| Absorbance 260 nm | 1.2 | - | 2 | 1.2 | - | 2 |
| Total phosphorus | 0.023 | 0.003 | 7 | 0.026 | 0.002 | 7 |
| Phosphate phosphorus | 0.009 | 0.002 | 8 | 0.012 | 0.003 | 8 |
| Diss. org. phosphorus | 0.016 | 0.001 | 6 | 0.018 | 0.002 | 7 |
| Total nitrogen | 0.70 | 0.03 | 7 | 0.73 | 0.02 | 7 |
| Ammonium nitrogen | 0.038 | 0.005 | 7 | 0.057 | 0.017 | 7 |
| $NO_2 + NO_3$ nitrogen | 0.026 | 0.004 | 6 | 0.028 | 0.005 | 6 |
| Diss. org. nitrogen | 0.67 | 0.02 | 7 | 0.66 | 0.09 | 7 |
| Calcium | 4.3 | - | 1 | 4.3 | - | 1 |
| Magnesium | 0.9 | - | 1 | 0.9 | - | 1 |
| Fe | 0.7 | - | 1 | 0.7 | - | 1 |
| Cl | 0.9 | - | 1 | 0.9 | - | 1 |
| $SO_4$ | 6.8 | - | 1 | 6.8 | - | 1 |
| DOC | 25.3 | 1.5 | 6 | 25.1 | 0.3 | 6 |
| POC | 0.14 | 0.04 | 7 | 0.12 | 0.03 | 7 |

Nuclepore filter into sterile 500 ml bottles and subsequently labelled with 5.55–7.40 MBq $^3H$-(methyl)thymidine for 13–20 hours in the incubation tank. The following morning 80–150 *Daphnia* from the experimental tank were added to bottles with labelled bacteria. After 10–12 min the experiment was stopped by filtering animals on a net (mesh size 70 $\mu$m) and narcotizing them with carbonated water. The numbers of *Daphnia* in different size classes were counted as described above (for other details of the liquid scintillation procedure see Kankaala, 1988). Blanks were made using 20 glutardialdehyde-killed animals in bottles with 100 ml water with $^3H$ (methyl)thymidine labelled bacteria as described earlier. The activity of the blank animals was 0.1–5% of that of the experimental animals.

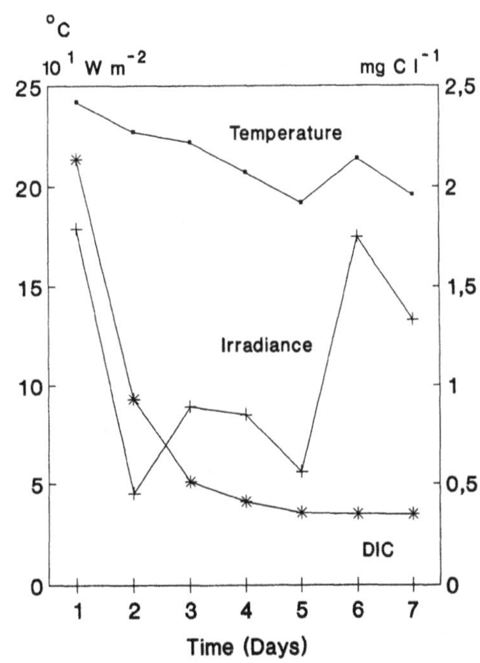


*Fig. 1.* The change in the daily maximum temperature and dissolved inorganic carbon (DIC) in the tank with *Daphnia*, and in solar irradiance (above the acrylic plastic cover) during the experiment.

## Results

### *Abiotic variables*

Because of the cloudy weather, irradiance was low (mean 109 W $m^{-2}$ $d^{-1}$) but higher at the beginning and at the end of the experiment (Fig. 1).

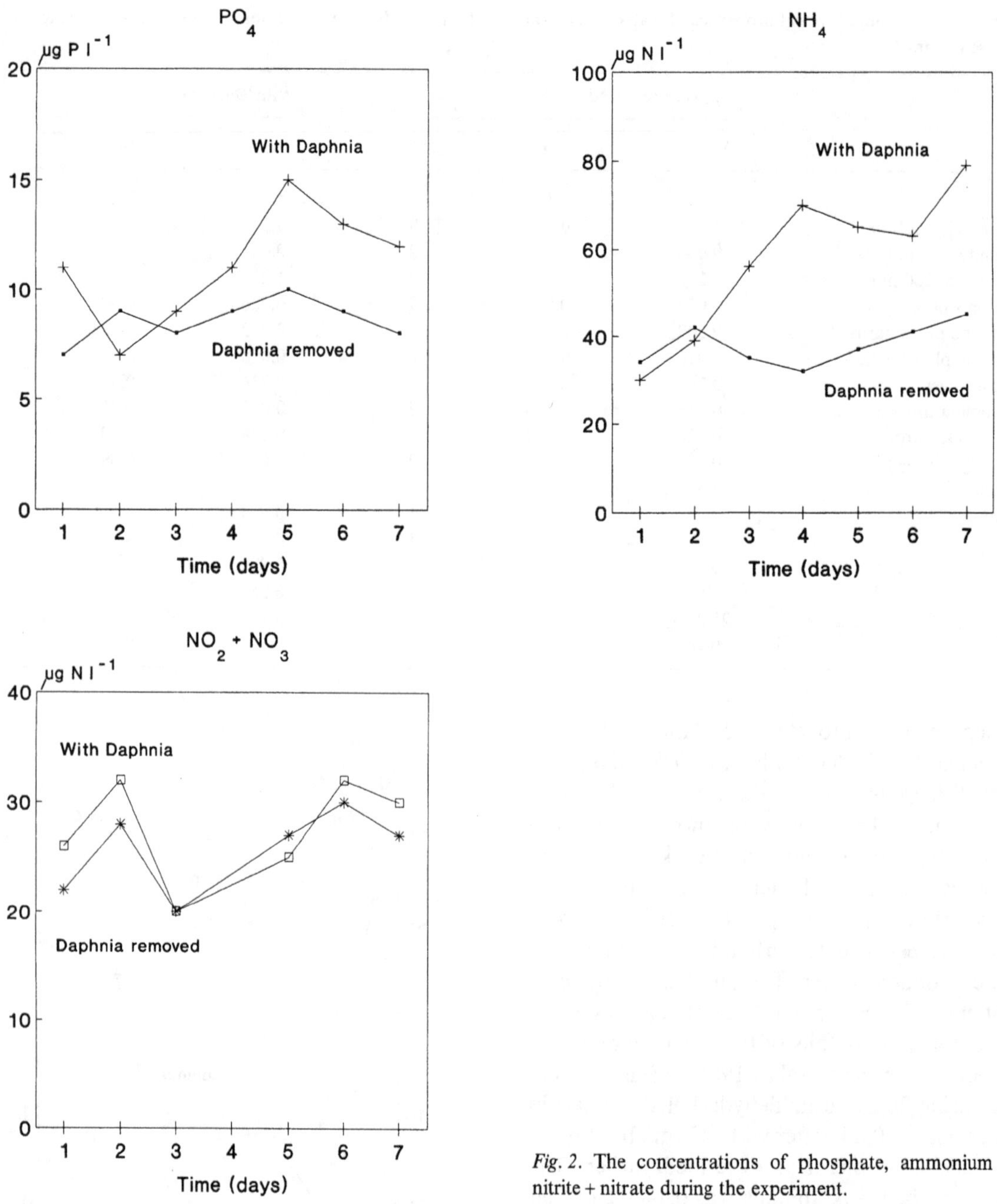


*Fig. 2.* The concentrations of phosphate, ammonium and nitrite + nitrate during the experiment.

Condensation on the under side of the acrylic plastic cover during rainy days further reduced the irradiance at the water surface in the tanks by *ca.* 50%. The mean water temperature was 20.3 °C (SD = 1.1) with a daily range of 0.5–4.0 °C, which corresponds well the diel range observed in the lake. The concentration of dissolved organic carbon (DOC) was high and no significant changes were observed during the experiment (Table 1). Dissolved inorganic carbon (DIC) decreased rapidly during the first three days in both tanks (Fig. 1). The concentrations of phosphate and ammonium increased in the tank with *Daphnia*, but remained at about the same

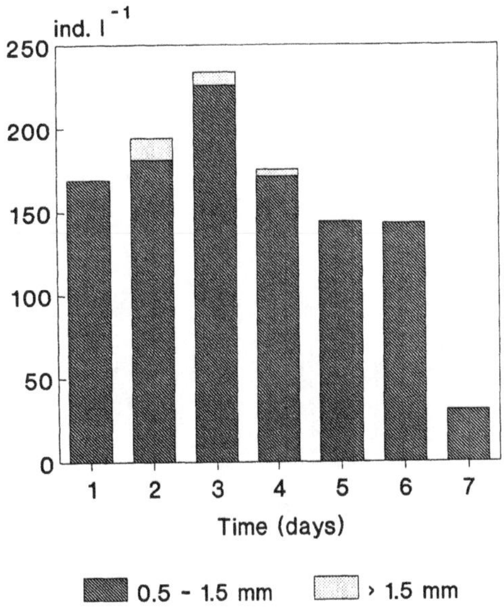


*Fig. 3.* The densities of two size classes of *Daphnia longispina* in the tank with *Daphnia* during the experiment.

level in the tank where *Daphnia* was removed (Fig. 2). No difference was found in the concentration of nitrate + nitrite between the tanks.

*Dynamics of plankton assemblage*

At the beginning of the experiment *D. longispina* was very abundant, *ca.* 200 ind. $l^{-1}$ (Fig. 3), but its density had decreased to 30 ind. $l^{-1}$ by the end of the experiment. The proportion of large individuals (> 1.5 mm) was < 10% of the population and decreased to none at the end of the experiment.

The density of autotrophic phytoplankton was consistently low (< 120 cells $ml^{-1}$) in both tanks. More than 90% of the phytoplankton biomass comprised *Cryptomonas* spp., *Mallomonas costata*, and *M. akrokomos*. The densities of these species (Fig. 4) as well as phytoplankton biomass (Fig. 5) seemed to decrease steadily in the presence of *D. longispina* while in the other tank no significant decrease was found during the experiment. A similar trend was observed in the concentration of chlorophyll *a* (Fig. 6). A Prymnesiophycean *Scourfieldia cordiformis* and a Chlorophycean *Chlamydomonas* sp. appeared in low densities in both tanks.

After the removal of *Daphnia*, heterotrophic flagellates (no autofluorescence of chlorophyll, length < 3 μm) increased exponentially for two days to a maximum of 5100 cells $ml^{-1}$ and were soon followed by a maximum (1400 cells $ml^{-1}$) of a choanoflagellate *Desmarella moniliformis* (Fig. 4). Thus grazing by *Daphnia* seemed to control the density of heterotrophic flagellates. Also a non-autofluorescent, and presumably heterotrophic, crysophycean *Chromulina* sp. increased in numbers after the removal of *Daphnia*, but for an unknown reason its growth rate was lower than in the presence of *Daphnia*. The number of ciliate protozoans was very low (< 10 cells $ml^{-1}$) in both tanks.

The biomass of bacteria was about four times higher than that of algae (Fig. 5). The sum of bacterial and algal carbon biomasses, estimated from microscopic counts, was on average 71–75% of particulate organic carbon (POC). The proportion of detritus (the difference between POC and carbon in organisms) showed a slight increase towards the end of the experiment (Fig. 7).

*Production and respiration of plankton*

The daily primary production was very low (2–19 mg C $m^{-3}$ in 10 h) in both tanks, being highest on the first day (Fig. 5). A slight increase was observed during the last two days in the tank with *Daphnia*, which may have been due to the nutrient concentration (Fig. 2) having increased above that in the other tank. Due to the rapid attenuation of light in the dark coloured water, the primary production at 0.5 m depth was only 30–64% of that measured at 0.05 m depth. The proportion of exudates (passing through a 0.2 μm filter) in the total phytoplankton primary production averaged 22% (mean with SD 1.10 ± 0.86 mg C $m^{-3}$) in the tank without *Daphnia* and 24% (mean 1.48 ± 1.67 mg C $m^{-3}$) in the tank with *Daphnia*. Respective proportions obtained with 0.02 μm filters were 5.9% and 6.5%. The latter results are based on medians, since in

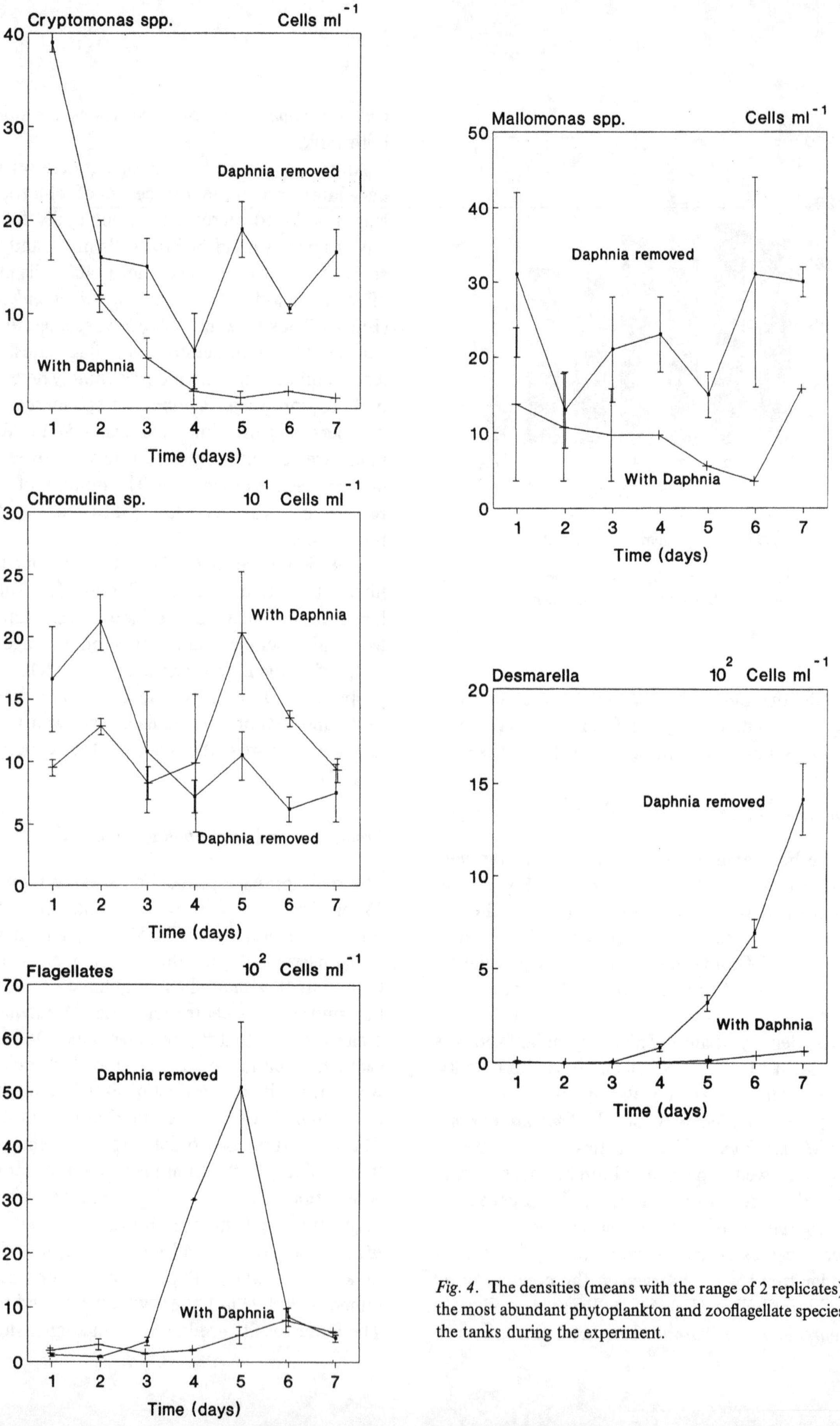


*Fig. 4.* The densities (means with the range of 2 replicates) of the most abundant phytoplankton and zooflagellate species in the tanks during the experiment.

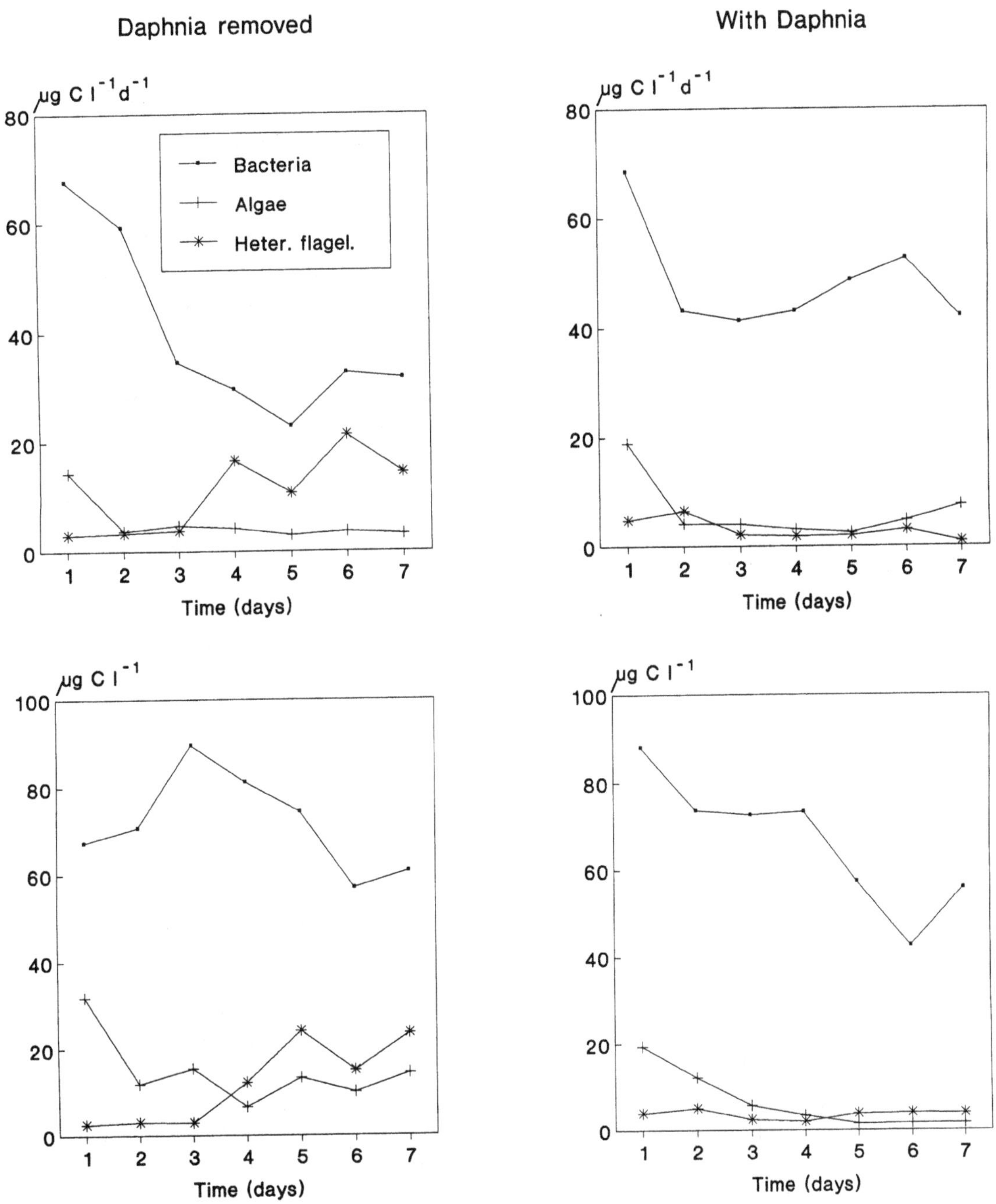


*Fig. 5.* Production and biomass of bacteria, autotrophic phytoplankton and heterotrophic flagellates in the two tanks during the experiment.

the early phase of the experiment there were some unreasonably high values caused by leakage of seals, when using the filtration equipment with the hard Anopore filters. On average 83% (range 65–91%) of daily irradiance was measured during the 10 h incubations. Daily primary production correlated significantly with irradiance ($r = 0.596$, $p < 0.05$).

In the tank with *Daphnia* the biomass and production of autotrophic and heterotrophic

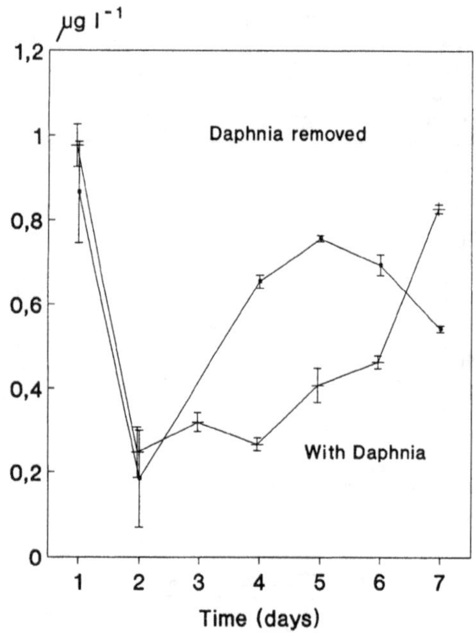


*Fig. 6.* The concentration of chlorophyll *a* (means with the range of 2 replicates) in the two tanks during the experiment.

phytoplankton and zooflagellates remained low (Fig. 5). After the removal of *Daphnia*, the biomass and daily production of heterotrophic flagellates exceeded those observed for autotrophic algae. In both tanks the production of bacteria was about eight times the primary production (Fig. 5) indicating that it was the main driving force of the food chains. At the surface the production of bacteria was about twice that at 0.5 m depth (Table 3) which suggests light dependence of the production of substrates used by bacteria.

*Table 2.* Average growth rates (divisions $d^{-1}$) and generation times (h) of dominant algae, zooflagellates and bacteria.

| | *Daphnia* removed | | With *Daphnia* | |
|---|---|---|---|---|
| | div. $d^{-1}$ | h | div. $d^{-1}$ | h |
| Cryptomonas sp. | 0.68 | 24 | 0.97 | 17 |
| Mallomonas costata | 0.79 | 21 | 0.98 | 17 |
| Chromulina sp. | 0.81 | 21 | 0.28 | 59 |
| Desmarella moniliformis | 1.62 | 10 | 1.54 | 11 |
| Flagellates <3 µm | 1.62 | 10 | 1.66 | 10 |
| Bacteria | 0.5 | 33 | 0.6 | 28 |

With the exception of the first day, the production in the tank with *Daphnia* remained at almost the same level during the whole experiment. Before the elimination of *Daphnia*, bacterial production was similar in both tanks, but afterwards it was consistently lower in the tank without *Daphnia* (Fig. 5). This may have been due to released dissolved inorganic and/or organic nutrients in the presence of *Daphnia*. The growth rates of bacteria were markedly lower than those of flagellates (Table 2) and there was no difference between the two tanks.

Respiration of plankton stayed almost at the same level during the experiment and there was no significant difference (Wilcoxon's rank test) between the two tanks (mean 70 mg C $m^{-3}$ $d^{-1}$ with *Daphnia* removed and 63 mg C $m^{-3}$ $d^{-1}$ with *Daphnia*). Nor was there any significant difference in <1 µm fraction respiration (means 86 mg C $m^{-3}$ $d^{-1}$ and 94 mg C $m^{-3}$ $d^{-1}$, respectively) between the tanks (Fig. 8). In this fraction the respiration was higher at the beginning of the experiment, but the combined data from both tanks shoved no significant difference (Wilcoxon's test, $P = 0.055$) between total and <1 µm fraction (bacterial) respiration (Fig. 8). The latter showed significant correlation ($r = 0.744$, $P < 0.01$) with the production of bacteria. The proportion of respiration in the gross bacterial production (net growth + respiration in <1 µm fraction) was on average 70% in the tank with *Daphnia* removed and 66% in the tank containing *Daphnia* (Table 3). The consistently negative community net production (Fig. 9), even at the surface of the water column where light cannot have been limiting primary production, showed clearly that during the experiment heterotrophic processes were dominating. Again there was no significant difference between the two tanks (Wilcoxon's test).

### *Biotic interactions*

*Daphnia longispina* cleared algae at a rate about three times higher (mean 0.30 ml ind.$^{-1}$ $h^{-1}$) than bacteria (mean 0.09 ml ind.$^{-1}$ $h^{-1}$). The

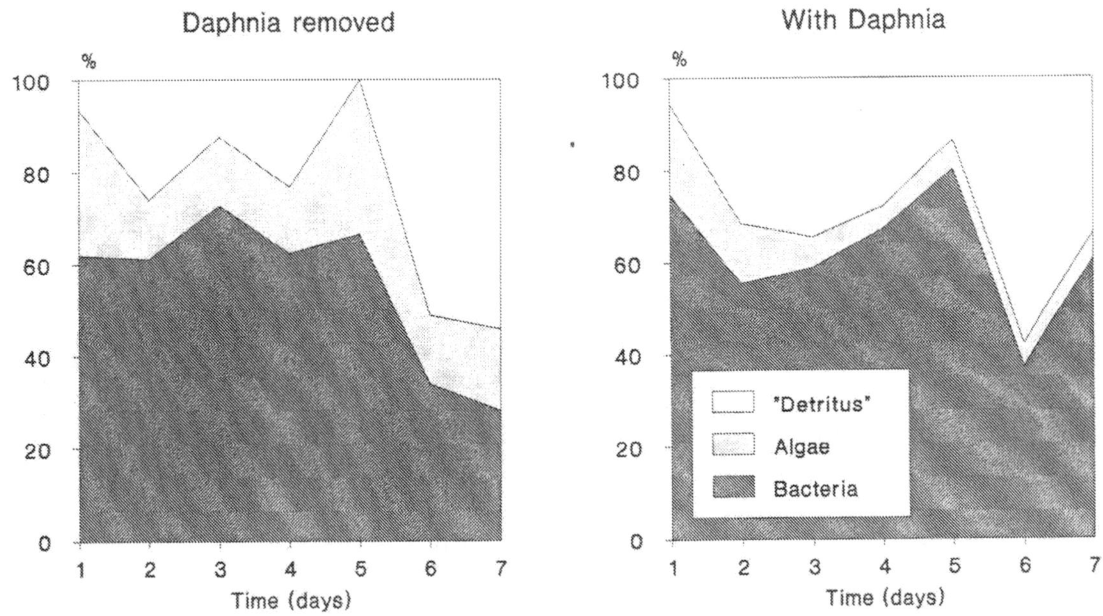


*Fig. 7.* Estimated proportions of detritus, algae and bacteria in POC in the two tanks during the experiment.

highest ratio between these clearance rates (11.2) was observed on the third day of the experiment, when large (> 1.5 mm) *Daphnia* individuals were still in good condition. During the first three days *Daphnia* cleared daily > 100% of the water volume (more details are shown in James & Salonen, 1992). Thus, in the tank with *Daphnia* the density and biomass of autotrophic algae and heterotrophic flagellates were radically affected. However, the intensive grazing by *Daphnia* did not so clearly affect phytoplankton primary production (Fig. 5). The growth rates of the major phytoplankton species (Table 2) were of the same order as *Daphnia* grazing.

Because of the low algal biomass, bacteria were a more important food source for *Daphnia* after the first two days of the experiment (Fig. 10). In the tanks where *Daphnia* were present throughout the experiment they contributed 35 ± 7% (mean with SD) of the total grazing on bacteria, and in the other tank 39% before the removal of *Daphnia* (Fig. 11). *Daphnia* evidently suffered from poor food conditions, because the population nearly collapsed at the end of the experiment.

Heterotrophic flagellates were the most important grazers of bacteria. According to the experiments carried out with fluorescent 0.57 μm latex beads, bacterial sized particles were ingested by

*Table 3.* Mean bacterial net production and respiration during the tank experiment. BB – Bacterioplankton biomass; GP – gross production (net production + respiration in < 1 μm fraction).

| Tank | Depth m | Net production | | Respiration | |
|---|---|---|---|---|---|
| | | $\mu g\ C\ l^{-1}\ d^{-1}$ | % of BB | $\mu g\ C\ l^{-1}\ d^{-1}$ | % of GP |
| *Daphnia* removed | 0.05 | 53 | 74 | 87 | 63 |
| | 0.5 | 26 | 37 | 85 | 76 |
| With *Daphnia* | 0.05 | 65 | 98 | 86 | 57 |
| | 0.5 | 32 | 48 | 101 | 76 |

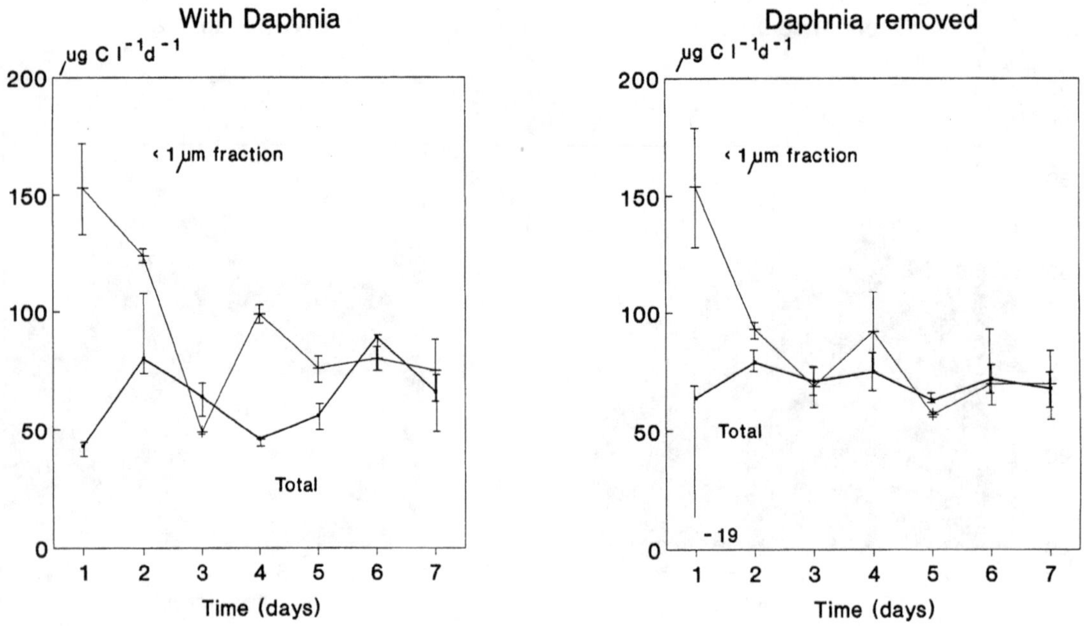


*Fig. 8.* Respiration of total plankton (medians for 4 replicate determinations; with ranges) and of the $<1\ \mu m$ fraction (means for 0 and 0.5 m samples) in the two tanks during the experiment.

*Desmarella moniliformis*, *Paraphysomonas* sp., *Lobomonas* sp. and small ($<3\ \mu m$) unidentified flagellates. The latter and *D. moniliformis* became particularly abundant after the elimination of *Daphnia* (Fig. 4). The ciliate *Strombidium* sp. also

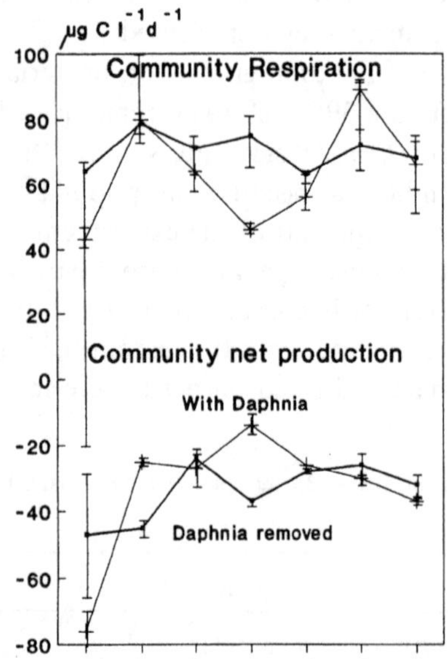


*Fig. 9.* Community respiration (medians of 4 replicates with the range) and community net production (with the ranges between 0 and 0.5 m samples) in the two tanks during the experiment.

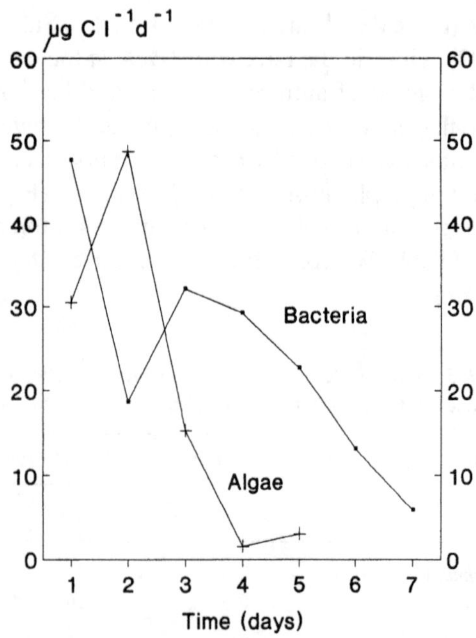


*Fig. 10.* The grazing of bacteria and algae by *Daphnia longispina* during the experiment.

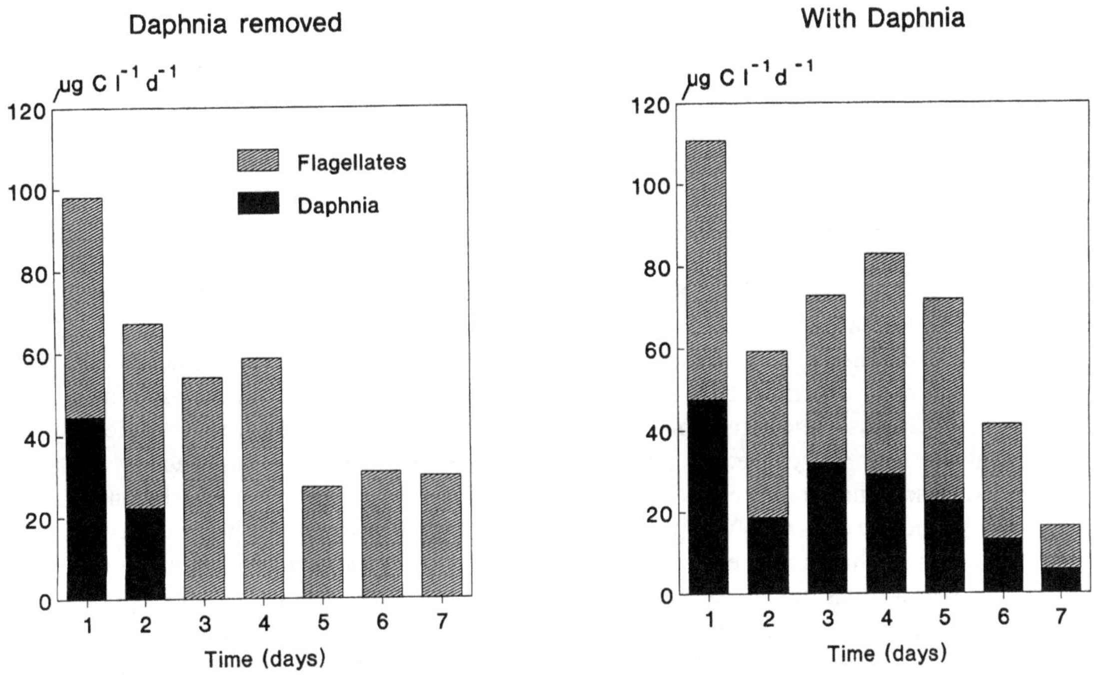


*Fig. 11.* The grazing of bacteria by flagellates and *Daphnia longispina* in the two tanks during the experiment.

ingested fluorescent beads, but the role of ciliates in bacterial grazing was probably small since they occurred only in very low numbers. On average flagellates daily grazed 132% of bacterial production in both tanks. Bearing in mind some uncertainties in the results (particularly volume to carbon conversion factors) the grazing on bacteria by flagellates and *Daphnia* seemed to balance the net production of bacteria. The growth rate of flagellates correlated rather well with the number of bacteria they grazed (Fig. 12).

## Discussion

Our experimental design eliminated the complex vertical stratification and allowed study of most of the components in the simple planktonic food chains and carbon fluxes between different organisms with biologically relevant time resolution. Thus different fluxes could be related to each other, allowing evaluation of the results at the

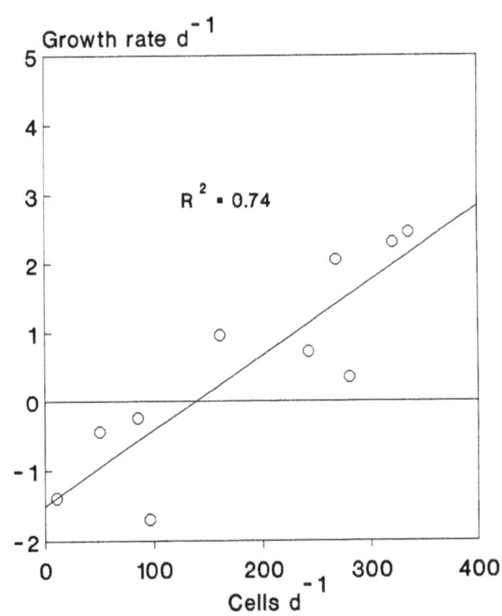


*Fig. 12.* The dependence of the growth of bacterivorous flagellates (*Desmarella moniliformis*, *Chromulina* sp. and <3 μm flagellates) on the number of grazed bacteria during the experiment.

whole community level. This kind of approach makes it easier to identify the limitations in the results as well to draw reliable conclusions.

The major criticism of this and other enclosure experiments with isolated water bodies is whether the results adequately represent the natural conditions in the lake. In this experiment the isolation of epilimnetic water from the underlying anoxic hypolimnion may have particularly affected the results. The rapid decrease in the concentration of DIC (Fig. 1) during the experiment (similar to that found in the similar experiment in spring by Salonen *et al.*, 1992) was the clearest indication of that. It was certainly not caused by primary production, because community net production was negative. Although excess $CO_2$ was probably lost to air, it is unlikely that the gas exchange rate between water and air was enhanced in the experiment, because the tank was covered by a plate with only *ca.* 10 mm air gaps at its edges. Thus the decrease in DIC probably reflects a rapid decrease in $CO_2$ production or cessation of vertical $CO_2$ flux from hypolimnion. In a similar way, nutrient fluxes from the hypolimnion could have beet cut, affecting primary production. In addition, fluxes between anoxic hypolimnion and oxic epilimnion might regenerate organic substrates or reduced elements essential for epilimnetic bacteria. This might have reduced bacterial respiration leading to a decrease in DIC. Our results provide some indication that the production of bacteria and phytoplankton decreased at the very beginning of the experiment (Fig. 5), consistent with the hypothesis that vertical fluxes from hypolimnion are important for plankton.

Another effect of the isolation of epilimnetic water is that vertical migrations of organisms between epilimnion and hypolimnion were prevented. Continuous occupation of the warm epilimnion may not be normal for *Daphnia* and may have led to higher energy consumption in the experiment than in the lake. Thus the insufficient food availability for *Daphnia* might be explained by the fact that this species was no longer able to migrate into cold water close to the oxycline (see Salonen & Lehtovaara, 1992). In Mekkojärvi many algal species also migrate diurnally (e.g. *Cryptomonas*) or they are confined to anoxic hypolimnion (e.g. *Scourfieldia cordiformis*, *Chlamydomonas*) (Arvola *et al.*, 1992), thus being able to escape or reduce the grazing pressure by *Daphnia* and also to improve their access to nutrients (Salonen *et al.*, 1984). In this experiment the migration of edible algal species to the hypolimnion was not possible and these species must have had short enough generation times or some other mechanism to survive heavy grazing by *Daphnia*.

Despite such potential criticism, the results show convincingly that in our experiment heterotrophic processes required considerably more energy than could be supplied by autotrophic plankton. High respiration in relation to low phytoplankton primary production was among the strongest indications of the dominance of heterotrophy. Although phytoplankton primary production was estimated using the technically reliable whole water modification of the $^{14}C$ method and respiration was measured directly as an increase of the end product ($CO_2$) of the mineralization, the interpretation of the ratio between these variables is still uncertain. There are a number of problems in the determination of primary production (e.g. Peterson, 1980), and respiration results may be biased. In both cases the dependence of results on the duration of incubation may be important. Respiration, which is artificially determined in darkness, may particularly suffer from the elimination of some substances with rapid turnover. On the other hand, as the numbers of zooplankters in respiration bottles are probably lower than in the lake, the reduction of grazing may increase respiration of a rapidly growing bacterial population. The results from 1 $\mu$m filtrate (bacterial respiration) are most prone to this possibility and indeed, some results were clearly higher than total respiration. Although this difference may also be due to increased concentration of substrates available for bacteria after mechanical damage to some organisms during the 1 $\mu$m filtration, the highest difference at the beginning of the experiment suggests that the flux of substrates from hypolimnion might have been important. Nevertheless, the order of magnitude difference between the respiration and primary production of plankton is so high as to surely represent a real trend in the epilimnetic water from Mekkojärvi. This conclusion is in agreement with the results of Kuuppo-Leinikki & Salonen (1992) who found continuously low (48–80%) saturation of oxygen in the upper epilimnion of Mekkojärvi. Moreover, in our experiment respiration was probably underestimated (see later), because very abundant and large *Daphnia longispina* probably partly evaded the sampling into the small bottles used in the determination of respiration. In two other humic headwater lakes Salonen *et al.* (1983) also found respiration to be considerably higher than phytoplankton primary production in the uppermost 2 m water column. Thus the discrepancy between primary production and respiration in our experiment was probably not due to any transient phenomenon, but a common situation in summer in that lake. It seems that only for a short period in spring is phytoplankton primary production dominating in Mekkojärvi (e.g. Salonen *et al.*, 1992).

Although a rather high percentage, the exuded primary production could support only about one percent of carbon needed for bacterial production and respiration. This proportion would be even less if the 0.02 $\mu$m filtrate, which probably better controls the passage of only truly dissolved organic matter, is taken as an estimate of exuded organic matter instead of the usual 0.2 $\mu$m filtrate. However, both exudation results are minimum estimates, since the uptake by bacteria during the incubation was not measured. Since the total primary production would have supported only 5% of the gross bacterial production, it seems to be completely out of question that the production of bacteria in the experiment was based on phytoplankton primary production. In other small humic lakes Jones & Salonen (1985) reached a similar conclusion. The difference between algal primary production and bacterial production was so large that the difference can hardly be explained by errors in conversion factors. Even the removal of high predation by zooplankton in the second tank did not result in markedly higher phytoplankton primary production. This might suggest that at the rather high temperature prevailing during the experiment algae were not able to compete favourably with bacteria for some essential elements. Predation of phytoplankton by heterotrophic flagellates is an unlikely explanation, since the dominating large species (particularly *Mallomonas* species with long spines) may not be efficiently grazed.

Heterotrophic flagellates were the main grazers of bacteria and were further eaten by *Daphnia longispina*. However, the latter was an almost equally important bacterial grazer. The proportion of bacteria (35%) it grazed was much higher than the <1–6% estimated by Sanders *et al.* (1989) for cladocerans in eutrophic Lake Oglethorpe. *D. longispina* is known as a 'high efficiency bacteria feeder' (cf. Geller & Müller, 1981) and is typical of humic lakes (Børsheim *et al.*, 1988; Hessen *et al.*, 1989). Particularly when the availability of algae is low, bacteria can be an important food source for *D. longispina* (Kankaala, 1988) Børsheim & Olsen (1984) have made similar observations on *D. pulex*. Juvenile *Daphnia* harvest bacteria and algae with almost equal efficiency, but adults (>1.5 mm) clear algal size particles at 2–10 times higher rate than bacteria (Børsheim & Andersen, 1987; Kankaala, 1988). Thus, juveniles and adults can have different food niches along the size spectrum of food particles.

The food limitation encountered by *Daphnia longispina* at the end of the experiment seems likely reason for the decrease of its population during the experiment. The biomass of *Daphnia* was not determined directly, but from other results we can estimate *ca.* 6 $\mu$g C as the mean size of animals, which would give *ca.* 1000 $\mu g\ C\ l^{-1}$ in *Daphnia* in the tank. This biomass is an order of magnitude higher than that of bacteria. The calculation of respiration of *Daphnia* using the size related equation of Lampert (1984) yields respiration of *ca.* 200 $\mu g\ C\ l^{-1}\ d^{-1}$ which is more than twice the total production of its food organisms (Fig. 5). Thus at the temperature prevailing in the experiment, zooplankton respiration would have been higher than that of bacteria. Although the filtering efficiency of bacteria by zooplankton is lower than that of flagellates, harvesting the former might still be an efficient way to get food, because respiration losses inevitable in the intermediate flagellate step are avoided. Although rough, the above calculation also makes it clear that the food consumption of *Daphnia* cannot be adequately supplied by phytoplankton. On the other hand, such high grazing pressure must to a great extent regulate species composition of phytoplankton. Hence only species with a high growth rate or with protective morphology or behaviour could survive under such circumstances. The dominating phytoplankton species can probably cope with high grazing by their spines (*Mallomonas costata* and *M. akrokomos*) or behavioural (*Cryptomonas* sp.) adaptations (Arvola *et al.*, 1992).

When primary production of phytoplankton was insufficient to sustain respiration of plankton, the only alternative energy source was the high concentration of detrital organic matter (see also Salonen & Hammar, 1986). Hessen *et al.* (1989, 1990) estimated that in a humic lake 46–82% of zooplankton carbon originated from detrital particles, which in turn comprised 73–74% of POC. In this study the estimated proportion of detrital particles was on average 25–29% of POC, but using lower end carbon conversion factors for algae and bacteria (see methods in Salonen *et al.*, 1992) it would have been as high (74–75%) as that observed by Hessen *et al.* (1989, 1990). We consider it inadequate to draw conclusions about the role of detrital particulate carbon using only indirect results obtained as a difference, because such an approach has a tendency to accumulate all errors in an unpredictable way.

Bacteria seem to be crucial in utilizing DOM and channelling it further in the food chain (cf. Tulonen *et al.*, 1992; Tranvik, 1988, 1990), although the direct use of colloidal DOM by heterotrophic and mixotrophic flagellates may also be possible (Sherr, 1988). In humic lakes, allochthonous DOM is channelled to zooplankton directly through bacteria, and indirectly through a bacteria-flagellate chain. In Mekkojärvi, and probably also in other equally humic lakes, the DOM-bacteria-flagellate food chain seems to be very important and is similar to the microbial loop described in the marine waters (Azam *et al.*, 1983). However, there is a basic difference between the two environments. In marine waters the microbial loop is largely maintained by release of DOM from phytoplankton or from higher trophic level organisms of grazing food chains. In humic lakes DOM is mainly allochthonous and thus, in spite of possibly significant loss by respiration, serves a food supply additional to that of primary production. Therefore it would be more appropriate to call this DOM-bacteria-flagellate food chain a microbial link rather than a microbial loop. The latter term should only be used when allochthonous sources of DOM are negligible.

## Conclusions

The results of this experiment clearly show that heterotrophic processes can dominate in the plankton of highly humic lakes. Among microorganisms bacteria were responsible for most respiration, but the share of *Daphnia longispina* in the total respiration was also high. The dense *Daphnia* population regulated nanoplankton species composition and population sizes and also grazed markedly on bacteria. However, the elimination of *Daphnia* did not greatly affect the grazing of bacteria, because heterotrophic flagellates took its place. Since the production of bacteria was not based on phytoplankton primary production, they serve as an additional and quantitatively important source of food for higher trophic levels, being able to supply food for high grazing by heterotrophic flagellates and cladoceran zooplankton. As most of the total DOM is probably not available for microorganisms, in further studies particular attention should be paid to factors regulating the regeneration of biologically available matter from allochthonous DOM. The importance of vertical fluxes in steeply stratified lakes such as Mekkojärvi also clearly deserves further investigation.

## Acknowledgements

This study was supported by the Academy of Finland and Maj and Tor Nessling Foundation.

We thank R. Ilola and J. Vainionpää for making the chemical determinations and R. Jones for checking the language of the manuscript.

## References

Arvola, L., K. Salonen, P. Kankaala & A. Lehtovaara, 1992. Vertical distributions of bacteria and algae in a steeply stratified humic lake under high grazing pressure from *Daphnia longispina*. Hydrobiologia 229: 253–269.

Azam, F., T. Fenchel, J. G. Field, J. S. Gray, L. A. Meyer-Reill & F. Thingstad, 1983. The ecological role of water-column microbes in the sea. Mar. Ecol. Progr. Ser. 10: 257–263.

Børsheim, K. Y. & S. Andersen, 1987. Grazing and food selection by crustacean zooplankton compared to production of bacteria and phytoplankton in a shallow Norwegian mountain lake. J. Plankton Res. 9: 367–379.

Børsheim, K. Y. & Y. Olsen, 1984. Grazing activities by Daphnia pulex on natural populations of bacteria and algae. Verh. int. Ver. Limnol. 22: 644–648.

Børsheim, K. Y., S. Andersen, G. H. Johnsen, E. O. Kambestad & S. Norland, 1988. Primary and bacterial production compared to growth and food requirements of *Daphnia longispina* in Lake Kvernavatnet, west Norway. J. Plankton Res. 10: 921–939.

Geller, W. & H. Müller, 1981. The filtration apparatus of Cladocera: filter mesh-sizes and their implications on food selectivity. Oecologia (Berl.) 49: 316–321.

Hessen, D. O., 1985. The relation between bacterial carbon and dissolved organic humic compounds in oligotrophic lakes. FEMS Microbiol. Ecol. 31: 215–223.

Hessen, D. O., T. Andersen & A. Lyche, 1989. Differential grazing and resource utilization of zooplankton in a humic lake. Arch. Hydrobiol. 114: 321–347.

Hessen, D. O., T. Andersen & A. Lyche, 1990. Carbon metabolism in a humic lake: Pool sizes and cycling through zooplankton. Limnol. Oceanogr. 35: 84–99.

James, M. & K. Salonen, 1992. Zooplankton-phytoplankton interactions and their importance in the phosphorus cycle of a polyhumic lake. Arch. Hydrobiol. (in press).

Johansson, J.-Å., 1983. Seasonal development of bacterioplankton in two forest lakes in central Sweden. Hydrobiologia 101: 71–88.

Jones, R. I. & L. Arvola, 1984. Light penetration and some related characteristics in small forest lakes in southern Finland. Verh. int. Ver. Limnol. 22: 811–816.

Jones, R. I. & K. Salonen, 1985. The importance of bacterial utilization of released phytoplankton photosynthate in two humic forest lakes in southern Finland. Holarct. Ecol. 8: 133–140.

Kankaala, P., 1988. The relative importance of algae and bacteria as food for Daphnia longispina (Cladocera) in a polyhumic lake. Freshwat. Biol. 19: 285–294.

Kuuppo-Leinikki, P. & K. Salonen, 1992. Bacterioplankton in a small polyhumic lake with an anoxic hypolimnion. Hydrobiologia 229: 159–168..

Lampert, W., 1984. The measurement of respiration. In J. A. Downing & F. H. Rigler (eds.), A manual on methods for the assessment of secondary productivity in fresh waters. IBP Handbook 17, Blackwell Scientific Publications, Oxford, 501 pp.

Moran, M. A. & R. E. Hodson, 1990. Bacterial production on humic and nonhumic components of dissolved organic carbon. Limnol. Oceanogr. 35: 1744–1756.

Peterson, B. J., 1980. Aquatic primary productivity and the $^{14}C$-$CO_2$ method: a history of the productivity problem. Annu. Rev. Ecol. Syst. 11: 359–385.

Salonen, K. & T. Hammar, 1986. On the importance of dissolved organic matter in the nutrition of zooplankton in some lake waters. Oecologia (Berl.) 68: 246–253.

Salonen, K. & S. Jokinen, 1988. Flagellate grazing on bacteria in a small dystrophic lake. Hydrobiologia 161: 203–309.

Salonen, K. & A. Lehtovaara, 1992. Migrations of haemoglobin-rich *Daphnia longispina* in a small, steeply stratified, humic lake with an anoxic hypolimnion. Hydrobiologia 229: 271–288.

Salonen, K., R. I. Jones & L. Arvola, 1984. Hypolimnetic phosphorus retrieval by diel vertical migrations of lake phytoplankton Freshwat. Biol. 14: 431–438.

Salonen, K., L. Arvola, T. Tulonen, T. Hammar, T-R. Metsälä, P. Kankaala & U. Münster, 1992. Planktonic food chains of a highly humic lake. I. A mesocosm experiment during the spring primary production maximum. Hydrobiologia 229: 125–142.

Salonen, K., K. Kononen & L. Arvola, 1983. Respiration of plankton in two small, polyhumic lakes. Hydrobiologia 101: 65–70.

Sanders, R. W., K. G. Porter, S. J. Bennett & A. E. DeBiase, 1989. Seasonal pattern of bacterivory by flagellates, ciliates, rotifers, and cladocerans in a freshwater planktonic community. Limnol. Oceanogr. 34: 673–687.

Sherr, E. B., 1988. Direct use of high molecular weight polysaccharide by heterotrophic flagellates. Nature 335: 348–351.

Tranvik, L., 1988. Availability of dissolved organic carbon for planktonic bacteria in oligotrophic lakes of differing humic content. Microb. Ecol. 16: 311–322.

Tranvik, L., 1989. Bacterioplankton growth, grazing mortality and quantitative relationship to primary production in a humic and a clearwater lake. J. Plankton Res. 11: 985–1000.

Tranvik, L., 1990. Bacterioplankton growth on fractions of dissolved organic carbon of different molecular weights from humic and clear waters. Appl. envir. Microbiol. 56: 1672–1677.

Tulonen, T., K. Salonen & L. Arvola, 1992. Effects of different molecular weight fractions of dissolved organic matter on the growth of bacteria, algae and protozoa from a highly humic lake. Hydrobiologia 229: 239–252.

*Hydrobiologia* **229**: 159–168, 1992.
*K. Salonen, T. Kairesalo & R.I. Jones (eds.),*
*Dissolved Organic Matter in Lacustrine Ecosystems: Energy Source and System Regulator.*



# Bacterioplankton in a small polyhumic lake with an anoxic hypolimnion

Pirjo Kuuppo-Leinikki[1] & Kalevi Salonen*
*Lammi Biological Station, University of Helsinki, SF-16900 Lammi, Finland (* author for correspondence);*
[1]*present address: Tvärminne Zoological Station, SF-10900 Hanko, Finland*

*Key words:* bacterioplankton, polyhumic lake, photosynthetic bacteria

## Abstract

Bacterioplankton biomass and dark fixation of inorganic carbon were measured in the highly humic (water colour up to 550 mg Pt $l^{-1}$) and acidic lake, Mekkojärvi. Strong thermal and chemical stratification developed in the water column early in spring and led rapidly to anoxia in the hypolimnion, which extended to less than 1.0 m from the surface. In the epilimnion only small bacteria were abundant. In the anoxic zone both the abundance and the mean size of bacteria were considerably higher than in the epilimnion. These differences are thought to be the result of different grazing pressure from zooplankton in the two zones. In late summer a high concentration of bacteriochlorophyll *d* in the upper hypolimnion indicated a high density of photosynthetic bacteria. Bacterial biomass was similar to that of phytoplankton in the epilimnion, but 23 times higher in the whole water column. In August, dark fixation of inorganic radiocarbon in the anaerobic zone was 51% of the total $^{14}C$-incorporation and the contribution of light fixation was only 5.4%. In the polyhumic Mekkojärvi, bacterioplankton was evidently a potentially significant carbon source for higher trophic levels, but bacterioplankton production could not be supported by phytoplankton alone. Allochthonous inputs of dissolved organic matter probably support most of the bacterial production.

## Introduction

A high concentration of dissolved organic matter (DOM) is typical of small, sheltered lakes in the northern boreal zone. Humic compounds have physical and chemical characteristics which directly or indirectly affect aquatic organisms. One of the most important factors is the strong absorption of light (Eloranta, 1978; Jones & Arvola, 1984). In spring, after melting of the ice, the absorption of light in a thin water layer results in rapid development of steep thermal and chemical stratification of the water column. In small lakes with sufficient depth this leads to incomplete vernal overturn (e.g. Salonen *et al.*, 1984). In humic lakes the euphotic layer is shallow (Jackson & Hecky, 1980; Arvola & Rask, 1984) and phytoplankton production may be restricted by light availability. However, utilization of the high concentration of DOM by decomposers means that bacterial biomass and total plankton respiration can be high in relation to phytoplankton biomass and production (e.g. Rodina, 1967; Salonen, 1981a). As a result of high decomposition, an anoxic hypolimnion is regularly found in small humic lakes (Salonen *et al.*, 1984).

This work was part of the ecosystem study of a small, sheltered, polyhumic headwater lake, Mekkojärvi. Its aim was to assess the potential role of bacteria for the food chains of Mekkojärvi.

Special attention was paid to the likely consequences for bacterioplankton of the steep stratification of the water column.

## Study site

Mekkojärvi (Fig. 1) is a small polyhumic (colour 300–550 mg Pt $l^{-1}$), naturally acidic (pH 4.6–6.2) lake in southern Finland and with sometimes incomplete spring mixing (Salonen *et al.*, 1984). The lake is surrounded by a floating Spaghnum mat and therefore has no shallow littoral. In summer the water column is vertically steeply stratified (Table 1).

## Material and methods

Sampling was performed weekly between May and August, 1984 at the middle of the lake (Fig. 1). Water samples were taken from the surface to 2.5 m depth at 0.25 m intervals using a plastic tube connected to a bottle to which vacuum was applied with a hand pump. Oxygen and temperature were measured in the field using a combined probe (Yellow Springs Instruments). Water colour was determined from the absorbance at 420 nm with a spectrophotometer (Hitachi 101). The attenuation of light in the water column was calculated (Hutchinson, 1957) as

$$I_z = I_0\, e^{-n'z},$$

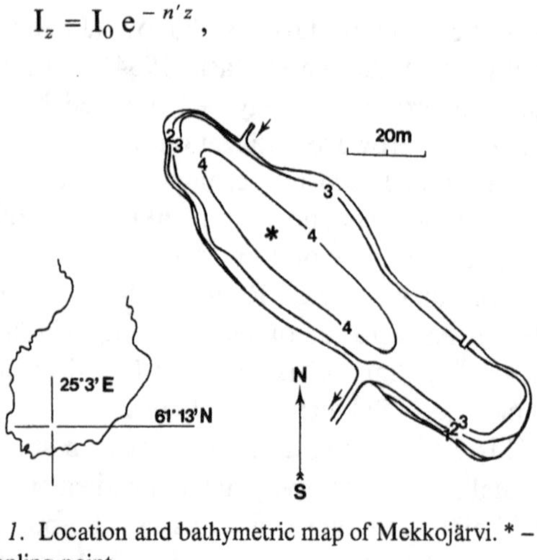


*Fig. 1.* Location and bathymetric map of Mekkojärvi. * – the sampling point.

*Table 1.* Some physical and chemical characteristics of Mekkojäri at two depths. The values are means of the results obtained between May and August, 1984.

| | 0 m | 2.5 m |
|---|---|---|
| Colour (mg Pt $l^{-1}$) | 458 | 334 |
| pH | 5.1 | 6.1 |
| Conductivity (mS $m^{-1}$ 20 °C) | 4.2 | 7.1 |
| Alkalinity (meq $l^{-1}$) | 0.025 | 0.420 |
| $NO_2 + NO_3 - N$ (µg $l^{-1}$) | 4.2 | 3.0 |
| $NH_4$-N (µg $l^{-1}$) | 21 | 518 |
| Total-N (µg $l^{-1}$) | 853 | 1378 |
| $PO_4$-P (µg $l^{-1}$) | 11 | 160 |
| Total-P (µg $l^{-1}$) | 32 | 192 |

where $I_z$ = incident light at depth $z$; $I_o$ = light at the surface; $n'$ = extinction coefficient of light, which was corrected by the water colour at each depth (Jones & Arvola, 1984).

Inorganic nutrients were determined with an AKEA autoanalyzer immediately after return to the laboratory. For phosphate we used the molybdate method (Murphy & Riley, 1962) and $NO_3 + NO_2$-nitrogen was determined as nitrite after the reduction in a cadmium-copper column (Wood *et al.*, 1967). $NH_4$-nitrogen was determined with an ammonium electrode (Orion Model 95–10). Total nitrogen and phosphorus of unfiltered water samples were determined after persulphate digestion (Koroleff, 1979).

For determination of chlorophyll, 1 liter of water was filtered through a Whatman GF/C filter and stored at −18 °C. The chlorophylls were extracted with 94 or 99.9% ethanol at +6 °C for 24 h (Arvola, 1981), and measured with a Hitachi 101 or Shimadzu UV-240 spectrophotometer using wavelengths of 665 and 750 nm for chlorophyll *a*, and 654 nm for bacteriochlorophyll *d* (Stanier & Smith, 1960; Takahashi & Ichimura, 1970).

Phytoplankton primary production and inorganic carbon dark fixation were measured with a $^{14}$C-technique (Vollenweider, 1969; Sorokin & Kadota, 1972; Overbeck, 1979). Close interval $^{14}$C-incorporation measurements were made using a sampler (Fig. 2) designed to complete all steps required in the determination *in situ* at the

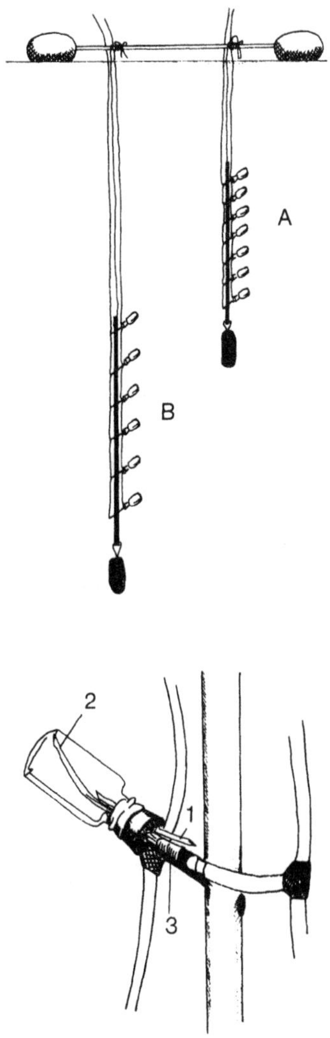


*Fig. 2.* Design of the device for making inorganic carbon uptake experiments in the hypolimnion. Upper panel – the two devices suspended in the water; lower panel – detailed drawing of the incubation bottle. 1 – Water inlet; 2 – air and water outlet; 3 – syringe for radiocarbon solution.

depth of sampling. Thus the organisms were not influenced by abnormal light intensity, and contamination of anoxic water by oxygen could be avoided. The dark $^{14}C$-fixation in the hypolimnion was measured one day later, because the close interval sampling/incubation system was available for only one vertical series. Washed and precombusted (4 h at + 500 °C) scintillation vials (20 ml) were attached to the rubber caps, and 0.5–1.0 ml $NaH^{14}CO_3$-solution was added into the injection syringes. After estimation of the depth at which the water became anoxic (using YSI oxygen meter), two samplers were slowly lowered into the anaerobic hypolimnion and the samplers were tied to a buoy. On the lower sampler the vials were located at 0.2 m and on the upper one at 0.1 m intervals (Fig. 2). The vials were first rinsed ca. 5 min by applying vacuum so that lake water flowed through the vials. Then $NaH^{14}CO_3$-solution was injected into the vials by applying pressure in the tubes attached to the syringes. The incubation was started at ca. 11.00 and continued for 24 h. After the incubation, the samplers were removed and immediately protected from direct sunlight. The scintillation vials were detached from the samplers and 25% glutaraldehyde added (final conc. 0.25%). After mixing, the vials were wrapped in aluminum foil and put in crushed ice. In the laboratory 0.5 ml water was injected under the surface of 0.5 ml of a carbon dioxide absorption solution (1 part ethanolamine, 7 parts ethyleneglycol monomethyl ether; Burnison & Perez, 1974) to determine the amount of added radioactivity. Finally the liquid volume in the vial was filled to 8 ml with deionised water. After this, inorganic carbon was measured with a carbon analyser (Salonen, 1981b) with an exhaust tube led outdoors to dispose of radioactive carbon dioxide. The incorporated radioactivity was measured after the addition of HCl into an 8 ml subsample and bubbling for 20 minutes (Schindler *et al.*, 1972). The radioactivities were measured with an LKB-Wallac Ultrobeta 1210 Liquid Scintillation Counter using Lumagel (Lumac) as scintillant.

Subsamples for the microscopic counting of bacteria were taken from the vials immediately after the determination of dissolved inorganic carbon and prepared according to Bergström *et al.* (1986), using acriflavine stain. One preparation from each sample was made the day after sampling and the slides were stored at room temperature protected from light. Earlier experience had shown that the fluorescence of acriflavine lasts for months under such conditions. Bacteria were enumerated within one week of sampling using a Nikon Optiphot epifluorescence microscope (light source HBO 100 mercury lamp, excitation 430–490 nm, objective UV-F100 N.A. 1.30 and final magnification 1250 ×). Bacteria were counted from 10 randomly chosen rectangular fields (until June 12, the number of fields was 10–40) from the center of each filter. Bacterial aggregates were sometimes observed, but they were not discriminated in the counts. The total number of cells counted on each filter was 100–300.

The linear dimensions of one hundred cells on each filter were measured with a calibrated ocular micrometer. We chose an estimate of 0.25 pg C $\mu m^{-3}$ (range in literature from 0.11 (Nagata, 1986) to 0.56 (Bratbak, 1985), respectively) to convert bacterial volume to carbon. Phytoplankton carbon biomasses were estimated from chlorophyll using a carbon to chlorophyll ratios 27 and 67, which is the range given by Riemann *et al.* (1989).

## Results and discussion

### *Physical and chemical properties*

Mekkojärvi lost its ice cover at the beginning of May and thermal stratification then developed rapidly (Fig. 3). During summer, the temperature fluctuated between 12.3 and 19.6 °C at the surface and between 3.0 and 7.3 °C below 1.5 m. Due to the shallow epilimnion, diel changes in the surface temperature of Mekkojärvi can be 10 °C in extreme cases in midsummer. Therefore the epilimnetic temperatures reported here are nearer to the minimum temperature, which occurs early in the morning, than the maximum temperature, which is generally observed at the end of the afternoon.

After ice melt, oxygen in the hypolimnion was rapidly depleted and the euphotic zone also remained undersaturated. Even during the May primary production maximum, the oxygen saturation at the surface varied between 60 and 80%. In June, July and August the respective values were 55–83, 52–80 and 48–65% indicating continuous dominance of heterotrophic processes. The anoxic layer reached its highest position in August (Fig. 3).

During the summer inorganic nutrients accumulated in the anaerobic hypolimnion (Fig. 4) where nitrates and nitrites were probably partly reduced to ammonium. The concentration of $NO_2 + NO_3$ nitrogen varied from 0 to 20 $\mu g\,l^{-1}$ (mean 3 $\mu g\,l^{-1}$), ammonium from 7 to 810 $\mu g\,l^{-1}$ (mean 193 $\mu g\,l^{-1}$) and inorganic phosphorus from 7 to 240 $\mu g\,l^{-1}$ (mean 64 $\mu g\,l^{-1}$) in the

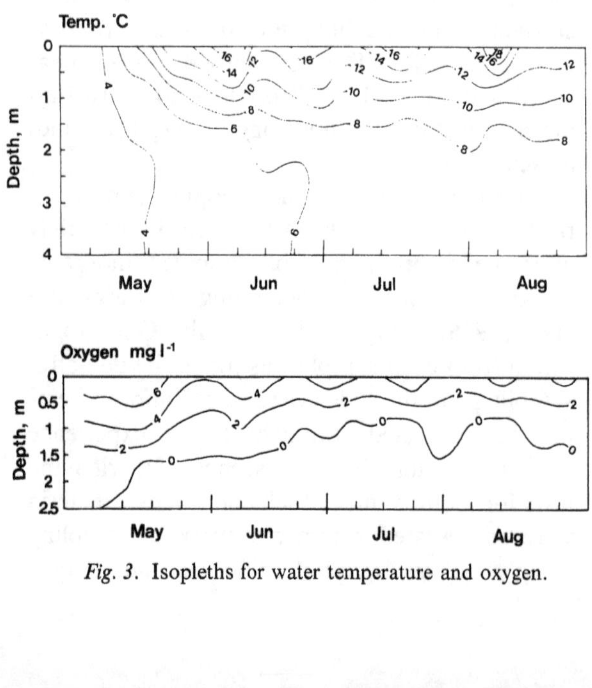


*Fig. 3.* Isopleths for water temperature and oxygen.

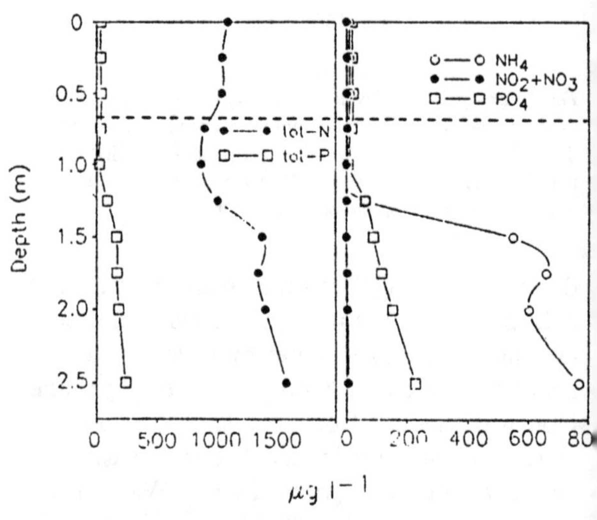


*Fig. 4.* Example of the vertical distribution of nitrogen (as N) and phosphorus (as P) on 14 August 1984. The oxycline is indicated by the dashed line.

water column during the summer. In the aerobic epilimnion the concentrations of $NO_2 + NO_3$ nitrogen were 0 to 18 $\mu g\,l^{-1}$ (mean 6 $\mu g\,l^{-1}$), $NH_4$ nitrogen 10 to 74 $\mu g\,l^{-1}$ (mean 28 $\mu g\,l^{-1}$) and phosphate phosphorus 5 to 36 $\mu g\,l^{-1}$ (mean 15 $\mu g\,l^{-1}$). Also total nitrogen (range 720–1300 $\mu g$ N $l^{-1}$) and phosphorus (20–250 $\mu g$ P $l^{-1}$, respectively) concentrations increased towards the bottom by the end of the summer. However, it is noticeable that low concentration of inorganic nutrients extended well below the epilimnion (Fig. 4).

The increase in water colour (Fig. 5) towards the end of the summer indicated increased input of DOM from the catchment area. Consequently the attenuation of light in the water column was high (Fig. 5) limiting the 1% level of surface light to a depth of 0.75–1.0 m.

*Vertical distribution of chlorophylls*

The concentration of chlorophyll *a* in 0–2.0 m water column decreased during May from 19–31 $\mu g\,l^{-1}$ to 1.1–5.5 $\mu g\,l^{-1}$ in the epilimnion (Fig. 6). After the depletion of oxygen in the 1.5–2 m water layer (Fig. 3), bacteriochlorophyll *d* developed in the hypolimnion (Fig. 6) indicating the existence of green sulphur bacteria. The maximum concentration of bacteriochlorophyll *d* developed at 1.5–1.75 m where the concentration reached 800–1000 $\mu g\,l^{-1}$. Light intensity at that depth was <0.03% of the surface light. At the lower boundary of the ca. 1 m thick maximum of green sulphur bacteria, the cells were probably limited by light, while the upper boundary was possibly controlled by the availability of $H_2S$ and cladoceran grazing. The occurrence of green sulphur bacteria rather than other types in Mekkojärvi is probably due to the strong predominance of red light (a result of the very brown colour of the water) in the anoxic hypolimnion, low light intensity and low concentration of hydrogen sulphide (e.g. Bergstein *et al.*, 1979; Parkin & Brock, 1980; 1981a; 1981b; Mazumder & Dickman, 1989). Parkin & Brock (1980a) found only bacteriochlorophyll *d* in humic Mary lake, where the proportion of surface light available for bacterial photosynthesis was of the same order (<0.02%) as in Mekkojärvi. Compared with the concentrations of bacteriochlorophyll *d* observed in Mary lake and in the other 5 lakes studied by Parkin & Brock (1980), the concentrations in Mekkojärvi were high. Our bacteriochlorophyll *d* values may be in error because we applied the formula given for acetone extraction (Takahashi & Ichimura, 1970) although we used ethanol as a solvent. However, Takahashi & Ichimura (1968) in Jap-

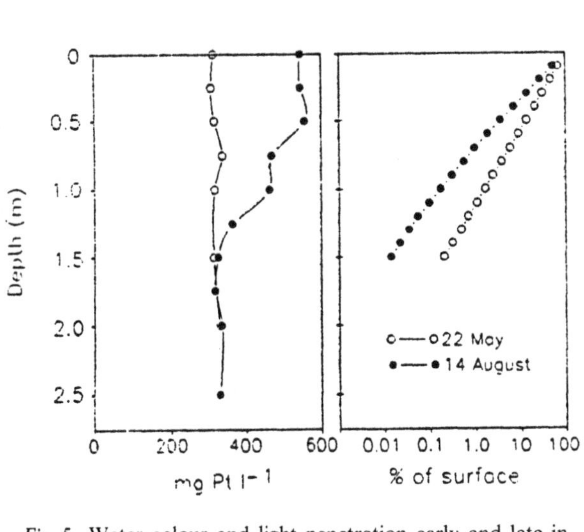


*Fig. 5.* Water colour and light penetration early and late in summer.

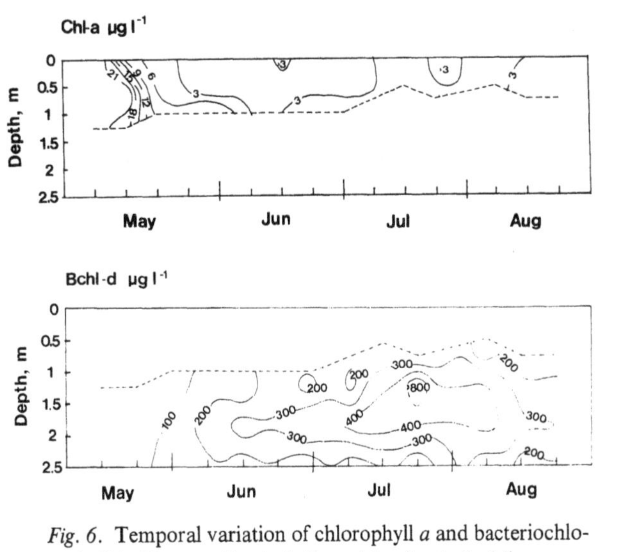


*Fig. 6.* Temporal variation of chlorophyll *a* and bacteriochlorophyll *b*. The oxycline is indicated by the dashed line.

anese lakes and Lawrence *et al.* (1978) in a meromictic salt lake observed similarly high concentrations of bacteriochlorophyll *d*. Small bacterial biomass in relation to bacteriochlorophyll *d* concentration may also suggest overestimation of pigment, but this kind of comparison is so dependent on the volume to carbon conversion factor as to be of little real value. In general there was no clear relationship between the bacterial biomasses and the concentrations of bacteriochlorophyll *d*.

*Bacterial density and biomass*

In Mekkojärvi the number of bacteria varied between $2.3 \times 10^6$ and $3.9 \times 10^7$ cells $ml^{-1}$ and the highest numbers were observed in the anoxic hypolimnion. These densities are higher than in acidic lakes with somewhat lower humic content (Salonen, 1981a; Andersson, 1983; Johansson, 1983). Bacterial abundance in lake water seems to correlate positively with humic concentration (Hessen, 1985b; Tranvik, 1988) and this might explain the high density of bacteria in Mekkojärvi.

The mean volume of bacteria varied from 0.089 to 0.295 $\mu m^3$ being within the range reported from humic lakes (e.g. Salonen, 1981a: 0.09–0.25 $\mu m^3$; Johansson, 1983: 0.10–0.35 $\mu m^3$; Hessen, 1985b: 0.076–0.144 $\mu m^3$). Small cells dominated in the aerobic epilimnion while in the anaerobic hypolimnion the proportion of large bacteria increased (Fig. 7). In spite of their high density, the contribution of small cells to the total bacterial biomass was low. In the anoxic hypolimnion, where bacteria were larger and more abundant, bacterial biomasses were several times higher than in the epilimnion (Fig. 8). Consequently epilimnetic bacteria contributed only 7–31% to the bacterial biomass of the whole water column. In the middle

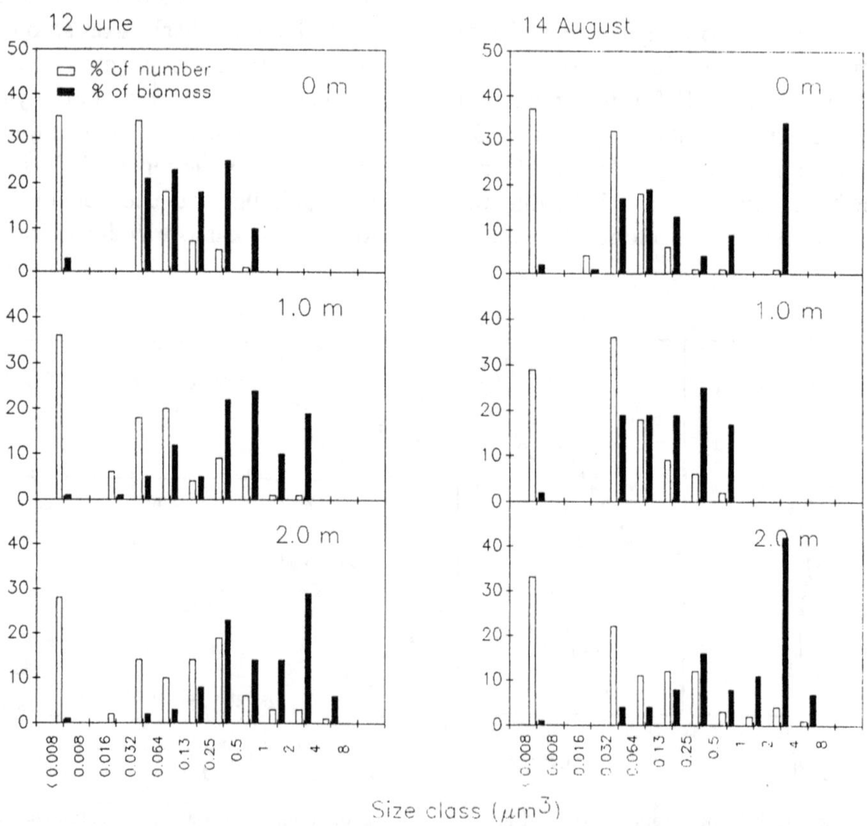


*Fig. 7.* Size distributions of bacteria at two dates.

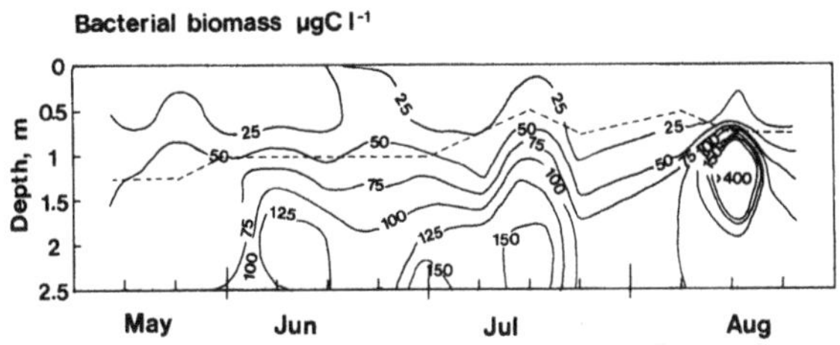


*Fig. 8.* Temporal variation of bacterial biomass in the water column. The oxycline is indicated by the dashed line.

of August bacterial biomass increased remarkably at the depth of maximum concentration of bacteriochlorophyll *d*.

### $^{14}$*C-incorporation*

In August, the inorganic carbon uptake at the surface of the aerobic epilimnion was below 100 mg C $m^{-3}$ $d^{-1}$ in light and reduced rapidly, along with decreasing light, in deeper water (Fig. 9). The measured phytoplankton primary production was within the range of values observed in other humic lakes of the same area (Arvola, 1983; Rask *et al.*, 1986). In the anaerobic hypolimnion the maximum values were similar, up to 90 mg C $m^{-3}$ $d^{-1}$ in darkness and 63 mg C $m^{-3}$ $d^{-1}$ in light. In the hypolimnion both dark and light incorporation of inorganic carbon showed similar vertical distributions. In the second experiment dark results were consistently higher than light results, possibly because the light and dark incubations were not simultaneous. The differences between light and dark $^{14}$C-incorporation and between different dates suggest that, in spite of the rather low temperature in the hypolimnion, the activity of the anaerobic microbial community can exhibit appreciable fluctuations. This aspect should be investigated in view of possible diel periodicity of bacterial photosynthesis. Bacterial photosynthesis was so low, and dark inorganic carbon uptake so fluctuating that it could not have been estimated by successive incubations.

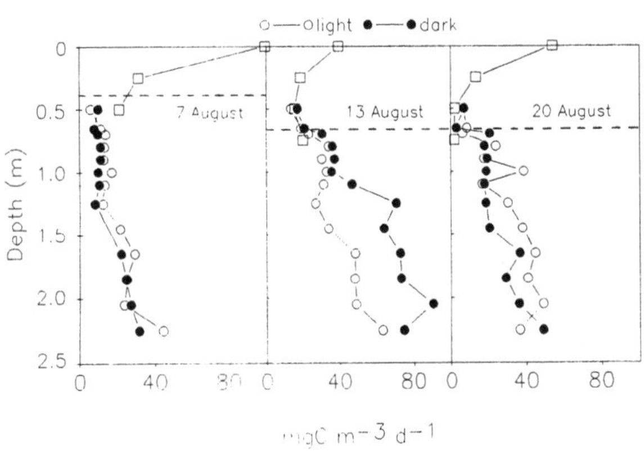


*Fig. 9.* Light and dark fixation of inorganic carbon in August. The oxycline is indicated by the dashed line. Light fixation determined in the epilimnion (open squares) using a more traditional vacuum sampling through a tube and incubation in glass bottles is shown for comparison.

High incorporation in the hypolimnion reflects various chemosynthetic and anaerobic processes (Sorokin, 1965). In general, incorporation of $^{14}$C in the hypolimnion followed the same pattern as the bacterial biomass. During summer, the plankton of Mekkojärvi became increasingly 'heterotrophic', as the illuminated and aerobic epilimnion became shallower. Because of the much higher depth of the anaerobic hypolimnion than that of the epilimnion, areal dark fixation of inorganic carbon in anoxic water was quantitatively of the same order as, or higher than, the primary production of epilimnetic phytoplankton. In different bacterial groups the proportion of cell carbon originating from inorganic carbon in water varies according to metabolic type. For aerobic bacteria the range is typically only 3–5% (Sorokin, 1965; Overbeck, 1979), whereas for chemoautotrophic and photosynthetic bacteria it can be the sole carbon source (Sorokin, 1965; Hansen, 1983). Thus it is impossible to convert dark inorganic carbon uptake in the anoxic hypolimnion to values directly comparable with phytoplankton primary production. In the literature, estimates of the photosynthetic bacterial contribution to total planktonic primary production in humic and eutrophic lakes range from 0.26–6.3% (Parkin & Brock, 1980; 1981a; Steenbergen & Korthals, 1982). In meromictic clear water lakes bacterial photosynthesis may contribute as much as 50–80% of primary production (e.g. Lawrence *et al.*, 1978; Biebl & Pfennig, 1979).

## General discussion

High bacterial biomass compared to primary production (e.g. Rodina, 1969; Salonen, 1981a; Hessen, 1985b) and phytoplankton biomass (Johansson, 1983) is typical of humic lakes. In the whole water column, bacterial biomass exceeded phytoplankton biomass on average 23-fold, whereas in the oxic epilimnion there was practically no difference (respective ratio ca. 2). In Mekkojärvi, where the coloured water attenuates light rapidly, algae might be adapted to low light intensity and hence may have developed higher chlorophyll concentrations than in those lakes from which the chlorophyll to carbon ratio used in this study (Riemann *et al.*, 1989) was obtained. If so, the difference between the estimated bacterial and algal biomasses would have been even higher.

The mortality of bacteria (Servais *et al.*, 1985) can be a consequence of spontaneous cell lysis, infection by bacteriophages (Sherr, 1989; Børsheim *et al.*, 1990), or grazing. In the first two cases, autotrophic bacteria would be indifferent, but heterotrophic bacteria would act as a 'sink' in the energy transfer of the system (Ducklow *et al.*, 1986). In Mekkojärvi the small size of epilimnetic bacteria most probably results from size selective grazing by heterotrophic nanoflagellates (e.g. Andersson *et al.*, 1983) and daphnids (Peterson *et al.*, 1978; Kankaala, 1988). Large cell size in the anaerobic layer is probably due to different metabolic bacterial groups, high concentration of inorganic and organic nutrients and low grazing pressure. Grazing by phagotrophic nanoflagellates has been suggested to be the predominant loss mechanism for bacterioplankton (e.g. Fenchel, 1982; Sanders *et al.*, 1989; Tranvik, 1989). Salonen & Jokinen (1988) studied grazing by heterotrophic nanoflagellates in the epilimnion of Mekkojärvi using latex beads as tracers and found phagotrophic nanoflagellates to be major consumers of epilimnetic bacteria, clearing daily ca. 22% of the water column. Their estimated clearance may, however, be an underestimate caused by the methods (Sieracki *et al.*, 1987).

Cladocerans are also able to feed on bacterioplankton in lakes (Riemann & Bosselmann, 1984). The dominant species in Mekkojärvi, *Daphnia longispina*, is a very efficient filter feeder which, particularly when young, is also able to collect small bacterial size particles (Brendelberger & Geller, 1985; Hessen, 1985a). According to Kankaala (1988), *D. longispina* consumed 3–48% of bacterial biomass daily in the epilimnion of Mekkojärvi; she concluded, however, that the pathway via bacterivorous flagellates (e.g. Sanders & Porter, 1990) is a more important link to zooplankton than direct grazing on bacteria. This conclusion is sensitive to the size of the factor used to convert bacterial volumes to carbon. Kankaala used a conservative conversion factor of 0.121 pg $\mu m^{-3}$ and the use of a higher factor (e.g. 0.35, Bjørnsen, 1986) would have made direct utilization of bacteria by zooplankton seem much more important. Irrespective of the pathway by which bacterial production enters zooplankton, the extremely high biomass of *D. longispina* in Mekkojärvi (Salonen & Lehtovaara, 1991) suggests that phytoplankton primary production is insufficient to support zooplankton growth and hence that bacteria play an important role in the planktonic food chains of Mekkojärvi.

Direct grazing estimates for protozoans in anaerobic waters have not generally been performed (see Sanders & Porter, 1986). Pigments of photosynthetic bacteria have sometimes been observed in the guts of zooplankters (Takahashi & Ichimura, 1968; Sorokin, 1970) and Mazumder & Dickman (1989) demonstrated that *Daphnia* can control the upper boundary of phototrophic bacterial population in a meromictic lake. In Mekkojärvi, the dense population of *D. longispina* typically stays near to the oxycline, and may dive into the upper anoxic hypolimnion (Salonen & Lehtovaara, 1991) to feed on bacterioplankton. Taking into account the proportionally high volume of anaerobic water in small humic lakes with shallow stratification, hypolimnetic production and its utilization may well prove important for the productivity of such lakes.

## Acknowledgements

This study was supported by the Academy of Finland and the University of Helsinki. We wish to thank H. Kuosa for comments on the manuscript and R. I. Jones for the revision of English.

## References


Andersson, I. B., 1983. Bacterioplankton in the acidified Lake Gårdsjön. Hydrobiologia 101: 59–64.

Andersson, A., U. Larsson & Å. Hagström, 1986. Size-selective grazing by a microflagellate on pelagic bacteria. Mar. Ecol. Prog. Ser. 33: 51–57.

Arvola, L., 1981. Spectrophotometric determination of chlorophyll *a* and phaeopigments in ethanol extractions. Ann. bot. fenn. 18: 221–227.

Arvola, L., 1983. Primary production and phytoplankton in two small, polyhumic forest lakes in southern Finland. Hydrobiologia 101: 105–110.

Arvola, L. & M. Rask, 1984. Relations between phytoplankton and environmental factors in a small, spring-meromictic lake in Southern Finland. Aqua fenn. 14: 129–138.

Bergstein, T., Y. Henis & B. Z. Cavari, 1979. Investigations on the photosynthetic sulfur bacterium Chlorobium phaeobacterioides causing seasonal blooms in Lake Kinneret. Can. J. Microbiol. 25: 999–1007.

Bergström, I., A. Heinänen & K. Salonen, 1986. Comparison of acridine orange, acriflavine, and bisbenzimide stains for enumeration of bacteria in clear and humic waters. Appl. envir. Microbiol. 51: 664–667.

Biebl, H. & N. Pfennig, 1979. Anaerobic $CO_2$ uptake by phototrophic bacteria. A review. Arch. Hydrobiol. Beih. Ergebn. Limnol. 12: 48–58.

Bjørnsen, P. K., 1986. Automatic determination of bacterioplankton biomass by image analysis. Appl. envir. Microbiol. 51: 1199–1204.

Børsheim, K. Y., G. Bratbak & M. Heldal, 1990. Enumeration and biomass estimation of planktonic bacteria and viruses by transmission electron microscopy. Appl. envir. Microbiol. 56: 352–356.

Bratbak, G., 1985. Bacterial biovolume and biomass estimations. Appl. envir. Microbiol. 49: 1488–1493.

Brendelberger, H. & W. Geller, 1985. Variability of filter structures in eight *Daphnia* species: mesh sizes and filtering areas. J. Plankton Res. 7: 473–486.

Burnison, B. K. & K. T. Perez, 1974. A simple method for the dry combustion of $^{14}C$-labeled materials. Ecology 55: 899–902.

Ducklow, H. W., D. A. Purdie, P. J. LeB Williams & J. M. Davies, 1986. Bacterioplankton: a sink for carbon in a coastal marine plankton community. Science 232: 865–867.

Eloranta, P., 1978. Light penetration in different types of lakes in Central Finland. Holarct. Ecol. 1: 362–366.

Fenchel, T., 1982. Ecology of heterotrophic microflagellates. IV. Quantitative occurrence and importance as bacterial consumers. Mar. Ecol. Prog. Ser. 9: 35–42.

Hansen, T. A., 1983. Electron donor metabolism in phototrophic bacteria. In J. G. Ormerod (ed.), The phototrophic bacteria: Anaerobic life in the light. Studies in Microbiology 4. Blackwell, Oxford: 76–95.

Hessen, D. O., 1985a. Filtering structures and particle size selection in coexisting Cladocera. Oecologia (Berl.) 66: 368–372.

Hessen, D. O., 1985b. The relation between bacterial carbon and dissolved humic compounds in oligotrophic lakes. FEMS Microbiol. Ecol. 31: 215–223.

Hutchinson, G. E., 1957. A treatise on limnology, 1. J. Wiley & Sons, N.Y., 1015 pp.

Jackson, T. A. & R. E. Hecky, 1980. Depression of primary production by humic matter in lake and reservoir waters of boreal forest zone. Can. J. Fish. aquat. Sci. 37: 2300–2317.

Johansson, J.-A., 1983. Seasonal development of bacterioplankton in two forest lakes in Central Sweden. Hydrobiologia 101: 71–88.

Jones, R. I. & L. Arvola, 1984. Light penetration and some related characteristics in small forest lakes in Southern Finland. Verh. int. Ver. Limnol. 22: 811–816.

Kankaala, P., 1988. The relative importance of algae and bacteria as food for *Daphnia longispina* (Cladocera) in a polyhumic lake. Freshwat. Biol. 19: 285–296.

Koroleff, F., 1979. Meriveden yleisimmät kemialliset analyysimenetelmät. Meri 7: 47–49.

Lawrence, J. R., R. C. Haynes & U. T. Hammer, 1978. Contribution of photosynthetic green sulphur bacteria to total primary production in a meromictic saline lake. Verh. int. Ver. Limnol. 20: 201–207.

Mazumder, A. & M. D. Dickman, 1989. Factors affecting the spatial and temporal distribution of phototrophic sulfur bacteria. Arch. Hydrobiol. 116: 209–226.

Murphy, J. & J. P. Riley, 1962. A modified single solution method for the determination of phosphate in natural waters. Analyt. chim. Acta 27: 31–36.

Nagata, T., 1986. Carbon and nitrogen content of natural planktonic bacteria. Appl. envir. Microbiol. 52: 28–32.

Overbeck, J., 1979. Dark $CO_2$ uptake – biochemical background and its relevance to *in situ* bacterial production. Arch. Hydrobiol. Beih. Ergebn. Limnol. 12: 68–47.

Parkin, T. B. & T. D. Brock, 1980. Photosynthetic bacterial production in lakes: the effects of light intensity. Limnol. Oceanogr. 25: 711–718.

Parkin, T. B. & T. D. Brock, 1981a. Photosynthetic bacterial production and carbon mineralization in a meromictic lake. Arch. Hydrobiol. 91: 366–382.

Parkin, T. B. & T. D. Brock, 1981b. The role of phototrophic bacteria in the sulfur cycle of a meromictic lake. Limnol. Oceanogr. 26: 880–890.

Peterson, B. J., J. E. Hobbie & J. F. Haney, 1978. Daphnia grazing on natural bacteria. Limnol. Oceanogr. 23: 1039–1044.

Rask, M., A. Heinänen, K. Salonen, L. Arvola, I. Bergström & M. Liukkonen, 1986. The limnology of a small, naturally acidic, highly humic forest lake. Arch. Hydrobiol. 106: 351–371.

Riemann, B. & S. Bosselmann, 1984. Daphnia grazing on natural population of bacteria. Verh. int. Ver. Limnol. 22: 795–799.

Riemann, B., P. Simonsen & L. Sondergaard, 1989. The carbon and chlorophyll content of phytoplankton from various nutrient regimes. J. Plankton Res. 11: 1037–1046.

Rodina, A. G., 1969. Bacterial population of humified lakes. Mikrobiologiya 38: 531–537.

Salonen, K., 1981a. The ecosystem of oligotrophic Lake Pääjärvi, 2. Bacterioplankton. Verh. int. Ver. Limnol. 21: 448–453.

Salonen, K., 1981b. Rapid and precise determination of total inorganic carbon and some gases in aqueous solutions. Wat. Res. 15: 403–406.

Salonen, K. & T. Hammar, 1986. On the importance of dissolved organic matter in the nutrition of zooplankton in some lake waters. Oecologia (Berlin) 68: 246–253.

Salonen, K. & S. Jokinen, 1988. Flagellate grazing on bacteria in a small dystrophic lake. Hydrobiologia 161: 203–209.

Salonen, K. & A. Lehtovaara, 1992. Migrations of haemoglobin-rich *Daphnia longispina* in a small, steeply stratified, humic lake with an anoxic hypolimnion. Hydrobiologia 229: 271–288.

Salonen, K., L. Arvola & M. Rask, 1984. Autumnal and vernal circulation of small forest lakes in Southern Finland. Verh. int. Ver. Limnol. 22: 103–107.

Sanders, R. W. & K. G. Porter, 1986. Use of metabolic inhibitors to estimate protozooplankton grazing and bacterial production in a monomictic eutrophic lake with an anaerobic hypolimnion. Appl. envir. Microbiol. 52: 101–107.

Sanders, R. W. & K. G. Porter, 1990. Bacterivorous flagellates as food resources for the freshwater crustacean zooplankter Daphnia ambigua. Limnol. Oceanogr. 35: 188–191.

Sanders, R. W., K. G. Porter, S. J. Bennett & A. E. DeBiase, 1989. Seasonal patterns of bacterivory by flagellates, ciliates, rotifers and cladocerans in a freshwater planktonic community. Limnol. Oceanogr. 34: 673–687.

Schindler, D. W., R. V. Schmidt & R. Reid, 1972. Acidification and bubbling as an alternative to filtration in determining phytoplankton production by the $^{14}C$ method. J. Fish. Res. Bd Can. 29: 1627–1631.

Servais, P., G. Billen & J. Vives-Rego, 1985. Rate of bacterial mortality in aquatic environments. Appl. envir. Microbiol. 49: 1448–1454.

Sherr, E., 1989. Aquatic viruses: And now, small is plentiful. Nature, Lond. 340: 429–430.

Sieracki, M. E., L. W. Haas, D. A. Caron & E. J. Lessard, 1987. Effect of fixation on particle retention by microflagellates: underestimation of grazing rates. Mar. Ecol. Prog. Ser. 38: 251–258.

Sorokin, Yu. I., 1965. On the trophic role of chemosynthesis and bacterial biosynthesis in water bodies. Mem. Ist. ital. Idrobiol. 18 (suppl.): 187–205.

Sorokin, Yu. I., 1970. Interrelations between sulfur and carbon turnover in meromictic lakes. Arch. Hydrobiol. 66: 391–446.

Sorokin, Yu. I. & H. Kadota, 1972. Techniques for the assessment of microbial production and decomposition in freshwaters. IBP Handbook 23. Blackwell, Oxford, 112 pp.

Stanier, R. Y. & J. H. C. Smith, 1960. The chlorophylls of green bacteria. Biochim. Biophys. Acta 41: 478–484.

Steenbergen, C. L. M., H. J. Korthals, 1982. Distribution of phototrophic organisms in the anaerobic and microaerobic strata of Lake Vechten (The Netherlands). Pigment analysis and role in primary production. Limnol. Oceanogr. 27: 883–895.

Takahashi, M. & S. Ichimura, 1968. Vertical distribution and organic matter production of photosynthetic sulfur bacteria in Japanese lakes. Limnol. Oceanogr. 13: 644–655.

Takahashi, M. & S. Ichimura, 1970. Photosynthetic properties and growth of photosynthetic sulfur bacteria in lakes. Limnol. Oceanogr. 15: 920–944.

Tranvik, L., 1988. Availability of dissolved organic carbon for planktonic bacteria in oligotrophic lakes of differing humic content. Microb. Ecol. 16: 311–322.

Tranvik, L., 1989. Bacterioplankton growth, grazing mortality and quantitative relationship to primary production in a humic and a clearwater lake. J. Plankton Res. 11: 985–1000.

Vollenweider, R. A., 1969. A manual on the methods for measuring primary production in aquatic environments. IBP Handbook 12. Blackwell, Oxford, 214 pp.

Wood, E. D., F. A. J. Armstrong & F. A. Richards, 1967. Determination of nitrate in sea water by cadmium-copper reduction to nitrite. J. Mar. Biol. Ass. U.K. 47: 23–31.

*Hydrobiologia* **229**: 169–180, 1992.
*K. Salonen, T. Kairesalo & R.I. Jones (eds.),*
*Dissolved Organic Matter in Lacustrine Ecosystems: Energy Source and System Regulator.*




# Microbial control of dissolved organic carbon in lakes: research for the future

John E. Hobbie
*The Ecosystems Center, Marine Biological Laboratory, Woods Hole, MA 02543, USA*

*Key words:* dissolved organic carbon, methods, stable isotopes, bacteria

## Abstract

The dissolved organic carbon (DOC) of lakes dominates any budget of organic carbon in these systems. Limnologists are still limited by techniques and particularly by the lack of measurements of rates of microbial transformation and use of this DOC.

There are now four different approaches to the study of the microbial control of DOC in lakes. The first is through measurements of the total DOC. Recent advances in measurement with high temperature combustion will likely lead to higher and more consistent measurements in freshwaters than previously. It is possible that a biologically active fraction may be identified. The second approach is through measurements of microbial incorporation and respiration of $^{14}C$-labeled organic matter. The kinetics of this process are well known but advances in measurement of the size of the substrate pool are still being made. The third approach is to use bacterial growth in batch or continuous flow experiments in order to understand how much of the total DOC can be decomposed by microbes. The assay in this approach may be microbial growth (thymidine incorporation, biomass changes) or change in the DOC (total concentrations, specific compounds, or fractions of the DOC by molecular weight). These methods are promising but are not developed enough for routine use. For example, growth measurements in the laboratory are all subject to experimental artifacts caused by changes in the DOC and in the bacterial populations. Finally, the fourth approach is through the use of isotopes of the natural DOC. In the sea this approach has given the age of the bulk DOC ($^{14}C$ data). In freshwaters it has great potential for differentiating between bacterial use of terrestrial DOC vs. use of algal-derived DOC ($^{13}C$ data). Stable isotopes are also useful for experimentally labeling DOC produced by algae and following the use of this material by bacteria.

## Introduction

The interactions between the dissolved organic carbon (DOC) in lakes and the bacteria remains an unanswered question beset by technical problems. On the one hand there are the problems of measurement of the origin and composition of the DOC; as chemical techniques improve we detect low levels of just about every possible compound plus the ill-defined high molecular weight humic and fulvic acids. On the other hand there are problems of growing the heterotrophic bacteria responsible for the decomposition of the DOC, of determining the growth rates of the heterotrophs *in situ*, and of finding out the controls of this decomposition. Despite the difficulties, we do know enough to state that the DOC of lakes dominates the input budget of carbon and energy. In many lakes there is the potential that as a carbon source DOC from land is more important than the primary productivity of algae. It is now time to go beyond describing potentials and use state-of-the-art techniques to determine the actual importance of the DOC and heterotrophic bacteria in controlling the structure and function of lake organisms.

The emphasis in this review is on microbial control of DOC in lakes. However, I believe that there is little difference between the general features of microbial ecology in lakes and in oceans (Hobbie, 1988). For this reason, many of my examples and ideas are drawn from ocean studies because they apply equally well to lake studies. Another belief is that the whole field of microbial ecology is still an immature and new branch of science. One bit of evidence is that the entire field of microbial ecology is still techniques limited, so much so that each technique advance has produced a large step up in our understanding. Thus, this review is organized around techniques, what various techniques have told us and what the new techniques will be.

There are now four different approaches to the study of the microbial control of DOC in lakes. The first is through measurements of the total DOC. Here, the concentration of DOC is measured in the lake and in the input and outlet streams or over a seasonal cycle to understand the changes wrought by microbes. The second is through measurements of microbial incorporation and respiration of $^{14}C$-labeled organic matter. This could be total DOC or individual compounds. This approach produces a turnover time of individual compounds of the DOC or of a fraction such as the DOC from algae but has not yet produced information on the total use. The third approach is to use bacterial growth in batch or continuous flow experiments in order to understand how much of the total DOC can be decomposed by microbes. The assay in this approach may be microbial growth (thymidine incorporation, biomass changes) or change in the DOC (total concentrations, specific compounds, or fractions of the DOC by molecular weight). Finally, the fourth approach is through the use of isotopes of the natural DOC. In the sea this approach has given the age of the bulk DOC ($^{14}C$ data). In freshwaters it has great potential for differentiating between bacterial use of terrestrial DOC vs. use of algal-derived DOC ($^{13}C$ data).

**Measurements of total DOC concentration**

The DOC in natural waters has been measured by wet oxidation (persulfate or photo-oxidation with UV), by dry combustion (drying, then combustion), or by direct injection into a tube in a furnace. The most common methods, persulfate and direct injection, appeared to agree and most of our understanding has been based upon results of these methods. The few 'strange' results (Gordon & Sutcliffe, 1973; Romankevich, 1984) for ocean samples were many-fold higher and until recently were mostly ignored. There was no evidence of errors in freshwater.

The single conclusion that arises from measurements of the total DOC concentration in lakes is that there is relatively small change with depth and season. For example, Wetzel (1975) measured concentrations throughout the year in Lawrence Lake and found mostly 5–7 mg DOC $l^{-1}$. In Mirror Lake, the range was very small, 2.4–3.0 mg DOC $l^{-1}$ (Likens *et al.*, 1985). We now understand that the internal cycling of DOC that goes on in a lake is intense and accounts for much more carbon than is present at any one time or than enters the lake in streams. In Mirror Lake, for example, Cole (1985) measured an input of 12 g C $m^{-2}$ $y^{-1}$ from outside the lake and an internal source of more than 20 g C $m^{-2}$ $y^{-1}$. Thus, we have learned little about the microbial control of DOC by this DOC measurement approach. A new method has now appeared that bears investigation and has the potential for providing information on the biotically available DOC.

For ocean studies the revolution has come through the new method developed by Sugimura & Suzuki (1988). As summarized in Williams & Druffel (1988), the method consists of injecting 200 $\mu$l of filtered, acidified seawater onto a catalyst (3% Pt on $Al_2O_3$) at 680 °C and measuring the $CO_2$ by IR absorption. The results (Fig. 1) indicate that this high temperature oxidation method (HTCO) gives up to 500% higher concentrations, especially in the upper waters. Are these numbers real or is there some artifact? One suggestion is that other gases produced in the HTCO might appear to be $CO_2$ in the IR analyzer but Saupe *et al.* (1990) have confirmed similar measurements with a technique that uses a dried sample and closed-tube combustion with a Pt catalyst followed by a gas separation and measurement of the $CO_2$ as pressure (a mass spectrometer technique). At the present time there is still caution in the oceanographic community but a growing number of scientists now are obtaining similar results in their own laboratories.

The first reaction of ecologists to these unexpected data was to assume that the newly found DOC was not biologically available. In fact the reverse seems to be true, although it is not clear why decomposer organisms can use a portion of the DOC that persulfate oxidation can not break down. For example, in Fig. 1 the DOC data from the HTCO are inversely proportional to the apparent oxygen utilization (AOU), a measure of the oxidation of organic matter and reduced ions such as ammonium. From this, Sugimura & Suzuki (1988) concluded that the decomposition of organic matter accounted for the distribution of the AOU in the ocean. A later paper is less cautious and Suzuki *et al.* (in press) suggest that the difference between the HTCO results and the persulfate results is due to a biologically available fraction. The most striking evidence comes from Suzuki & Tanoue (1991) (Fig. 2). Here, the difference between the HTCO method and the persulfate method is large in the surface waters of the open ocean but is reduced to about a 20% difference inside a coral reef. This type of difference is also found between near-shore (20%) and offshore waters off China (Suzuki *et al.*, in press).

My interpretation of the data is that in the euphotic zone of very oligotrophic waters the bacteria are not active enough to immediately break down all of a certain fraction of the DOC being produced by biotic processes (photosynthesis of algae, sloppy feeding of zooplankton, decomposition of algae) and organic matter accumulates. When the water parcel is transported out of the euphotic zone into deeper waters, the fraction is slowly broken down and not replenished (which gives the inverse relationship DOC:AOU). In biologically richer coastal waters, the bacterial breakdown keeps pace with production and little of this fraction accumulates. Similarly, a coral reef is also an area of high microbial activity so that the break-down of this fraction of the DOC

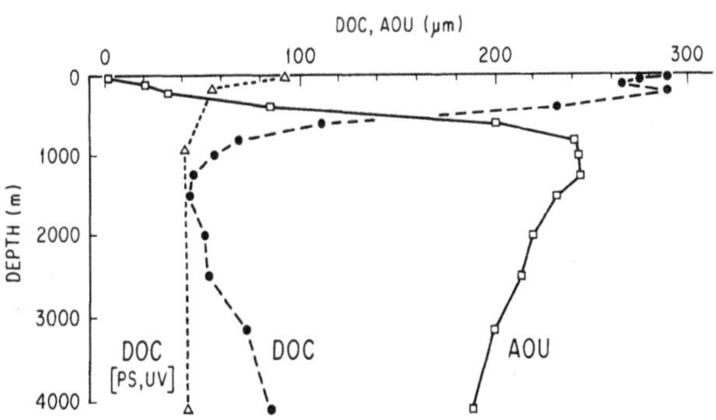


*Fig. 1.* Profiles of DOC and apparent oxygen utilization (AOU) as measured by Sugimura & Suzuki (1988) and persulfate and UV data (PS + UV) from Williams & Druffel (1987). The figure has been redrawn from Figures 1 and 2 of Williams & Druffel (1988).

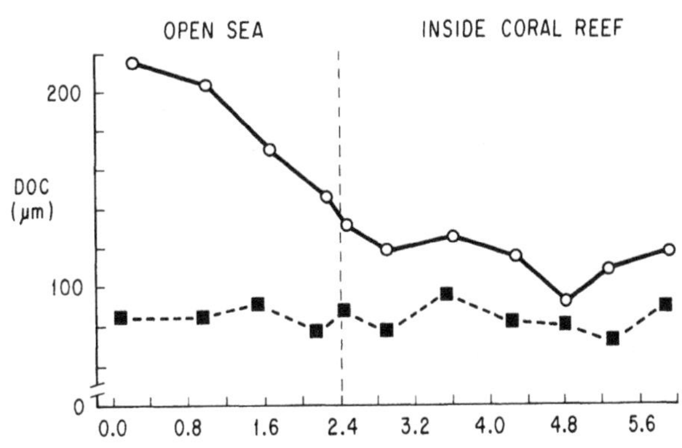


*Fig. 2.* The surface water amounts of DOC outside and inside a coral reef (Majuro Island) in the Pacific (redrawn from Suzuki & Tanoue 1991). The HTCO values (O--O) are from the high temperature combustion oxidation and the PS values (■--■) from persulfate oxidation.

is suddenly accelerated when open ocean water passes over a reef.

How much of this oceanic scenario applies to freshwater? The evidence so far suggests some cause for concern, but that errors will be closer to 20% than to 300%. One study (Sakamota & Miyasaka, 1987) compared results from a high temperature (Beckman 915B, 950 °C) analyzer with those from a persulfate-UV analyzer (Astro 1800). Almost all organic compounds gave equal efficiencies of oxidation in both machines. The only exceptions were humic and fulvic acids where the persulfate-UV method results were 30–60% those of the high temperature machine. Lake and stream water had the same result (persulfate-UV:high temperature combustion = 49%). Suzuki believes that the Pt catalyst is the key to this method, so the Beckman and Suzuki methods may not be comparable. He also has some (unpublished) data for samples from the freshwater section of the Delaware River (USA) where he found that results of the persulfate method were 89% lower than those of the HTCO method. In contrast to the Sakamoto & Miyasaka (1987) data, Lee & Wakeham (in press) report that P. M. Williams thinks that the UV and the persulfate oxidation techniques do oxidize humic and fulvic acids well.

I conclude that comparisons must be made between the persulfate methods and the new Suzuki method for freshwater samples. If the differences found in oceanic waters are caused by a slowdown of bacterial decomposition in oligotrophic systems, then perhaps only oligotrophic lakes will show these extreme differences. Most freshwaters have high bacterial activity compared with that of the oligotrophic oceans but there may still be the 20% error found by Suzuki even in productive waters. It is also possible that there is a temperature related seasonal asynchrony between production and uptake; that is, it is well known that photosynthesis is relatively insensitive to temperature and that large algal blooms occur at cold temperatures. Bacteria, in contrast, are usually sensitive to temperature and, as a result, are not very active below 10 °C. Therefore, it may be worth seeing if there is a springtime, cold water buildup of the biologically available fraction as measured by Suzuki's method. By measuring the release of $^{14}$C-DOC during photosynthesis experiments, Bell & Kuparinen (1984) found just such a buildup during the spring algal bloom (3.5 °C) in Lake Erken; 44 $\mu$g DOC $l^{-1}$ $d^{-1}$ was released by algae but bacteria only took up 1.7 $\mu$g DOC $l^{-1}$ $d^{-1}$.

### *Microbial incorporation and respiration of $^{14}$C-labeled organic matter*

The first insight into the dynamics of the bacteria-DOC relationship came from the techniques using the uptake of $^{14}$C-labeled organic compounds. Parsons & Strickland (1962) suggested that the kinetic approach could be used; it was modified for freshwater by Wright & Hobbie (1966). Today there are two radioisotope approaches. The first is a tracer approach which involves adding labeled substrate to water samples at concentrations well below those found in nature and using the fractional uptake to calculate a turnover rate ($t$) (Williams & Askew, 1968). The second is a kinetic approach which measures the uptake of added isotope at several concentrations of substrate and calculates $t$, a maximum velocity of uptake ($V_m$), and a value for the half-saturation constant plus substrate ($K + S$). Figure 3 illustrates a theoretical example with two separate types of bacteria (or one type with two different

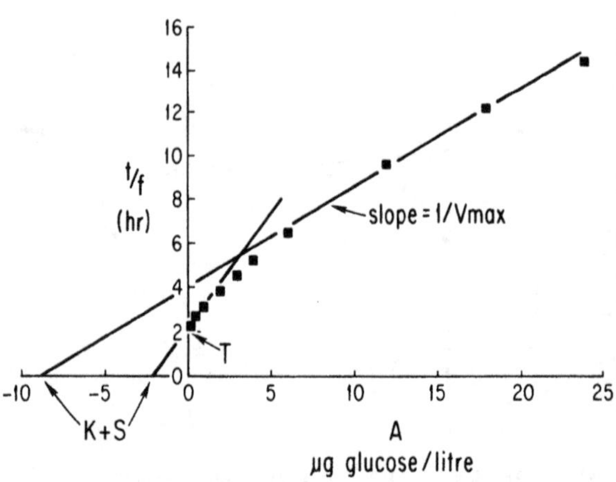


*Fig. 3.* The inverse of the fractional uptake (incubation time divided by fractional uptake) of added $^{14}$C-glucose (A) by two bacterial uptake systems. The example is imaginary. Reprinted courtesy of Academic Press from Hobbie (1990).

transport systems). Only the points close to the natural substrate levels of 2 $\mu$g glucose $l^{-1}$ should be used for the estimates. But errors can also arise if too low concentrations are used (Fig. 4). Here, Fuhrman & Ferguson (1986) have added such low levels that a different physiological property of the membrane is likely being measured by the three points closest to the ordinate. They measured 4.5 $\mu$M serine in the water sample (Table 1), quite close to the $K + S$ extrapolation of the line if the bottom two points are ignored. These techniques measure the turnover of these pools of specific organic compounds even though the concentrations could not be measured. The general topic of the dynamics of these pools has been reviewed by Wright (1984).

It is by now well known that the bacteria of lakes are capable of turnover rates and take up all of the substrate, such as specific sugars, amino acids, and acetate, in a few hours or in hundreds of hours. These turnover rates are directly related to the general productivity of the lake. The next step is to combine either of these radioisotope methods with an analysis of the concentration of substrate in order to calculate the actual rate of substrate uptake into bacteria (Table 1). In the last decade the HPLC has made a number of analyses of individual compounds possible at the nM level without concentration of the water (Lee & Wakeham, in press). For example, Lindroth & Mopper (1979) developed a technique for amino acids and R. E. Hodson of the University of Georgia (personal communication) has developed one for carbohydrates in natural waters (a separation by HPLC and detection by pulsed amperometry on a Dionex Ion Chromatograph). These new methods are a great improvement over some of the non-specific methods for dissolved carbohydrates or amino acids. These earlier methods have given data that yield diel cycling rates of carbohydrates that are 100 times the isotope methods (e.g., Burney *et al.*, 1981). The earlier experiments need to be repeated with the new chemical methods (see Wright, 1984, for a discussion).

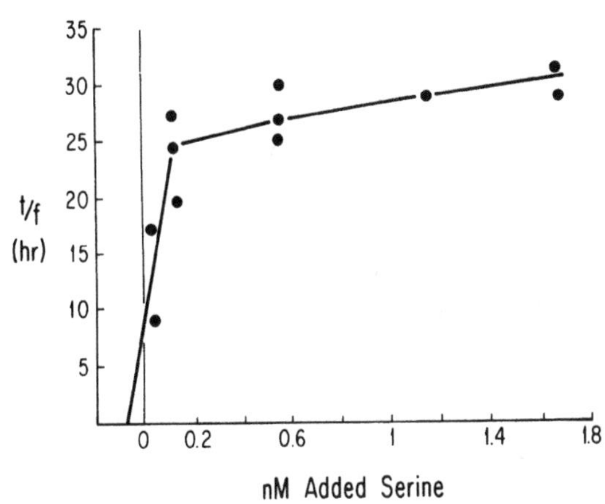


*Fig. 4.* The uptake of $^{3}$H-serine by bacteria in a water sample from New York Bight in Feb 1984. The figure is modified from Fuhrman & Ferguson (1986) and reprinted from Hobbie (1990) courtesy of Academic Press.

*Table 1.* Concentrations of amino acids and estimates of $K + S$ and turnover times at various stations in Long Island Sound, NY (extracted from Fuhrman & Ferguson 1986).

| Date | Amino acid | $K + S$ nM | HPLC conc. nM | Turnover h | Use rate $\mu$g C $l^{-1}$ $h^{-1}$ |
|---|---|---|---|---|---|
| Feb 1984 | Glutamic | 5.6 | 2.9 | | |
| | Serine | 1.4 | 4.5 | | |
| | Alanine | 6.6 | 3.0 | | |
| 2 Jun 1984 | Alanine | 6.9 | 4.2 | 0.4 | 0.38 |
| 7 Jun | Alanine | 7.7 | 14.5 | 0.9 | 0.58 |
| 15 Jun | Alanine | 31.4 | 5.1 | 0.8 | 0.23 |
| 29 Jun | Alanine | 9.2 | 7.4 | 1.2 | 0.22 |

One recent finding of interest has been Fuhrman & Ferguson's (1986) demonstration that there is close agreement between the $K + S$ estimates and the HPLC measures of concentration of amino acids (Table 1). This finding is in contrast to those of earlier workers who sometimes found that the $K + S$ was smaller than the measured substrate. I conclude (in agreement with Wright, 1984) that earlier chemical methods were not as precise as the authors believed and that the conclusion of earlier authors that a portion of the chemically-measured substrate pool was not biologically available is not correct.

The research using the $^{14}$C-labeled organic compounds has been concerned with the bacterial removal of specific compounds from solution. The pools of these compounds are virtually constant in size; that is, uptake is apparently balanced by supply rate (Wright, 1984). A logical next question is the source of these compounds constantly entering the pools of compounds. Are they released from the bursting of algal cells after death or from leaking or excretion during photosynthesis? Or are they being formed from the decomposition of macromolecules such as proteins? Hollibaugh & Azam (1983) examined the mechanisms of protein utilization by bacteria in seawater and found that radiolabeled protein was decomposed by a particle-bound enzyme system, presumably on the surface of bacteria cells. The decomposition products are taken up by bacteria and may even be transported as peptides (Kirchman & Hodson (1986) demonstrated transport of dipeptides such as phenyanyl-phenylalanine). The question of source has just begun to be studied.

A different approach uses the $^{14}$C-DOC produced by phytoplankton as a tracer of microbial utilization. One of the best studies of this type, by Larsson & Hagström (1982) in the Baltic, concluded that of the yearly phytoplankton production of 160 g C m$^{-2}$, 26 g C m$^{-2}$ (16%) was released as DOC. Bacterial production was estimated to be 38 C m$^{-2}$ from all sources. The entire pool of DOC, 6 mg DOC l$^{-1}$, is 1000 times larger than the exudate pool (Fig. 5). Bacteria take up each day most of the DOC produced during that day. The conclusion from this study was that most of the bacterial production came from the algal DOC and a smaller fraction came from DOC from other sources. Wiebe & Smith (1977) also concluded that the consumption of the algal-produced DOC was so tightly coupled to consumption that there would be little or no accumulation (a turnover time of 8.6–9.6 h).

## Bacterial growth in batch or continuous flow experiments

The total amount of DOC available to the microbes can also be determined by measuring the growth of bacteria in nature and translating that number to DOC removed or by growing bacteria in batch or continuous culture and measuring the

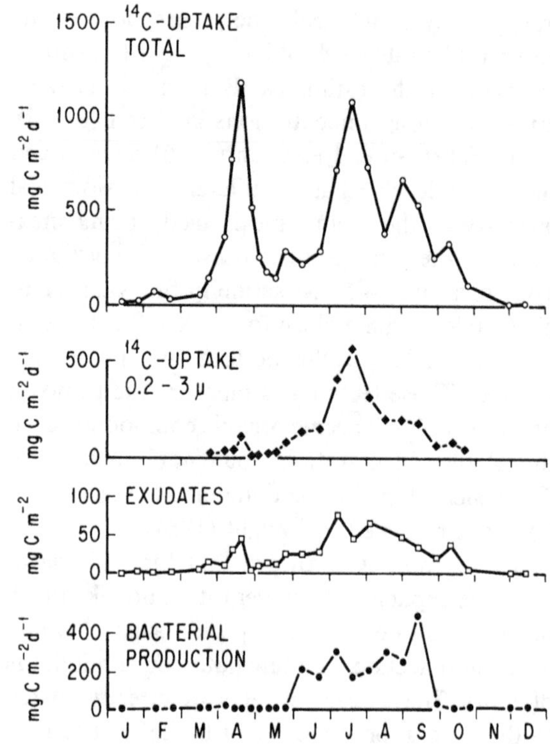


*Fig. 5.* Carbon-14 uptake in different size fractions, production of algal exudates and bacterial production in the trophic layer in the Baltic, 1978 (modified from Larsson & Hagström 1982).

DOC removed or amount of respiration and bacterial growth.

Culture studies have only become feasible in recent years. One problem was that bacteria cultured in the laboratory would not grow at the low substrate concentrations found in nature. Even when great care was taken to obtain cultures from nature and to use continuous culture techniques, the bacteria would not grow at low substrate concentrations (Jannasch, 1969). Another problem was that natural bacteria do not grow in plate culture so there was no way to count the bacteria that might grow. Not until the development of direct counting of bacteria with the epi-illuminated fluorescent microscope was it learned that bacteria grew well in natural waters. The secret may be in the use of the natural mix of many species instead of pure culture (Hagström *et al.*, 1984). In this approach, growth is studied by filtering natural lake or seawater through a 0.2 μm pore sized membrane filter, inoculating the filtrate with a small amount of 0.6 μm filtrate, and allowing the bacteria to grow over several days.

A seawater culture experiment carried out by Ammerman *et al.* (1984) showed doubling times for the microbes of 9–18 h, similar to that in the original seawater. The bacteria reached a density of $2 \times 10^6$ $ml^{-1}$ before stopping growth and removed about 5.3% of the DOC (estimated from number and size of the bacteria). The authors concluded that there were enough organic nutrients in seawater to support 1–2 doublings of the microbial assemblage.

In freshwaters, Tranvik (1988) and Tranvik & Höfle (1987) studied the bacterial removal of DOC in clear and humic lakes through the use of batch cultures. In a series of oligotrophic humic lakes in Sweden, about 9.5% of the DOC was consumed while in two lakes in Germany 15–22% of the DOC was consumed (Fig. 6). The bacteria seemed equally versatile in the two lakes as the growth rates were equal even when the sources and inocula were switched (Fig. 6). One unexpected finding was that the number of bacteria increased with increasing concentrations of DOC in the series of lakes (Tranvik, 1988); in other words, the percentage of available DOC was the same in spite of the range of DOC concentrations (5–31 mg DOC $l^{-1}$).

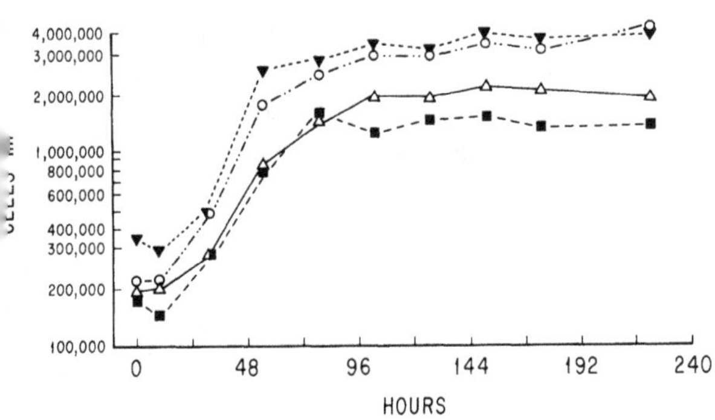


*Fig. 6.* Bacterial cell counts in batch cultures from a humic and a clear lake (modified from Tranvik & Höfle (1987)). The open circles are humic medium and humic inoculum, the inverted triangles are humic medium and clear-water inoculum, the squares are clear-water medium and clear-water inoculum and the open triangles are clear-water medium and humic inoculum.

A completely different approach is to measure over the entire year the growth of bacteria in short-term incubations (up to an hour), and then calculate the amount of DOC consumed from the bacterial growth rates. There are two techniques for the direct measurement of bacterial growth; both make use of the incorporation of radiolabeled precursors into vital components of bacteria that must be synthesized before division. The first of these techniques (Fuhrman & Azam, 1982) measures the rate of incorporation of $^3$H-thymidine into DNA. Bacteria do incorporate externally supplied thymidine through a salvage pathway and the amount of thymidine incorporated may be translated into bacterial DNA and then into bacterial biomass. The rates of production measured in nature with this technique are reasonable and for the most part fit with ecological constraints such as the total amount of organic matter available. However, the technique as presently developed is not perfect and must be used with care. One problem is that sometimes there is an increase in DNA without any change in the number or biomass of the bacteria. Another problem is in the translation from the amount of DNA produced to the bacterial biomass. So far, it has not been possible to determine the actual amount of DNA in the small bacteria that live in the plankton so the factors used are based on laboratory cultures of bacteria. There have been attempts to calibrate the technique using natural bacteria but often the calibration factors are not believable (as pointed out by Hobbie & Cole, 1984). A third problem is that some of the $^3$H from the thymidine ends up in protein. Quimby *et al.* (1990) found that 50% of the label appeared in protein in Chesapeake Bay while more than 90% appeared in DNA in the open ocean. The result of these and other problems is that the thymidine technique must be used cautiously; it is not yet well enough developed to be used for calculating the amount of DOC removed.

The second promising technique measures the incorporation of $^3$H-leucine into bacterial protein. Kirchman *et al.* (1985) pointed out that most of the heterotrophic bacteria take up 10 nM leucine and incorporate it directly into protein. Recent measurements by Simon & Azam (1989) confirm that leucine is a constant fraction of the total amino acids in bacterial protein and that protein is a constant percent of bacterial dry weight and carbon. Thus, leucine incorporation may be translated into increase in carbon (growth). A careful comparative study of the leucine and thymidine techniques by Ducklow (Quimby *et al.*, 1990, personal communication) showed that when an algal bloom occurred the bacteria first incorporated a great deal of leucine (presumably as they built up their enzymes), then incorporated thymidine (as they built up their DNA), then reproduced. My conclusion is that leucine incorporation is a very promising technique and may replace the thymidine method.

In order to complete the translation of bacterial growth into DOC used, the growth yield of the bacteria must be known. This is important because a 20% yield means that 2 $\mu$moles of bacteria carbon are produced by the decomposition of 10 $\mu$moles of DOC while a 50% yield means that 5 moles of each are produced. Presently the field is split. Measurements of the fate of $^{14}$C-labeled organic compounds, such as amino acids, sugars, and acetate, reveal that 50–60% of the carbon is incorporated (summarized in Williams, 1984). In contrast, measurements of growth of a small inoculum of native bacteria in lake and estuary water after filtration reveals a 20% efficiency (Tranvik, 1988; Bjornsen, 1986). This topic is discussed further below.

## Isotopes of natural DOC

Microbes grow on the DOC in lakes and thereby control the chemical composition and quantity of the DOC. But DOC in lakes comes from both autochthonous and allochthonous sources so a fundamental question for study must be not only the amount but also the source of the DOC used by the planktonic microbes. One source, the DOC released from phytoplankton, is believed by some to be sufficient for most of the bacterial growth. When data from 70 studies of lakes and oceans were compared, Cole *et al.* (1988) found a significant correlation between bacterial production and phytoplankton production. Bacterial production of biomass averages 20% of phytoplankton production so the numbers are reasonable. In one specific case of eutrophic lakes in southern Sweden, for example, DOC from phytoplankton fueled 32 to 95% of the bacterial production (Coveney, 1982). Is this reliance on algal carbon always the case or do the bacteria also use DOC from land or from emergent vegetation? Some of the answers may come from measurements of the distribution of the stable isotope $^{13}$C (see Peterson & Fry (1987) for background information on stable isotope use in ecological studies).

Stable carbon isotopic ratios in the bacteria may be used as tracers of the source of bacterial production if there are two photosynthetic sources that have different ratios. These differences arise because plants derive $CO_2$ from sources with different $\delta^{13}C$ values (aquatic or atmospheric) and because different pathways for carbon fixation (C-3 or C-4) yield different $\delta^{13}C$ values. The example in Fig. 7 comes from an experiment in an estuary but the principle is the same for freshwater systems. The main problem is to collect and measure only the bacterial $\delta^{13}C$ in a water sample where many small non-living particles are the same size as bacteria. One way to do this is through a growth bioassay (Peterson *et al.*, 1990). Here, Coffin *et al.* (1989) collected a water sample from a Massachusetts estuary and filtered it through a 0.2 $\mu$m filter to remove the particulate matter. Then, 1% of the original sample water was filtered through a 1.0 $\mu$m pore-sized filter and added as an inoculum; the bacteria were allowed to grow for a day or more before being filtered onto GF/F filters for isotope analysis.

The results from the estuarine study (Fig. 7) show that bacteria do, indeed, reflect the $\delta^{13}C$ of their food whether it is glucose or leachate from oak leaves. Bacteria grown in the unknown DOC from the river, the top set of points, show a strong influence of either upland plant or marsh grass DOC depending upon the tide and time of year. However, it must be realized that the $\delta^{13}C$ values for the bacteria are one step removed from values for the bacteria in nature. It is not possible to

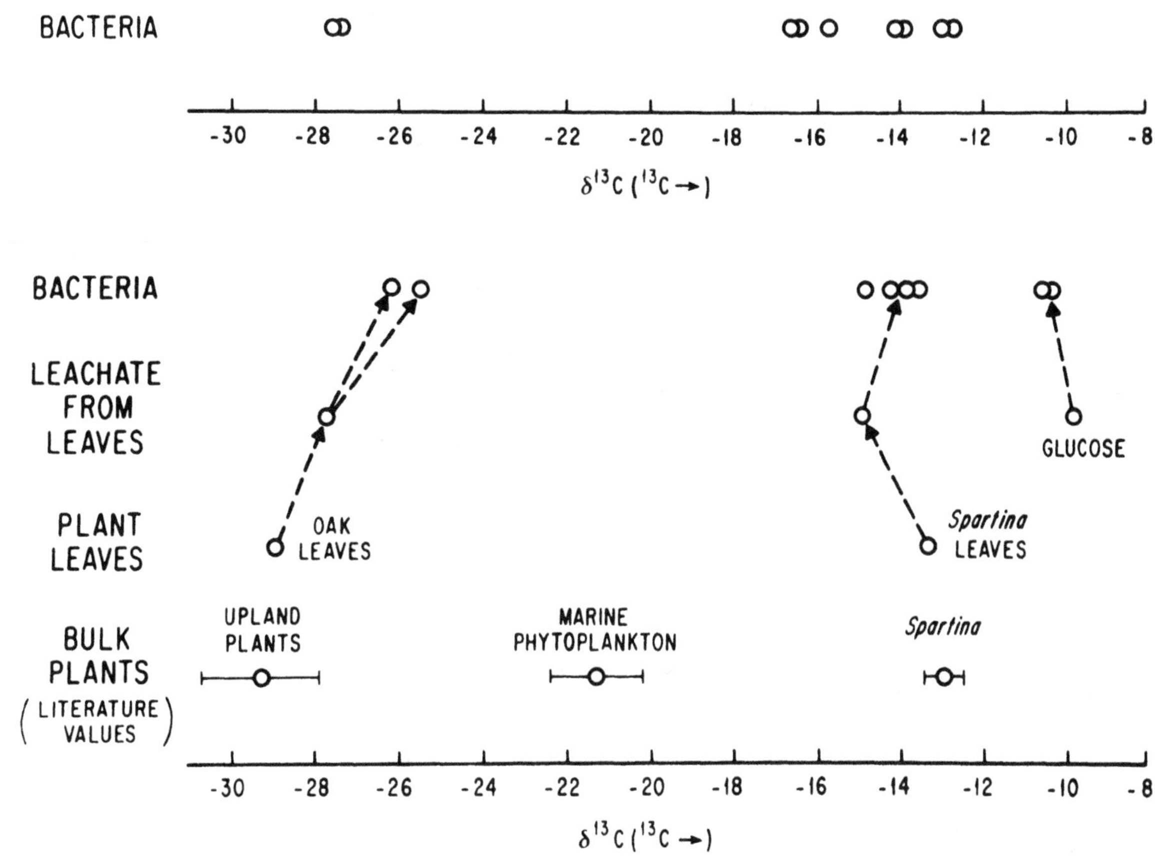


*Fig. 7.* Above: $\delta^{13}C$ values for bacteria grown on DOC from the Parker River estuary. Below: $\delta^{13}C$ values for bacteria grown on Parker River water with added substrates of glucose, leaf leachate from land plants (C-3), and *Spartina* leaves (a saltmarsh C-4 grass). Reprinted from Coffin *et al.* (1989) with the permission of the American Society of Limnology and Oceanography.

separate the bacteria from small particles and colloidal DOC by filtration, but are there other ways?

One promising way to obtain the $\delta^{13}C$ of the natural bacteria is through extraction and concentration of the bacterial DNA and RNA. Coffin *et al.* (1990) have developed a method in which water samples are filtered through a 1 $\mu$m pore-sized filter, the particles concentrated from 50 l to 1 l with a tangential flow filtration, the particles centrifuged at 10 000 g, and the nucleic acids extracted with chloroform. After extraction, the nucleic acids were precipitated with ethanol. Testing of the nucleic acids with a molecular probe (16S rRNA hybridization) gave the relative proportion of eucaryotic and procaryotic nucleic acids; less than 10% was from eucaryotes.

The nucleic acid extraction and the growth bioassay measurements gave generally similar results for an estuary in Florida (Table 2). In the waters near the Gulf Breeze Lab and near Range Point, the marine sources of carbon are dominant. There is some indication that the Range Point samples, near the seagrass beds ($\delta^{13}C$ of −3 to −13‰), are slightly heavier. The samples from the Perdido Estuary show the influence of terrestrially derived carbon ($\delta^{13}C$ of −31 to −28‰) and of a pulp mill whose effluent enters the estuary. Obviously the bacteria were making use of some of the terrestrially-derived DOC.

An experimental manipulation of a Baltic Sea mesocosm gave further insights into the role of DOC from different sources. In this experiment, run in plastic bags containing 1 $m^3$ of water, aged DOC from leaf leachate (hardwoods) was added

*Table 2.* The $\delta^{13}C$ values (‰) for bacteria concentrated from bioassay experiments with estuarine waters and for nucleic acids extracted from the same water samples (from Coffin *et al.*, 1990). All samples are from estuaries close to the Gulf Breeze Environmental Research Laboratory, Pensacola, Florida.

| Sample location | Growth bioassay | Nucleic acids |
|---|---|---|
| Gulf Breeze Lab, photoplankton major source | - 24.2<br>- 22.3 | - 22.6<br>- 21.1 |
| Range Point, seagrass source | - 20.5<br>- 21.4 | - 20.2<br>- 21.2 |
| Perdido Estuary, upstream of pulp mill | - 26.8 | - 26.4 |
| Perdido Estuary, near pulp mill | - 31.9 | - 27.5 |
| Perdido Estuary, 5 km down stream | - 29.9 | - 27.9 |

to increase the ambient DOC by 5.4 mg DOC $l^{-1}$ (total = 11.4 mg DOC $l^{-1}$) and $^{13}C$-bicarbonate was added to increase the $\delta^{13}C$ of the dissolved inorganic carbon (DIC) to +292‰. Nutrients were also added and one bag was darkened. Both bags were incubated in the Baltic near the Askö Laboratory of the University of Stockholm from 11–30 August 1989 (Fig. 8).

One source of carbon was the phytoplankton algae. After 13 days, the phytoplankton algae had picked up the $^{13}C$ from the DIC and were quite heavy. As a result, the POC in the light bag had a $\delta^{13}C$ of +42‰. It is not possible to tell exactly what the $\delta^{13}C$ of the algae was because of the presence of the non-labeled particulate matter in the sample. A minimum value comes from the copepods, which feed on the algae. The high value of a copepod sample on 30 August indicates that the algae were at least at +74‰. In the future, a chlorophyll extraction will be used for algae.

The other source was the DOC from the Baltic and from terrestrial hardwood leaves. The $\delta^{13}C$ of this aged leaf material was −28‰, a typical value for land plants. The values for Baltic DOC are unknown at this point. A best guess is that it will be a little light, perhaps −26 to −27‰, because of the inflow of DOC in rivers.

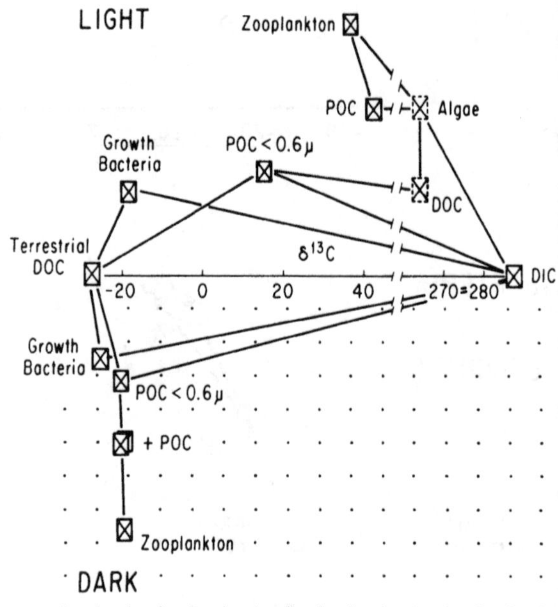


*Fig. 8.* The $\delta^{13}C$ values (‰) of various carbon pools in two (one light, one dark) 1 $m^3$ bag experiments in the Baltic. Both bags received extra DOC (about double) and both received added $^{13}C$-bicarbonate. The experiment began on 11 August 1989 and the samples were collected on 24 August.

In the light bag experiment (Fig. 8), the bacteria probably took up the algal DOC and their $\delta^{13}C$ was heavy (around +70‰?). There is some evidence for this from the POC of the particles less than 0.6 μm which was +15‰. However, when the algae were filtered off and bacteria allowed to regrow, the $\delta^{13}C$ of the bacteria was −18 to −20‰. It appears that there is very little of the algal DOC left for the regrowth and the bacteria are using a combination of a small amount (less than 10%) of the algal DOC and a large amount of leaf + Baltic DOC.

In the dark bag experiment (Fig. 8), the copepods, the POC, and the bacteria after regrowth are all similar in their $\delta^{13}C$ to the leaf and Baltic DOC pools. The very heavy DIC with a $\delta^{13}C$ of +292‰ had no apparent influence on the bacteria and copepods.

The implication of these results from the bag experiments are the following:

1. Bacteria of the plankton are so tightly coupled in their growth to the release of DOC from phy-

toplankton that nearly all this DOC is taken up within a few hours. This confirms the work of Larsson & Hagström (1982) and many others.
2. In the absence of algal DOC, bacteria do make use of the recalcitrant DOC pool which here is leaf DOC plus the background Baltic DOC.
3. Growth yield studies such as those of Bjornsen (1986) and Tranvik & Höfle (1987) uncouple any algal production of DOC and thus measure the growth yield on the recalcitrant DOC only; they miss the growth on phytoplankton-produced DOC. It is possible that bacteria in the plankton use both sources at different efficiencies but the close positive correlation between bacterial and algal production argues that almost all of the bacterial growth comes from the algal DOC. Measurements need to be made of the $\delta^{13}C$ of bacterial nucleic acids in algal blooms and in experiments with increased $\delta^{13}C$-bicarbonate to find out the predominant source of bacterial carbon in nature.
4. Bacteria use little of the allochthonous DOC of lakes.

## References


Ammerman, J. W., J. A. Fuhrman, A. Hagström & F. Azam, 1984. Bacterial growth in seawater. I. Growth kinetics and cellular characteristics in seawater cultures. Mar. Ecol. Prog. Ser. 18: 31–39.

Bell, R. T. & J. Kuparinen, 1984. Assessing phytoplankton and bacterioplankton production during early spring in Lake Erken, Sweden. Appl. envir. Microbiol. 48: 1221–1230.

Bjornsen, P. K., 1986. Bacterioplankton growth yield in continuous seawater cultures. Mar. Ecol. Prog. Ser. 30: 191–196.

Burney, C. M., P. G. Davis, K. M. Johnson & J. M. Sieburth, 1981. Dependence of dissolved carbohydrate concentrations upon small scale nanoplankton and bacterioplankton in the western Sargasso Sea. Mar. Biol. 65: 289–296.

Coffin, R. B., B. Fry, B. J. Peterson & R. T. Wright, 1989. Carbon isotopic compositions of estuarine bacteria. Limnol. Oceanogr. 34: 1305–1310.

Coffin, R. B., D. J. Velinsky, R. Devereux, W. A. Price & L. A. Cifuentes, 1990. Stable carbon isotope analysis of nucleic acids to trace sources of dissolved substrate used by estuarine bacteria. Appl. envir. Microbiol. 56: 2012–2020.

Cole, J., 1985. Decomposition. In G. E. Likens (ed.) An Ecosystem Approach to Aquatic Ecology. Springer-Verlag, New York, Berlin, Heidelberg, Tokyo: 302–310.

Cole, J. J., S. Findlay & M. Pace, 1988. Bacterial production in fresh and saltwater ecosystems: A cross-system overview. Mar. Ecol. Prog. Ser. 43: 1–10.

Coveney, M. F., 1982. Bacterial uptake of photosynthetic carbon from freshwater phytoplankton. Oikos 38: 8–20.

Fuhrman, J. A. & F. Azam, 1982. Thymidine incorporation as a measure of heterotrophic bacterioplankton production in marine surface waters: Evaluation and field results. Mar. Biol. 66: 109–120.

Fuhrman, J. A. & R. L. Ferguson, 1986. Nanomolar concentrations and rapid turnover of dissolved free amino acids in seawater: Agreement between chemical and microbiological measurements. Mar. Ecol. Prog. Ser. 33: 237–242.

Gordon, D. C. & W. H. Sutcliffe, 1973. A new dry combustion method for the simultaneous determination of total organic carbon and nitrogen in seawater. Mar. Chem. 1: 231–244.

Hagström, Å., J. W. Ammerman, S. Henrichs & F. Azam, 1984. Bacterioplankton growth in seawater, II. Organic matter utilization during steady-state growth in seawater cultures. Mar. Ecol. Prog. Ser. 18: 41–48.

Hobbie, J. E., 1988. A comparison of the ecology of planktonic bacteria in fresh and salt water. Limnol. Oceanogr. 33: 750–764.

Hobbie, J. E., 1990. Measuring heterotrophic activity in plankton. In R. Grigorova & J. R. Norris (eds), Methods in Microbiology, Volume 22. Academic Press, London: 235–250.

Hobbie, J. E. & J. J. Cole, 1984. Response of a detrital foodweb to eutrophication. Bull. mar. Sci. 35: 357–363.

Hollibaugh, J. T. & F. Azam, 1983. Microbial degradation of dissolved proteins in seawater. Limnol. Oceanogr. 28: 1104–1116.

Jannasch, H. W., 1969. Estimations of bacterial growth rates in natural waters. J. Bact. 99: 156–160.

Kirchman, D. L. & R. E. Hodson, 1986. Metabolic regulation of amino acid uptake in marine waters. Limnol. Oceanogr. 31: 339–350.

Kirchman, D. L., E. K'Ness & R. Hodson, 1985. Leucine incorporation and its potential as a measure of protein synthesis by bacteria in natural aquatic systems. Appl. envir. Microbiol. 49: 599–607.

Larsson, U. & Å. Hagström, 1982. Fractionated phytoplankton primary production, exudate release, and bacterial production in a Baltic eutrophication gradient. Mar. Biol. 67: 57–70.

Lee, C. & S. G. Wakeham. Organic matter in the water column. Future research challenges. Mar. Chem. (in press).

Likens, G. E., J. S. Eaton & N. M. Johnson, 1985. Physical and chemical environment. In G. E. Likens (ed.) An Ecosystem Approach to Aquatic Ecology. Springer-Verlag, New York, Berlin, Heidelberg, Tokyo: 89–108.

Lindroth, P. & K. Mopper, 1979. High performance liquid chromatographic determination of subpicomole amounts of amino acids by pre-column fluorescence derivatization with o-phthaldialdehyde. Analyt. Chem. 51: 1667–1674.

Parsons, T. R. & J. D. M. Strickland, 1962. On the production of particulate organic carbon by heterotrophic processes in sea water. Deep Sea Res. 8: 211–222.

Peterson, B. J. & B. Fry, 1987. Stable isotopes in ecosystem studies. Ann. Rev. Ecol. Syst. 18: 293–320.

Peterson, B. J., B. Fry, M. Hullar, S. Saupe & R. Wright, 1990. A stable isotope approach for tracing sources of DOC used by estuarine bacteria. EOS 71(2): 190.

Quimby, H., H. W. Ducklow, E. R. Peele & K. A. Glowersen, 1990. 3H-thymidine incorporation and bacterial production in Chesapeake Bay and the North Atlantic Ocean. EOS 71: 109.

Romankevich, E. A., 1984. Geochemistry of organic matter in the ocean. Springer-Verlag, Berlin, 334 p.

Sakamoto, T. & T. Miyasaka, 1987. TOC analysis. Study confirming the accuracy of a method for measuring TOC by wet oxidation. Ultrapure Water, December 1987: 24–31.

Saupe, S., M. Hullar, B. Fry & B. J. Peterson, 1990. A method for determining carbon isotopic compositions of DOC. EOS 71(2): 190.

Simon, M. & F. Azam, 1989. Protein content and protein synthesis rates of planktonic marine bacteria. Mar. Ecol. Prog. Ser. 51: 201–213.

Sugimura, Y. & Y. Suzuki, 1988. A high-temperature catalytic oxidation method for the determination of non-volatile dissolved organic carbon in seawater by direct injection of a liquid sample. Mar. Chem. 24: 105–121.

Suzuki, Y. & E. Tanoue, 1991. Dissolved organic carbon enigma – Implications for ocean margins. In R. F. Mantoura, J. M. Martin & R. F. Wollast (eds), Ocean Margin Processes in Global Change, Report of the Dahlem Workshop, Berlin, March 18–23, 1990. Wiley, New York.

Suzuki, Y., E. Tanoue & Y. Sugimura. Distribution of dissolved organic carbon in subtropical and tropical Pacific oceanic waters. Deep-Sea Res. (in press).

Tranvik, L. J., 1988. Availability of dissolved organic carbon for planktonic bacteria in oligotrophic lakes of differing humic content. Microb. Ecol. 16: 311–322.

Tranvik, L. J. & M. Höfle, 1987. Bacterial growth in mixed cultures on dissolved organic carbon from humic and clear waters. Appl. envir. Microbiol. 53: 482–488.

Wetzel, R. G., 1975. Limnology. W.B. Saunders Company, Philadelphia, London, Toronto, 743 p.

Wiebe, W. J. & D. F. Smith, 1977. Direct measurement of dissolved organic carbon release by phytoplankton and incorporation by microheterotrophs. Mar. Biol. 42: 213–223.

Williams, P. J. leB., 1984. Bacterial production in the marine food chain: The emperor's new suit of clothes. In M. J. R. Fasham (ed.) Flows of Energy and Materials in Marine Ecosystems. Plenum Press, New York, London: 271–299.

Williams, P. J. leB. & C. Askew, 1968. A method of measuring the mineralization by micro-organisms of organic compounds in sea-water. Deep Sea Res. 15: 365–375.

Williams, P. M. & E. R. M. Druffel, 1987. Radiocarbon in dissolved organic matter in the central North Pacific Ocean. Nature 330: 246–248.

Williams, P. M. & E. R. M. Druffel, 1988. Dissolved organic matter in the ocean: Comments on a controversy. Oceanography 1: 14–17.

Wright, R. T., 1984. Dynamics of pools of dissolved organic carbon. In J. E. Hobbie & P. J. leB. Williams (eds) Heterotrophic Activity in the Sea. Plenum Press, N.Y. and London: 121–154.

Wright, R. T. & J. E. Hobbie, 1966. The use of glucose and acetate by bacteria and algae in aquatic ecosystems. Ecology 47: 447–464.

*Hydrobiologia* **229**: 181–198, 1992.
*K. Salonen, T. Kairesalo & R.I. Jones (eds.),*
*Dissolved Organic Matter in Lacustrine Ecosystems: Energy Source and System Regulator.*




# Gradient-dominated ecosystems: sources and regulatory functions of dissolved organic matter in freshwater ecosystems

Robert G. Wetzel
*Department of Biology, The University of Alabama, Tuscaloosa, Alabama 35487-0344, U.S.A.*

## Abstract

The emergent wetland and littoral components of the land-water zone are functionally coupled by the amounts and types of dissolved organic matter that are released, processed, transported to, and then further processed within the recipient waters. Operational couplings and integrations in freshwater ecosystems occur along physical and metabolic gradients of a number of scales from micrometer to kilometer dimensions. The operation and turnover of the microbial communities, largely associated with surfaces, generate the metabolic foundations for material fluxes along larger-scale gradients.

Because of the predominance of small, shallow freshwater bodies, most dissolved organic carbon (DOC) of lacustrine and riverine ecosystems is derived from photosynthesis of higher plants and microflora associated with detritus, including sediments, and is only augmented by photosynthesis of phytoplankton. As the dissolved organic compounds generated in the wetland and littoral interface regions move toward the open-water regions of the ecosystems, partial utilization effects a selective increase in organic recalcitrance. Even though DOC from allochthonous and from interface sources is more recalcitrant than that produced by planktonic microflora, decomposition of the much larger interface quantities imported to the pelagic zone dominates *ecosystem* decomposition. The observed high sustained productivity of the land-water interface zone results from extensive recycling of essential resources (nutrients, inorganic carbon) and conservation mechanisms. On the average in lakes and streams, greater than 90 percent of the decomposition in the *ecosystem* is by bacteria utilizing DOM from non-pelagic sources of primary productivity. In addition to direct mineralization of DOC from non-pelagic sources, many of the organic compounds function indirectly to influence metabolism. New evidence is presented to demonstrate formation of complexes between humic and fulvic organic acids and extracellular enzymes. These complexes inhibit enzyme activity and can be transported within the ecosystem. The complex can be decoupled by mild ultraviolet photolysis with regeneration of enzyme activity in displaced locations.

## Introduction

When requested to evaluate the functional couplings of the land-water interface zone with the dissolved organic matter of freshwater ecosystems, I was first confronted with the extent to which our understanding has changed since I did a similar review twice at approximately decade intervals in the past (Wetzel, 1968, 1975, 1983). Clearly much progress has resulted from the intensive interest in dissolved organic matter, what this large reservoir of chemical energy does in fresh waters, and how these compounds affect metabolism of the biota at many levels of organization within diverse ecosystems. My ensuing remarks are applicable to all aquatic ecosystems, marine and freshwater, lotic and lacustrine. Although the loadings to and the structure of these ecosystems differ, they function similarly.

Despite much progress, the traditional conventional philosophy of food web structure and operation still largely prevails, in which phytoplankton is purported to provide the organic matter needed to drive the ecosystem, and support higher trophic levels and decomposition. This scenario is certainly only a partial and variable truth, and in many cases and times of the year the phytoplankton route of energy flux is minor. I have provided recently a major treatment of the functional importance of the biota of zones between land and water (Wetzel, 1990a). There I argued that ecosystem analyses of both standing and flowing fresh waters have ignored the functional significance of the biota of the land-water interface zones to the whole ecosystem metabolism and operation. Further, much of the nutrient loading, cycling, and recycling within many fresh waters is controlled by the metabolism of the wetland and littoral macrophytes and associated epiphytic and detrital microflora. Lastly, the high productivity of these interface wetland and littoral plants generates very large amounts of organic matter that can be transported to the open water regions and effect major changes to pelagic metabolism. That organic matter is relatively recalcitrant to rapid decomposition and enters the ecosystems largely in dissolved forms. The large but slow decomposition of this dominating soluble organic pool drives ecosystem metabolism and provides a stability that could not usually prevail with the much more ephemeral biota of the trophic dynamic structure components of freshwater ecosystems (Wetzel, 1975, 1983, 1984, 1990a).

The ensuing remarks offer a number of statements and hypotheses about the manner in which aquatic ecosystems operate. Supporting evidence is strong for many statements; in other cases evidence is suggestive and in need of much more rigorous and extensive investigation. In many cases, for various reasons, strongly supporting evidence is simply ignored; I argue vigorously that the realities of complex gradient-dominated ecosystems cannot be avoided if we are to manage aquatic ecosystems effectively and prudently.

## Sources and decomposition of dissolved organic matter

### *Gradient-dominated freshwater ecosystems*

In order to evaluate the sources and decomposition of organic matter in aquatic ecosystems, one must appreciate the fundamental properties of gradients within these systems. *The operational couplings and integrations in freshwater ecosystems are driven by gradients at both physical and metabolic levels.*

Physical gradients abound at many spatial scales, from ecosystem chemical components to microgradients at cell-water boundary layers. The sources of organic matter clearly depend upon primary productivity within and adjacent to the aquatic ecosystem. Net organic productivity of primary producers along the gradient from land to open-water pelagic zones has been emphasized previously (Fig. 1). Clearly the primary productivity of emergent aquatic plants and epiphytic algae of submersed plants is manifold greater than that of plant communities on land or of phytoplankton of the pelagic zone. These areal comparisons must be interpreted in terms of the areas within the freshwater ecosystems occupied by the different plant communities. A prevailing but commonly incorrect assumption in limnology is that the pelagic area is much larger than the land-water interface regions, and thus the phytoplankton functions as the low but collectively dominant source of organic matter to lakes as well as among certain streams, rivers, and reservoirs.*

The drainage basin often provides the dominant loading of organic matter to streams and rivers. What portion of the 'drainage basin' is within the wetland-littoral category is highly variable with different geomorphological and flood-stage conditions (Wetzel, 1990a). The extent of soil saturation and inundation is critical to the efficacy of inorganic and organic chemical loadings to the recipient streams. Floodplain macro-

* Reference is made to surface phenomena. High loadings of dissolved organic matter from volcanic hydrothermal processes, as is apparently the case in Lake Tanganyika (Degens & Ittekkot, 1983), is not the typical case in most fresh waters.

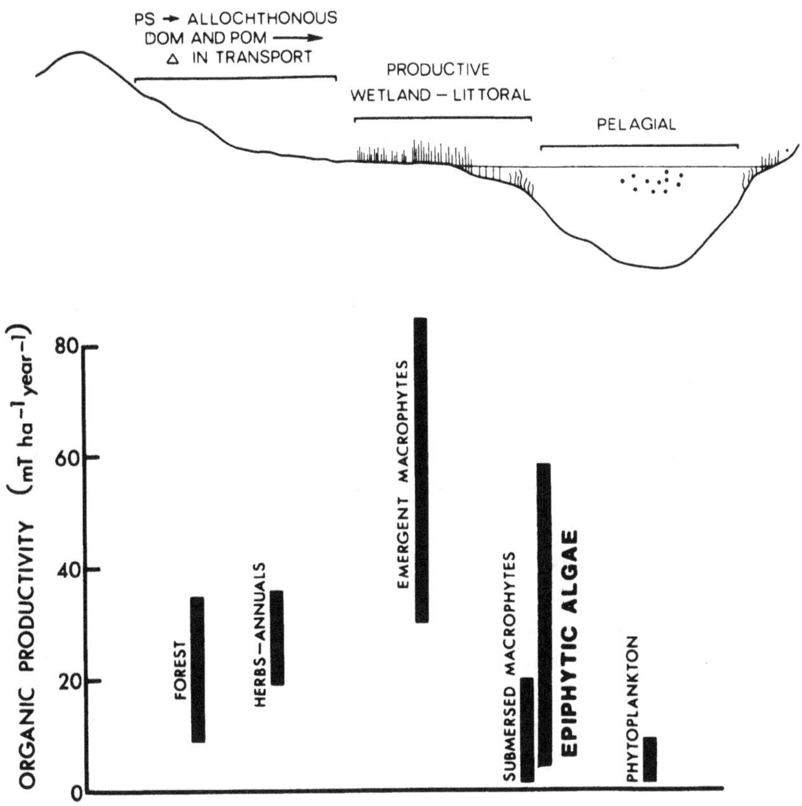


*Fig. 1.* The lake ecosystem, with the drainage basin of terrestrial synthesis (PS) of organic matter, movement of nutrients and dissolved (DOM) and particulate (POM) organic matter in surface and groundwater flows toward the lake basin, and chemical and biotic alteration of these materials en route, particularly as they pass through the productive and metabolically active wetland-littoral zones of the lake *per se* (net organic productivity in metric tons per hectare per year). (Modified from Wetzel, 1983).

phyte productivity and the metabolism of microflora associated with these large deposits of particulate organic detritus regulate the fluxes of much of the organic matter to the river waters.

The interface zone of reservoirs shifts away from emergent plants toward dominance by the littoral zone of floating, floating-leaved, and submersed macrophytes, and attendant epiphytic and epipelic microflora (Wetzel, 1990a, 1990b). The more productive emergent macrophytes of wetlands adjacent to reservoirs are often restricted or excluded because artificial fluctuations in water levels are often too great for physiological tolerances of desiccation and/or flooding of the plants. Emergent wetlands are therefore commonly restricted to the inflow regions of reservoirs where water tables at the interface zones are moderately stable.

Extant lake basin morphology is obviously critical to the development of the emergent wetland and littoral zone plant interface (Wetzel, 1983, 1990a). The gradient of slope from the land to the water must be sufficiently small to allow organic accumulation and sedimentation as substrata for wetland and littoral plant colonization, as well as penetration of light to tissues of rooted plants and their attached microbial communities. The mean depth of an average lake is less than half (0.46) of its maximum depth (Håkanson, 1983). Thus, if depths are modest, most lakes have a large proportion of their basin in shallow waters and potentially available for colonization by rooted or attached photosynthetic biota of much greater productivity than phytoplankton.

*Area and depth of freshwater bodies*

Lacustrine morphometry has been evaluated elsewhere (Wetzel, 1983, 1990a); I summarize only the salient points here. Most lake basin depressions were formed by geological processes that have resulted in predominantly shallow basins. As a result, most standing fresh waters are shallow with mean depths much less than 10 meters (Fig. 2). In large areas of the Northern Hemisphere millions of permanent water bodies exist with mean depths of less than three meters. Net sedimentation rates per volume are nearly always greater in shallow lake basins than in deeper lakes; in deeper waters, organic matter is exposed to high redox conditions for long periods of time and decomposition is often more intensive. Most of these shallow basins can support macrophytic vegetation over a large proportion of their basins. Surrounding drainage basins are commonly flat and emergent wetlands can extend nearly continuously in areas between the lakes. In arctic and subarctic regions, mean depths of literally millions of basins are frequently less than one meter (Hobbie, 1980).

The ratio of the wetland and littoral land-water interface zones to the pelagic (cf. modified definitions of Hutchinson in Wetzel, 1983) permits estimation of the relationship of areas of the wetland and littoral regions to the pelagic zone (Fig. 3). On the basis of world distribution, the littoral zone dominates over the pelagic (low P/L ratios) among most standing water ecosystems (Wetzel, 1990a). Addition of the wetland components to the littoral zone results in very low ratios (P/W + L) for most standing water bodies. A similar very low ratio often exists among flowing waters because the floodplains are functioning as major wetland and littoral components of most river ecosystems.

*Fate of organic carbon of the interface zones*

Upland terrestrial and wetland/littoral aquatic plants provide a dominant source of organic matter loading to the collective aquatic ecosystem. Much of the organic matter produced by the terrestrial and macrophytic aquatic plants enters a particulate detrital pool; herbivory of the plants is consistently very low ($< 10\%$). Decomposition of the particulate detrital pools undergoes marked successional and temporal stages and is regulated by numerous physical (particularly temperature), chemical (redox, nutrient availabilities), and biotic (microfloral successional patterns) factors (cf. Wetzel, 1983, 1990a). Transport of the detrital particulate organic matter away from sites

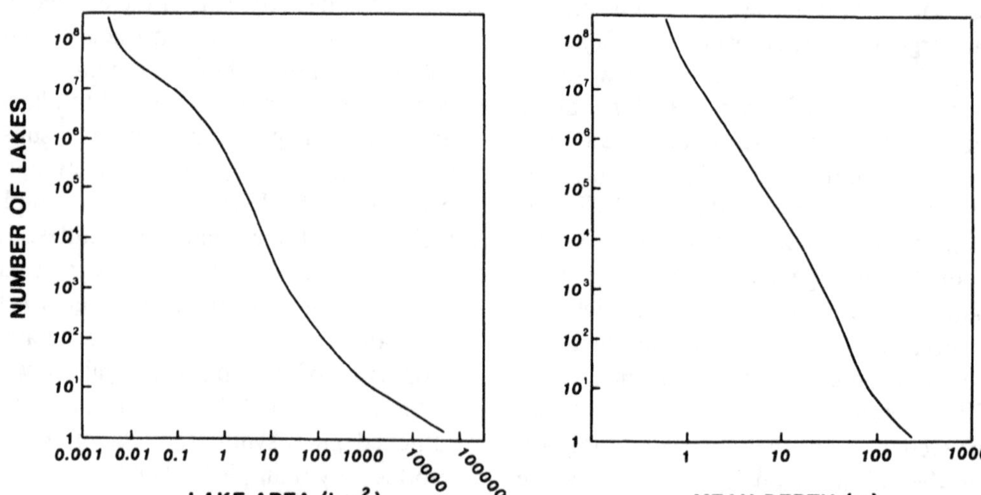


*Fig. 2.* Number of lakes of the world in relation to lake area and approximate mean water depth. (From Wetzel, 1990a).

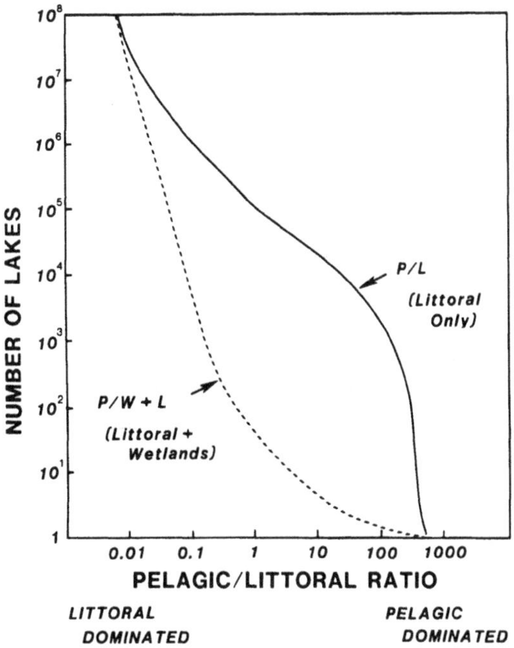


*Fig. 3.* Number of lakes of the world in relation to ratios of pelagic to littoral and wetland regions. (From Wetzel, 1990a).

of immediate production is usually small; most of the particulate organic matter remains in deposits undergoing decomposition near sites of production along the gradient of productivity. Although erosive molar action by inflowing water can occur during and following major precipitation events, particularly in river floodplains, under most conditions inflowing water passes through the particulate detrital deposits.

The dominant exportation of organic matter from the plant detrital masses is as dissolved organic compounds from various stages of decomposition. As these dissolved organic compounds pass through the detrital masses within the wetlands and through the submersed macrophytes and their epiphytic microbial communities, a selective removal of organic compounds occurs. In addition to biotic utilization, certain compounds (e.g., amino compounds) can adsorb to inorganic particles, such as to clays and calcium carbonate. As a result, the recalcitrance of the dissolved organic carbon compounds from upland and wetland-littoral detrital sources increases en route to pelagic regions. Because most lakes are small and shallow, the loading of organic carbon, largely in dissolved forms, from the wetland and littoral communities is significant, usually exceeding by several fold to several orders of magnitude that produced by phytoplankton in the open water pelagic zone. As a result, *most dissolved organic carbon of the lacustrine ecosystem is derived from photosynthesis of higher plants and microflora associated with detritus, including sediments, and is only augmented by photosynthesis of phytoplankton.*

I maintain that it is incorrect to state that this DOC is 'allochthonous'. Although a portion is from external upland sources, that DOC loaded from wetlands and littoral communities is truly autochthonous and produced as part of the lacustrine or lotic ecosystems. To state that this interface organic matter loading is functionally distinct from that produced by planktonic photosynthesis is artificial, and erroneously decouples the integration of these ecosystem components.

Although algae can also synthesize relatively recalcitrant humic compounds, synthesis is much greater among structural tissues of higher plants. Nearly all of the loading of humic compounds to fresh waters is as dissolved organic compounds from partial decomposition of higher plant tissues. Decomposition of these compounds occurs largely *at surfaces in the particulate detritus of wetlands, in sediments*, particularly in the shallow areas and hyporheic areas in streams, *and among epiphytic microflora* in the littoral zones.

## *Processing of organic matter in wetland and littoral zones*

Wetland and littoral zones are invariably depositional regions. Sedimentation of *particulate* organic matter is a prevalent characteristic of depositional regions that usually exist along reduced elevational gradients among most sites of land-water intersection. Rates of decomposition of this organic matter are markedly influenced by the hydrology of the regions. In most natural freshwater ecosystems, drainage inflows are diffuse and meandering, which permits sedimentation and allows sufficient time for appreciable degradation of particulate organic matter before potential displacement or permanent interment. However, the sedimentation of detritus results in well-known reducing conditions and lowered rates of degradation. Despite these conditions, the amounts of organic matter produced and loaded to these interface zones are very large; rates of decomposition are very high within the plant communities as many of the perennial plants undergo constant turnover and senescence of part of their foliage and rooting tissues. Intensive decomposition occurs both among the standing dead tissue as well as in the surficial portions of the complex deposits of particulate detritus.

In most wetlands of the world, the slope gradients are so modest that surface hydrological flows are diffuse and erosion is small. Hence, water from precipitation events tends to migrate slowly over and through the detrital masses of decaying organic matter, rather than physically transporting the particles. During major precipitation events in which the absorptive, retentive capacities of the wetland vegetation and detrital masses for water are exceeded, some movements of particulate organic matter occurs. However, its dispersion is not great in lacustrine ecosystems because of the rapid dissipation of energy. Therefore, the common deltaic accumulations occur at inlet mouths (e.g. Håkanson, 1983; Godshalk & Wetzel, 1984). Interception of particulate organic matter from both inlet and diffusive runoff within emergent and submersed littoral vegetation is major and results in rapid, accelerated deposition (e.g., Lundqvist, 1927; Godshalk & Wetzel, 1984; Moeller & Wetzel, 1988). Beyond the submersed vegetation at greater depths, an erosional zone is common in lacustrine ecosystems depending on the depths to which surficial mixing occurs.

Exportation of organic matter from the land-water interface zone is largely as *dissolved* organic matter. These compounds are decomposed selectively en route. For example, mineralization of proteins by wetland sediment microflora conformed to kinetics of polymer degradation (Cunningham & Wetzel, 1989). Turnover times of amino acids were very much faster (1.6–3.7 h) than those of proteins (13–69 h). The latter have appreciably more time for transport and movement to other parts of the ecosystem. This example is perhaps representative of the type of differential degradation of complex products released in soluble form from the large amounts of living and decaying macrophytes.

Large amounts of humic substances are released from decaying macrophytes, particularly from the very productive emergent and floating-leaved plants with high relative amounts of lignified and cellulosic tissues. These dissolved organic compounds are selectively removed by metabolic and adsorptive processes as they pass through the submersed macrophyte-epiphytic complex en route to the recipient lake or stream (e.g., Larson, 1978; Mickle & Wetzel, 1978a, 1978b, 1979; Wetzel, 1990a). The process essentially *effects a selective increase in recalcitrance of the dissolved organic compounds as they move toward the open water regions of the ecosystem.*

### *Decomposition of DOC in the open pelagic zones*

A frequent assumption of many contemporary researchers is that the dissolved organic compounds entering from the interface regions of aquatic ecosystems are so recalcitrant to decomposition that most bacterial metabolism must depend upon the smaller quantities of more labile sources synthesized and released by the phytoplankton, possibly coupled with the 'microbial loop' (*vide infra*).

Dissolved organic matter (DOM) loaded from interface zones functions in many ways. Recalcitrance of DOC that has been selectively metabolized en route to the pelagic zone is relative. The DOC decomposes at rates of 1 to 5 per cent per day (Saunders, 1972; Wetzel, 1984). Although slow, these rates occur with growth efficiencies of 25 to 50 per cent by bacteria (i.e., bacterial biomass production per amount of DOC consumed) (e.g. De Haan, 1977; Tranvik, 1988, 1989; Tranvik & Höfle, 1987). Hence, as was first emphasized nearly two decades ago (Wetzel *et al.*, 1972;

Wetzel, 1975, 1979), *although DOC from interface and allochthonous sources is more recalcitrant than that produced by microflora, the decomposition of large quantities imported to the pelagic zone dominates ecosystem* decomposition of DOC.

The paramount point is that this DOC is decomposed, though slowly, and as such provides a major energetic subsidy to the ecosystem and augments that produced planktonically. This DOC provides a major source of energy to drive lacustrine and lotic ecosystems. With a few exceptions, synthesis of organic carbon by phytoplankton is totally inadequate to support (i) extant bacterial metabolism of pelagic, littoral, and sediment communities, and (ii) extant rates of pelagic bacterial decomposition and recycling of inorganic nutrients. Protozoan decomposition of *particulate* organic matter can accelerate decomposition rates and recycling fluxes (e.g., Sherr & Sherr, 1989; Kairesalo & Saukkonen, 1990). In the pelagic zone, however, most decomposition is by free-living bacteria and fungi that utilize *dissolved* organic compounds.

*Rates of DOC loading and utilization*

Natural concentrations of DOC are approximately proportional to bacterial growth and productivity (e.g., Wetzel, 1983; Cole *et al.*, 1988), and this relationship is also true in heavily-stained brown-water lakes (Tranvik, 1988, 1989). Evidence is consistent in indicating that bacterioplankton are often chronically carbon-substrate limited in the pelagic of natural waters. Detailed comparative analyses of the uptake kinetics for nutrients by planktonic bacteria and algae indicate that phytoplankton utilize both phosphate and dissolved organic phosphorus compounds, particularly at high substrate concentrations, and that bacterial utilization of phosphorus is often limited by the availability of carbon substrates (Coveney & Wetzel, 1988, 1989, 1991; Cotner & Wetzel, 1991b).

The 'microbial loop', that is a utilization of DOC by bacteria, consumption and decomposition of bacteria by protozoans, and a release of dissolved organic compounds and $CO_2$, is now recognized as a common energetic cycle in the pelagic of most fresh waters (e.g., Sherr & Sherr, 1989; Carrick *et al.*, 1991). Frequently 20 to 40 percent or greater of the bacterial productivity is consumed by the protozoans. Because most of the organic matter 'processing' by the microbial loop is simply decomposed without ever reaching the gut of higher animals, the bacterial-protozoan microbial loop has been termed an energetic waste. This myopic zoological bias is not acceptable. I emphasize that the planktonic bacterial utilization of DOC from all sources, planktonic as well as from the land-water interface photosynthetic producers, followed by autolytic death, dormancy, or subsequent decomposition by protozoan digestion represents a major carbon and energy flux route. This decomposition is not an energetic 'waste' to the *ecosystem*, but rather is essential for ecosystem stability particularly in relation to nutrient regeneration (Wetzel, 1983, 1984).

It is important to view the loadings of DOC to the lake *ecosystem* with a realistic perspective. Many of the conceptual treatments view the loadings to enter the ecosystem in large individual masses, whereupon decomposition progresses in the familiar 'batch' culture pattern (Fig. 4). In reality, however, the loadings of DOC differs considerably, particularly in relation to the sources. Four distinctions could be made (Fig. 5).

(a) DOC movements from upland soil sources to aquatic ecosystems are driven by precipitation events. Quantitatively large loadings often occur, but at infrequent intervals separated by hiatus periods of little or no loading.

(b) The greatest amounts of DOC usually emanate from the decomposition of the large amounts of organic matter produced in wetlands. The cohorts of dominant vegetation species are constantly shifting with new cohorts growing while old are senescing and decaying (Wetzel, 1990a). The reservoir of particulate detrital materials is very large and in many stages of senescence and decomposition. As a result, DOC loadings are large and more or less continuous.

(c) Littoral floating-leaved and submersed

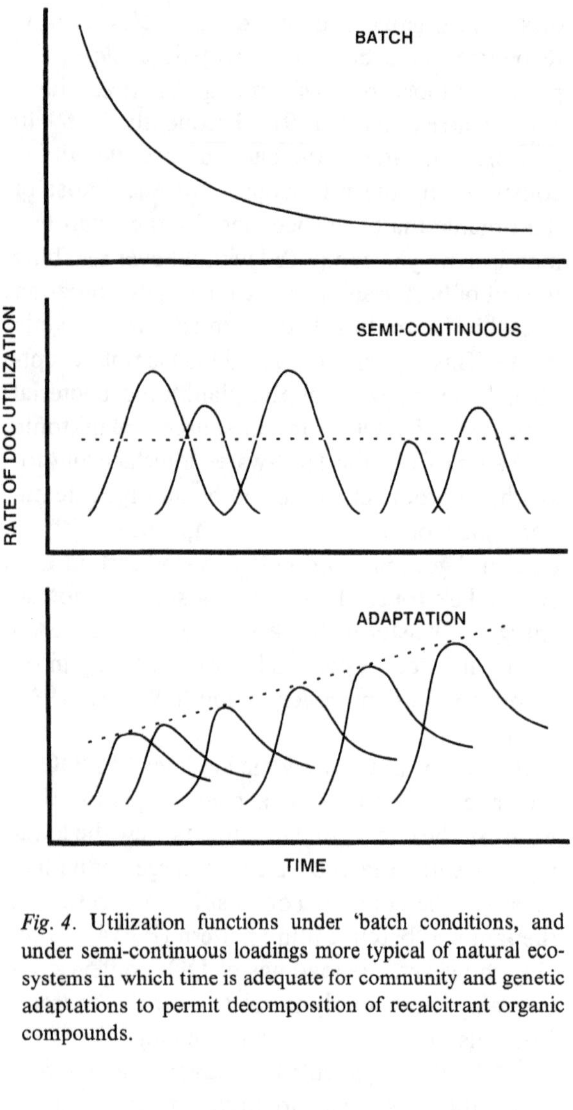


*Fig. 4.* Utilization functions under 'batch conditions, and under semi-continuous loadings more typical of natural ecosystems in which time is adequate for community and genetic adaptations to permit decomposition of recalcitrant organic compounds.

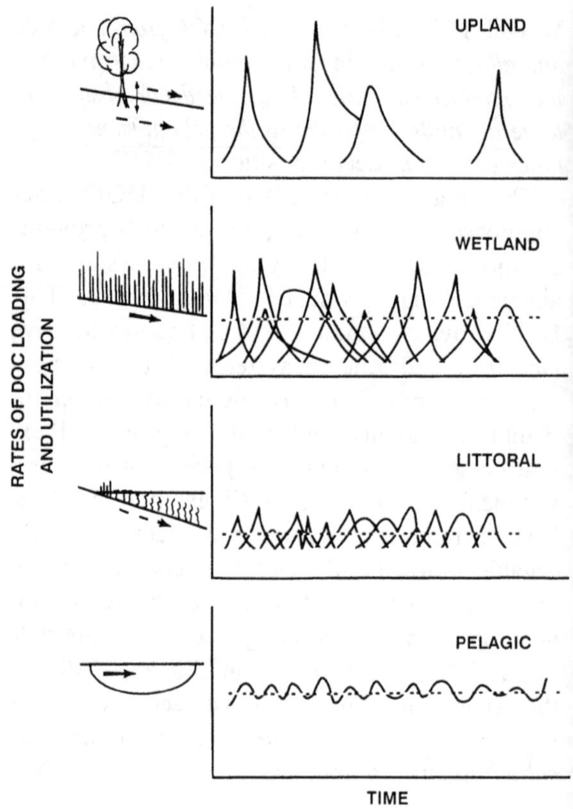


*Fig. 5.* Variations in the quantities and intervals of loading and decomposition of DOC from different portions of freshwater ecosystems.

macrophytes, as well as the extremely productive epiphytic algal communities, as the wetland communities, provide a high production of DOC that is loaded to the ambient environment in large amounts more or less continuously. Wind-induced water currents assist in moving and dispersing DOC within the basins.

(d) Loading of DOC from extracellular releases from phytoplankton is commonly assumed to be a major source for bacteria in the pelagic zone. The quantities produced are low, however, and much evidence indicates that this source is nearly always inadequate (by 50 to nearly 100%) to meet the needs of bacterial productivity of the ecosystem (pelagic, benthic). Even though methods at this time exhibit high variance, productivity and carbon budgets in the pelagic zone rarely balance. DOC substrates from other sources play a major role in bacterial metabolism in the pelagic and dominate in the inshore regions (e.g., Borsheim *et al.*, 1988; Tolstoy, 1988; Tranvik, 1988, 1989; Cotner & Wetzel, 1990a).

Arguments that decomposition of allochthonous and wetland/littoral DOC is low in the pelagic zone are poorly substantiated. Much evidence is to the contrary, namely that although slow, the decomposition quantity is large. One could argue that with the more or less continuous loadings from highly variable sources, a semi-continuous loading of DOM of different qualities leads to constant reinforcement of enzymatic synthesis (Fig. 4). In addition, the continual loading reinforcements should lead to bacterial genetic adaptations and, from the ecosystem standpoint, an enhancement of efficiencies of decomposition of the more recalcitrant organic compounds.

*Gradients from the wetlands-littoral to the pelagic*

Numerous studies have shown that gradients exist between the land-water interface communities and the pelagic zone in both inorganic nutrients as well as dissolved organic carbon compounds (cf. review of Wetzel, 1983, 1990a; Kairesalo & Matilainen, 1988; Craft *et al.*, 1988; Cotner & Wetzel, 1991a). Clearly the interface regions have higher organic matter concentrations, greater bacterial metabolism, high release of dissolved organic matter, and greater recycling and utilization rates of mineralized nutrients. Even the synthesis and release of regulatory compounds and essential organic constituents (e.g., cyclic AMP, B-complex vitamins) are more abundant in the littoral areas of greater productivity and decomposition (e.g., Francko & Wetzel, 1982; Kurata *et al.*, 1979).

Decomposition by bacteria is frequently carbon limited, and often limited by the availability of nutrients, particularly phosphorus in fresh waters. In the freshwater *ecosystem*, however, most decomposition is by bacteria associated with *surfaces*. Decomposition is several orders of magnitude greater in biofilm communities on particulate detritus, particularly of sediments (epipelic) and macrophyte (epiphytic) communities, than in dilute planktonic realms (reviewed in Wetzel, 1990a; cf. also van Loosdrecht *et al.*, 1990). As a result, *On the average in lakes and rivers, over 90 percent of the decomposition occurring in the ecosystem is by bacteria utilizing DOM from non-pelagic sources of primary productivity.*

The high rates of decomposition and mineralization among these surface-associated microbial communities is possible because of the increased juxtaposition of the algal and bacterial populations. A host of compensatory physiological mechanisms are coupled to permit intensive *recycling* of both carbon and nutrients. All physical and biotic couplings are organized and functionally operating to minimize losses to the surrounding waters and pelagic zone. *The only way that the observed high sustained productivity of the land-water interface zones can be maintained is to evolve simultaneous recycling and conservation mechanisms. These conservation mechanisms occur at all gradient levels: biochemical, physiological, community, and ecosystem integrations.*

These relationships have been discussed in detail recently by Wetzel (1990a). To summarize, these distinctions in fresh waters focus largely upon an integrated community of microbes when attached, in which recycling and retention mechanisms predominate (Fig. 6). Growth is accommodated by acquisition of new resources, with maintenance being retained by recycling nutrients and carbon while minimizing losses from the communities. In the pelagic environment, individuals are patchily dispersed which minimizes *effectiveness* of recycling and increases losses and displacement of organic matter and nutrients from sites of production (Fig. 6).

*Changes with enhanced ecosystem productivity*

Some years ago I discussed the generalized relationships of primary productivity of submersed and emergent macrophytic flora, attached algae, and phytoplankton of a series of whole lake ecosystems with increasing fertility (Fig. 7) (Wetzel & Hough, 1973; Wetzel, 1975, 1983). Initially under nutrient limitations of oligotrophic lake ec-

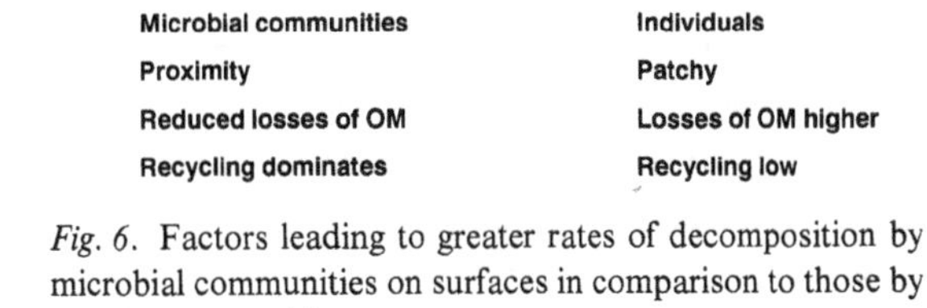


*Fig. 6.* Factors leading to greater rates of decomposition by microbial communities on surfaces in comparison to those by pelagic communities.

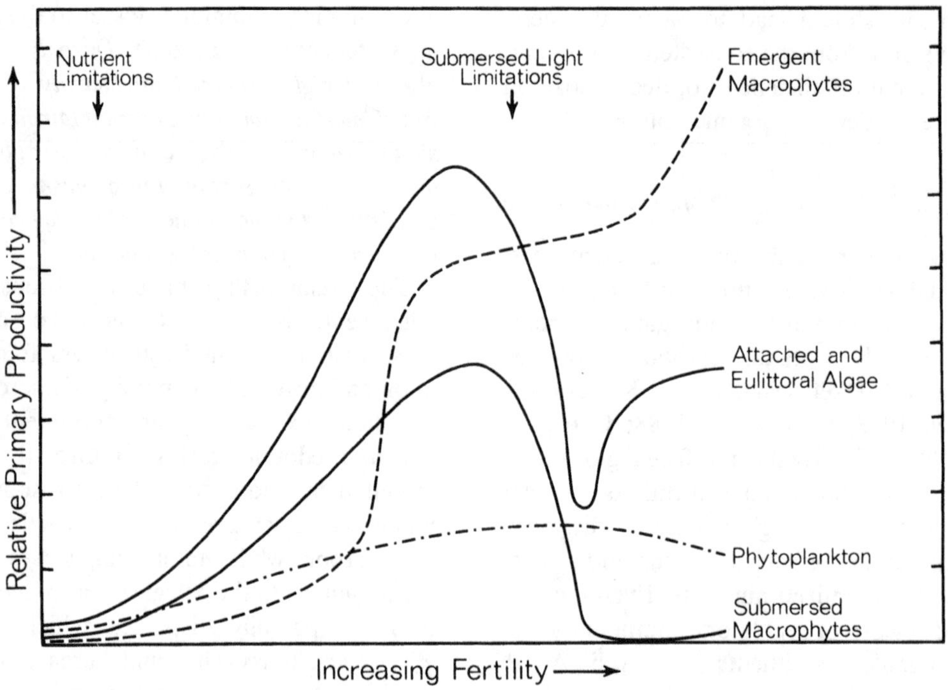


*Fig. 7.* Generalized relationships of primary productivity of submersed and emergent macrophytic flora, attached algae, and phytoplankton of lakes of increasing fertility of the whole lake ecosystem. (After Wetzel, 1983).

osystems, total productivity is low and approximately equally contributed by the different primary producers. Accompanying increased nutrient loading, submersed macrophytes and particularly attached microflora increase to very high levels of productivity. With further nutrient loading, phytoplankton productivity can increase markedly until dense populations and increased epiphytic periphyton attenuate light penetration sufficiently to restrict growth of submersed macrophytes. Productivity declines precipitously at this point and phytoplanktonic growth cannot be increased significantly beyond that point. As emergent vegetation colonizes greater proportions of the lake ecosystems, its high productivity is commonly accompanied by increased attached algae and littoral flora (Fig. 7).

Within this generalized sequence of primary productivity, the relative contributions of phytoplankton, littoral producers, dark and chemosynthetic $CO_2$ fixation, and allochthonous sources of dissolved organic carbon were evaluated along a gradient of lakes of increasing fertility in the progression of dominating pelagic and littoral plants (Fig. 8) (Wetzel, 1975, 1983). The major contributions of the littoral and wetland vegetation to

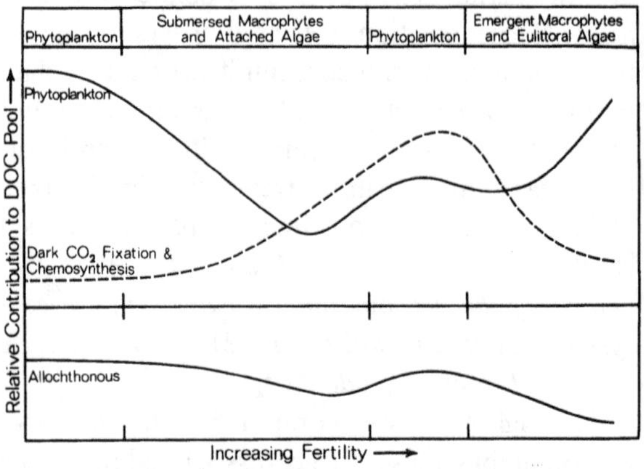


*Fig. 8.* Generalized relative contributions of phytoplankton, littoral producers, dark and chemosynthetic $CO_2$ fixation, and allochthonous sources of dissolved organic carbon (DOC) to lakes of increasing fertility in the progression of dominating pelagic and littoral flora. Upper panel of generalized changes in productivity by autotrophic flora refers to those detailed in Fig. 6. (After Wetzel, 1983).

the DOC pool of the whole of lacustrine ecosystems are apparent. In 'humic lakes', in which the water is heavily stained with dissolved organic compounds, the loadings of DOC from allochthonous and particularly wetlands is very large.

The sequence of increasing ecosystem productivity in the transitions along a range of lake ecosystems from low to very high littoral/wetland dominated productivity (Fig. 7) can also be interpreted in term of the loadings of particulate and dissolved organic detritus (Fig. 9). Although the total detritus increases markedly, the increasing syntheses and loadings from emergent aquatic macrophytes results in reducing conditions not conducive to rapid decomposition. Conversions of POC to DOC are reduced, although the totals of both categories increase markedly. Because of the increased organic matter loadings from the high littoral and wetland productivity in the later stages, the proportions of organic matter processing by bacterial and fungal microflora increases markedly as well. As a consequence, the amount of more labile DOC increases as a percentage of the whole, and overall recalcitrance declines. Both the syntheses of organic matter and the rates of total decomposition increase in these advanced stages.

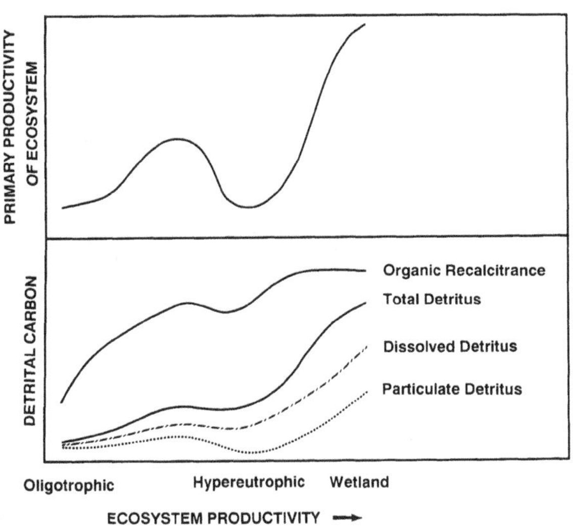


*Fig. 9.* Generalized alterations in quantities and recalcitrance of particulate, dissolved, and total detritus as the ecosystem productivity changes or is viewed along a gradient of freshwater ecosystems of increasing system productivity.

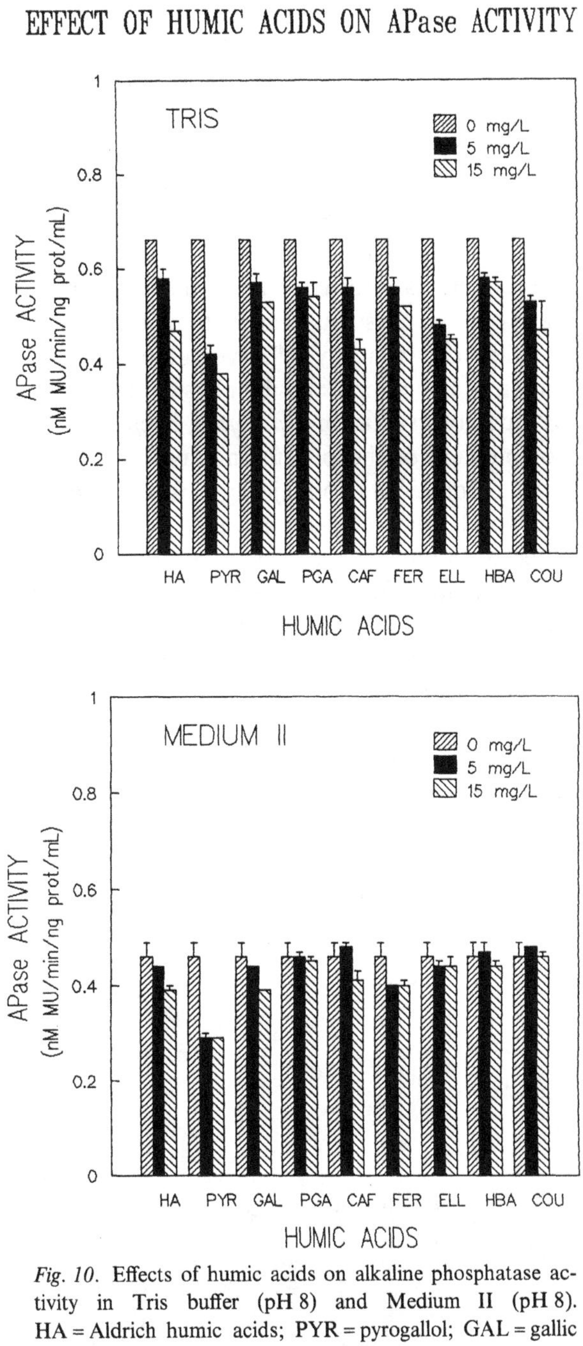


*Fig. 10.* Effects of humic acids on alkaline phosphatase activity in Tris buffer (pH 8) and Medium II (pH 8). HA = Aldrich humic acids; PYR = pyrogallol; GAL = gallic acid; PGA = polygalacturonic acid; CAF = caffeic acid; FER = ferulic acid; ELL = ellagic acid; HBA = *p*-hydroxybenzoic acid; COU = *p*-coumaric acid.

### *Complexation of DOM with extracellular enzymes*

Recent investigations have demonstrated an additional important coupling of the higher plants of the wetland/littoral interface zones to the metabolism of pelagic organisms. Studies have demonstrated the inhibitory effects of dissolved organic compounds produced by and released from living, senescing, and decaying wetland and littoral macrophytes (Wetzel, 1990a, 1991, this paper). The humic compound-enzyme complex inactivates the enzymatic activity and the complex can be transported to and relocated in other parts of the ecosystem. Reactivation of enzymatic activities can occur by selective microbial degradation of the complex, or by physical photolysis by ultraviolet radiation (*vide infra*). Because of the commonality of phosphorus limitations to the primary productivity of a majority of fresh waters of the world, much of the initial research has been directed to phosphatase activities. However, the complexation efficacy of DOM of higher plant origins applies variously to many other surface-bound and free extracellular enzymes (e.g., aminopeptidases, glucosidases, others). The results can be summarized as follows.

a) Limitations to growth by phosphorus availability invariably induces production of high exoenzymatic phosphatase activity among diverse populations of planktonic algae and bacteria. This phosphatase activity is both surface-bound to cells as well as secreted into the water (e.g., Wetzel, 1981, 1991; Stewart & Wetzel, 1982a).

b) Release of large amounts of dissolved organic compounds by wetland and littoral macrophytes is regulated seasonally by balances among photosynthetic production, shifts in nutrient availability, growth reductions and senescence of foliage, and hydrology (Wetzel, 1990a). The types of dissolved organic compounds, particularly polyphenols, that originate from lignin and cellulosic structural components of aquatic plants (cf. Conn, 1981; Aiken *et al.*, 1985; Thurman, 1985; Haslam, 1989; Suffet & MacCarthy, 1989; Perdue & Gjessing, 1990) are coupled tightly to plant characteristics (species, growth forms) and conditions of decomposition.

c) These relatively recalcitrant dissolved polyphenolic compounds from the plants complex chemically with many enzymes, particularly phosphatases, of both planktonic and attached algae and bacteria (Figs. 10, 11). This complexation oc-

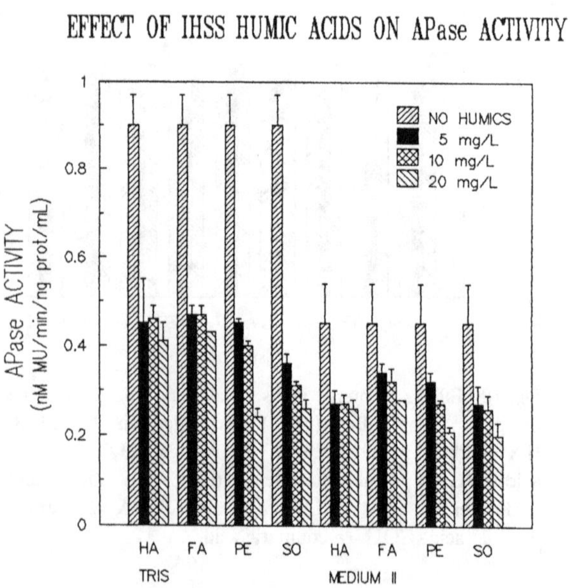


*Fig. 11.* Effect of purified humic compounds (International Humic Substances Society) on alkaline phosphatase activities. HA = Suwannee River humic acid; FA = Suwannee River fulvic acid; PE = peat humic acid; SO = soil humic acid.

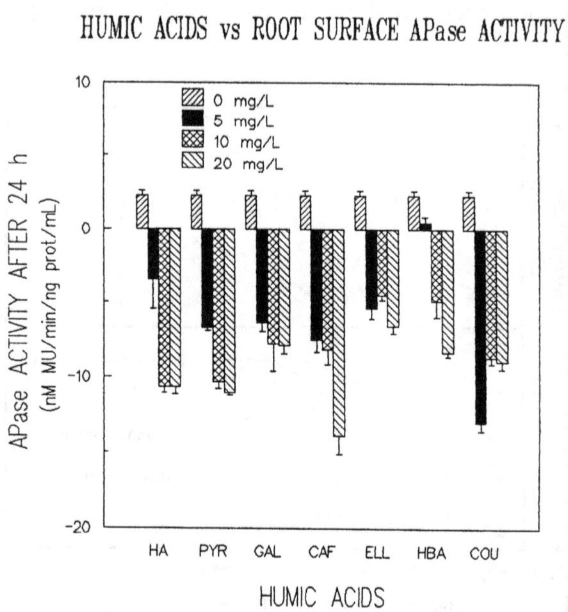


*Fig. 12.* Effects of humic acids on alkaline phosphatase activities of the root tissue of *Eleocharis acicularis* from Third Sister Lake, Washtenaw County, Michigan. Symbols as in Fig. 10.

curs both among free enzymes as well as those bound to cell surfaces.

d) Because this complex binds and inactivates extracellular phosphatase activities, cells experiencing phosphorus limitations will be required to expend greater amounts of energy to synthesize phosphatases to hydrolyze dissolved organic phosphorus compounds. Hence, with increased loadings of polyphenolic compounds to fresh waters, one observes an enhanced production of alkaline phosphatase activities (Stewart & Wetzel, 1982b; Wetzel, 1990a, 1991, in preparation). The enzymes are both competitively and non-competitively inhibited by these compounds. These results are in accordance with the frequent observations that the primary productivity of humic-stained fresh waters is consistently lower than rates in clear-water waters with comparable nutrient loadings and light availability (Borsheim *et al.*, 1988; Jones *et al.*, 1988; Tolstoy, 1988; Tranvik, 1989; Jones, 1990).

In addition, the polyphenolic compound inhibition of enzyme activities can be found at many sites. For example, polyphenolic compounds inhibit surface-bound phosphatase activities of the roots of emergent macrophytes (Fig. 12), as well as in soils (e.g., Blum & Shafer, 1988; Lähdesmäki & Piispanen, 1988) and digestive systems (Blytt *et al.*, 1988; Haslam, 1989).

e) The dissolved polyphenolic-enzyme complex can be transported with water, stored in an inactivated but viable state for long periods, either dissolved or possibly adsorbed to inorganic particles, and then reactivated in a displaced site. In this way, the dissolved polyphenolic compounds of macrophytes permit enzymes to be stored in a

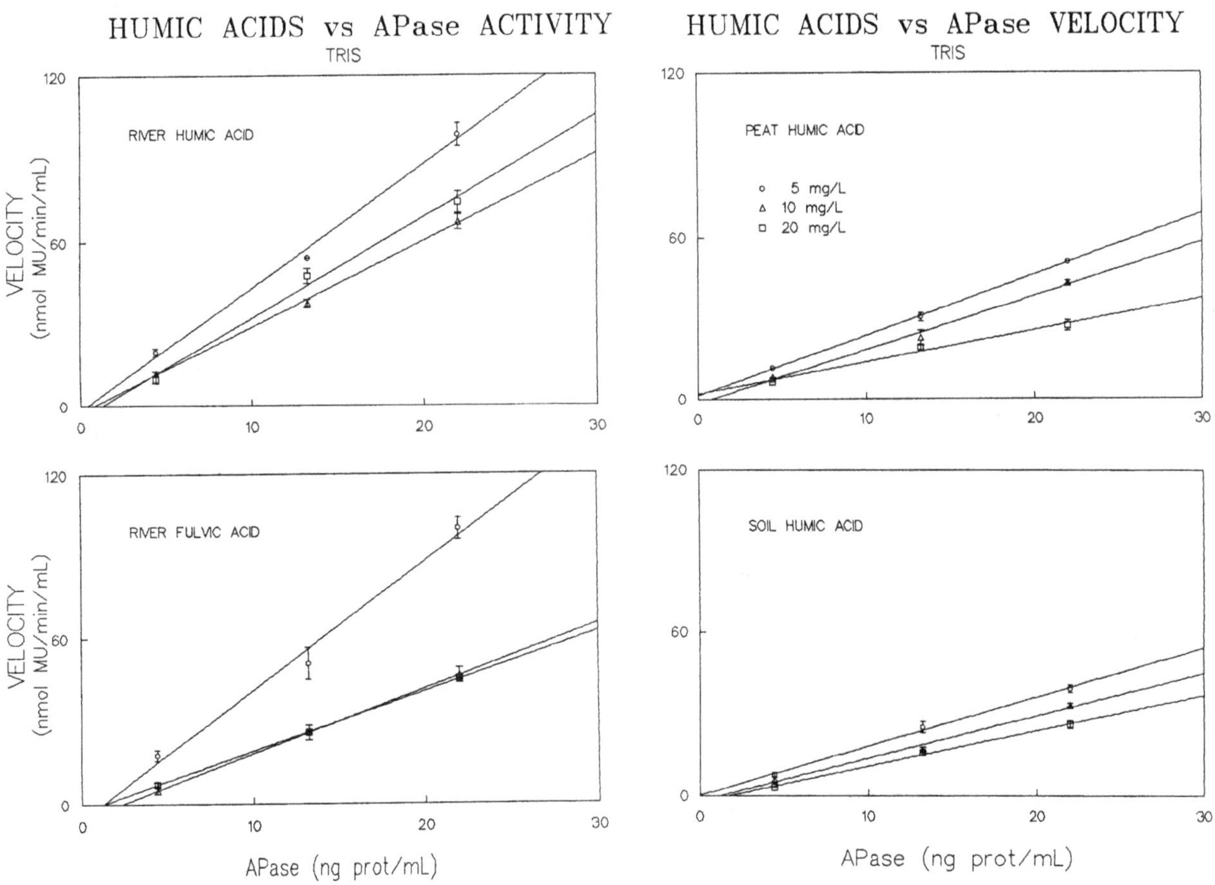


*Fig. 13.* Effects of purified natural humic acids on alkaline phosphatase activities.

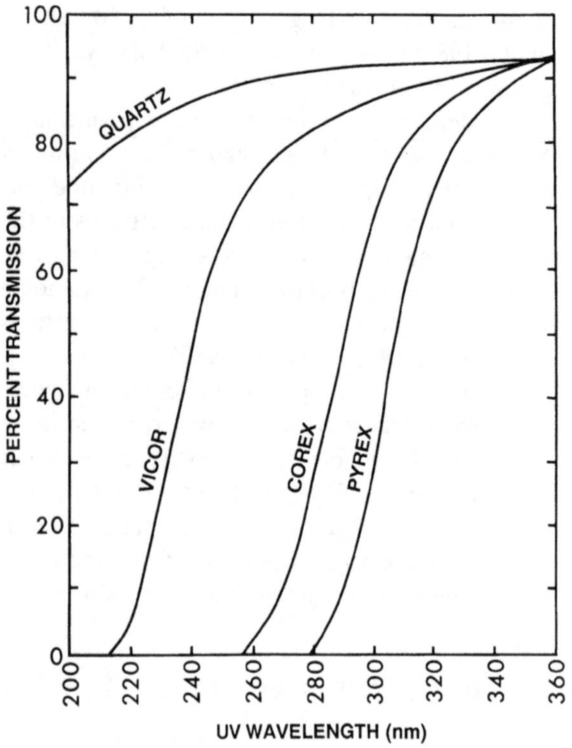


*Fig. 14.* Absorption of UV radiation by different glass types.

complexed form for long periods of time (even months or years) and later reactivated by partial bacterial hydrolysis or photolysis in distant locations within the ecosystem. In the sense of the whole ecosystem, this mode of operation is analogous to enzymatic functions in some higher organisms (e.g., liver, a leaf).

f) Non-destructive cleavage of the polyphenol-enzyme complex in order to release active enzyme could occur by a number of processes, both chemical and biotic. Photolysis of individual proteinaceous and humic (plant polyphenols) compounds by ultraviolet (UV) radiation is well known (e.g., McLaren & Shugar, 1964; Gjessing & Gjerdahl, 1970; Manny *et al.*, 1971; Strome & Miller, 1978; Stewart & Wetzel, 1981; Zepp *et al.*, 1981; Geller, 1986; Faust & Holgné, 1987; Guittonneau *et al.*, 1988).

The important operational question is whether the polyphenolic-enzyme complex can be photolysed gently by UV radiation in order to release still functional enzyme activity. As demonstrated above, a large number of low molecular weight phenolic acids (pyrogallol, hydroxybenzoic acid, gallic acid, ellagic acid), phenylpropanoids (ferulic acid, caffeic acid, coumaric acid), and polymers (polygalacturonic acid, humic acids and fulvic acids from several standard sources from natural waters purified by the International Humic Substances Society) all consistently formed complexes with enzymes and reduced enzymatic activities (Fig. 13).

The polyphenolic-enzyme complexes were then exposed to UV radiation of selective intensities using a high-intensity source (identical to that in Manny *et al.*, 1971). Direct exposure to the UV source caused nearly instantaneous loss of alkaline phosphatase (AP) activity (within seconds; data not shown). Selective reduction of the short and long-wavelength UV radiation with different types of glass (Fig. 14) resulted in selective photolysis of the polyphenol-enzyme complex. The slight increase in AP activity after two minutes by Vicor selection (Fig. 15) was much greater when further short wavelength UV was selectively removed by Corex glass (Fig. 16). Photolytic denaturation of the protein and polyphenolic degradation then followed upon longer exposure. When exposed to long-wavelength UV radiation only (Pyrex, Fig. 17), AP activity from the complex only began to increase after long periods of exposure. Photolytic separation of the humic acid-enzyme complexes by natural sunlight has also been demonstrated (Wetzel, unpublished). These results indicate clearly that the polyphenolic-enzyme complex can be fractured by mild UV radiation, as would be found in fresh waters, and that an increase in the enzymatic activity can occur as a result.

g) Many organic and inorganic compounds can compete for complexation sites on the proteinaceous enzyme molecules. As a result many modes of complexation and alterations of enzymatic reaction rates have been demonstrated.

h) Divalent cations also bond to organic acids (e.g., Hering & Morel, 1988). High cation concentrations in natural waters can reduce the reactivity of organic acids with proteins and enzymes. Enzyme activity would be less inhibited

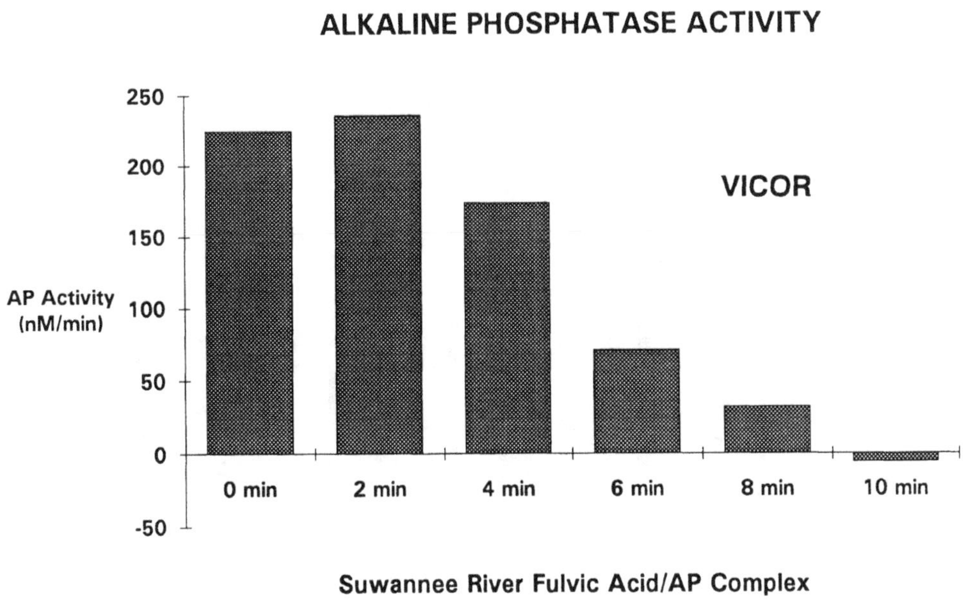


*Fig. 15.* Photolysis of fulvic acid-phosphatase complex by ultraviolet light through Vicor glass.

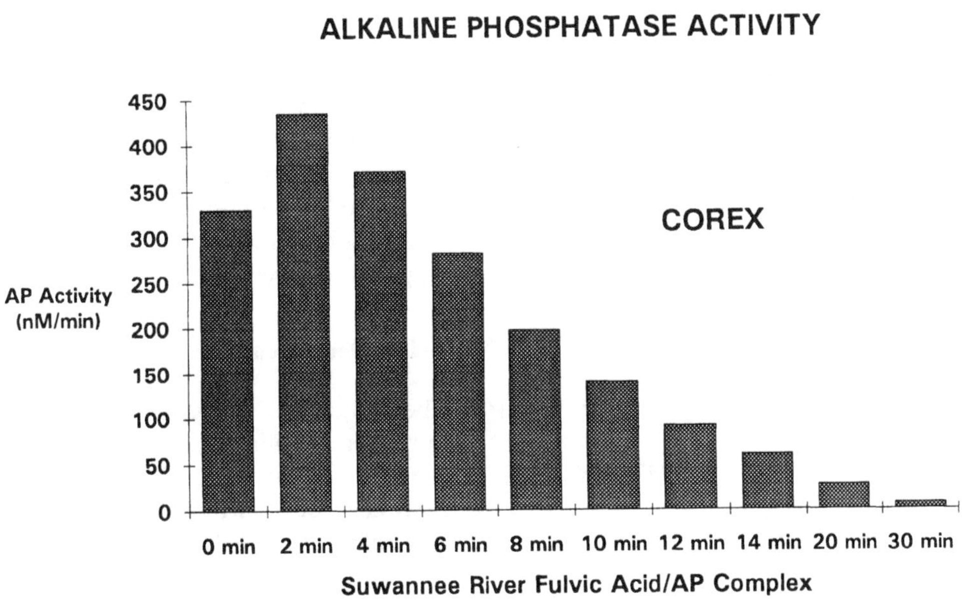


*Fig. 16.* Photolysis of fulvic acid-phosphatase complex by ultraviolet light through Corex glass.

by humic or other polyphenolic compounds in solutions with high divalent cation concentrations (Fig. 10, 11).

Among a series of waters of increasing salinity from soft to hard waters, productivity of the biota generally increases (cf. Wetzel, 1983). This increase is in part related to loading rates of essential nutrients, particularly nitrogen and phosphorus, from the drainage basin and internally from storage depositories (e.g., sediments). Causal relationships are not always that direct, but this sequence is commonly observed. Divalent cation

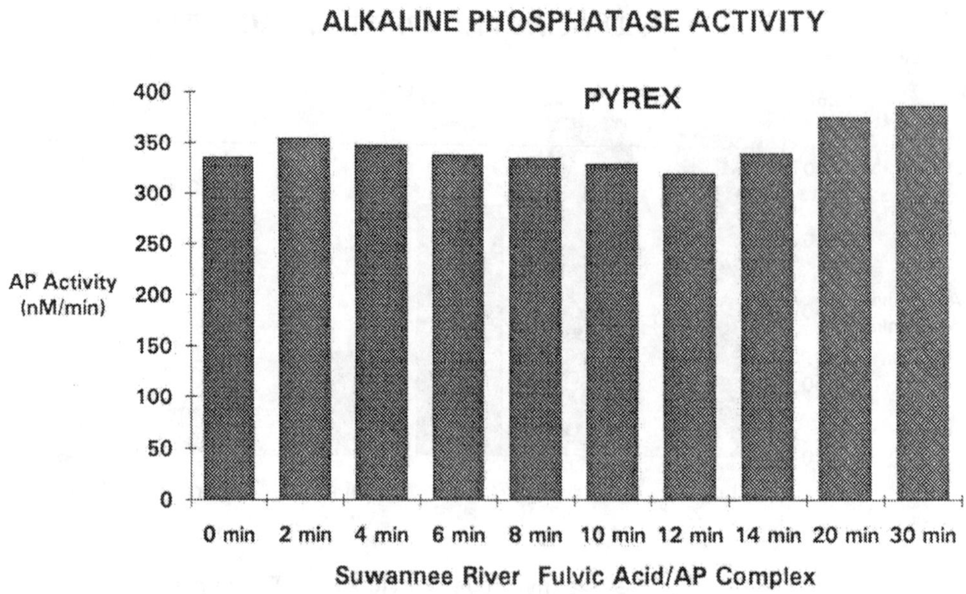


*Fig. 17.* Photolysis of fulvic acid-phosphatase complex by ultraviolet light through Pyrex glass.

concentrations increase markedly along the progression from soft to hard waters, low to higher productivity gradient, likely in part related to cationic suppression of humic compound interferences with extracellular enzymatic activities. The latter reduction would in turn permit enhanced use of nutrients bound in organic substrates (Fig. 18). In addition, as $CaCO_3$ precipitates in very hard waters, significant quantities of dissolved organic acids are adsorbed or coprecipitated with the nucleating carbonates (Otsuki & Wetzel, 1972, 1973; cf. also Steinberg & Kühnel, 1987; Dalton *et al.*, 1989).

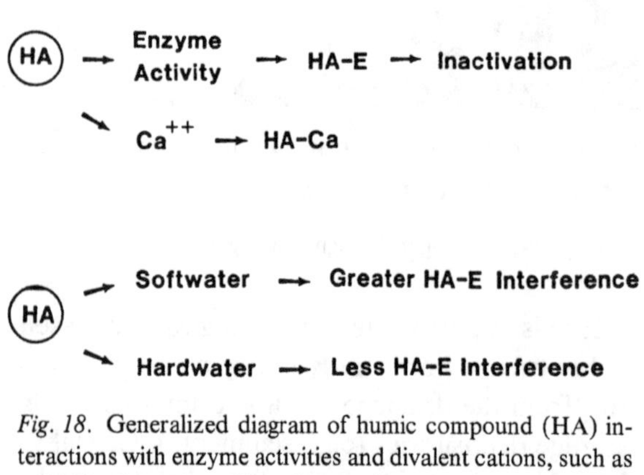


*Fig. 18.* Generalized diagram of humic compound (HA) interactions with enzyme activities and divalent cations, such as calcium along a gradient from soft to hard waters. (After Wetzel, 1991).

These mechanisms indicate how the dissolved organic compounds, particularly polyphenolic compounds originating from decomposition of wetland and littoral macrophytes, can result in major modifications of nutrient availability to, and metabolic pathways of, pelagic organisms of aquatic ecosystems.

## Acknowledgements

The author acknowledges the excellent subventions from the National Science Foundation and the U.S. Department of Energy. The technical assistance of K. Moore and Mark Dedmon is gratefully acknowledged.

## References

Aiken, G. R., D. M. McKnight, R. L. Wershaw & P. MacCarthy, 1985. Humic Substances in Soil, Sediment, and Water. Geochemistry, Isolation, and Characterization. J. Wiley & Sons, New York, 692 pp.

Blum, U. & S. R. Shafer, 1988. Microbial populations and phenolic acids in soil. Soil Biol. Biochem. 20: 793–800.

Blytt, H. J., T. K. Guscar & L. G. Butler, 1988. Antinutritional effects and ecological significance of dietary condensed tannins may not be due to binding and inhibiting digestive enzymes. J. Chem. Ecol. 14: 1455–1465.

Borsheim, K. Y., S. Andersen, G. H. Johnsen, E. O. Kambestad & S. Norland, 1988. Primary and bacterial production compared to growth and food requirements of *Daphnia longispina* in Lake Kvernavatnet, west Norway. J. Plankton Res. 10: 921–939.

Carrick, H. J., G. L. Fahnenstiel, E. F. Stoermer & R. G. Wetzel, 1991. Protozoan growth rates and trophic couplings in Lake Michigan. Limnol. Oceanogr. (In press).

Cole, J. J., S. Findlay & M. L. Pace, 1988. Bacterial production in fresh and saltwater ecosystems: A cross-system overview. Mar. Ecol. Progr. Ser. 43: 1–10.

Conn, E. E. (Ed.), 1981. The Biochemistry of Plants. Vol. 7. Secondary Plant Products. Academic Press, New York, 798 pp.

Cotner, J. B., Jr. & R. G. Wetzel, 1991a. Characterization of bacterial phosphatases from different habitats in a small, hardwater lake. In: R. J. Chrost, Editor. Exoenzymes in the Aquatic Environment. Developments in Hydrobiology (In press).

Cotner, J. B., Jr. & R. G. Wetzel, 1991b. Uptake of dissolved inorganic and organic phosphorus compounds by phytoplankton and bacterioplankton. Limnol. Oceanogr. (Submitted).

Coveney, M. F. & R. G. Wetzel, 1988. Experimental evaluation of conversion factors for the [$^3$H]thymidine incorporation assay of bacterial secondary productivity. Appl. environ. Microbiol. 54: 2018–2026.

Coveney, M. F. & R. G. Wetzel, 1989. Bacterial metabolism of algal extracellular carbon. Hydrobiologia 173: 141–149.

Coveney, M. F. & R. G. Wetzel, 1991. Nutrient effects on specific growth rate of bacterioplankton in oligotrophic lakewater cultures. Appl. environ. Microbiol. (In press).

Craft, C. B., S. W. Broome, E. D. Seneca & W. J. Showers, 1988. Estimating sources of soil organic matter in natural and transplanted estuarine marshes using stable isotopes of carbon and nitrogen. Estuar. coast. Shelf Sci. 26: 633–641.

Cunningham, H. W. & R. G. Wetzel, 1989. Kinetic analysis of protein degradation by a freshwater wetland sediment community. Appl. environ. Microbiol. 55: 1963–1967.

Dalton, B. R., U. Blum & S. B. Weed, 1989. Differential sorption of exogenously applied ferulic, *p*-coumaric, *p*-hydroxybenzoic, and vanillic acids in soils. Soil Sci. Soc. Am. J. 53: 747–762.

Degens, E. T. & V. Ittekkot, 1983. Dissolved organic matter in Lake Tanganyika and Lake Baikal – a brief survey. Mitt. Geol.-Paläont. Inst. Univ. Hamburg 55: 129–143.

De Haan, H., 1977. Effect of benzoate on microbial decomposition of fulvic acids in Tjeukemeer (the Netherlands). Limnol. Oceanogr. 22: 38–44.

Faust, B. C. & J. Holgné, 1987. Sensitized photooxidation of phenols by fulvic acid in natural waters. Envir. Sci. Technol. 21: 957–964.

Francko, D. A. & R. G. Wetzel, 1982. The isolation of cyclic adenosine 3′:5′-monophosphate (cAMP) from lakes of differing trophic status: Correlation with planktonic metabolic variables. Limnol. Oceanogr. 27: 27–38.

Geller, A., 1986. Comparison of mechanisms enhancing biodegradability of refractory lake water constituents. Limnol. Oceanogr. 31: 755–764.

Gjessing, E. T. & T. C. Gjerdahl, 1970. Influence of ultraviolet radiation on aquatic humus. Vatten 26: 144–145.

Godshalk, G. L. & R. G. Wetzel, 1984. Accumulation of sediment organic matter in a hardwater lake with special reference to lake ontogeny. Bull. mar. Sci. 35: 576–586.

Guittonneau, S., J. de Laat, M. Dore, J. P. Duguet & C. Bonnel, 1988. Etudes comparative de la degradation de quelques molecules aromatiques simples en solution aqueuse par photolyse UV et par photolyse du peroxyde d'hydrogene. Envir. Technol. Lett. 9: 1115–1128.

Håkanson, L., 1983. Principles of Lake Sedimentology. Springer-Verlag, New York, 316 pp.

Haslam, E., 1989. Plant Polyphenols. Vegetable Tannins Revisited. Cambridge Univ. Press, Cambridge, 230 pp.

Hering, J. G. & M. M. Morel, 1988. Humic acid complexation of calcium and copper. Envir. Sci. Technol. 22: 1234–1237.

Hobbie, J. E., Editor, 1980. Limnology of Tundra Ponds. Dowden, Hutchinson & Ross, Inc., Stroudsburg, 514 pp.

Jones, R. I., 1990. Phosphorus transformation in the epilimnion of humic lakes: Biological uptake of phosphate. Freshwat. Biol. 23: 323–337.

Jones, R. I., K. Salonen & H. de Haan, 1988. Phosphorus transformations in the epilimnion of humic lakes: Abiotic interactions between dissolved humic materials and phosphate. Freshwat. Biol. 19: 357–369.

Kairesalo, T. & T. Matilainen, 1988. The importance of low flow rates to the phosphorus flux between littoral and pelagial zones. Verh. int. Ver. Limnol. 23: 2210–2215.

Kairesalo, T. & P. Saukkonen, 1990. Thymidine incorporation by littoral and pelagial bacterioplankton in a mesohumic lake. Verh. int. Ver. Limnol. 24: 677–681.

Kurata, A., C. Saraceni & H. Kadota, 1979. The status of B group vitamins in macrophyte and pelagic zones of Lake Biwa. Mem. Ist. ital. Idrobiol. 37: 63–85.

Lähdesmäki, P. & R. Piispanen, 1988. Degradation products and the hydrolytic enzyme activities in the soil humification processes. Soil Biol. Biochem. 20: 287–292.

Larson, R. A., 1978. Dissolved organic matter of a low-colored stream. Freshwat. Biol. 8: 91–104.

van Loosdrecht, M. C. M., J. Lyklema, W. Norde & A. J. B. Zehnder, 1990. Influence of interfaces on microbial activity. Microbiol. Rev. 54: 75–87.

Lundqvist, G., 1927. Bodenablagerungen und Entwicklungstypen der Seen. In: A. Thienemann (Ed.), E. Schweizerbart'sche Verlagsbuchhandlung, Stuttgart. Die Binnengewässer 2: 1–122.

Manny, B. A., M. C. Miller & R. G. Wetzel, 1971. Ultraviolet combustion of dissolved organic nitrogen compounds in lake waters. Limnol. Oceanogr. 16: 71–85.

McLaren, A. D. & D. Shugar, 1964. Photochemistry of Proteins and Nucleic Acids. Pergamon Press, Oxford, 449 pp.

Mickle, A. M. & R. G. Wetzel, 1978a. Effectiveness of submersed angiosperm epiphyte complexes on exchange of nutrients and organic carbon in littoral systems. I. Inorganic nutrients. Aquat. Bot. 4: 303–316.

Mickle, A. M. & R. G. Wetzel, 1978b. Effectiveness of submersed angiosperm epiphyte complexes on exchange of nutrients and organic carbon in littoral systems. II. Dissolved organic carbon. Aquat. Bot. 4: 317–329.

Mickle, A. M. & R. G. Wetzel, 1979. Effectiveness of submersed angiosperm-epiphyte complexes on exchange of nutrients and organic carbon in littoral systems. III. Refractory organic carbon. Aquat. Bot. 6: 339–355.

Moeller, R. E. & R. G. Wetzel, 1988. Littoral vs profundal components of sediment accumulation: Contrasting roles as phosphorus sinks. Verh. int. Ver. Limnol. 23: 386–393.

Otsuki, A. & R. G. Wetzel, 1972. Coprecipitation of phosphates with carbonates in a marl lake. Limnol. Oceanogr. 17: 763–767.

Otsuki, A. & R. G. Wetzel, 1973. Interaction of yellow organic acids with calcium carbonate in fresh water. Limnol. Oceanogr. 18: 490–493.

Perdue, E. M. & E. T. Gjessing (eds.), 1990. Organic Acids in Aquatic Ecosystems. John Wiley & Sons, Chichester, 345 pp.

Saunders, G. W., 1972. The transformation of artificial detritus in lake water. Mem. Ist. ital. Idrobiol. Suppl. 29: 261–288.

Sherr, B. & E. Sherr, 1989. Trophic impacts phagotrophic Protozoa in pelagic foodwebs. In T. Hattori, Y. Ishida, Y. Maruyama, R. Y. Morita, and A. Uchida (eds.), Recent Advances in Microbial Ecology. Jap. Sci. Soc. Press, Tokyo: 388–393.

Strome, D. J. & M. C. Miller, 1978. Photolytic changes in dissolved humic substances. Verh. int. Ver. Limnol. 20: 1248–1254.

Steinberg, C. & W. Kühnel, 1987. Influence of cation acids on dissolved humic substances under acidified conditions. Wat. Res. 21: 95–98.

Stewart, A. J. & R. G. Wetzel, 1981. Dissolved humic materials: Photodegradation, sediment effects, and reactivity with phosphate and calcium carbonate precipitation. Arch. Hydrobiol. 92: 265–286.

Stewart, A. J. & R. G. Wetzel, 1982a. Phytoplankton contribution to alkaline phosphatase activity. Arch. Hydrobiol. 93: 265–271.

Stewart, A. J. & R. G. Wetzel, 1982b. Influence of dissolved humic materials on carbon assimilation and alkaline phosphatase activity in natural algal-bacterial assemblages. Freshwat. Biol. 12: 369–380.

Suffet, I. H. & P. MacCarthy (eds.), 1989. Aquatic Humic Substances. Influence on Fate and Treatment of Pollutants. American Chemical Soc., Washington, DC, 864 pp.

Thurman, E. M., 1985. Organic Geochemistry of Natural Waters. Martinus Nijhoff/Dr W. Junk Publ., Dordrecht, 497 pp.

Tolstoy, A., 1988. Predicted and measured annual primary production of phytoplankton – examples from some Swedish lakes. Arch. Hydrobiol. 113: 381–404.

Tranvik, L. J., 1988. Availability of dissolved organic carbon for planktonic bacteria in oligotrophic lakes of differing humic content. Microb. Ecol. 16: 311–322.

Tranvik, L. J., 1989. Bacterioplankton growth, grazing mortality and quantitative relationship to primary production in a humic and clearwater lake. J. Plankton Res. 11: 985–1000.

Tranvik, L. J. & M. G. Höfle, 1987. Bacterial growth in mixed cultures on dissolved organic carbon from humic and clear waters. Appl. envir. Microbiol. 53: 482–488.

Wetzel, R. G., 1968. Dissolved organic matter and phytoplankton productivity in marl lakes. In Symposium on biogenic Metabolism in Fresh Waters, Chemistry and Microbiology. Mitteilungen int. Ver. Limnol. 14: 261–273.

Wetzel, R. G., 1975. Limnology. W.B. Saunders Co., Philadelphia, 743 pp.

Wetzel, R. G., 1979. The role of the littoral zone and detritus in lake metabolism. In G. E. Likens, W. Rodhe, and C. Serruya, (eds.), Symposium on Lake Metabolism and Lake Management. Arch. Hydrobiol. Beih. Ergebn. Limnol. 13: 145–161.

Wetzel, R. G., 1981. Longterm dissolved and particulate alkaline phosphatase activity in a hardwater lake in relation to lake stability and phosphorus enrichments. Verh. int. Ver. Limnol. 21: 337–349.

Wetzel, R. G., 1983. Limnology. 2nd Edition. Saunders College Publ., Philadelphia, 860 pp.

Wetzel, R. G., 1984. Detrital dissolved and particulate organic carbon functions in aquatic ecosystems. Bull. Mar. Sci. 35: 503–509.

Wetzel, R. G., 1990a. Land-water interfaces: Metabolic and limnological regulators. Edgardo Baldi Memorial Lecture. Verh. int. Ver. Limnol. 24: 6–24.

Wetzel, R. G., 1990b. Reservoir ecosystems: Conclusions and Speculations. In K. Thornton, B. L. Kimmel and F. E. Payne (eds.), Reservoir Limnology. Academic Press, New York: 227–238.

Wetzel, R. G., 1991. Extracellular enzymatic interactions in aquatic ecosystems: Storage, redistribution, and interspecific communication. In R. J. Chrost (ed.), Extracellular Enzymes in Aquatic Ecosystems. Springer-Verlag, New York: 6–28.

Wetzel, R. G. & R. A. Hough, 1973. Productivity and role of aquatic macrophytes in lakes. An assessment. Pol. Arch. Hydrobiol. 20: 9–19.

Wetzel, R. G., P. H. Rich, M. C. Miller & H. L. Allen, 1972. Metabolism of dissolved and particulate detrital carbon in a temperate hard-water lake. Mem. Ist. ital. Idrobiol. Suppl. 29: 185–243.

Zepp, R. G., G. L. Baughmann & P. F. Schlotzhauer, 1981. Comparison of photochemical behavior of various humic substances in water. I. Sunlight induced reactions of aquatic pollutants photosensitized by humic substances. Chemosphere 10: 109–117.

*Hydrobiologia* **229**: 199–224, 1992.
*K. Salonen, T. Kairesalo & R.I. Jones (eds.),*
*Dissolved Organic Matter in Lacustrine Ecosystems: Energy Source and System Regulator.*



# Littoral-pelagial interchange and the decomposition of dissolved organic matter in a polyhumic lake

Timo Kairesalo, Anja Lehtovaara & Pentti Saukkonen
*Lammi Biological Station, University of Helsinki, SF-16900 Lammi, Finland*

*Key words:* aquatic moss, *Warnstorfia*, *Daphnia longispina*, bacterial production, leucine uptake, phenol metabolism, decomposition, polyhumic lake

## Abstract

The small, polyhumic lake, Mekkojärvi (southern Finland), is bordered by a moss vegetation zone (*Warnstorfia* and *Sphagnum* species) which provides a habitat-rich and productive environment for many planktonic and periphytic animals. Impacts of moss on the metabolism of bacterioplankton, phytoplankton and zooplankton in polyhumic water were investigated in laboratory throughflow systems. Growing *Warnstorfia* (together with epiphytic algae and bacteria) suppressed the production of planktonic algae but had no clear effect on leucine uptake, and hence bacterial production, or on the decomposition of humic substances. Phenol uptake and mineralization rates, however, were lower in the littoral water than in the pelagial water. Excretion of organic carbon by *Warnstorfia*, algae or *Daphnia longispina* (the predominant crustacean in the pelagial water) provided only a minor contribution to bacterial production; therefore, a major contribution had to be from humic substances. A bacterial production efficiency of 31–38% could account for the microbial respiration in the water. The results indicated that bacterial, or detrital matter (originating largely from the littoral zone), could not obviate the need for algal food, and that a great deal of particulate matter in the water was poor or useless food for *Daphnia*. In all, the bulk of dissolved organic matter in Lake Mekkojärvi was biochemically highly recalcitrant. Our results indicate that humic substances (from watershed or littoral area) which, through bacterial degradation, enter the planktonic food web of the lake are mainly lost through respiration by microorganisms.

## Introduction

Littoral vegetation zones are the most productive areas in lakes. Areal production of macrophytes and the associated algal communities may be orders of magnitude higher than that of the pelagial phytoplankton. However, the fates and fluxes of the organic matter synthesized by littoral primary producers are poorly known, although the question of the importance of littoral DOM, released either by growing macrophytes or through microbial degradation, was addressed two decades ago by Wetzel & Allen (1972). In particular, the fate of littoral DOM is of great significance when considering the contributions of allochthonous and autochthonous DOM to the lake's total metabolism.

In Finland, many sheltered forest lakes are especially rich in humic materials originating from the drainage area and from the mossgrown, peaty littoral zone. Mosses (*Warnstorfia*, *Sphagnum*) may form thick mats extending from the shoreline out to the open water area, eventually covering the lake and turning it to a bog.

A small, polyhumic lake, Mekkojärvi, has been an object of plankton studies for several years (Kankaala & Eloranta, 1987; Salonen & Jokinen, 1988; Kankaala, 1988; Arvola & Kankaala, 1989; Jones, 1990). The moss, *Warnstorfia procera* (Ren. & H. Arn.) (Tuomikoski, 1973), is spreading out to the lake and possibly having an increasingly important influence on the quantity and quality of DOM in the lake water and hence also affecting the whole lake ecosystem. In addition, *Warnstorfia* clumps and mats, while colonizing the open water area, make the area more productive with plenty of new microhabitats for a great variety of micro- and macro-organisms, and leading towards the dominance of *Sphagnum* and higher vegetation (Glime *et al.*, 1982).

In this study we have tested the hypothesis that littoral primary producers (*Warnstorfia* + associated algal communities), while releasing newly synthesized, labile organic compounds into the ambient water, also enhance the degradation of more recalcitrant humic substances, and thus accelerate the energy flow up to cladoceran production. Thus, the impacts of *Warnstorfia* on the production and decomposition processes in the ambient water, the first steps of the succession towards a bog ecosystem, were assessed in this study.

## Materials and methods

### Description of the lake

Mekkojärvi is a small (0.35 ha), polyhumic (300–500 mg Pt $l^{-1}$) forest lake at Lammi in southern Finland. The lake, with a maximum depth of 4.3 m and mean depth of 2 m, is steeply stratified in the summer; a narrow thermocline separates the 0.5–1 m thick, oxic epilimnion from the anoxic hypolimnetic water. A peaty moss bed, with sedges and grasses emerging from the shoreline, borders the open water area (Fig. 1). Details of the lake's chemical characteristics have been presented elsewhere (e.g. Arvola & Kankaala, 1989; Münster *et al.*, 1989).

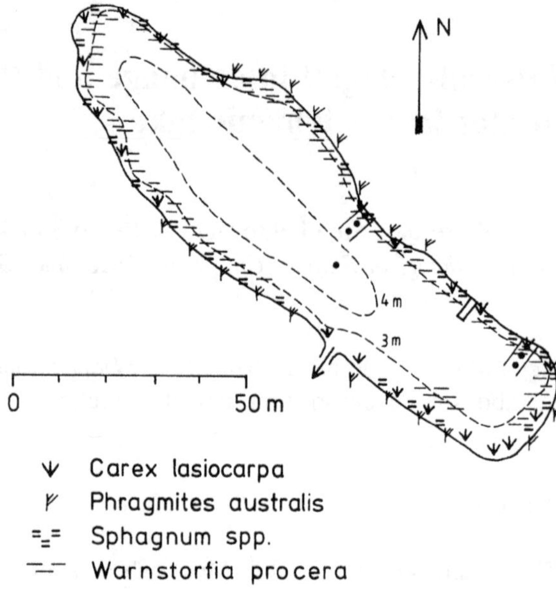


*Fig. 1.* Depth contours, macrophyte zonation and the sampling transects and points in the polyhumic lake Mekkojärvi.

### Field sampling

Field samples of plankton were collected on 21 May, 1 July and 18 August, 1987. Samples were taken along two transects from the shoreline out to the open water area: from among the moss vegetation (Lit. 1); from the outer edge of the moss bed (Lit. 2); from the open water area, 3–4 m outside the moss bed (Pel. 1); and from the mid area of the lake, from 0–2 m water column (Pel. 2). A detailed study of the littoral vegetation was made in August 1987.

Samples from among the moss bed were taken using an acrylic plastic tube (bottom area 54.8 $cm^2$, length 85 cm), or with a smaller tube (bottom area 15 $cm^2$), and pelagial samples with a Sormunen tube sampler (volume 6.4 l, length 1 m). Bulked (12–25 l) water samples were treated as follows. (1) A 0.6–1.9 l subsample was sieved onto a 25 $\mu$m net, preserved in 4% formaldehyde, and later used for microscopical (Wild M5) identification and enumeration of rotifers. (2) The rest of the sample water was sieved onto an 80 $\mu$m net, preserved in 4% formaldehyde, and later used for microscopical identification and enumeration of crustaceans and other meta- and macrofauna. Quantitative sampling of animals living among and attached to the floating moss material proved too difficult; therefore, only qualitative samples were taken by enclosing moss clumps into containers. In the laboratory, animals were carefully washed from the moss and preserved in 4% formaldehyde for microscopical analysis.

*Experimental design for the laboratory studies*

The three laboratory experiments, in May, June and July (1989), were all carried out under controlled temperature and light conditions using a throughflow system (Fig. 2). Water and moss samples for the May and June experiments were taken from the eastern corner of the lake a few days prior to each experiment; 'littoral water' (total volume 100–120 l) from among the moss vegetation using an acrylic plastic tube (diameter 4.4 cm), or a manual waterpump (boatpump), and 'pelagial water' (400–500 l) from the uppermost 20 cm of the open water area using 8 l enclosures. Moss samples were taken from among a dense patch of the *Warnstorfia*-bed. In July, 'littoral water' (total volume 400–500 l) was taken as described earlier, whereas 'pelagial water' (total volume 100–120 l) was taken from the 0–1 m water column using a Sormunen tube sampler (volume 7 l).

Littoral and pelagial waters, transported in large tanks into the laboratory, were filtered through 25 $\mu$m nets (except *Daphnia*-water in the July experiment; cf. Fig. 18) and mixed well before being introduced either as throughflowing water into the 500 l darkened and cooled (6–8 °C) store aquarium (lined with polyethylene), or as test waters into 25 l (acid-washed) glass community bottles. The community bottles were hung in two aquaria at a water temperature of 20 ± 0.5 °C. Water in the store aquarium, as well as in each community bottle, was stirred with a 12 cm long, two-blade propeller connected to a 12 V electric pump; the aquarium water was stirred (100 rpm) all the time while the water in the community bottles with an on/off interval of 15 min./45 min. During the experiments a 16 h/8 h light/dark period was used. The laboratory lamps provided white light with a photon flux density of 350 $\mu$E m$^{-2}$ s$^{-1}$ on the surface of the aquarium water.

Throughflowing water from the store aquarium ran through a peristaltic pump (Ismatec MV-GES) in separate, darkened silicon tubes into each community bottle with an average flow rate

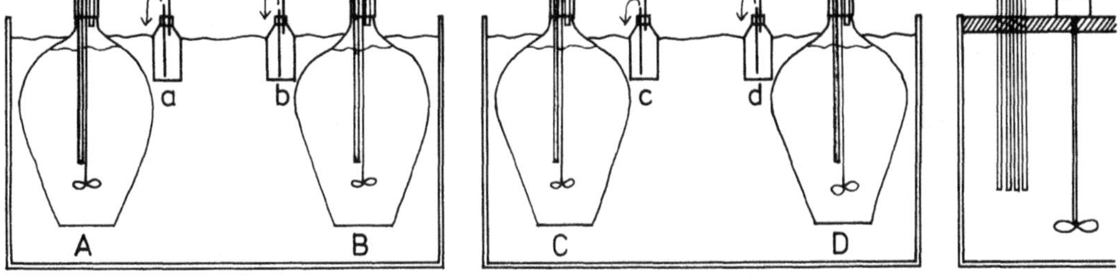


*Fig. 2.* Schematic representation of the throughflow system, the capital letters (A–D) indicating the community bottles and the small letters (a–d) the corresponding sample bottles (see text for further details).

of 5.5 ml min$^{-1}$; the renewal time in the bottles was thus about 3 days. Water from the community bottles then discharged by another peristaltic pump (Ismatec IPS-12) through 1 l, darkened sample bottles (acid washed glass bottle).

## Analytical methods

### *Sample bottles*

Each of the four 1 l sample bottles was replaced every day at the same time (at 8–9 a.m., i.e. 10–11 hours from the start of the light period) and treated, together with samples from the store aquarium, as follows.

(1) A 5 ml subsample was taken from undisturbed sample bottles with a syringe and needle, and dissolved inorganic carbon (DIC) was analysed as $CO_2$ in an infrared gas analyser according to Salonen (1981).

(2) 3 transparent and 2 darkened glass scintillation vials (20 ml) were filled with each sample water for measurement of algal production by the $^{14}C$-method. 30 $\mu$l $NaH^{14}CO_3$-solution ($1.11 * 10^4$ Bq) was added into each vial. After 3.5 h incubation in the aquarium conditions, total organic $^{14}C$ ($TO^{14}C$) was determined from 8 ml subsamples with Lumagel scintillation cocktail after removal of $^{14}CO_2$ by acidification with 0.1 N HCl and bubbling with air for 10–15 min.

(3) For dissolved organic carbon (DOC), 20 ml subsamples were filtered through 0.2 $\mu$m Nuclepore filters; 10 ml was used for rinsing the PE-plastic tube in which the DOC sample was then frozen for later analysis of carbon using high temperature (980 °C) combustion with oxygen (Salonen, 1979).

(4) For chlorophyll-*a*, 30 ml subsamples were filtered through Whatman GF/C-filters which were then frozen and kept in darkness in separate plastic tubes for fluorometric analysis.

(5) For particulate organic carbon content (POC), 30 ml subsamples were filtered through precombusted (500 °C for 6 hours) Whatman GF/F-filters, which were then kept frozen in plastic tubes for later analysis of carbon by the high temperature combustion technique (Salonen, 1979).

(6) For dissolved inorganic nutrients ($PO_4$, $NO_3 + NO_2$, $NH_4$), 50–100 ml of sample water were filtered through 0.2 $\mu$m Millipore filters, and the molybdate reactive phosphate, nitrate + nitrite (after reduction of nitrate to nitrite in a cadmium-copper column) and ammonium were determined according to Murphy & Riley (1962), Wood *et al.* (1967), and Solorzano (1969), respectively, using an AKEA automatic analysing system (Datex).

(7) Total nutrient concentrations were determined later from frozen subsamples after alkaline peroxosulphate oxidation in an autoclave (Koroleff, 1979).

For measuring heterotrophic activity in the water, three different methods were used.

(8) Total community respiration was measured as an increase of DIC in 50 ml Jena glass bottles after 24 h incubation in darkness in a 20 °C water bath.

(9) For measuring the incorporation of leucine by bacteria L-[U-$^{14}$C]leucine (specific activity, 11.5–12.7 GBq/mmol, Amersham) was added into precombusted (500 °C for several hours) test tubes (5 ml triplicates + 2 blanks containing 250 $\mu$l 39% formaldehyde) at a final concentration of 100 nM (Fig. 3). After 60 min incubation in a 20 °C water bath, 5 ml of 10% TCA were added and then for determining the incorporation of $^{14}$C-leucine in cold-TCA-insoluble fraction, the tubes were kept in crushed ice/water mixture for 20–30 min waiting for filtering of the samples through 0.2 $\mu$m Sartorius 1137-filters. To eliminate contamination of the filters with non-incorporated, radioactive label, the filters were rinsed carefully – before and after filtering the samples – with cold, unlabelled 600 nM leucine-solution. The radioactivities of the filters were then determined with 10 ml Lumagel scintillation cocktail, and with 0.2 ml ethylene-glycol-monomethylether for dissolving the filter. Bacterial production was calculated using the conversion factors of $1.64 \times 10^{17}$ cells per mol leucine (Chin-Leo & Kirchman, 1988) and 360 fg C $\mu m^{-3}$ (Tulonen *et al.*, unpublished; average cell volume 0.025 $\mu m^3$).

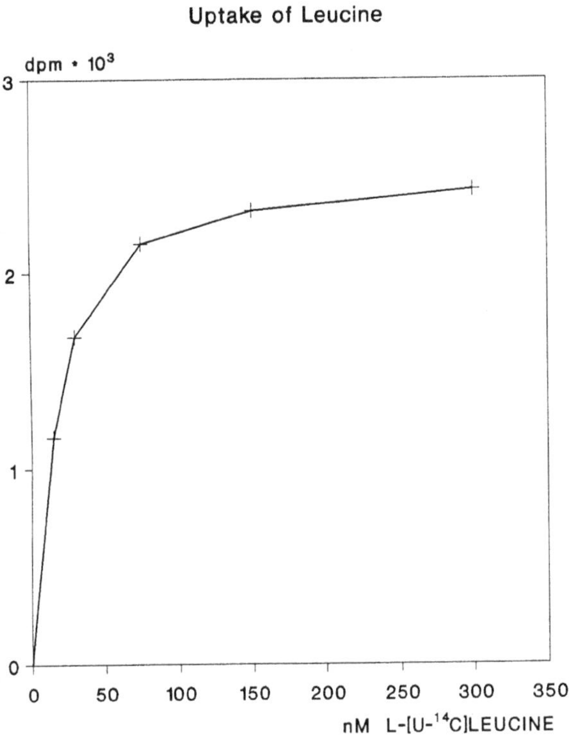


*Fig. 3.* Dependence of leucine uptake on leucine concentration in Mekkojärvi water.

(10) For measuring the uptake and mineralization rates of phenol by bacteria, [U-$^{14}$C]phenol (specific activity 4.48 GBq/mmol, Amersham) was added into each test bottle (9 ml bottle; 8 ml triplicates + 1 blank containing 250 $\mu$l 39% formaldehyde) at a final concentration of 100 nM. The bottles were then sealed tightly with parafilm and a rubber stopper and incubated in a 20 °C water bath for 2–2.5 hours. Incubation was stopped by inserting the bottles into crushed ice/water mixture. 2 ml of the sample was used for assessing the mineralization of phenol to $CO_2$; using a syringe and needle, subsamples were taken through parafilm and $^{14}CO_2$ was absorbed into 9 ml of ethanolamine/ethanol (v/v 1:7) solution in a scintillation vial connected to the outflow of the DIC analysing system. The rest of the sample (6 ml) was filtered through Sartorius 1137-filters and the radioactivities of the filters, after a careful rinsing with MQ-water, were assessed with 10 ml of Lumagel scintillation cocktail ( + 0.2 ml ethylene-glycol-monomethylether for dissolving the filter).

(11) 100 ml subsamples for bacterial and algal counts were preserved with 1 ml of glutaraldehyde in plastic bottles. Bacteria were counted using the acriflavine epifluorescence counting method (Bergström *et al.*, 1986), and algae were identified in an inverted microscope using the Utermöhl-technique.

### *Community bottles*

In May and June, replicate community bottles were filled with the < 25 $\mu$m littoral water. 250 fresh, 5 cm long tips of *Warnstorfia* thalli were then introduced into two community bottles, the *Warnstorfia*-bottles (cf. Fig. 5), and 50 similar tips in May and 127 tips in June were frozen immediately for the assessment of the initial plant biomass. The growth of *Warnstorfia* was determined as an increase in carbon biomass during the experiment. Carbon percentage of the dried, homogenized and weighed plant material was assessed using high temperature combustion in oxygen; the evolved $CO_2$ was analysed in an infrared gas analyser.

In July, the community bottles were filled with the pelagial water; for replicate Control-bottles (cf. Fig. 18), the water was first filtered through a 25 $\mu$m net, while replicate *Daphnia*-bottles (cf. Fig. 18) contained the unmodified lake plankton.

In the May and June experiments, pelagial water (< 25 $\mu$m), and in the July experiment littoral water (< 25 $\mu$m), provided the throughflowing water. In order to maintain and enhance biological processes (to simulate the introduction of nutrient-rich hypolimnetic water into the epilimnion), nutrient enrichment with $NaHCO_3$, $KH_2PO_4$ and $KNO_3$ (cf. Figs. 5–7; Figs. 18–21) was made on the third or fourth day from the start of the experiment.

Different molecular weight fractions (> $10^4$ dalton; $10^3$–$10^4$ dalton; < $10^3$ dalton), in the beginning and end of the May and June experiments, were separated using tangential flow membrane ultrafiltration (Pellicon, Millipore). DOC concentration of the different fractions was determined using the high temperature combustion method (Salonen, 1979).

*Excretion of organic carbon (EOC) by* Warnstorfia

EOC by *Warnstorfia* was measured using 'fully labelled' plants which had been grown in transparent, sealed glass bottles in $<0.2\ \mu m$, enriched (with $NaHCO_3$, $KH_2PO_4$ and $KNO_3$) and continuously stirred Mekkojärvi water, under constant conditions of light (300–400 $\mu E\ m^{-2}\ s^{-1}$), temperature (18–20 °C) and $DI^{14}C/DI^{12}C$-ratio for several weeks, from 1.5–1 cm long tips up to about 5 cm long plants. Five replicate measurements of EOC were made. In each experiment, 3 labelled, washed plants were introduced into fresh Mekkojärvi culture medium (with $NaH^{14}CO_3$) in a 200 ml transparent glass bottle under the described light and temperature conditions. EOC by *Warnstorfia* was measured as an increase in $TO^{14}C$ (see the determination of total algal production) in the water during the 2–3 hour incubation. EOC was calculated as:

$$\frac{EO^{14}C}{TO^{14}C \text{ (in the plants)}} \times TOC \text{ (in the plants)}$$

and

$$\frac{EO^{14}C}{\text{assimilated } DI^{14}C} \times DIC$$

*EOC by* Daphnia

EOC by *Daphnia* was measured using both 'fully labelled' and 'nonlabelled' animals which had been fed, respectively, either a $^{14}C$-labelled *Chlamydomonas*/bacteria suspension or an unlabelled but otherwise similar suspension, in sealed glass bottles in Mekkojärvi water for 2–3 months (under the constant laboratory conditions; see above). EOC measurements were made in a throughflow system. 5–7 individuals (length 1.2–1.5 mm, biomass 2.5–3.5 $\mu g\ C\ ind^{-1}$; rinsed with filtered Mekkojärvi water) from both culture populations were introduced into two separate Pasteur-pipettes, a third pipette without animals being a reference. The lower end of each pipette was wrapped with a 100 $\mu m$ net. Two experiments were carried out with a continuous throughflow rate of 1 ml $min^{-1}$. Both experiments started with a throughflow of filtered ($<0.2\ \mu m$) Mekkojärvi water, followed by the labelled *Chlamydomonas*-suspension ($9–10 * 10^3$ cells $ml^{-1}$), then by the unlabelled algal suspension and, finally, by filtered lake water. Outflow water from each pipette was sampled separately in 9 ml (i.e. 9 min) fractions into 20 ml glass vials, 4 ml of that volume was used for the measurement of $TO^{14}C$; 5 ml was filtered through 0.2 $\mu m$ Millipore-filters and subsequently used for the measurement of $DO^{14}C$. Carbon and $^{14}C$ contents of the dried test animals were assessed separately using the high temperature combustion method and subsequent absorption of $^{14}CO_2$ into ethanolamine.

## Results

*Field studies*

The moss bed, surrounding the whole lake and mainly formed by *Sphagnum* species in the inner zone and by *Warnstorfia* species in the outer zone, may extend 5–10 m out from the shoreline (Fig. 1). In the inner part, within the peaty *Sphagnum* bed, some rooted plants, particularly *Carex lasiocarpa* Ehrh. and *Phragmites australis* (Cav.) Trin. ex Steudel, emerge and form narrow stands.

The density of rotifers, cladocerans and copepods together with other animal groups (nematods, oligochaetes, chironomids, water mites) was almost always highest in the inner littoral zone; only in May were copepods as plentiful in the other zones (Fig. 4).

The most abundant littoral rotifers were *Rotaria* spp., *Conochilus* spp., and *Trichocerca* spp.. *Keratella testudo* Ehr., which was also abundant in the littoral, was clearly the dominant rotifer in the pelagial area. *Cyclops strenuus* Fischer and *Eucyclops serrulatus* Fischer were the dominant

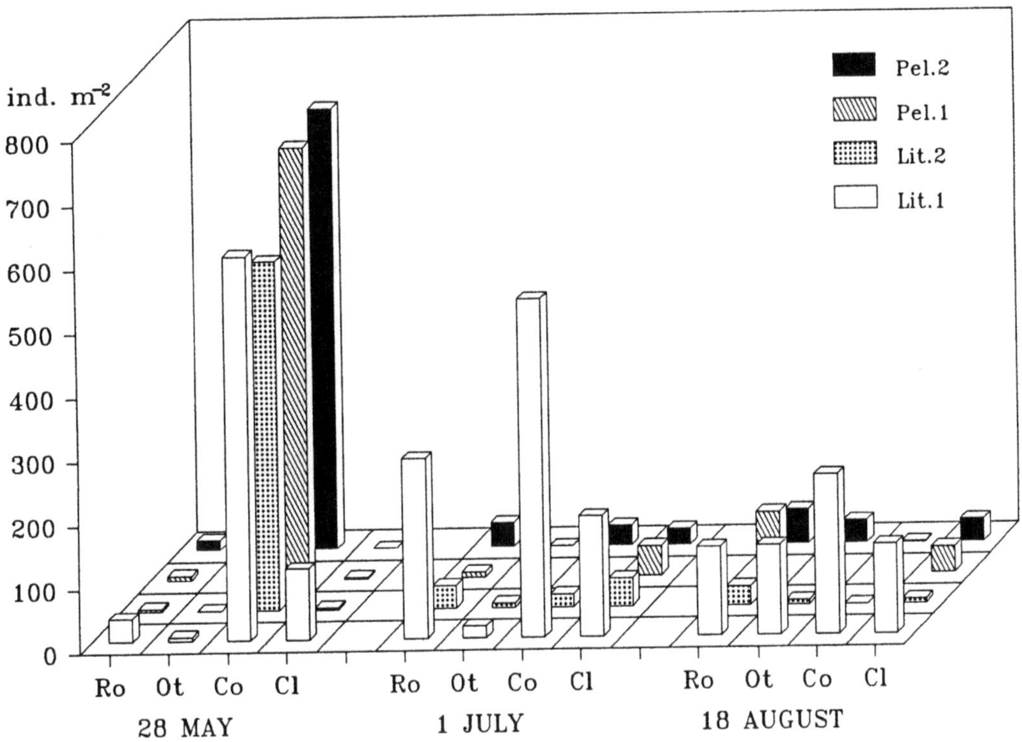


*Fig. 4.* Densities (ind $m^{-2}$) of different animal groups (Ro = Rotatoria; Co = Copepoda; Cl = Cladocera; Ot = Others) in two littoral (Lit. 1, Lit. 2) and two pelagial (Pel. 1, Pel. 2) zones in May, in June–July, and in August.

copepods. Especially in July–August, *Daphnia longispina* O. F. Müller predominated in the pelagial zones, whereas in the littoral zones it was replaced by chydorids (*Acroperus harpae* Baird, *Alonella excisa* Fischer, *Chydorus sphaericus* O. F. Müller). *Polyphemus pediculus* L. was most abundant at the outer edge of the moss bed.

*Laboratory experiments in May and June*

*Chemical parameters*

Available phosphate and nitrate (including possible nitrite) were used efficiently in the community bottles during both experiments, since no distinct increase in their concentrations could be measured despite a substantial load of these nutrients with the throughflowing water. However, the concentrations of total nutrients, although being first higher in *Warnstorfia*-bottles (especially in May) than in Control-bottles, achieved a quite equal level in all community bottles by the end of the experiments (Figs. 5 and 6). A part of the nutrients was lost through sedimentation or through uptake by attached algae and bacteria on the glass surfaces of the community bottles; this loss was assumed to be equal in the *Warnstorfia*-bottles and in the Control-bottles.

DIC started to increase in the community bottles after the enrichment of the throughflowing water; the difference between *Warnstorfia*-bottles and Control-bottles was attributable to the uptake of DIC by the moss and epiphytic algae. In May, the increase of DIC was linear to the end of the experiment, but in June, saturation was achieved five days after the start of the enrichment (Fig. 7). The load of DIC ($NaHCO_3$) also raised pH in the throughflowing water by one unit, from 5.9 to 6.9 and, subsequently, pH in the community bottles increased from the initial pH of 5.9 up to 7–7.5 at the end of the experiment.

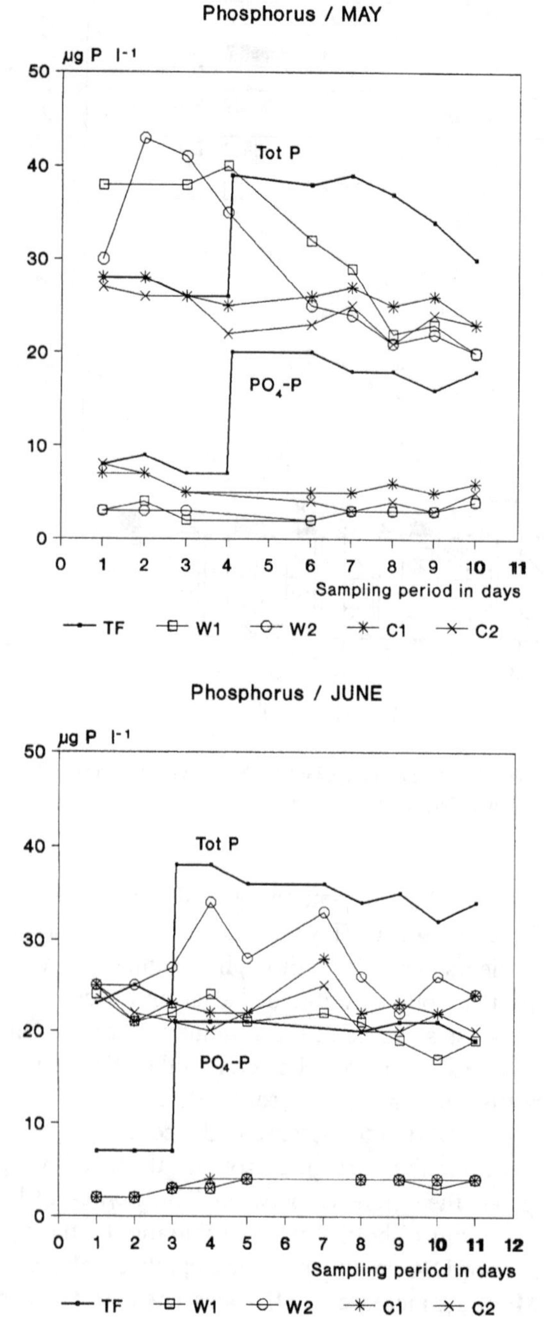


*Fig. 5.* Total and phosphate phosphorus concentrations ($\mu g\ P\ l^{-1}$) in the throughflowing water (TF), and the different community waters (W1–2, with *Warnstorfia*; C1–2, controls without *Warnstorfia*) in the May and June experiments.

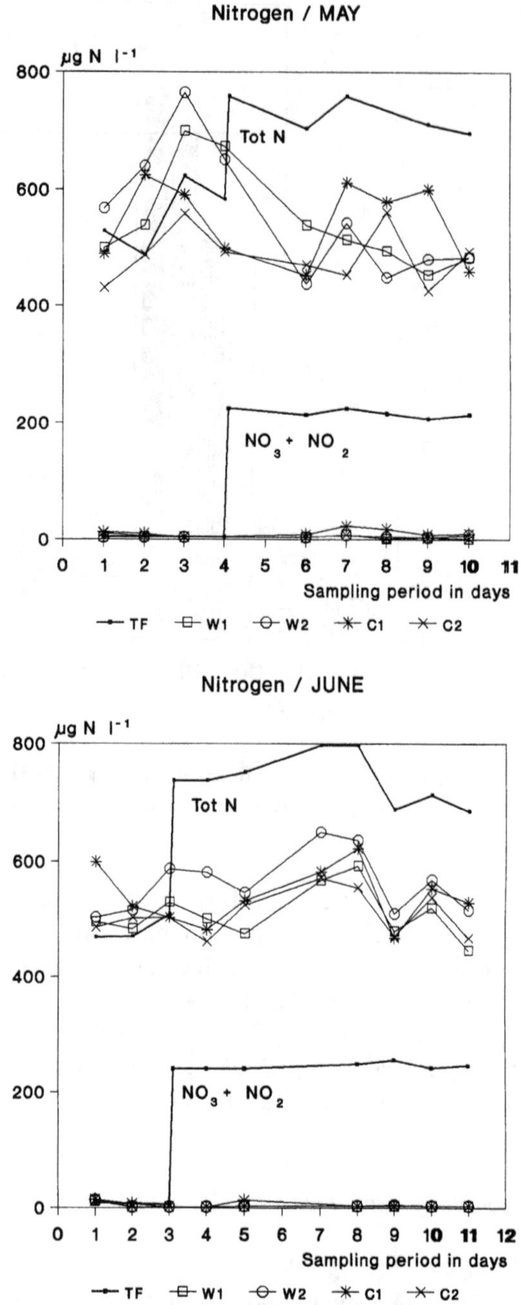


*Fig. 6.* Total and nitrate (+nitrite) nitrogen concentrations ($\mu g\ N\ l^{-1}$) in the May and June experiments (for symbols see Fig. 5).

*Microbial metabolism*

In May, microbial ($<25\ \mu m$) dark respiration was initially distinctly higher in *Warnstorfia*-bottles than in Control-bottles; thereafter, the microbial respiration was enhanced by the nutrient enrichment more in Control-bottles than in *Warnstorfia*-bottles (Fig. 8). In June, microbial respiration was

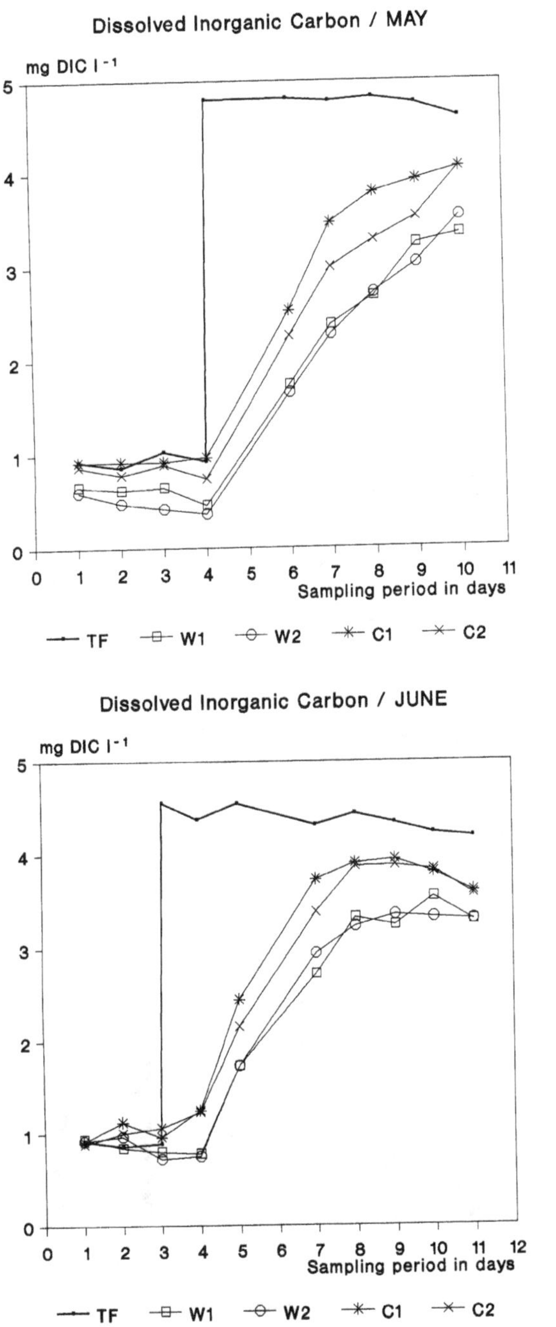


*Fig. 7.* Dissolved inorganic carbon concentrations (mg C $l^{-1}$) in the May and June experiments (for symbols see Fig. 5).

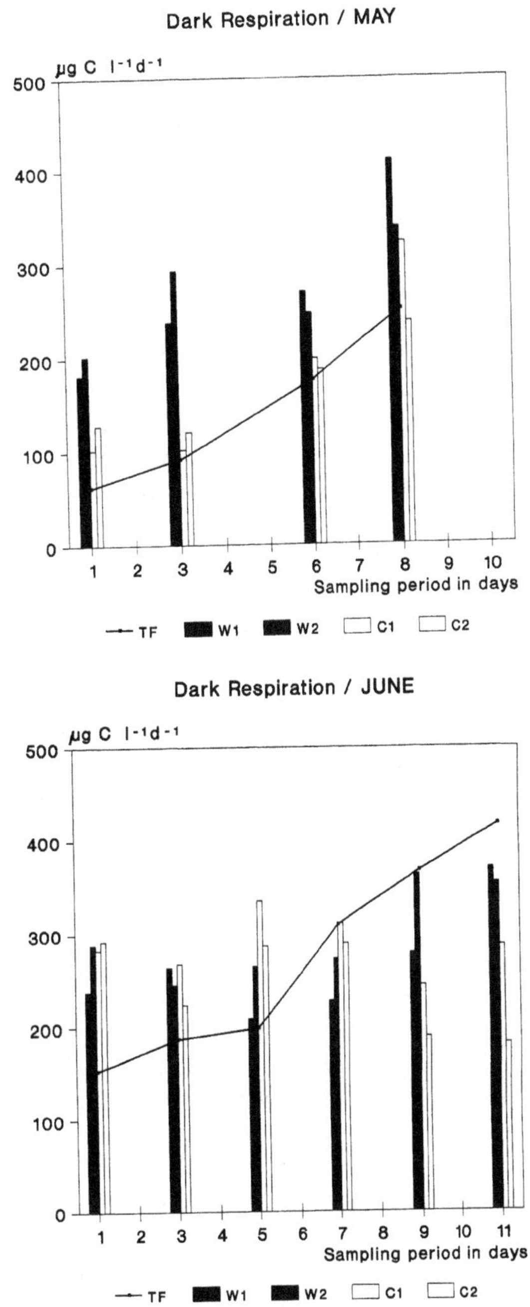


*Fig. 8.* Microbial (<25 μm) dark respiration rates (μg C $l^{-1}$ $d^{-1}$) in the May and June experiments (for symbols see Fig. 5).

initially at the same level in both systems, and at first again responded faster in Control-bottles to the enrichment with nutrients. However, at the end of the experiment the respiration rate in *Warnstorfia*-bottles, as well as that in the store aquarium, clearly exceeded that in Control-bottles (Fig. 8).

Planktonic algal production, although varying

widely between the duplicate systems and reaching a much higher level in June than in May, developed in a similar way during both experiments. Prior to the nutrient enrichment the algal production was higher in *Warnstorfia*-bottles, while at the end of the experiment it was higher in Control-bottles. Planktonic algae completely lost their photosynthetic capability in the dark store aquarium after 10 days in May, but within 4 days in June (Fig. 9).

During both experiments, uptake of leucine in the community bottles was stimulated only for a few days by the enrichment with nutrients. No such stimulation, however, was measured in the throughflowing water. In May, the highest uptake rate was measured on almost every day in each of the *Warnstorfia*-bottles, while in June the highest rates were measured in Control-bottles subsequently to the nutrient-enrichment; on other days the uptake rates were quite similar in both systems (Fig. 10).

During both experiments, but especially in May, the uptake and mineralization of phenol was distinctly higher in the (throughflowing) pelagial water than in the littoral (community) waters. The lowest rates in all systems were measured at the end of each experiment. The enrichment with nutrients had no clear influence on phenol metabolism (Fig. 11). When including all the samples in the May experiment, 14.9–19.7% of the phenol taken up, on average, was mineralized to $CO_2$ during the incubation time, while in June the percentages were higher, averaging 43.1–50.5%.

*Bacterio- and phytoplankton, and POC*

Density and biomass (volume) of bacteria were, on average, higher in Control-bottles than in *Warnstorfia*-bottles, and the enrichment of the throughflowing water did not cause any clear change in the densities or biomasses of bacteria in the community bottles, or in the throughflowing water (Figs. 12 and 13).

During both experiments, chlorophyll-a concentration was first higher in *Warnstorfia*-bottles and at the end higher in Control-bottles, but remained relatively low ($<5\ \mu g\ l^{-1}$) throughout.

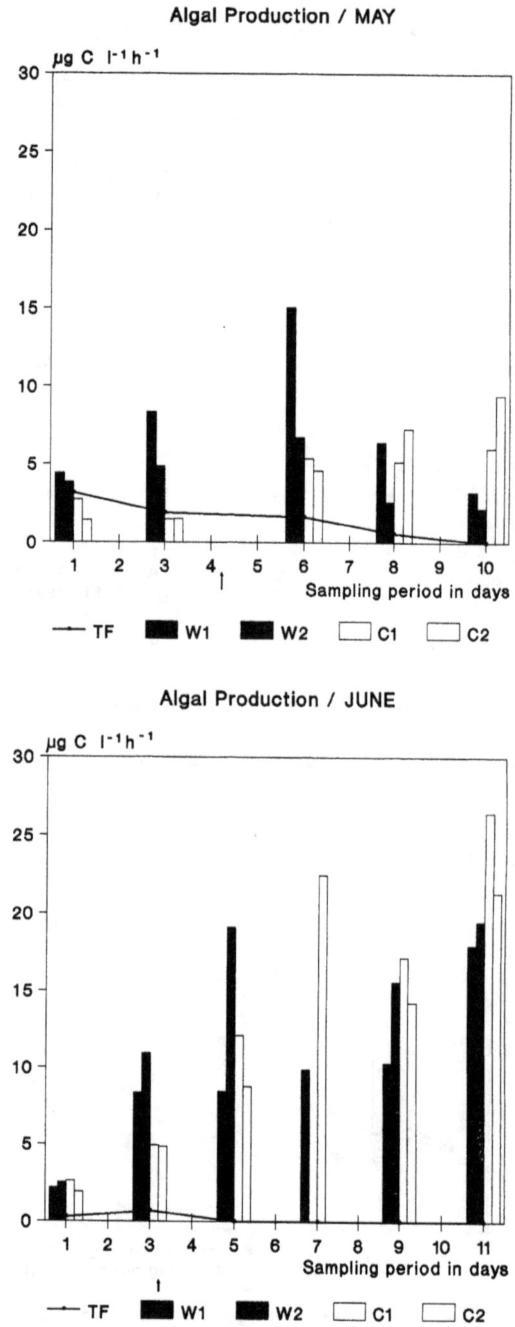


*Fig. 9.* Phytoplankton production rates ($\mu g\ C\ l^{-1}\ h^{-1}$) in the May and June experiments (for symbols see Fig. 5).

In the dark store aquarium, planktonic algae lost their chlorophyll (Fig. 14). Small (3–10 $\mu m$) flagellates dominated most samples (densities being up to $2 * 10^4$ cells $ml^{-1}$) together with *Cryp-*

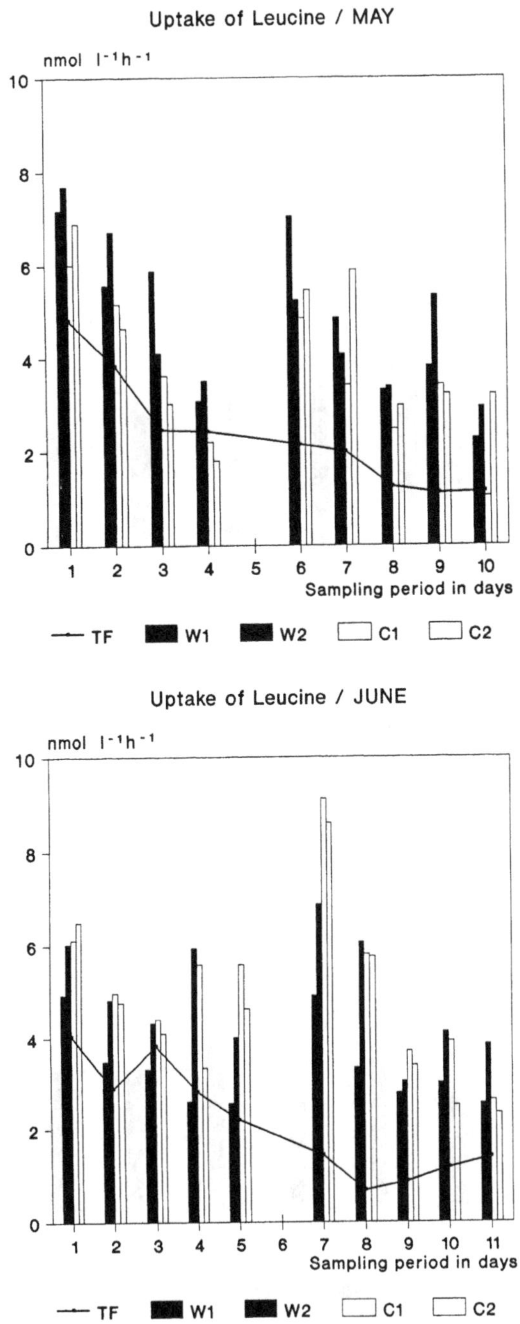


*Fig. 10.* Leucine uptake rates (nmol $l^{-1} h^{-1}$) in the May and June experiments (for symbols see Fig. 5).

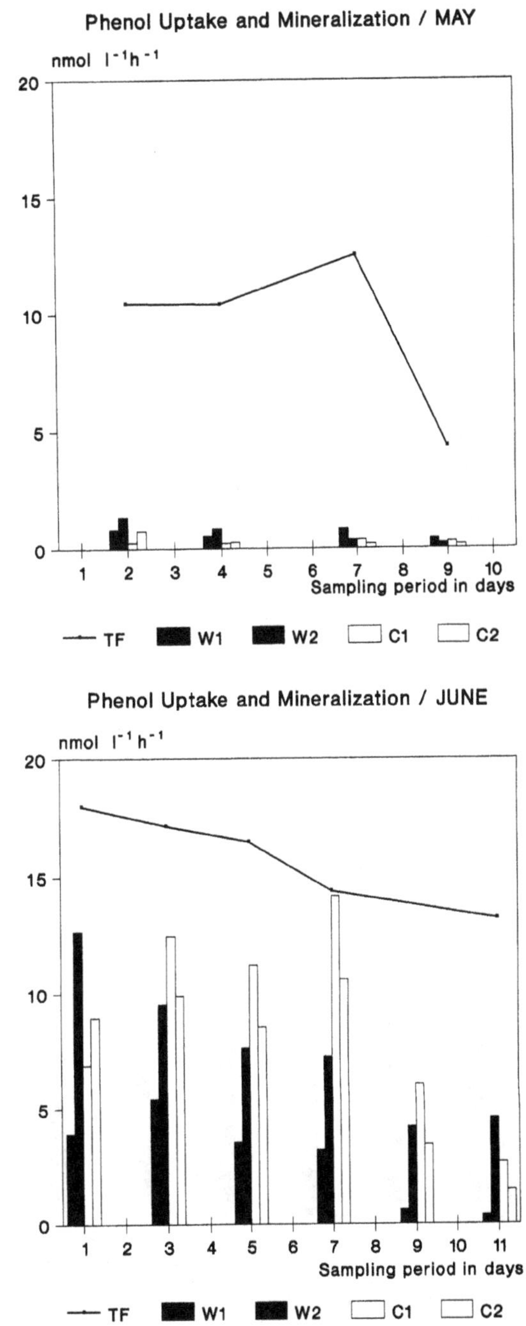


*Fig. 11.* Phenol uptake and mineralization rates (nmol $l^{-1} h^{-1}$) in the May and June experiments (for symbols see Fig. 5).

*tomonas*, *Chlamydomonas*, *Tabellaria* and *Synedra* species. *Cryptomonas* was especially numerous among detritus 'clumps'.

The high POC-concentrations in *Warnstorfia*-bottles in May were mainly attributable to moss detritus (Fig. 15). Detached moss tissue also evidently contributed to the chlorophyll-a concentrations (Fig. 14).

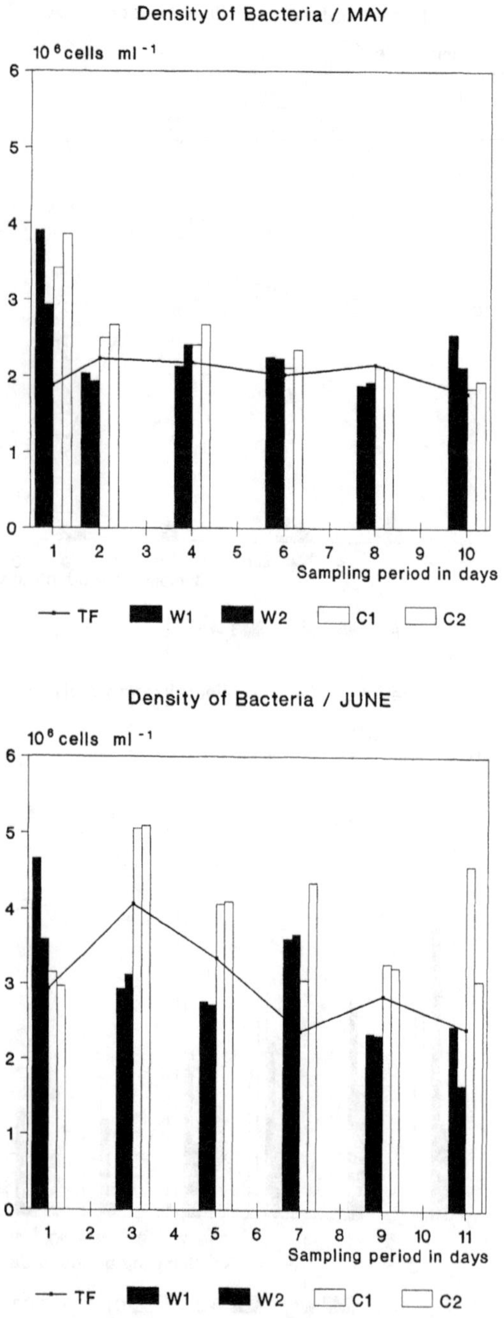


*Fig. 12.* Bacterial densities ($10^6$ cells $ml^{-1}$) in the May and June experiments (for symbols see Fig. 5).

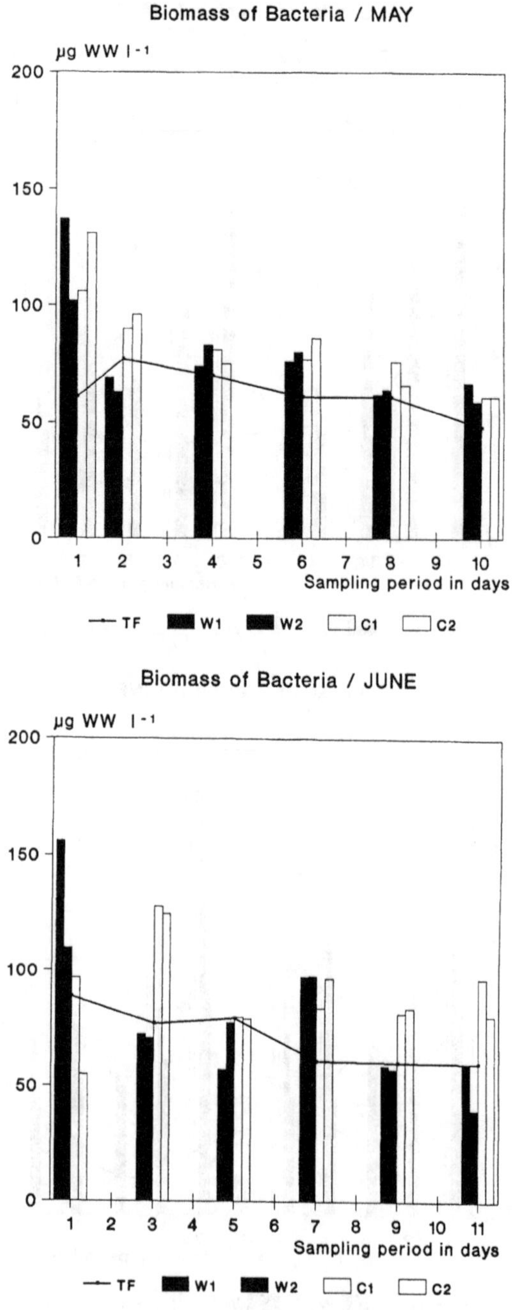


*Fig. 13.* Bacterial biomasses ($\mu g$ WW $l^{-1}$) in the May and June experiments (for symbols see Fig. 5).

*Production of* Warnstorfia *and its epiphytes*

Dry weights and the carbon percentages (43.2–44.2 ± 1.0–4.9%; range of the median ± SD) gave an average daily growth of *Warnstorfia* from 1.70 to 4.30 mg C during the May and June experiments (Table 1). The difference in DIC-concentration between Control- and *Warnstorfia*-bottles, which was assumed to be solely attributable to carbon fixation of the moss community (*Warnstorfia* + epiphytes), gave higher values for average

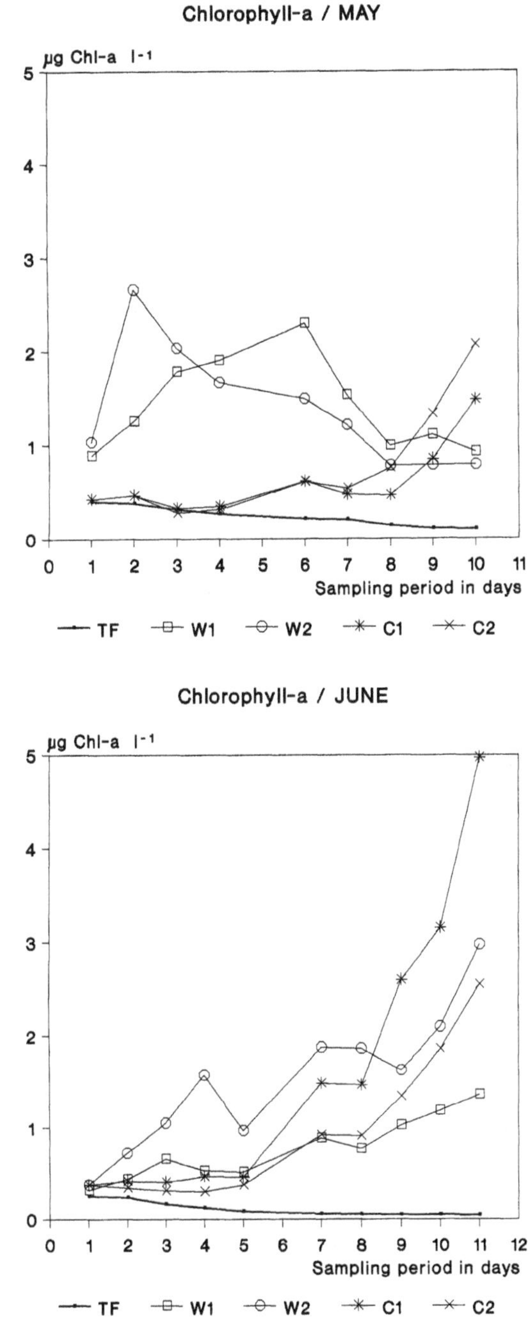


*Fig. 14.* Chlorophyll-*a* concentrations ($\mu$g Chl-*a* $l^{-1}$) in the May and June experiments (for symbols see Fig. 5).

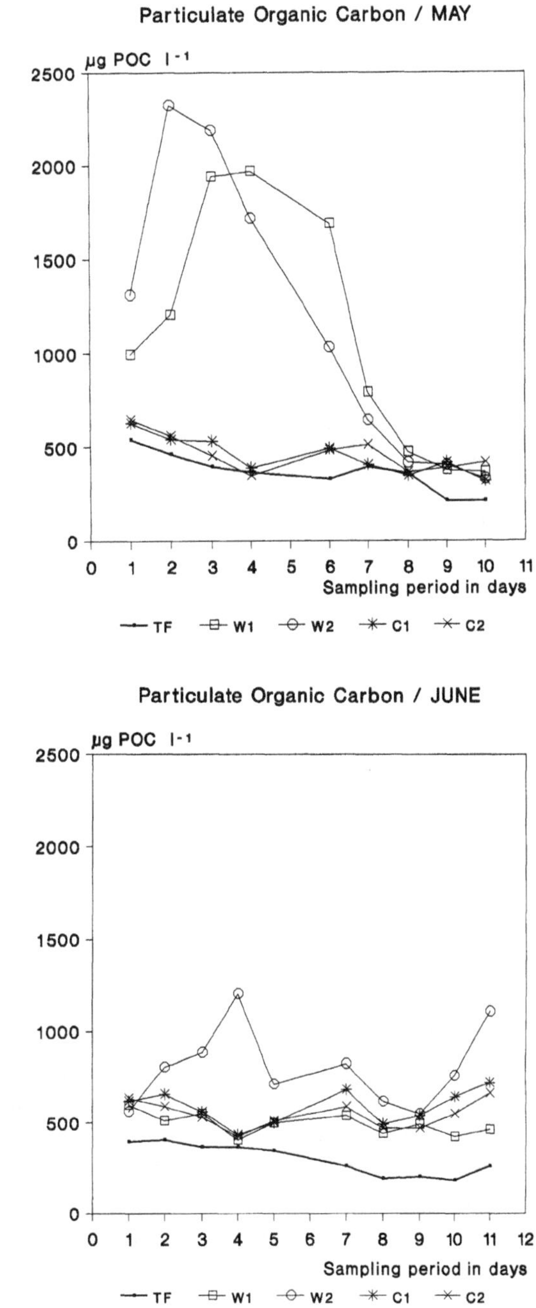


*Fig. 15.* Particulate organic carbon concentrations ($\mu$g C $l^{-1}$) in the May and June experiments (for symbols see Fig. 5).

daily production of 4.27 to 4.82 mg C. The production of epiphytes on moss surfaces ranged from 0.52 to 1.71 mg C, i.e. 10–50% of the total production of the moss community, if it is assumed that the difference between the growth of *Warnstorfia* and the DIC consumption was attributable only to the epiphytic algae. At the end of the experiment, diatoms, such as *Tabellaria* spp., *Synedra* spp., and *Cocconeis* spp., together with

*Table 1.* Average daily rates for carbon metabolism (mg C $d^{-1}$) in the community bottles (25 l) in the May and June experiments before (B) and after (A) nutrient enrichment (see text and figures for further details).

| | | W1 | W2 | C1 | C2 |
|---|---|---|---|---|---|
| MAY | | | | | |
| *Warnstorfia* | (B) | 1.5 | 2.1 | – | – |
| + Epiphytes prod. | (A) | 3.7 | 3.9 | – | – |
| *Warnstorfia* EOC | (B) | 0.1 | 0.1 | | |
| | (A) | 0.2 | 0.2 | – | – |
| Phytoplankton prod. | (B) | 2.8 | 1.9 | 0.9 | 0.7 |
| | (A) | 3.5 | 1.6 | 2.4 | 3.1 |
| Bacterioplankton prod. | (B) | 5.0 | 5.1 | 4.0 | 3.8 |
| | (A) | 4.0 | 3.9 | 2.8 | 2.7 |
| Microbial respiration | (B) | 5.7 | 6.7 | 2.8 | 3.4 |
| | (A) | 9.2 | 7.9 | 7.0 | 5.7 |
| JUNE | | | | | |
| *Warnstorfia* | (B) | 1.2 | 1.0 | – | – |
| + Epiphytes prod. | (A) | 2.8 | 2.8 | – | – |
| *Warnstorfia* EOC | (B) | 0.1 | 0.1 | – | – |
| | (A) | 0.1 | 0.1 | – | – |
| Phytoplankton prod. | (B) | 2.2 | 2.9 | 1.6 | 1.5 |
| | (A) | 5.1 | 7.8 | 8.5 | 6.4 |
| Bacterioplankton prod. | (B) | 3.6 | 4.7 | 4.8 | 4.8 |
| | (A) | 2.9 | 4.5 | 4.8 | 4.1 |
| Microbial respiration | (B) | 6.8 | 7.2 | 7.5 | 7.0 |
| | (A) | 7.3 | 8.5 | 7.9 | 6.4 |

some filamentous green algae, such as *Oedogonium* sp., were found attached on the moss thalli.

The enrichment of the throughflowing water greatly enhanced the production of the moss community. In May, the average production rate increased by 75–165%, from 92–134 up to 234–244 $\mu g\,C\,h^{-1}$, and in June by 136–175%, from 63–73 up to 172–173 $\mu g\,C\,h^{-1}$.

### *EOC by* Warnstorfia

EOC by *Warnstorfia* was 4.8–1.3% $h^{-1}$ of the photosynthetically fixed carbon, or $6.21 \pm 2.47 * 10^{-4}\%\ h^{-1}$ (mean ± SE; $n = 5$) of the total carbon biomass, which means 6.21 $\mu g$ C per g org. C $h^{-1}$. Thus, in May the EOC ranged from 8.7 to 12.9 $\mu g\,C\,h^{-1}$ in the community bottles according to the DIC consumption rates, or

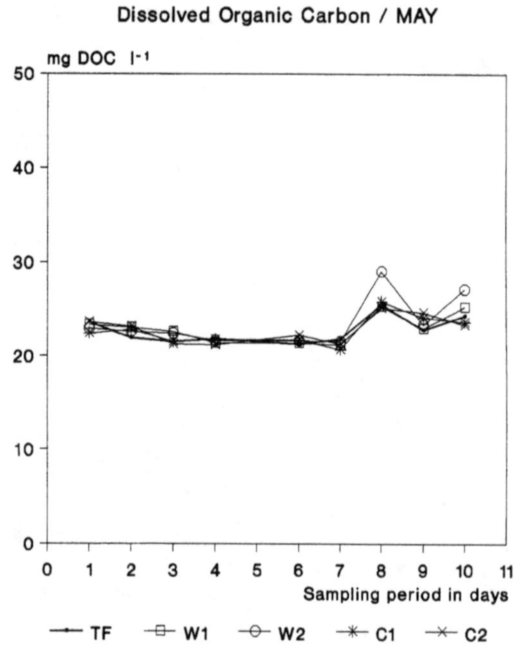


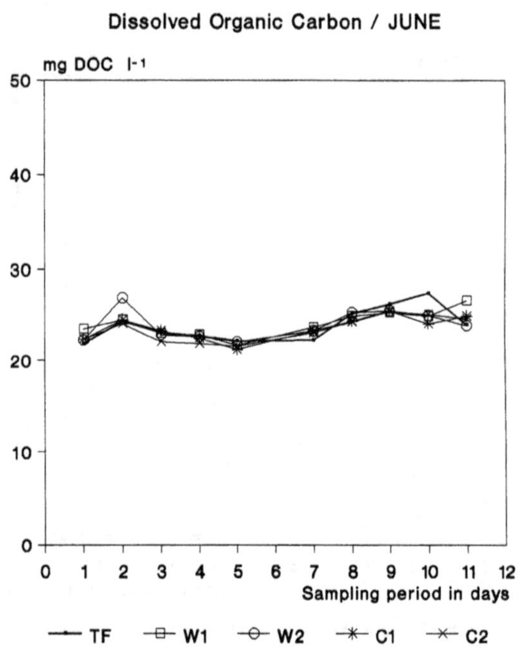


*Fig. 16.* Dissolved organic carbon concentrations (mg C $l^{-1}$) in the May and June experiments (for symbols see Fig. 5).

from 4.3 to 5.0 $\mu g\,C\,h^{-1}$ according to the moss biomasses (biomasses: W1 687.6; W2 801.5 mg C). In June, the EOC rates were respectively, 5.1–6.0 $\mu g\,C\,h^{-1}$, or 3.8–4.1 $\mu g\,C\,h^{-1}$ (biomasses: W1 659.2 mg C; W2 608.5 mg C).

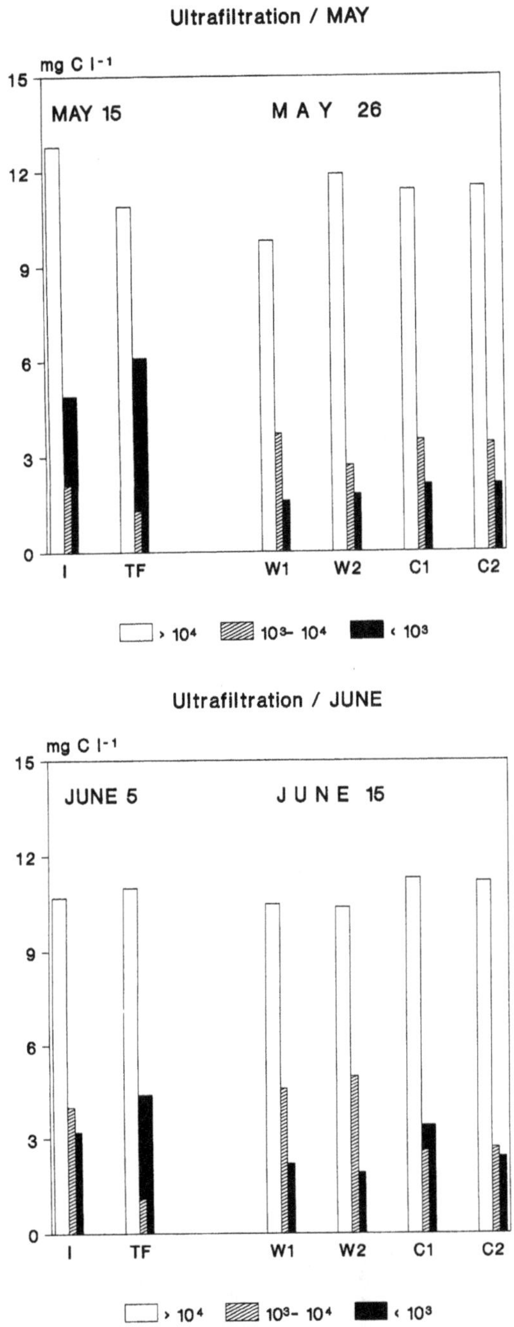


*Fig. 17.* Carbon concentrations (mg C $l^{-1}$) in different molecular weight fractions ($>10^4$ dalton; $10^3$–$10^4$ dalton; $<10^3$ dalton) of ultrafiltrated samples in the beginning (I, initial; for other symbols see Fig. 5) and in the end of the May and June experiments.

*DOC*

DOC concentration remained constant in all waters throughout the experiments (Fig. 16). However, in the molecular weight fractions there appeared a distinct change during the experiments; the lowest fraction ($<10^3$ dalton) decreased in all systems while the second lowest fraction ($10^3$–$10^4$ dalton) tended to increase (Fig. 17). However, up to 25% of the total DOC concentration was lost in the ultrafiltration treatments which may have biased these results.

## *Laboratory experiment in July*

*Chemical parameters*

The presence of large zooplankton in *Daphnia*-bottles produced slightly lower concentrations of total nutrients and higher concentrations of inorganic nutrients, especially of $NH_4$ (Figs. 18–20). However, the load of nutrients with the throughflowing water caused only a slight increase in the nutrient concentrations in the community bottles.

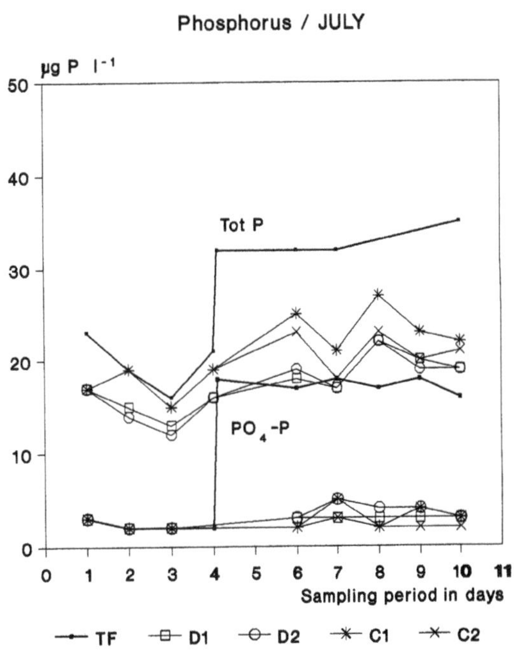


*Fig. 18.* Total and phosphate phosphorus concentrations (µg P $l^{-1}$) in throughflowing water (TF) and in different community waters (D1–2, complete plankton with *Daphnia*; C1-2, $<25$ µm controls without *Daphnia*).

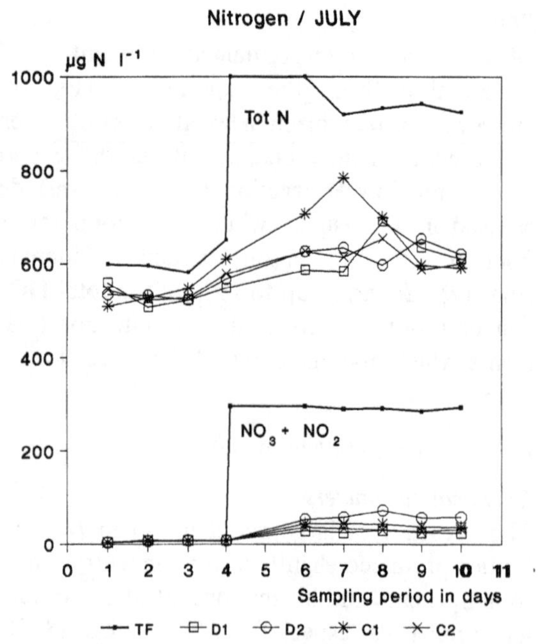


*Fig. 19.* Total and nitrate ( + nitrite) nitrogen concentrations ($\mu g\ N\ l^{-1}$) in the July experiment (for symbols see Fig. 18).

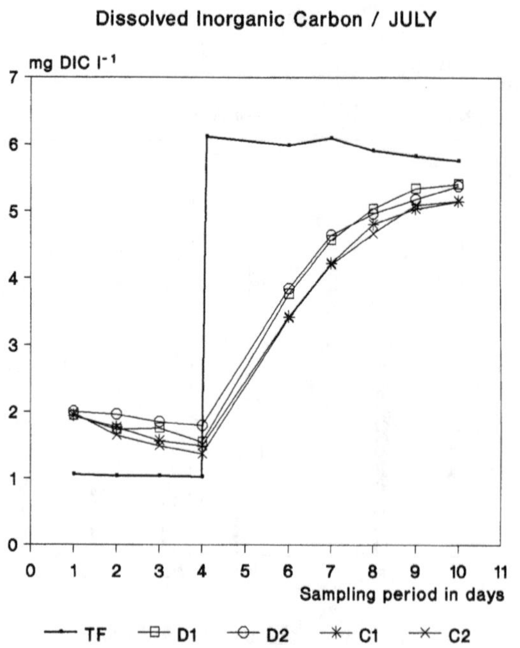


*Fig. 21.* Dissolved inorganic carbon concentrations ($mg\ C\ l^{-1}$) in the July experiment (for symbols see Fig. 18).

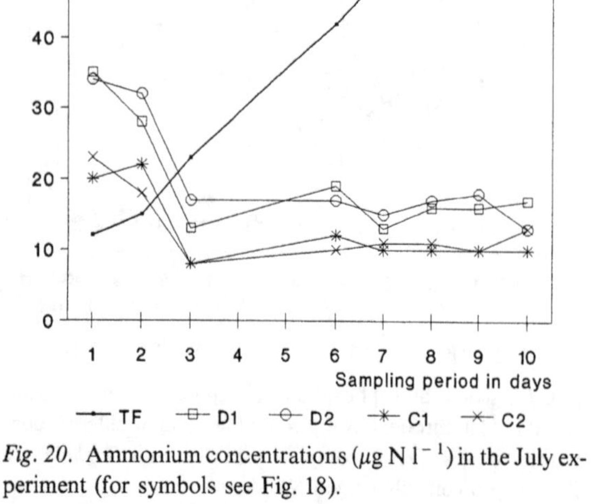


*Fig. 20.* Ammonium concentrations ($\mu g\ N\ l^{-1}$) in the July experiment (for symbols see Fig. 18).

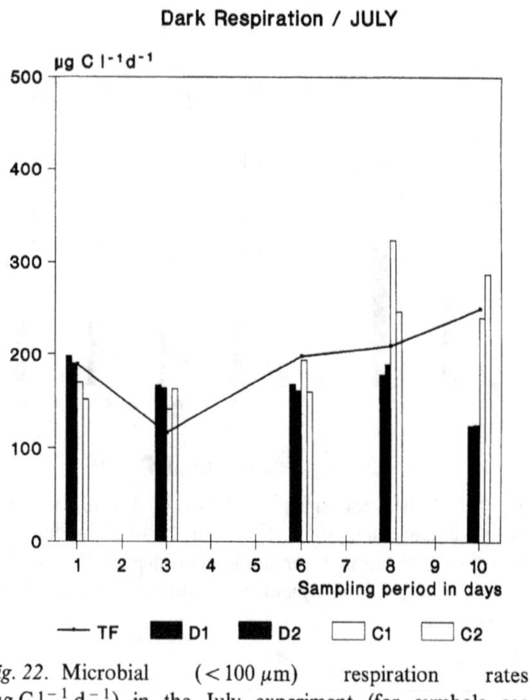


*Fig. 22.* Microbial ($< 100\ \mu m$) respiration rates ($\mu g\ C\ l^{-1}\ d^{-1}$) in the July experiment (for symbols see Fig. 18).

The higher respiration rate in *Daphnia*-bottles, primarily due to the presence of *Daphnia*, resulted in higher DIC concentrations (Fig. 21; photosynthetic carbon fixation by phytoplankton contributed <10% to the observed difference). Since respiration by large zooplankton ('*Daphnia*') was calculated as the difference between DIC-concentrations in *Daphnia*-bottles and Control-bottles, after first excluding the photosynthetic carbon fixation by planktonic algae, the results would have underestimated the respiration of *Daphnia* by ignoring the contribution of grazed algae.

*Microbial metabolism*

Dark respiration in Control-bottles, being initially a little lower than in *Daphnia*-bottles, increased during the course of the experiment, and by the end of the experiment distinctly exceeded that in the *Daphnia*-bottles (Fig. 22). The respiration rates in the community waters in July (pelagial water, Fig. 22) were, on average, lower than those in May or in June (littoral water, Fig. 8).

Planktonic algal production (Fig. 23) was very low in *Daphnia*-bottles while in the Control-bottles

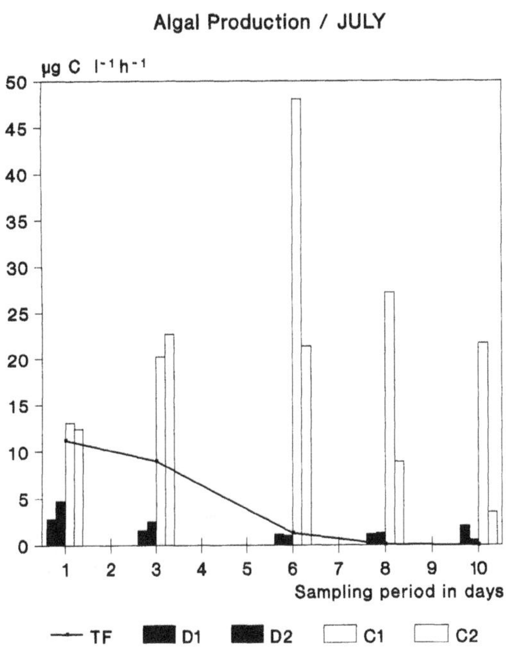


*Fig. 23.* Phytoplankton production rates ($\mu g\, C\, l^{-1}\, h^{-1}$) in the July experiment (for symbols see Fig. 18).

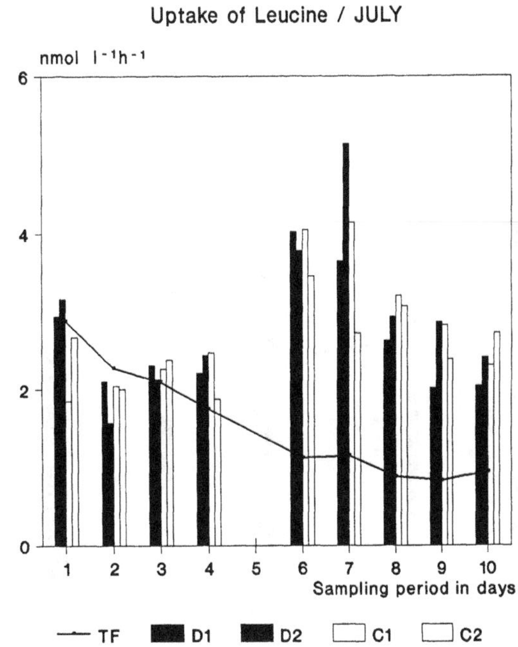


*Fig. 24.* Leucine uptake rates ($nmol\, l^{-1}\, h^{-1}$) in the July experiment (for symbols see Fig. 18).

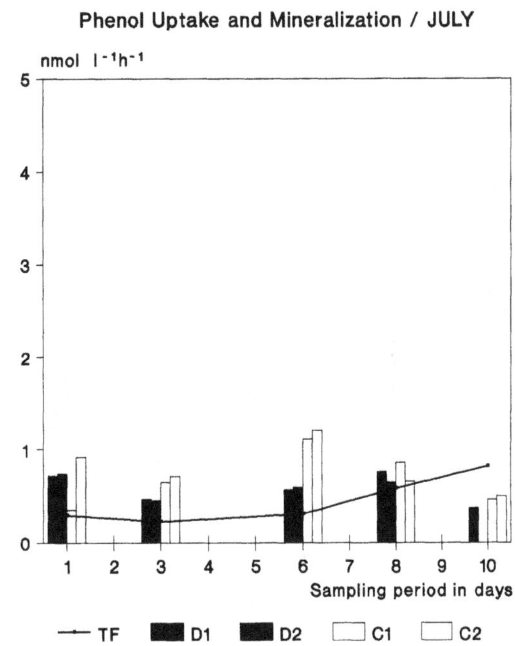


*Fig. 25.* Phenol uptake and mineralization rates ($nmol\, l^{-1}\, h^{-1}$) in the July experiment (for symbols see Fig. 18).

it reached the same level as in June (Fig. 9), but with a high deviation between the replicate systems. In the throughflowing water, the photosynthetic capability of planktonic algae, as in May and June, was lost during the course of the experiment (Fig. 23).

Leucine uptake rates developed equally in both systems. The enrichment with nutrients, as in May and June, enhanced the uptake rate but only for a few days (Fig. 24). In July, however, the incorporation of leucine remained distinctly lower than that in May or in June (Fig. 10).

Uptake and mineralization of phenol in the community waters was on average a little higher in Control-bottles than in *Daphnia*-bottles (Fig. 25), but was much lower than in June, and resembled the situation in May (Fig. 11). On average, 35.8–41.4% of the phenol taken up was mineralized to $CO_2$ during the incubation period. Disregarding the final sampling day, phenol metabolism was again lower in the littoral (throughflowing) water than that in the pelagial (community) waters.

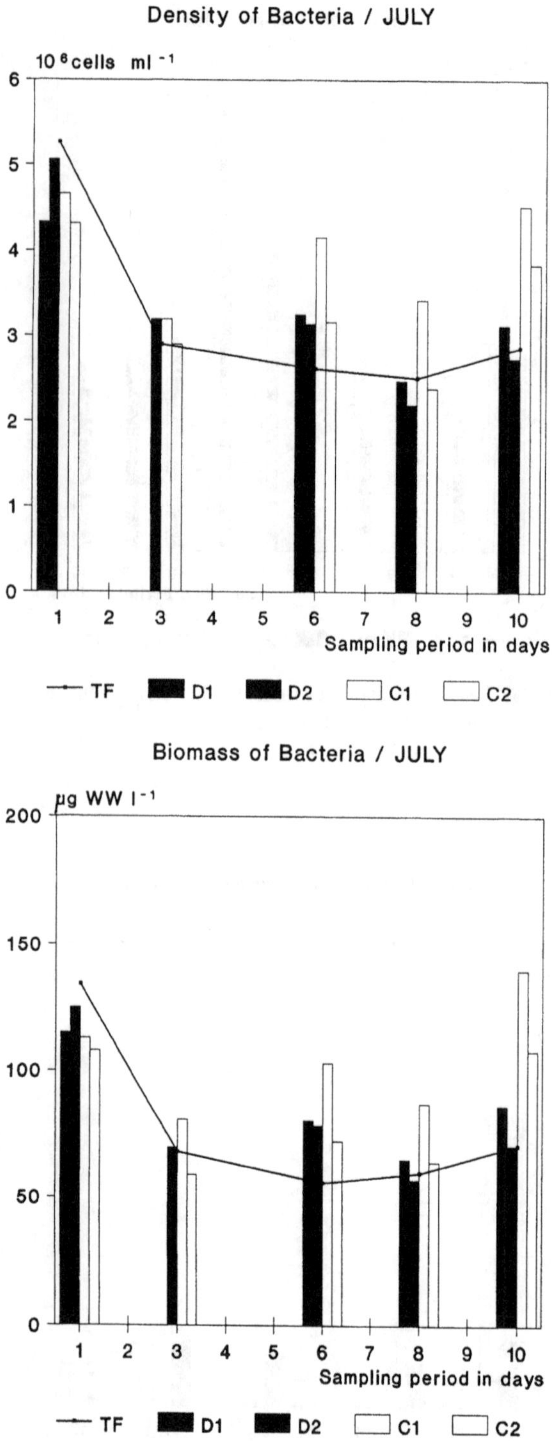


*Fig. 26.* Densities ($10^6$ cells $ml^{-1}$) and biomasses ($\mu g$ WW $l^{-1}$) of bacteria in the July experiment (for symbols see Fig. 18).

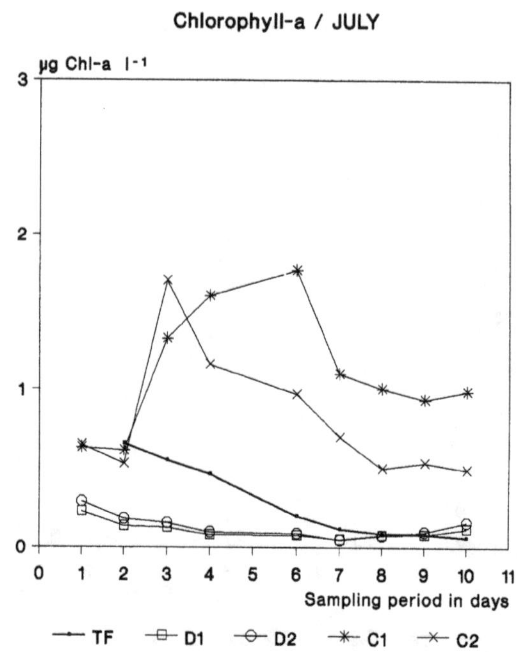


*Fig. 27.* Chlorophyll-*a* concentrations ($\mu g$ Chl-*a* $l^{-1}$) in the July experiment (for symbols see Fig. 18).

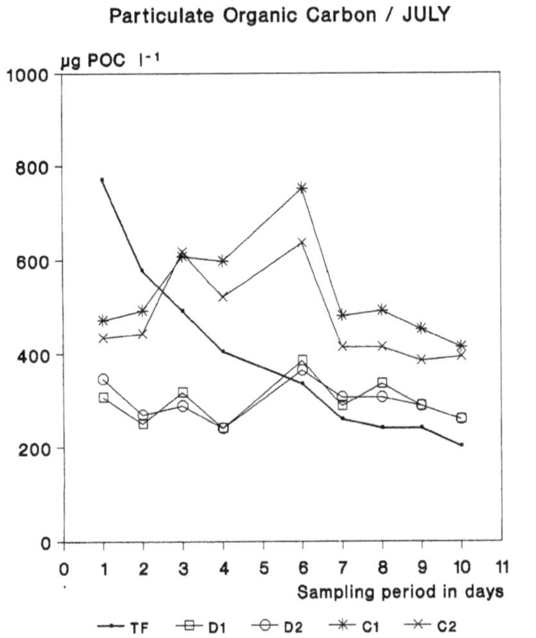


*Fig. 28.* Particulate organic carbon concentrations ($\mu g\ C\ l^{-1}$) in the July experiment (for symbols see Fig. 18).

*Bacterio- and phytoplankton, and POC*

Density and biomass (volume) of bacteria at first developed similarly in both systems, but by the end of the experiment distinctly higher densities and biomasses were found in Control-bottles (Fig. 26).

In the chlorophyll-*a* concentration there was a clear difference between the two systems, due to grazing by larger zooplankton on planktonic algae (Fig. 27). In *Daphnia*-bottles, chlorophyll-a concentration dropped to near zero during the first half of the experiment. Replicate Control-bottles deviated greatly, chlorophyll concentration being higher in C1-bottle, in which higher algal production rates were also measured (Fig. 23).

Small (3–10 µm) flagellates dominated most samples. At the end of the experiment *Cryptomonas* sp. became the dominant algal in Control-samples, whereas in *Daphnia*-samples there were only small flagellates and detrital material left.

The impact of grazing by zooplankton also became evident as a difference in POC concentrations between *Daphnia*- and Control-bottles (Fig. 28). That difference was used for calculating

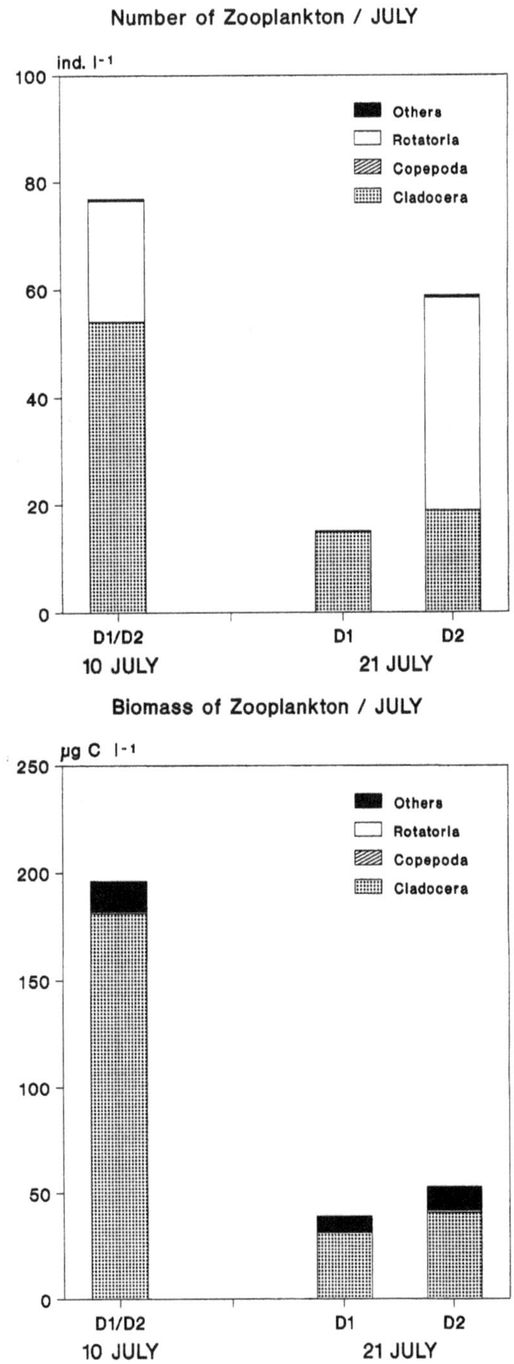


*Fig. 29.* Densities ($ind\ l^{-1}$) and biomasses ($\mu g\ C\ l^{-1}$) of different animal groups in the beginning and in the end of the July experiment (for symbols see Figs. 4 and 18).

the total grazing rate of zooplankton ('*Daphnia*') for the first 4 days of the experiment as an average difference in POC concentration in *Daphnia*-

and Control-bottles, when first excluding the contribution of different development of algal POC (POC:Chl-*a* = 33:1) in the *Daphnia*- and Control-bottles. (Development of the POC-concentrations during the latter half of the experiment made the grazing estimates meaningless).

*Zooplankton*
During the experiment, the numbers of *Daphnia* decreased in the community bottles (Fig. 29). However, *Daphnia* always contributed most to the total animal biomass. The number of rotifers (*Rotaria* spp.) increased in the D2-bottle but they had a negligible impact on the total biomass. The few Hydracarina and *Chaoborus* individuals in the samples clearly raised the animal biomasses (Fig. 29).

*DOC*
DOC concentration was again constant throughout the experiment (Fig. 30). However, the concentrations in Control-bottles (with higher densities of bacteria; Fig. 26) were generally a little lower than those in *Daphnia*-bottles, or than that in the throughflowing water.

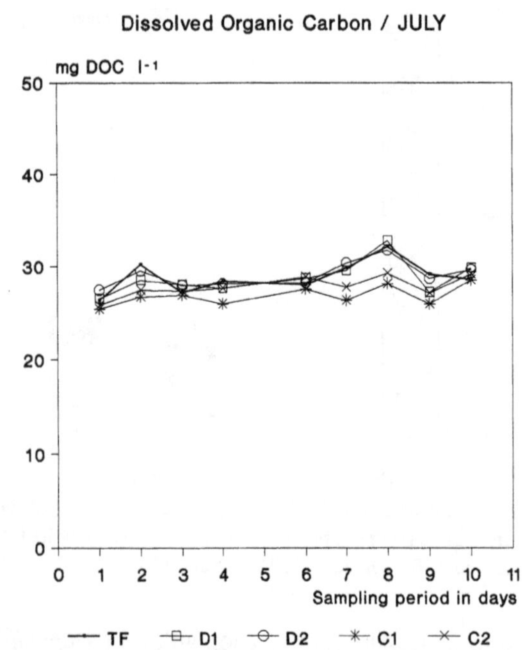


*Fig. 30.* Dissolved organic carbon concentration (mg C $l^{-1}$) in the July experiment (for symbols see Fig. 18).

*EOC by* Daphnia
'True' EOC by *Daphnia* could be calculated only at the start or the end of the experiments since, due to incomplete ingestion of algae by *Daphnia*, $DO^{14}C$ concentration increased drastically subsequent to the introduction of the radioactive algal suspension into the test tubes (Fig. 31). During the supply of unlabelled algal suspension, or filtered lake water, EOC by the 'fully labelled animals' was 0.13–0.27% $h^{-1}$ (mean 0.2%) of the total radioactivity or carbon content in the animals. Thus, 54 *Daphnia* individuals per litre (a similar density as in the lake sample in July; biomass 181 $\mu g$ C $l^{-1}$) would increase the ambient DOC content by 0.2–0.5 $\mu g$ C $h^{-1}$ (Table 2). However, the 'sloppy feeding' of the cladocerans may very significantly increase the release rates.

The results also revealed that the EOC may be

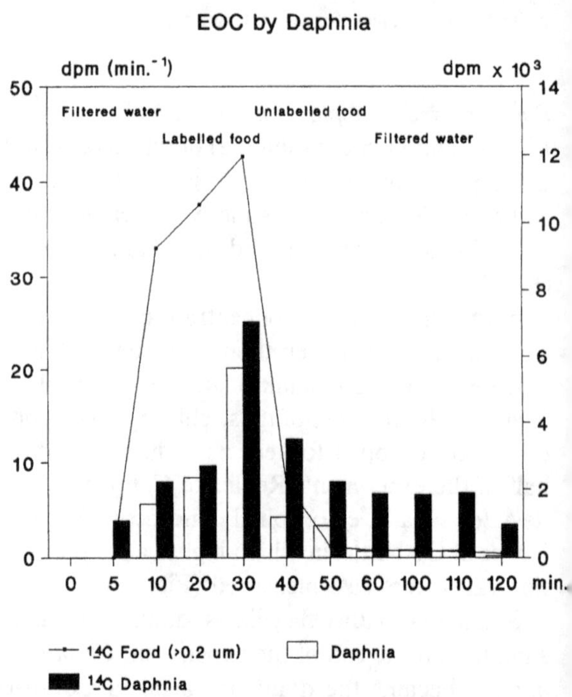


*Fig. 31.* Release of dissolved radiolabelled organic carbon by *Daphnia longispina* (white and black bars; dpm $min^{-1}$; left axis) when fed with different, labelled or unlabelled (line; $10^3$ dpm $min^{-1}$; right axis) suspensions. (After the two hour experiment the radioactivity of the 'prelabelled animals' averaged $26.6 \times 10^3$ dpm $ind^{-1}$, and that of the 'nonlabelled animals' $1.5 \times 10^3$ dpm $ind^{-1}$; see text for further details).

*Table 2.* Average daily rates for carbon metabolism (mg C $d^{-1}$) in the community bottles (25 l) in the July experiment before (B) and after (A) nutrient enrichment (*'Daphnia'* includes the whole zooplankton; see text and figures for further details).

| JULY | | D1 | D2 | C1 | C2 |
|---|---|---|---|---|---|
| *'Daphnia'* grazing | (B) | 1.3 | 1.6 | – | – |
| | (A) | ? | ? | – | – |
| *'Daphnia'* respiration | (B) | 1.3 | 2.4 | – | – |
| | (A) | 2.2 | 2.1 | – | – |
| *Daphnia* EOC | (B) | 0.2 | 0.2 | – | – |
| | (A) | 0.1 | 0.1 | – | – |
| Phytoplankton prod. | (B) | 1.0 | 1.6 | 7.2 | 7.6 |
| | (A) | 0.6 | 0.4 | 14.0 | 4.9 |
| Bacterioplankton prod. | (B) | 2.2 | 2.1 | 2.1 | 2.1 |
| | (A) | 2.6 | 3.2 | 3.1 | 2.7 |
| Microbial respiration | (B) | 4.9 | 4.8 | 4.2 | 4.3 |
| | (A) | 4.2 | 4.3 | 6.8 | 6.2 |

substantially underestimated in short-term experiments without prelabelled animals (Fig. 31).

## Discussion

### *Impacts of* Warnstorfia *on plankton metabolism*

Submerged macrophytes may affect plankton metabolism directly (1) by competing with algae for nutrients and light and (2) by providing particulate and dissolved organic matter for degradation and mineralization by bacteria and fungi, or indirectly (3) by providing substratum for epiphytic communities (algae, bacteria, protozoans) which then interact both with the macrophytes and with planktonic organisms by competing for resources from the ambient water. On the other hand, epiphytic communities may also be an important source of food and energy for some planktonic organisms (e.g. Downing, 1981).

The moss bed bordering Mekkojärvi harbours an abundant planktonic and periphytic fauna. Indeed, the high densities of the animals living among the moss vegetation suggested that the production of those animal communities exceeded that of the pelagial community (cf. Fig. 4). In such a heterogeneous, habitat-rich environment, the species diversity, especially within the inner littoral zone, was much higher than in the open water environment. For instance, while the pelagial area was dominated by only one cladoceran species, *Daphnia longispina*, there were several abundant cladocerans, such as *Acroperus harpae*, *Alona affinis*, and *Chydorus sphaericus*, coexisting within the littoral area. Also the number and diversity of predators (belonging to Copepoda, Odonata, *Notonecta*, Trichoptera, Ephemeroptera) among the moss vegetation was high and indicative of high primary and secondary production from the lower trophic levels of the littoral food chains to meet the energy requirement of the predators.

The laboratory experiments in May and June, demonstrated that *Warnstorfia* also grows well when submerged. In May, production of the *Warnstorfia* community (*Warnstorfia* + epiphytes) was on average the same or higher than that of phytoplankton; in June, however, planktonic algal production exceeded that of the *Warnstorfia* community both before and after the nutrient enrichment (Table 1). Part of the 'planktonic' production, however, was attributable to detached moss pieces (and epiphytes), as seen from the higher initial chlorophyll-a and POC concentrations in *Warnstorfia*-bottles than in Control-bottles (Figs. 14 and 15). The enrichment of nutrients into the community bottles more than doubled the production of the moss community, while a suppression was observed in the production of phytoplankton (compared to Control-bottles) (Fig. 9). That suppression could not be due to shading by moss material since the production measurements were made in a separate aquarium, in equal light conditions for all the plankton samples. In any case, the amount of moss in the community bottles was too low to produce effective shading during the experiment. The experiments therefore demonstrated that the submerged moss community can compete effectively with phytoplankton for nutrients and hence could also benefit from nutrients upwelling from the anoxic hypolimnion of Mekkojärvi. That phenomenon will be still more pronounced in the field conditions where *Warnstorfia* is colonizing the water surface; in submerged conditions epiphytic algae would evidently respond faster to the nutrient enrichment.

Microbial dark respiration seemed to be enhanced to some extent by the moss (Fig. 8). However, leucine uptake rates, and hence bacterial production rates, were higher in *Warnstorfia*-bottles only during the first half of the May experiment; at other times, there was no clear difference between the two systems. If the conversion factors adopted gave true figures for the bacterial production, then a 31–38% bacterial production efficiency would entirely explain the average dark respiration (Table 1). A higher percentage would, however, yield a more realistic figure; for instance, a 50% bacterial production efficiency (cf. Bratbak, 1985) would account for about half of the total dark respiration.

EOC by *Warnstorfia* could quantitatively contribute only little to bacterial metabolism in the ambient water, particularly as a great deal of the EOC was most likely utilized by epiphytic bacteria and algae (Søndergaard, 1983; Kairesalo *et al.*, 1991). However, phenol uptake and mineralization rate was almost always (in May, June and July) much higher in pelagial water than in littoral water. This suggests that in the pelagial water environment, with lower availability of newly synthesized organic matter by the littoral primary producers, bacteria had a higher capacity to utilize the more recalcitrant organic compounds.

## *The role of* Daphnia *in the plankton metabolism*

In *Daphnia*-bottles, representing natural summer plankton density and composition in the pelagial of Mekkojärvi, *Daphnia longispina* clearly had a key role, representing 92–98% of the cladoceran numbers and biomass, and about 65% of the total plankton biomass (other animals comprising about 5%, algae about 10% and bacteria about 20%; cf. Figs. 26, 27, and 29).

Light conditions in the aquarium experiments corresponded to the 0–20 cm depth zone on a cloudy day or to the 25–50 cm depth zone on a bright summer day, i.e. to the depth zone where *Daphnia* usually lives in the lake (Salonen & Lehtovaara, 1992). Nevertheless, in the experiments (Figs. 23 and 27), planktonic algae were overgrazed and could not produce enough food to maintain the *Daphnia* population under those conditions.

In *Daphnia*-bottles, the density and biomass of bacteria (20–40 $\mu g\ C\ l^{-1}$) remained constant after the start of the experiment. Since bacterial production rates in *Daphnia*- and Control-bottles did not differ significantly from each other (Table 2), the observed differences in the bacterial values at the end of the experiment suggested that the mortality of bacteria was higher in *Daphnia*-bottles than in Control-bottles. However, the increase in microbial respiration at the end of the experiment was probably attributable to the higher algal (and heterotrophic flagellate) numbers in Control-bottles, with the higher bacterial numbers being derived from the senescence of the algal material.

Kankaala (1988) demonstrated that bacteria may serve as a major food source for *Daphnia longispina* in Mekkojärvi when the availability of algal food is low. In our experiment, bacterial production (2–8 $\mu g\ C\ l^{-1}\ h^{-1}$; 55–215 $\mu g$ C bottle$^{-1}$ $h^{-1}$) alone would have been high enough to supply the respiration by zooplankton (Table 2). However, the grazing of particulate material in the community bottles (in any case an underestimate due to excluding fecal pellets) could not satisfy the animal respiration (Table 2). Thus much of the particulate material in the water (250–350 $\mu g$ POC $l^{-1}$; Fig. 28) was poor or useless food for *Daphnia*, and available bacteria and/or detrital matter could not supplant the need for algal food (ten times lower production rates were measured for *Daphnia* with a shortage of algal food; Kairesalo & Lehtovaara, unpublished). Accordingly, the observed decline in the *Daphnia* density in *Daphnia*-bottles, down to one third or fourth of the initial value during the 11 day laboratory experiment, was most probably caused by food limitation and starvation of the animals (cf. Hessen, 1989).

The laboratory experiments were carried out under constant 20 °C temperature conditions; consequently, the impact of temperature on respiration rate was not addressed in this study. However, in steeply stratified Mekkojärvi, summer surface temperature often exceeds 20 °C while the temperature at a depth of 50–100 cm may be only 10 °C (Kankaala, 1988). *Daphnia* might then derive some metabolic advantage from vertical migration between the warm and productive epilimnion (0–50 cm) and the cold and dark thermocline and uppermost hypolimnion (50–100 cm), i.e. between the grazing zone and the low-respiration zone, respectively. This kind of metabolic adaptation in the environment with low algal production (40–150 $\mu g\,C\,l^{-1}\,d^{-1}$ at 0–25 cm depth, unpublished results) together with relatively low predation pressure (from *Chaoborus* and *Notonecta*), might explain the observed high density and biomass of *Daphnia* in the pelagial water. The results of Stich & Lampert (1983) with *Daphnia hyalina* and *Daphnia galeata*, however, did not support the hypothesis that daphnids would gain some metabolic advantage from vertical migration. Therefore, the ecological value of the vertical migrations by *Daphnia longispina* in Mekkojärvi is still speculative and requires further study (cf. Salonen & Lehtovaara, 1992).

In the laboratory experiment, a major part (43–76%) of particulate material seemed to remain untouched by the pelagial zooplankton. Microscopic analysis revealed that this comprised mostly detrital material, such as fibrous material from the live moss bed or peaty remains (cf. Boers & Boon, 1988). Nevertheless, algal carbon could explain only a minor fraction (13–14%) of the grazed POC when assuming that chlorophyll corresponded to 3% of the algal carbon content (Kairesalo, 1984; Riemann *et al.*, 1989. Also, free bacteria seemed to be of minor or negligible importance as a food source for the zooplankton. Therefore, a major fraction of the grazed POC must have been mainly nutritionally poor detrital particles, although perhaps colonized by bacteria (cf. Tranvik & Sieburth, 1989), and hence nutritionally improved food for the cladocerans.

EOC by *Daphnia*, being 5–10% of bacterial respiration (regarding bacterial production efficiency as 50% and *Daphnia* density as 50 ind $l^{-1}$), was quantitatively of minor importance to bacterial metabolism (Table 2). However, as shown by the results (Fig. 31), the by-products of ingestion and digestion by *Daphnia* may release much organic material of algal origin into the dissolved carbon pool, making it more readily available for bacterial metabolism and production. This mechanism may be an important pathway of dissolved organic carbon from phytoplankton to bacteria (Lampert, 1978; Olsen *et al.*, 1986; Jumars *et al.*, 1989). However, regardless of the higher DOC concentrations in *Daphnia*-bottles than in Control-bottles (Fig. 30), bacterial production was not enhanced in our experiment. This difference in the bacterial activity was evident before and after the nutrient enrichment, which indicates that bacterial production and biomass were controlled rather by some factor independent of mineral nutrition.

Flagellates and ciliates, the most important bacteria grazers (e.g. Sherr & Sherr, 1984; Sanders *et al.*, 1989), played key roles in controlling bacterial biomass and hence also production. Unexpectedly, however, *Daphnia* seemed not to have a clear impact on bacterial numbers and biomasses via grazing of flagellates and protozoans.

Collapse of fecal particles during filtration through the 0.2 $\mu m$ Nuclepore-filters might have contributed to the observed increase of DOC in *Daphnia*-bottles. Olsen *et al.* (1986), however, found that most of the compounds released by *Daphnia longispina* were in small ($< 1.2\ \mu m$) fecal particles, which were resistant to mechanical treatment. In our experiment, the number of 0.5–1 $\mu m$, 'fecal pellets' (dark particles in a phase contrast microscope) in *Daphnia*-samples increased towards the end of the experiment (seen as higher POC concentration in *Daphnia*-bottles than in the throughflowing water; Fig. 28), also suggesting that they were relatively resistant to bacterial degradation. If so, and the feces are a sink rather than a source for bacterial metabolism and production in the epilimnetic water, the role of *Daphnia* in recycling of nutrients and carbon needs to be reconsidered.

*Production and utilization of DOM in Mekkojärvi*

Littoral-pelagial interchange in Mekkojärvi can be evaluated in respect to the carbon budget and metabolism in the lake.

Input of allochthonous organic matter, usually comprising >90% of dissolved fraction, has been regarded as the predominant source for the high organic carbon concentration and the dark brown colour of water in Mekkojärvi. Of the autochthonous sources, phytoplankton production, while restricted to the uppermost surface layer, remains low ($<10\ g\ C\ m^{-2}\ yr^{-1}$; Salonen & Jokinen, 1988) and therefore is of marginal importance in the total carbon budget of the lake. The living and peaty moss bed in the littoral zone, on the other hand, represents a large store of organic matter and evidently plays an important role in the carbon budget. For instance, the laboratory experiments demonstrated that water currents may release a substantial amount of particulate organic material from the littoral moss bed into the ambient water, thus providing an energy flow from the littoral to the pelagial area. Those detritus particles, however, seemed to contribute little directly to the zooplankton production. Therefore, most of the energy requirement for the maintenance of the dense *Daphnia longispina* population in the pelagial water, had to be fullfilled through other food chains, i.e. via autotrophic algal production and/or via a heterotrophic link from DOM and bacteria.

Our laboratory experiments indicated that the availability of algal food was of principal importance in regulating production of the *Daphnia* population in Mekkojärvi. Accordingly, such high population densities ($\geq 50\ ind\ l^{-1}$) cannot be achieved and maintained in the warm epilimnion without the availability of fresh algal food. Most of the planktonic algae were obligatory photoautotrophs, since within a few days of storage in the darkness they lost their capability to photosynthesize and died. Only some small (3–10 $\mu$m) flagellates remained alive in the dark store aquarium, presumably by grazing on bacteria (cf. Salonen & Jokinen, 1988).

Bacterial production rates were high (Tables 1 and 2). Since EOC by phytoplankton (assuming 10–20% of the production; Kairesalo & Saukkonen, 1990), *Warnstorfia* or *Daphnia* could account for only a minor portion of the carbon requirement for bacterial metabolism, the major part must have derived from the 'unidentified' DOM pool of the water, namely from humic substances. Differences in the ultrafiltration fractions in the beginning and in the end of the experiments indicated that the lowest molecular weight fraction ($<10^3$ dalton), in particular, provided a carbon source for bacteria. Interestingly, however, regardless of the relatively small decrease in the bacterial densities and biomasses (Figs. 12, 13 and 26) in the throughflowing water, the leucine uptake rate (Figs. 10 and 24), and hence bacterial production, fell to a low level. Thus, darkness and low temperature in the store aquarium (corresponding to the thermocline conditions in the lake) reduced the production potential of bacteria. Perhaps light, either directly (e.g. Kieber *et al.*, 1989) or indirectly through algal photosynthesis, enhanced the regeneration of labile organic compounds for bacterial metabolism thus maintaining their production potential at a high level. On the other hand, no clear relation between leucine uptake and algal production rates were found in the experiments. Unexpectedly also, our results did not support the idea that bacteria served as a significant food source for *Daphnia*, or as an essential link in the energy flow from humic substances up to metazoans (also Kairesalo & Lehtovaara, unpublished). Rather, the supplementary energy gained from humic substances seemed primarily to be consumed and lost already within the microbial link or loop.

In all, the bulk of dissolved organic matter in Mekkojärvi water was biochemically highly recalcitrant, remaining practically unavailable for bacterial degradation. The increase of pH by one unit caused no noticeable, direct or indirect deviation in the DOC concentration. Nor could any variation in DOC concentration be detected among the moss vegetation, where biological activities and turnover rates might have been an order of magnitude higher than in the pelagial water environment. Only the lower phenol metabolism in the littoral water (compared to pelagial water) implied some difference in the composition of DOC between the littoral and pelagial waters.

Metabolic impacts on the DOC pool, both in the pelagial and littoral waters, were efficiently masked by the 'humus' of the water. Regardless of the massive moss growth in the littoral zone and of the relatively low water currents (due to the sheltered location of the lake), DOM seemed to have a quite uniform concentration and distribution in the lake. That phenomenon supports the idea that (1) dissolved humic substances in the pelagial and littoral areas were decomposed very slowly, and that (2) they originated principally from the same source, or from similar sources, i.e. were either only allochthonous by origin, or rather, involved both allochthonous and autochthonous plant biomolecules, highly degraded bryophyte remains leached from both the mossgrown watershed and the mossgrown littoral areas.

Thus the extent to which the DOM pool in Mekkojärvi is autochthonous by origin, and to which seasonal fluctuations in the DOM pool are attributable to the growth and senescence of the littoral vegetation is still uncertain. Pyrolysis-GC-MS technique could be a powerful tool for addressing those questions (Gadel & Bruchet, 1987). Exoenzyme studies have also proved promising (Münster *et al.*, 1992; Wetzel, 1992), and will surely give more insight into metabolic interchange between the littoral and pelagial communities.

In conclusion, our results were not consistent with the hypothesis that the littoral primary producers, by producing newly synthetized, labile organic compounds into the ambient water, would significantly enhance the utilization of humic substances and thus the energy flow up to cladoceran production. Instead, our results were in accordance with those of Moran *et al.* (1988) that the fate of the humic substances (lignocellulosic carbon) that enter the food web of a detritus-based ecosystem, like Mekkojärvi, appears to be largely that of respiratory loss from microorganisms.

## Acknowledgements

This study was supported financially by the Academy of Finland, the Maj and Tor Nessling Foundation, and Lammi Biological Station, which is gratefully acknowledged. We thank Riitta Ilola and Jaakko Vainionpää for invaluable help in the laboratory, and Roger Jones for checking the language.

## References

Arvola, L. & P. Kankaala, 1989. Winter and spring variability in phyto- and bacterioplankton in lakes with different water colour. Aqua fenn. 19: 29–39.

Bergström, I., A. Heinänen & K. Salonen, 1986. Comparison of acridine orange, acriflavine, and bisbenzimide strains for enumaration of bacteria in clear and humic waters. Appl. Envir. Microbiol. 51: 664–667.

Boers, P. C. M. & J. J. Boon, 1988. Unmasking the particulate organic matter in a lake ecosystem: Origin and fate of POM in the shallow eutrophic Loodsdrecht Lakes (The Netherlands). Arch. Hydrobiol. Beih. Ergebn. Limnol. 31: 27–34.

Bratbak, G., 1985. Bacterial biovolume and biomass estimations. Appl. Envir. Microbiol. 49: 1488–1493.

Chin-Leo, G. & D. L. Kirchman, 1988. Estimating bacterial production in marine waters from the simultaneous incorporation of thymidine and leucine. Appl. Envir. Microbiol. 54: 1934–1939.

Downing, J. A., 1981. *In situ* foraging responses of three species of littoral cladocerans. Ecol. Monographs 51: 85–103.

Gadel, F. & A. Bruchet, 1987. Application of pyrolysis-gas chromatography-mass spectrometry to the characterization of humic substances resulting from decay of aquatic plants in sediments and waters. Wat. Res. 21: 1195–1206.

Glime, J. M., R. G. Wetzel & B. J. Kennedy, 1982. The effects of bryophytes on succession from alkaline marsh to Sphagnum bog. Am. Midl. Nat. 108: 209–223.

Hessen, D. O., 1989. Factors determining the nutritive status and production of zooplankton in a humic lake. J. Plankton Res. 11: 649–664.

Jones, R. I., 1990. Phosphorus transformations in the epilimnion of humic lakes: biological uptake of phosphorus. Freshwat. Biol. 23: 323–337.

Jumars, P. A., D. L. Penry, J. A. Baross, M. J. Perry & B. W. Frost, 1989. Closing the microbial loop: dissolved pathway to heterotrophic bacteria from incomplete ingestion, digestion and absorption in animals. Deep-Sea Res. 36: 483–495.

Kairesalo, T., 1984. The seasonal succession of epiphytic communities within an *Equisetum fluviatile* L. stand in Lake Pääjärvi, southern Finland. Int. Revue ges. Hydrobiol. 69: 475–505.

Kairesalo, T. & P. Saukkonen, 1990. Thymidine incorporation by littoral and pelagial bacterioplankton in a mesohumic lake. Verh. int. Ver. Limnol. 24: 677–681.

Kairesalo, T., G. S. Jónsson, K. Gunnarsson, C. Lindegaard & P. M. Jónasson, 1992. Metabolism and community dynamics within *Nitella opaca* (Charophyceae) beds in Thingvallavatn. Oikos 64.

Kankaala, P., 1988. The relative importance of algae and bacteria as food for *Daphnia longispina* (Cladocera) in a polyhumic lake. Freshwat. Biol. 19: 285–296.

Kankaala, P. & P. Eloranta, 1987. Epizooic ciliates (*Vorticella* sp.) compete for food with their host *Daphnia longispina* in a small polyhumic lake. Oecologia (Berlin) 73: 203–206.

Kieber, D. J., J. McDaniel & K. Moppet, 1989. Photochemical source of biological substrates in sea water: implications for carbon cycling. Nature 341: 637–639.

Koroleff, F., 1979. Methods for the chemical analysis of seawater. Meri 7: 1–60 (in Finnish).

Lampert, W., 1978. Release of dissolved organic carbon by grazing zooplankton. Limnol. Oceanogr. 23: 831–834.

Moran, M. A., T. Legovic, R. Benner & R. E. Hodson, 1988. Carbon flow from lignocellulose: a simulation analysis of a detritus-based ecosystem. Ecology 6: 1525–1536.

Murphy, J. & J. P. Riley, 1962. A modified single-solution method for the determination of phosphate in natural waters. Analyt. Chim. Acta 27: 31–36.

Münster, U., P. Einiö & J. Nurminen, 1989. Evaluation of the measurements of extracellular enzyme activities in a polyhumic lake by means of studies with 4-methylum-belliferyl-substrates. Arch. Hydrobiol. 115: 321–337.

Münster, U., P. Einiö, J. Nurminen & J. Overbeck, 1992. Extracellular enzymes in a polyhumic lake: important regulators in detritus processing. In K. Salonen, T. Kairesalo & R. I. Jones (eds), Dissolved Organic Matter in Lacustrine Ecosystems: Energy Source and System Regulator. Kluwer Academic Publishers.

Olsen, Y., K. M. Varum & A. Jensen, 1986. Some characteristics of the carbon compounds-released by *Daphnia*. J. Plankton Res. 8: 505–517.

Riemann, B., P. Simonsen & L. Stensgaard, 1989. The carbon and chlorophyll content of phytoplankton from various nutrient regimes. J. Plankton Res. 11: 1037–1045.

Salonen, K., 1979. A versatile method for the rapid and accurate determination of carbon by high temperature combustion. Limnol. Oceanogr. 24: 127–129.

Salonen, K., 1981. Rapid and precise determination of total inorganic carbon and some gases in aqueous solutions. Water Research 15: 403–406.

Salonen, K. & S. Jokinen, 1988. Flagellate grazing on bacteria in a small dystrophic lake. Hydrobiologia 161: 203–209.

Salonen, K. & A. Lehtovaara, 1992. Migrations of haemoglobin-rich *Daphnia longispina* in a small, steeply stratified humic lake with an anoxic hypolimnion. In K. Salonen, T. Kairesalo & R. I. Jones (eds), Dissolved Organic Matter in Lacustrine Ecosystems: Energy Source and System Regulator. Kluwer Academic Publishers.

Sanders, R. W., K. G. Porter, S. J. Bennett & A. E. DeBiase, 1989. Seasonal patterns of bactivory by flagellates, ciliates, rotifers, and cladocerans in a freshwater planktonic community. Limnol. Oceanogr. 34: 673–687.

Sherr, B. F. & E. B. Sherr, 1984. Role of heterotrophic protozoa in carbon and energy flow in aquatic ecosystems. In M. Klug & C. A. Reddy (eds), Current Perspectives in Microbial Ecology. Am. Soc. Microbiol. (Wash. D.C.).

Solorzano, L., 1969. Determination of ammonia in natural waters by the phenolhypochlorite method. Limnol. Oceanogr. 14: 799–801.

Søndergaard, M., 1983. Heterotrophic utilization and decomposition of extracellular carbon released by the aquatic angiosperm *Littorella uniflora* (L.) Aschers. Aquat. Bot. 16: 59–73.

Stich, H.-B. & W. Lampert, 1984. Growth and reproduction of migrating and non-migrating *Daphnia* species under simulated food and temperature conditions of diurnal vertical migration. Oecologia (Berlin) 61: 192–196.

Tranvik, L. J. & J. McN. Sieburth, 1989. Effects of flocculated humic matter on free and attached pelagic microorganisms. Limnol. Oceanogr. 34: 688–699.

Tuomikoski, R., 1973. New combinations in mosses; *Loeskypnum wickesii*, *Scorpidium vernicosum* and *Warnstorfia procera*. Ann. Bot. fenn. 10: 216.

Wetzel, R. G., 1992. Gradient dominated ecosystems: sources and regulatory functions of dissolved organic matter in freshwater ecosystems. In K. Salonen, T. Kairesalo & R. I. Jones (eds), Dissolved Organic Matter in Lacustrine Ecosystems: Energy Source and System Regulator. Kluwer Academic Publishers.

Wetzel, R. G. & H. L. Allen, 1972. Functions and interactions of dissolved organic matter and the littoral zone in lake metabolism and eutrophication. In Z. Kajak & A. Hillbricht-Ilkowska (eds), Productivity Problems of Freshwaters. Warszawa-Krakow 1972: 333–347.

Wood, E. D., F. A. J. Armstrong & F. A. Richards, 1967. Determination of nitrate in sea water by cadmium-copper reduction to nitrite. J. mar. biol. Ass. U.K. 47: 23–31.

*Hydrobiologia* **229**: 225–238, 1992.
*K. Salonen, T. Kairesalo & R.I. Jones (eds.),*
*Dissolved Organic Matter in Lacustrine Ecosystems: Energy Source and System Regulator.*




# Extracellular enzymes in a polyhumic lake: important regulators in detritus processing

U. Münster,[1] P. Einiö,[2] J. Nurminen[2] & J. Overbeck[1]
[1]*Max-Planck-Institute for Limnology, P.O. Box 165, D-2320 Plön, F.R.G.;* [2]*University of Helsinki, Lammi Biological Station, SF-16900 Lammi, Finland*

*Key words:* microbial activity, extracellular enzymes, regulators, detritus, humic matter, amino acids, sugars

## Abstract

Eight extracellular enzymes and their corresponding natural substrates were studied in an acid polyhumic lake. Highest activities were found for phosphatases (100–150 nmol $l^{-1}$ $h^{-1}$), glucosidase (70–120 nmol $l^{-1}$ $h^{-1}$) and aminopeptidases (20–30 nmol $l^{-1}$ $h^{-1}$). Significant relationships were found for natural polymeric substrate composition, variation and enzyme activities. Identified carbohydrates and amino acids contributed 1–5% to the DOC pool and are assumed to undergo significant processing by microbial glycosidases and aminopeptidases. Measured enzymes are partially modified in their activity by lake water acidity, temperature and humic matter. Extracellular enzymes are regarded as important regulators in microbial detritus processing and substrate utilization.

## Introduction

In many aquatic environments a major part of dissolved (DOM) and particulate organic matter (POM) is channelled through microheterotrophs to higher trophic levels (Sorokin, 1977; Williams, 1981; Pomeroy, 1974; Pomeroy & Wiebe, 1988). This 'microbial loop' has been now accepted as an important part of the food web in aquatic ecosystems (Azam *et al.*, 1983; Fenchel, 1987).

DOM contains many different solutes from which only 5–10% can be easily assimilated and metabolized by micro-heterotrophs (Overbeck, 1979). Of those solutes, the most important and frequently studied are dissolved free amino acids (DFAA) and carbohydrates (DFCHO), which comprise mostly monomeric and partially oligomeric compounds (Münster, 1984; Münster & Chróst, 1990). The other part of the DOM (80–90%) is polymeric in natura, e.g. polypeptides, polysaccharides, lipids, polynucleotides, other organophosphorus compounds, and recalcitrant polyphenolic humic substances (HS), (Münster, 1985; Steinberg & Münster, 1985; Münster & Chróst, 1990; Thurman, 1985).

The ratio between monomeric to polymeric substrates is in the range of 1:10 to 1:100 (Wetzel, 1983). Therefore, the polymeric compounds are the dominant organic solutes in the DOM pool. Monomeric, easily utilizable dissolved organic matter (UDOM) comprises predominately DFAA and DFCHO pools, and concentrations are in the nMOL to $\mu$Mol range (Thurman, 1985; Jørgensen, 1987; Münster & Chróst, 1990). Because of their efficient uptake systems, aquatic bacteria assimilate these monomeric solutes rapidly from the UDOM pool for energy, carbon, phosphorus and nitrogen requirements (Overbeck, 1979; Jørgensen, 1982; Azam & Cho, 1987). With respect to microbial growth, these monomers are rapidly exhausted and represents a first threshold concentration in the DOM pool (Jannasch, 1970; Sieburth, 1979). Bacterial growth also depends therefore, on the availability and utilization of polymeric substrates. Microplankton is known to be an efficient user of different kinds of organic matter, e.g. bacteria can synthesize extracellular enzymes for cleavage of polymeric substrates into monomeric and oligomeric solutes, with subsequent rapid uptake (Pollock, 1962; Rogers, 1961; Priest, 1984; Chróst *et al.*, 1989; Chróst, 1990; Münster, 1991). The ecological significance of such extracellular enzyme activities has been shown already in many aquatic environments (Hoppe, 1983; Hollibaugh & Azam, 1983; Ammerman & Azam, 1985; Billén, 1984; Meyer-Reil, 1986; Hoppe *et al.*, 1988; Chróst, 1989; Chróst *et al.*, 1989; Münster *et al.*, 1989; Münster, 1991). In some of these papers, correlations were found between heterotrophic uptake of [$^{14}$C]-glucose or [$^{3}$H]-leucine, bacterial production and exoenzyme activities (Hoppe, 1983; Hoppe *et al.*, 1988; Billén, 1984; Chróst *et al.*, 1989; Somville, 1984; Meyer-Reil, 1986). In most of these papers, the relations between added artificial, synthetic substrates and extracellular enzyme activities were tested. Fewer investigations have been made of the relation between natural substrates (e.g. polysaccharides, polypeptides) and the corresponding enzymes on the one hand and the relation between the enzymatic hydrolysis products (monomers) and the extracellular enzyme activities on the other hand (Chróst *et al.*, 1989; Münster, 1991).

We hypothesize that in most aquatic environments there should be a significant relationship between extracellular enzyme activities, their corresponding substrates (polymers) and their hydrolysis products (monmers). The validity of this hypothesis has already been verified in the eutrophic lake Plußsee (Chróst *et al.*, 1989). We intend to test this hypothesis also in a lake with high DOM and Humic Substance (HS) content, but probably lower UDOM content and we expect that polyphenolic compounds may modify enzyme activities partially, and environmental processes may affect their distribution patterns and activities. We consider these enzyme-substrate systems as important regulators for detritus processing and also for microbial food chain processes.

## Study site

Mekkojärvi is a head water lake, located in the Evo forest area in southern Finland (Fig. 1), near Lammi Biological Station. In this area (about 100 km$^2$) more then 100 lakes with different trophic state and HS content can be found. Most lakes are located in bedrock areas of Diorit and Gneis, covered by morainic sediments. Like other lakes in this area, Mekkojärvi is surrounded by forest with spruce and scots pine and the littoral zone is in close contact with 0.5 to 1 m thick peat layers.

Mekkojärvi contains high amounts of allochthonous detritus. The water is brown coloured

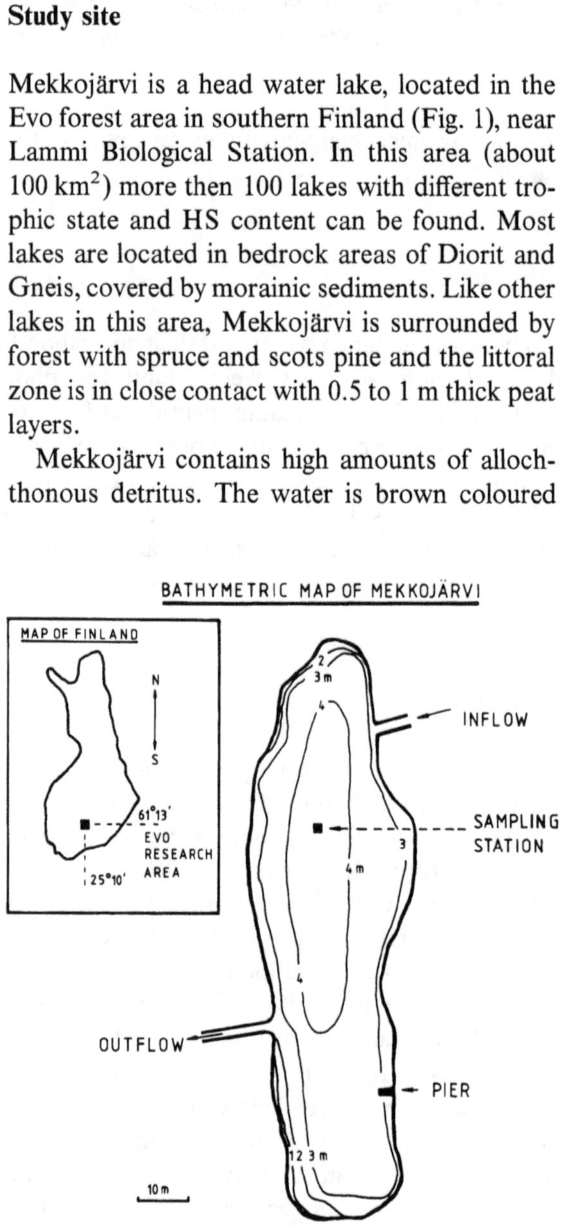


*Fig. 1.* Location of studies and bathymetric map of Mekkojärvi with sampling station, modified from Münster *et al.* (1989).

(300–600 mg Pt units $l^{-1}$) and after ice break the water warms rapidly. This water colour influences the light environment and compresses the trophogenic layer to a depth of 0.5–1 m. Rapid and steep light, nutrient, oxygen and temperature gradients can be measured (Jones & Arvola, 1984; Salonen, 1984). Because of this thin trophogenic zone, the most important nutrients ($PO_4^{3-}$, $NO_3^-$) are already rapidly depleted during spring bloom development (Arvola & Rask, 1984). The most important physical and chemical parameters are summarized in Table 1.

The dominant algae in polyhumic lakes like Mekkojärvi are motile, mostly flagellates like *Chlorophytes* and *Chrysophytes* (Ilmavirta, 1988). These can comprise more than 45% of total phytoplankton biomass in Mekkojärvi (Ilmavirta, 1983; Arvola, 1986). Intensive grazing by heterotrophic nanoflagellates (HNF) may also be expected. Salonen & Jokinen (1988) measured grazing rates by HNF in Mekkojärvi up to 40 bacteria cells $cell^{-1}$ $h^{-1}$. Phagotrophic species like *Ochromonas*, *Chromulina*, *Chrysococcus* and *Uroglena* may contribute up to 30–80% of phytoplankton biomass (Salonen & Jokinen, 1988).

## Methods

Lake water was sampled weekly from May until August 1986 and 1987 with a 30 cm long Patalas sampler in the trophogenic layer (0.2–1 m depth). Lake water was immediately filtered through a 100 $\mu$m plankton net (to remove larger zooplank-

*Table 1.* Basic limnological data of lake Mekkojärvi*, modified from Münster *et al.* (1991).

| Parameter | | Range | Average |
|---|---|---|---|
| Surface area | (ha) | – | 0.35 |
| Max. depth | (m) | – | 0.35 |
| Mean depth | (m) | – | 2.2 |
| Secchi depth | (m) | 0.3–0.5 | 0.4 |
| Mixing depth | (m) | 0.5–0.9 | 0.7 |
| Water color | (mg Pt $l^{-1}$) | 300–600 | 450 |
| Conductivity | ($\mu$S $cm^{-1}$) | 2–10 | 6 |
| pH | | 4.2–6.6 | 5.4 |
| $P_{tot}$ | ($\mu$g P $l^{-1}$) | 10–25 | 17 |
| $P-PO_4^{3-}$ | ($\mu$g P $l^{-1}$) | 2–12 | 7 |
| $N_{tot}$ | ($\mu$ N $l^{-1}$) | 500–1500 | 1000 |
| $N-NO_3^-$ | ($\mu$g N $l^{-1}$) | 20–80 | 50 |
| $N-NH_4^+$ | ($\mu$g N $l^{-1}$) | 10–70 | 40 |
| DIC (diss. inorg. carbon) | (mg C $l^{-1}$) | 0.1–0.5 | 0.3 |
| DOC (diss. organ. carbon) | (mg C $l^{-1}$) | 15–35 | 25 |
| DFAA (diss. free amino acids) | ($\mu$g C $l^{-1}$) | 3–110 | 56 |
| DFAA % C of DOC | | 0.02–0.6 | 0.3 |
| DCAA (diss. combined amino acids) | ($\mu$g C $l^{-1}$) | 160–850 | 505 |
| DCAA % C of DOC | | 0.85–3.2 | 2.2 |
| DFCHO (diss. free carbohydrates) | ($\mu$g C $l^{-1}$) | 1–50 | 25 |
| DFCHO % C of DOC | | 0.01–0.1 | 0.05 |
| DCCHO (diss. combined carbohydrates) | ($\mu$g C $l^{-1}$) | 0.5–100 | 50 |
| DCCHO % C of DOC | | 0.02–0.5 | 0.25 |
| Chlorophyll-a | ($\mu$g $l^{-1}$) | 5–30 | 18 |
| Bacteria numbers (AFDC) | ($10^6$ cells $ml^{-1}$) | 1–6 | 3 |
| Bacteria biomass | ($\mu$g C $l^{-1}$) | 10–40 | 25 |
| Heterotrophic activity | ($\mu$g glucose $l^{-1}$ $h^{-1}$) | 0.3–0.05 | 0.16 |

* Most physical and chemical data were taken from surface water (0–1 m) in 1986/87.

ton) into autoclaved plastic flasks (Nalgene Inc. USA), which were kept in coolers during sampling and transportation to the laboratory.

Nutrient concentrations (P-$PO_4^{3-}$, N-$NO_3^-$, $P_{tot}$, $N_{tot}$) were measured by standard methods as described by Arvola & Rask (1984) with an Akea auto analyzer. Dissolved organic carbon (DOC) was measured with a carbon analyzer (model Tocamaster, Beckman Inc. USA) with total combustion and IR-detection after lake water filtration through precombusted Whatman GF/F and prewashed Nuclepore 0.2 $\mu$m polycarbonate filters.

For chlorophyll measurements, lake water was filtered through precombusted glass fiber filter (Whatman GF/F). Filters were extracted with 10 ml of acetone for 10 hours at 4 °C in a refrigerator. Extracted chlorophylls were measured fluorimetrically according to Wetzel & Likens (1979), before and after acidification and corrected for phaeophytin. Chlorophyll concentrations were calculated from chlorophyll-*a* and phaeophytin-*a* calibration curves prepared with purified chlorophyll-*a*/phaeophytin-*a* from *Anacystis nidulans* (Sigma Chemie, FRG).

For bacterial counting, lake water was filtered through 0.2 $\mu$m Nuclepore filters, stained with acriflavine and counted by epifluorescence (Bergström *et al.*, 1986). According to this procedure, bacterial numbers are expressed as acriflavine direct counts (AFDC).

Extracellular enzyme assays were prepared according to Hoppe (1983) with small modifications as described by Münster *et al.* (1989). From freshly prepared 4-methylumbelliferyl (4-MUF) substrates and leucine-4-methyl-coumarinyl-7-amide (leu-MCA) stock solutions (10 mMol), 100 $\mu$l of 4-MUF- and leu-MCA substrate were added to 9.9 ml freshly sampled and 100 $\mu$m prefiltered lake water and incubated at *in-situ* temperature and lake water pH. Extracellular enzyme activities were followed fluorimetrically from enzymatic release of fluorescent 4-MUF and MCA from added non-fluorescent 4-MUF- and leu-MCA substrates. 4-MUF and MCA concentrations were calculated from freshly prepared 4-MUF and MCA calibration curves in filtered and preautoclaved lake water, to consider potential quenching or interference of 4-MUF- and MCA-fluorescence with HS (Münster *et al.*, 1989). Differentiation between acid- and alkaline phosphatase activities were achieved by measurements at two different pH values (pH = 5.5 and 8.2, adjusted with 0.5 mol borate buffer) in lake water.

The effect of polyvinylpyrrolidone (PVP) on enzyme activity was tested by adding soluble polyvinylpyrrolidone-15 (Serva Feinbiochemica, Heidelberg, FRG) to a final concentration of 3% (w/v) to lake water. After one hour of preincubation with PVP, enzyme activity was measured as described above.

## Substrate analysis

Delicate flagellates are common in the plankton of Mekkojärvi, so 10–20 ml lake water were filtered through precombusted Whatman GF/F and subsequently through prewashed Nuclepore polycarbonate filters (0.2 $\mu$m) without vacuum. This pre-tested procedure minimized probable cell leakage or disruption of flagellates during filtration. Filtered lake water was kept frozen (−25 °C) until analysis.

Carbohydrates were analyzed with gas liquid chromatography (glc) according to Münster (1985) before and after hydrolysis with 1 N HCl at 110 °C for 10 h under $N_2$ atmosphere in precombusted glass ampoules. Using this procedure, two carbohydrate fractions were analyzed: dissolved free carbohydrates (DFCHO) and dissolved combined carbohydrates (DCCHO). Amino acids were analyzed as ortho-phthaldialdehyde (OPA)-derivatives according to Lindroth & Mopper (1979) on a Spherisorb ODS-2 column (5 $\mu$m, 4.3 × 250 mm) with a Spectra-Physics HPLC system, model SP 8700, connected to a SP 4100 integrator. The solvent system was prepared from solvent-A (methanol, LiChrosolve, Merck) and solvent-B (0.03 Mol Na-Acetate, pH:6.55, with 10% methanol and 0.5% THF v/v). Amino acids were detected fluorimetrically with a Kratos SF 970 fluorescence detector, with wavelength settings of 225 nm at excitation and

410 nm band path filter at the emission side. With these wavelength settings, we have obtained higher sensitivities combined with lower background noise, according to HS interference on the HPLC column. Amino acids were analyzed before and after hydrolysis with freshly three-fold distilled 6 N HCl at 110 °C under nitrogen atmosphere in precombusted and sealed glass ampoules. By this procedure, we could differentiate between dissolved free (DFAA) and dissolved combined amino acids (DCAA).

## Results

### *Biomass indicators*

Chlorophyll concentration varied between 0.5–60 $\mu g\,l^{-1}$ (Fig. 2). This is in the upper range of more than 50 different lakes studied in this area (Arvola, 1986). Highest concentrations were found just after ice melting, which suggest that nutrient availability enables highest biomass production.

Number of bacteria (AFDC) varied between 1–7 $10^6$ cells $ml^{-1}$ (Fig. 2), but bacteria were small, with a mean volume of 0.04 $\mu m^3$. Calculated biomass varied from 10–40 $\mu g\,C\,l^{-1}$. Significant covariance between chlorophyll and number of bacteria (AFDC) was observed (Fig. 2).

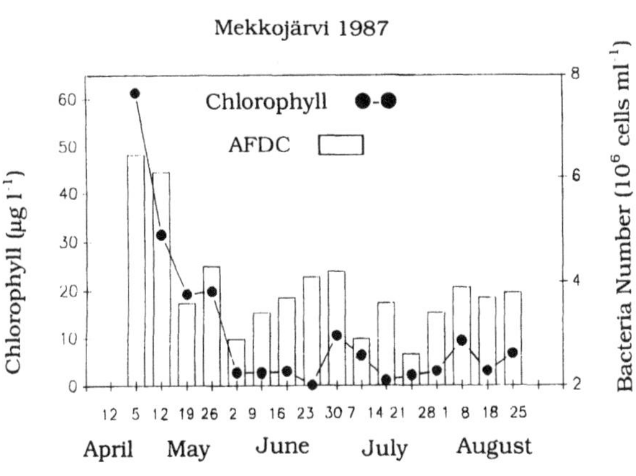


*Fig. 2.* Distribution of chlorophyll (filled circles) and bacterial number (AFDC, bars); modified from Münster (1991).

### *Composition of dissolved organic matter (DOM)*

Because of the large drainage basin and the close contact of the littoral with peat land, Mekkojärvi water is rich in HS, Dissolved organic carbon content (DOC) is therefore high and varies between 8–25 mg C $l^{-1}$ (Tab. 1). But the content of UDOM, such as dissolved free amino acids (DFAA) and carbohydrates (DFCHO), is low (Table 1). The carbon content of DFAA comprise only 0.02–0.6% of the DOC and the DFCHO even lower (0.01–0.1% of DOC, Table 1). This is less then the remaining part of DOC. The dissolved combined amino acids (DCAA) and carbohydrates (DCCHO) were found in a range of 0.3–3.2% (mean 2.2%) and 0.02–0.5% (mean 0.3%) of the DOC pool, respectively. We found consistently that the UDOM pool in Mekkojärvi was extremely small compared to eutrophic alkaline lakes (Münster, 1984, 1985; Münster & Chróst, 1990). Consequently the amount of protein comprised only 0.92–3.8% of DOC from Mekkojärvi. From these data it can be assumed that utilizable polymeric solutes like DCAA and DCHO in the DOM are also less available for heterotrophic assimilation.

### *Extracellular enzyme activity, variation and distribution*

Because of the low concentration of UDOM solutes, bacterial growth may also depend on the utilization and availability of polymeric substrates like DCAA and DCCHO. However, polymeric substrates have to be cleaved into monomers and oligomers by extracellular enzymes before they can be assimilated (Rogers, 1961). Only small molecules can enter microbial membranes and make contact with transport systems. Extracellular enzymatic substrate depolymerization may therefor become a rate-limiting step for growth of microheterotrophs.

We have followed eight different enzymes in Mekkojärvi at 0–1 depth (Table 2). Of these, phosphatases, glycosidases and aminopeptidase displayed the highest activities (Fig. 3). Highest

*Table 2.* Measured enzymes in Mekkojärvi.

| Enzyme | Substrate | Enzyme activity (nmol $l^{-1}$ $h^{-1}$) | |
|---|---|---|---|
| | | Range | Mean (± SD) |
| Phosphatase | MUF*–phosphate | 51–162 | 50.4 ± 28.2 |
| Leucine-aminopeptidase | MCA**–leucine | 6.5– 26 | 15.0 ± 7.5 |
| α-glucosidase | MUF-α-glucose | 0.1– 90 | 6.7 ± 6.1 |
| β-glucosidase | MUF-β-glucose | 4.4– 68 | 23.6 ± 13.5 |
| α-galactosidase | MUF-α-galactose | 0.1–117 | 11.1 – 4.5 |
| β-galactosidase | MUF-βgalactose | 0.1– 65 | 8.7 ± 10.6 |
| α-mannosidase | MUF-α-mannose | 0.1–102 | 6.9 ± 3.9 |
| β-mannosidose | MUF-β-mannose | 0.1– 81 | 6.2 ± 3.6 |

* MUF: 4-methylumbelliferyl-
** MCA: 4-methyl-coumarinyl-7-amide-

activities (100–150 nmol $l^{-1}$ $h^{-1}$) were measured for phosphatases and glycosidases in June and July. Glycosidase activity had maxima (70–120 nmol $l^{-1}$ $h^{-1}$) in July and August. Aminopeptidase measurements started later and fluctuated between 20–30 nmol $l^{-1}$ $h^{-1}$. Generally, a significant decrease of enzyme activity was measured from June to August. Among studied glycosidases (Table 3) β-glucosidase gave the highest activity. For all measured glycosidases the β-form displayed always higher values than α-form (Table 2). At present, we have no clear explanation for this result, and preliminary results from carbohydrate analysis in the DOM also gave no clear explanations. For all identified hexoses (glucose, galactose, mannose) only in 10–40% of measurements were the β-form of hexoses higher than the α-form. Because of analytical limitations with elucidating the complete structure of DOM and the included carbohydrates, we cannot clarify this phenomenon, but we assume that already during the analytical procedure (i.e. during the hydrolysis procedure) anomerization can occur at the C-1 position of analyzed hexoses (Pigman &

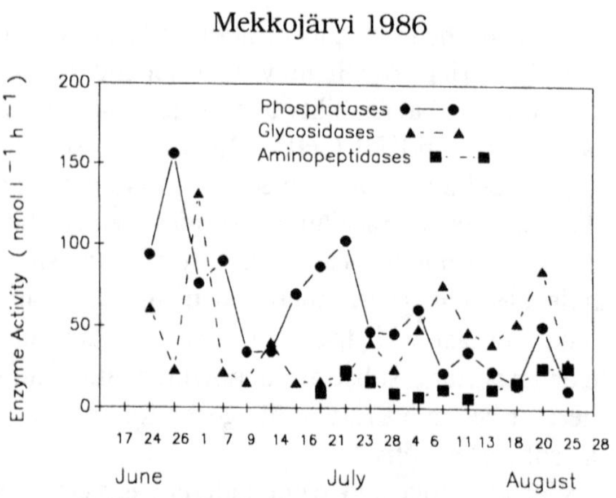


*Fig. 3.* Distribution of the three most active extracellular enzyme systems in Mekkojärvi (0–1 m depth): phosphatases (filled circles), glycosidases (filled triangles) and aminopeptidases (filled squares), modified from Münster (1991).

*Table 3.* α:β ratio of glycosidases activities in lake Mekkojärvi, 0–1 m depth, after Münster (1991).

| Sampling | α:β-ratio | | |
|---|---|---|---|
| Date | Glucosidase | Galactosidase | Mannosidase |
| 1–7–86 | 0.31 | 0.14 | 3.00 |
| 7–7–86 | 0.01 | 0.01 | 0.01 |
| 9–7–86 | 0.13 | 0.01 | 0.01 |
| 21–7–86 | 0.57 | 0.17 | 0.10 |
| 28–7–86 | 0.10 | – | – |
| 4–8–86 | 0.20 | – | – |
| 6–8–86 | 0.32 | 0.08 | – |
| 11–8–86 | 0.18 | 0.71 | – |
| 13–8–86 | 0.54 | – | 1.71 |
| 18–8–86 | 0.79 | 1.77 | 0.35 |
| 20–8–86 | 0.79 | 0.95 | 2.95 |
| 25–8–86 | 0.62 | 1.83 | 0.01 |
| 28–8–86 | 0.47 | 0.83 | 1.32 |

Anet, 1972). This means that physical-chemical reactions (hydrolysis and anomerization) can modify the structural composition ($\alpha:\beta$ ratio) of enzymatically and chemically cleaved polymeric carbohydrates.

*Substrate-enzyme interactions in Mekkojärvi*

The significance of extracellular enzyme-substrate interactions in natural aquatic habitats has already been demonstrated (Hoppe, 1983; Chróst *et al.*, 1989). However in polyhumic lakes with high HS content and partially acid character, such studies are scarce (Münster *et al.*, 1989; Münster, 1991). We assume that extracellular substrate depolymerization can contribute to the availability and dynamics of UDOM solutes. We selected two enzyme-substrate systems to evaluate their participation in DOM processing.

*The leucine-aminopeptidase-DFAA/DCAA system*
Although we incubated only one fluorogenic substrate (leucine-MCA) for measuring aminopeptidase activities, we expected a reasonable relationship to the DFAA and DCAA pool sizes and variations. The DFAA pool varied less than the DCAA pool size (Fig. 4) and the ratio of DFAA to DCAA concentration varies between 1:10 to 1:5, but with significantly higher dynamics and variation of the DCAA pool size. Both substrate concentrations are certainly regulated by release and uptake by microplankton. There is no significant relation between concentration of DFAA and measured aminopeptidase activity (Fig. 5a). However, the DCAA concentration was significantly correlated to aminopeptidase activity (Fig. 5b). A further detailed analysis of the relation of measured aminopeptidase activity and the composition of identified AA in the DFAA and DCAA pools in 1987 revealed that with this aminopeptidase enzyme assay, we have not measured a highly specific cleaving mechanism, but

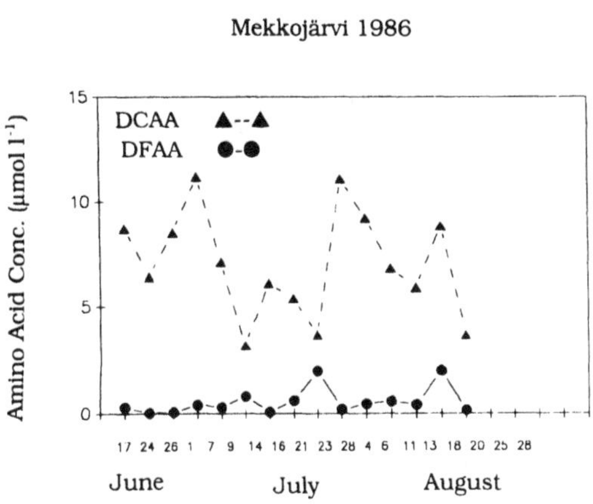


*Fig. 4.* Distribution and variation of dissolved free (DFAA, filled circles) and dissolved combined (DCAA, filled triangles) amino acids in Mekkojärvi (0–1 m depth).

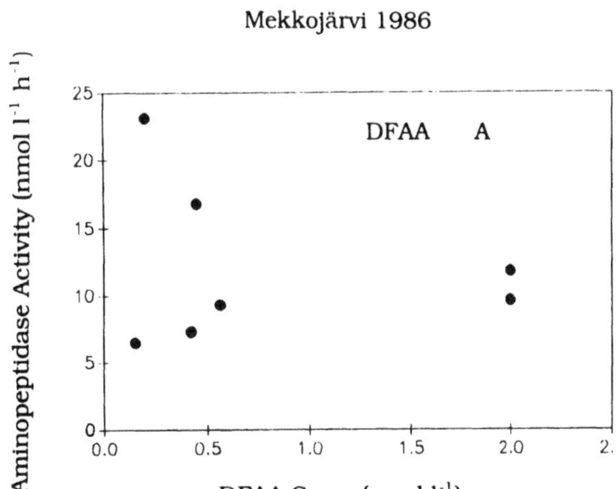


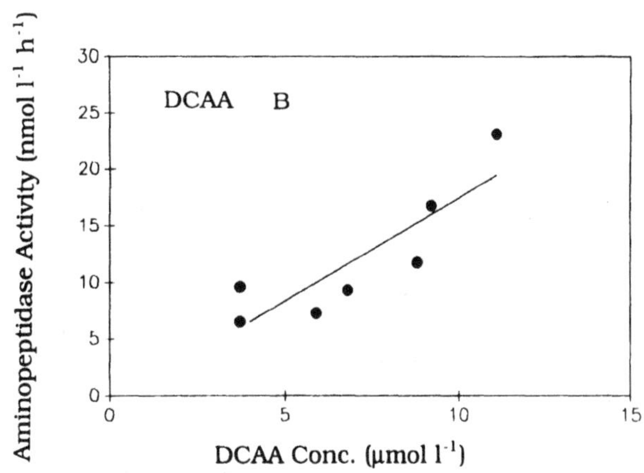


*Fig. 5.* Relationship between amino acids and aminopeptidase activity in Mekkojärvi water (0–1 m depth). A. Dissolved free amino acids (DFAA); B. Dissolved combined amino acids (DCAA, linear regression plot $y = 1.8257X - 0.7751$, $r^2 = 0.75459$, $n = 7$, $p < 0.0001$), after Münster (1991).

*Table 4.* Coefficients of correlations ($r^2$) between leucine-aminopeptidase and dissolved amino acids (daa) in lake Mekkojärvi, 1987, 0–1 m depth, after Münster (1991). $n=7$, $p<0.01$ in all cases.

| No. | DAA* | Correlation ($r^2$) | |
|---|---|---|---|
| | | DFAA** | DCAA*** |
| 1 | Asp | −0.70 | +0.87 |
| 2 | Glu | −0.56 | +0.87 |
| 3 | Asn | −0.50 | +0.88 |
| 4 | Ser | −o.59 | +0.81 |
| 5 | Gly | −0.79 | +0.78 |
| 6 | Thr | −0.78 | +0.87 |
| 7 | Arg | −0.62 | +0.77 |
| 8 | Tyr | −0.77 | +0.85 |
| 9 | Val | −0.77 | +0.65 |
| 10 | Meth | −0.53 | +0.61 |
| 11 | Phe | −0.61 | +0.81 |
| 12 | Ile | −0.73 | +0.78 |
| 13 | Leu | −0.74 | +0.93 |

* DAA : Dissolved amino acids.
** DFAA: Dissolved free amino acids.
*** DCAA: Dissolved combined amino acids.

rather unspecific peptide-cleaving mechanisms. Simple linear regression of DCAA with leucine-aminopeptidase activity from 1987 (Table 4) gave significant positive relationships for all DCAA (Münster, 1991). Measured aminopeptidase activity displayed, therefore, a less specific peptide-bond cleaving reaction in Mekkojärvi water.

*The glucosidase-DFCHO/DCCHO system*

From all measured carbohydrases in Mekkojärvi, about 50–70% of activity is represented by glucosidase activity. Because of the structural diversity of polysaccharides, we have measured $\alpha$- and $\beta$-glucosidases. There is a significant decrease of enzyme activity from June to August and the $\beta$-form is predominant except in June (Fig. 6). Maxima of activity (90–60 nmol $l^{-1}$ $h^{-1}$) were observed in June and July. As with the leucine-aminopeptidases system, we expected a positive relationship to natural substrate composition. The relation of DF-glucose to DC-glucose and their variation is shown in Fig. 7. Concentration of DF-glucose varied between 1–250 nmol $l^{-1}$ and surpassed (in some cases) that of DC-glucose,

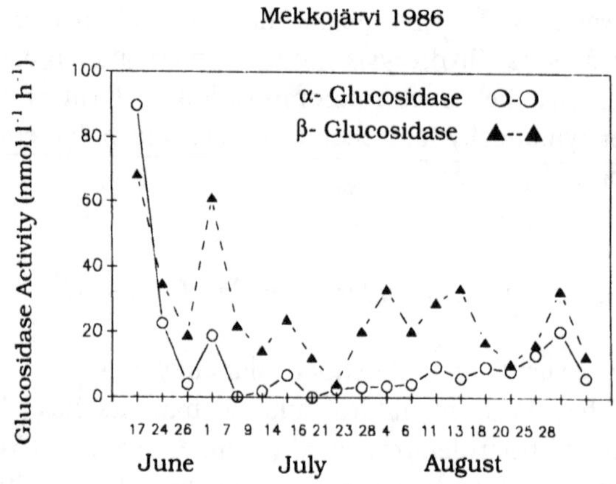


*Fig. 6.* Distribution and variation of glucosidase activity ($\alpha$-glucosidase open circles, $\beta$-glucosidase filled triangles) in Mekkojärvi water (0–1 m depth).

which varied between 50–500 nmol $l^{-1}$. It can be clearly seen that DF-glucose and DC-glucose maxima and minima are oscillating in an opposite pattern. The mechanisms and explanations for such patterns might be important in relation to the glucosidase activities. Therefore, the relationship of extracellular glucosidase activity to DF- and DC-glucose is plotted in Fig. 8. Surprisingly, a significant clear relationship exists between both glucose pools and the glucosidase activity. DF-glucose has a significant negative correlation and

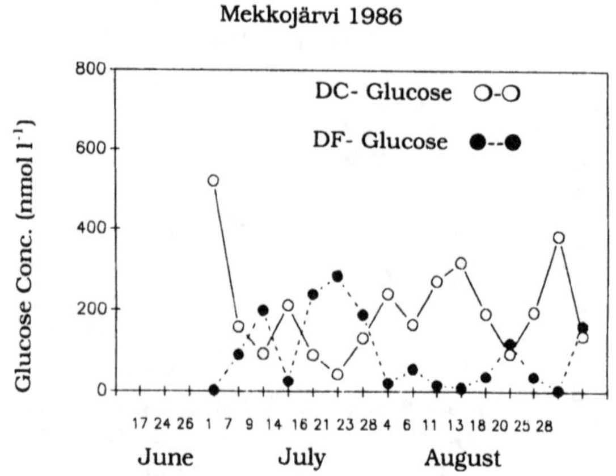


*Fig. 7.* Distribution and variation of dissolved free (DF-glucose, filled circles) and dissolved combined (DC-glucose, open circles) glucose concentration in Mekkojärvi water (0–1 m depth).

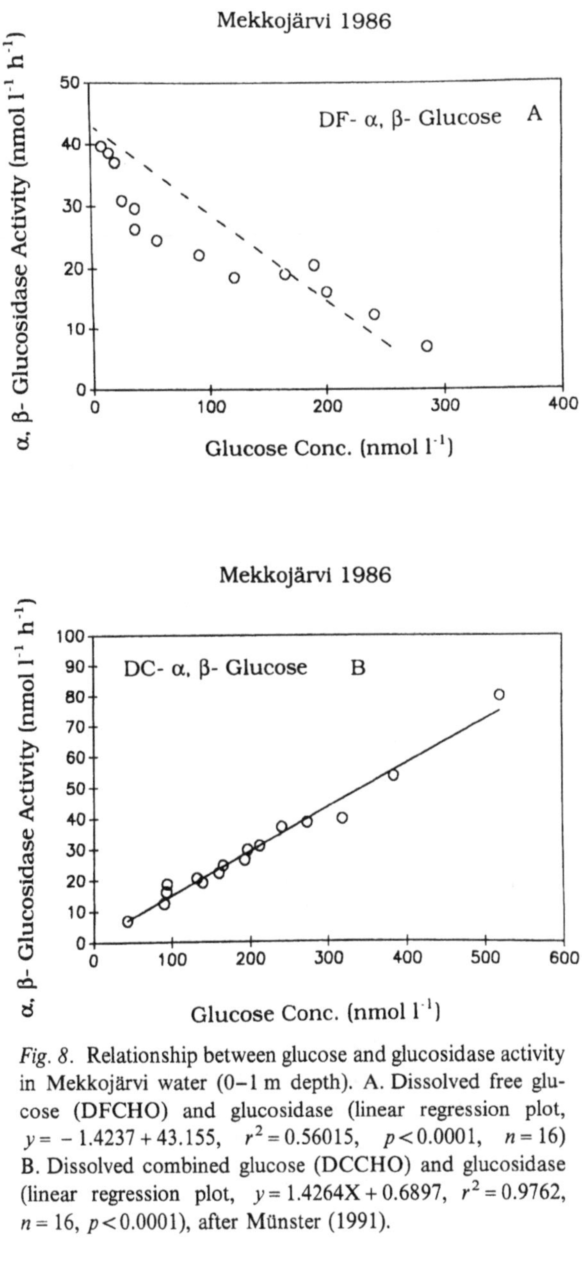


*Fig. 8.* Relationship between glucose and glucosidase activity in Mekkojärvi water (0–1 m depth). A. Dissolved free glucose (DFCHO) and glucosidase (linear regression plot, $y = -1.4237 + 43.155$, $r^2 = 0.56015$, $p < 0.0001$, $n = 16$) B. Dissolved combined glucose (DCCHO) and glucosidase (linear regression plot, $y = 1.4264X + 0.6897$, $r^2 = 0.9762$, $n = 16$, $p < 0.0001$), after Münster (1991).

DC-glucose a positive, similar to that found between DCAA and aminopeptidase. However, in contrast to the DFAA-aminopeptidase system DF-glucose gave a clear relation to the glucosidase activity. This deviation between both enzyme-substrate systems is not clearly understood, but it might be reasonable to suggest that glucosidases have a more specific reaction or cleaving mechanism, whereas leucine-aminopeptidase reacts unspecifically with many peptide bonds. It is not specifically inhibited by enzyme reaction products like DFAA (Table 4), whereas DF-glucose can inactivate the glucose-glucosidase system (Fig. 8a). We conclude that glucosidase activity is significantly regulated by product inhibition mechanisms, whereas leucine aminopeptidase was not significantly inhibited by the DFAA pool.

### *Environmental influences on extracellular enzyme activities*

Our knowledge about hydrolytic enzyme activities and their regulation in natural habitats is still rather limited. Enzyme characterization has mostly been done with purified enzymes. Such enzyme assays are recommended and useful for biochemical studies, but enzyme studies in an ecological context have to consider environmental aspects, temporal and spatial variability of activity, external influences like temperature, acidity, inhibition and immobilization reactions. This means that measurements have to be made under '*in-situ*' conditions. Therefore, we have studied extracellular enzyme activities under natural conditions, and have taken into account temporal and spatial variability of activity, effects of temperature, acidity, ion composition, inhibition and immobilization (Figs. 3, 6, 9).

#### *Temporal and spatial variation*

Enzymes clearly fluctuate in a temporal manner (Figs. 3, 6). We also expect spatial variation. This fluctuation is partially due to the oscillation of the whole lake metabolism. Many complex metabolic processes are responsible for such distribution patterns. The aim of such studies was to evaluate the different regulating parameters, which influence extracellular enzyme activities.

#### *Lake water acidity*

Acidity (pH) in small head water lakes may change according to the nature and composition of allochthonous material and may therefore influence the activity of extracellular enzymes. Five measured enzymes showed different activities at changing pH in lake water (Fig. 9). Phosphatases, and all three glycosidases (α-, β-glycosidases, exo-cellobiohydrolases) have their relative maxima of activity at pH = 5.4, which was the mean pH of Mekkojärvi water during the 1987 sampling season. The leucine-aminopeptidases had their maxima of activity at pH = 7.8. For phosphatases we measured two maxima, the larger in the acid region (pH = 5.4) and the smaller in the slight alkaline region (pH = 8.5). We have not followed specifically the contribution of the alkaline phosphatases to the processing of organic phosphorus compounds in Mekkojärvi, but we estimated from laboratory experiments that about 5–10% of total phosphohydrolases is represented by alkaline phosphatase in Mekkojärvi water.

*Temperature effects*

Studies on temperature dependence of extracellular enzymes in Mekkojärvi revealed that, in comparison with the pH dependency, the enzymes do not operate under optimal temperature conditions. For all measured enzymes, we found temperature optima at 35–38 °C (data not shown!). This is far from '*in situ*' conditions, but not very different from laboratory studies on purified enzymes.

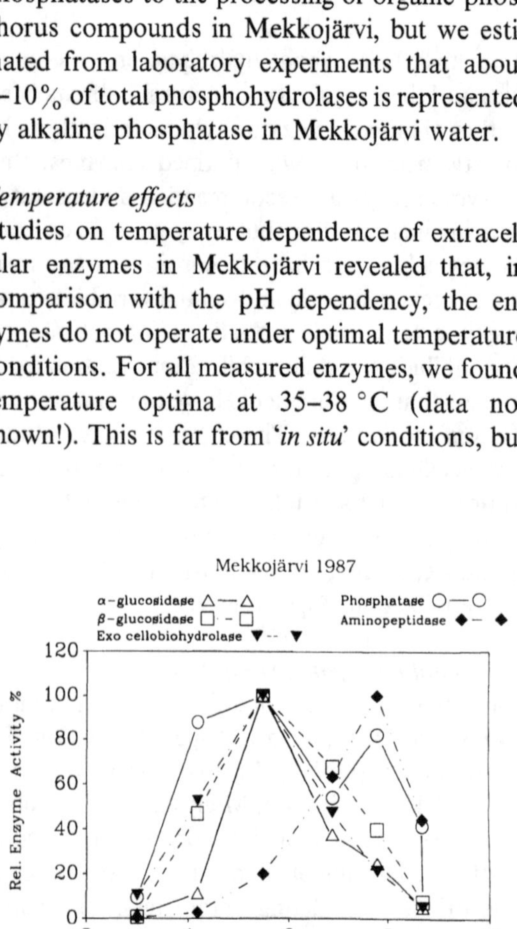


*Fig. 9.* pH dependence of phosphatases (open circles), amino-peptidases (filled diamonds), exo-cellobiohydrolases (filled inverse triangles), α-glucosidases (open triangles) and β-glucosidases (open squares) in Mekkojärvi water (0–1 m depth), after Münster (1991).

*Immobilization by HS*

Besides temperature and acidity, enzyme activities can be further modified and influenced by co-occurring organic or inorganic solutes. Because of the high amount of HS in Mekkojärvi water, we have tested the influence of HS on extracellular enzyme activities. According to Cannell *et al.* (1987, 1988a), microalgae can also produce protease and glycosidase inhibitors. These inhibitors have been characterized and identified as pentagalloylglucose derivatives (Cannell *et al.*, 1988b). The inhibitory mechanisms can be described as phenol-protein interactions (McManus *et al.*, 1985; Haslam, 1974). As HS contains polyphenolic compounds (Christman & Gjessing, 1983), such interactions may also occur in Mekkojärvi water. We have tried to evaluate such inhibitory interactions in the water samples by adding polyvinylpyrrolidone (PVP). PVP has been shown to react significantly with polyphenols via hydrogen bridge bonding (Olsson & Samuelson, 1974; Münster, 1985), which bind to the polyphenolic cores. If PVP is added to polyphenols, it immobilizes phenolic compounds. Such immobilized polyphenols no longer interact with proteins and decrease the inhibition of enzyme activities (Cannell *et al.*, 1988b). After adding PVP to Mekkojärvi water, enzyme assays displayed 10–23% higher activities when compared to untreated enzyme assays (Table 5). These results confirm our hypothesis that extracellular enzymes in natural habitats are partially modified in their activities.

*Table 5.* Extracellular enzyme activities before and after Polyvinylpyrrolidone (PVP) addition.

| Name of enzyme | Enzyme activity ($nmol\,l^{-1}\,h^{-1}$) | | |
|---|---|---|---|
| | Control | + PVP | % increase |
| Phosphatase | 185.2 | 212.9 | 14.9 |
| Leucine-aminopoptidase | 5.5 | 6.9 | 25.6 |
| β-glucosidase | 52.8 | 57.8 | 9.5 |

## Discussion

Studies on the role of extracellular enzymes in aquatic food web processes are relatively new and have only recently emphasized their importance in connection with microbial activities (Hoppe, 1983; Ammerman & Azam, 1985; Billén, 1984; Chróst, 1990; Wetzel, 1991; Münster, 1991). In recent studies on food chain structures in dystrophic lakes, it has been shown that heterotrophic activity may exceed that of autotrophy (Salonen, 1981; Winberg, 1980; Tranvik, 1989). The higher community respiration in polyhumic lakes has been attributed to additional DOM utilization by bacterioplankton (Salonen *et al.*, 1983). Allochthonous DOM, especially HS, was regarded as an additional carbon and energy source for bacterial growth (Salonen *et al.*, 1983; Salonen & Hammar, 1986; Johansson, 1983; Tranvik & Höfle, 1987; Tranvik, 1989). Simple UDOM solutes like radiolabelled monomeric sugars, amino acids, and also radiolabelled phenol have been tested as potential substrates for bacterial growth (Tranvik & Höfle, 1987), but these papers do not explain how the mechanisms operate or how rapidly bacterioplankton utilize natural polymeric and refractory substrates. Because of the complexity and dynamics of DOM in general and of HS in particular, there is still a need for information about the mechanisms and regulation of detritus processing (Wetzel, 1984; Münster & Chróst, 1990; Münster, 1991). UDOM solutes in Mekkojärvi do not comprise more than 0.01–0.35% of DOC (Table 1). Also, polymeric substrates like DCAA and DCCHO do not comprise more than 0.1% to 4% of DOC (Table 1) or 1–5% of dissolved organic nitrogen (DON). This is much lower in comparison to eutrophic lake water (Münster, 1985; Münster & Chróst, 1990; Jørgensen, 1987). We have to assume that bacterioplankton in Mekkojärvi are substrate-limited. Degradation and utilization of recalcitrant DOM, like HS, may contribute only slowly to the carbon demand of bacterioplankton. There is convincing evidence that natural bacterial assemblages in polyhumic lakes may have higher metabolic potentials for degradation of aromatic and polyphenolic compounds than those from clear water lakes (De Haan, 1974, 1977; Tranvik & Höfle, 1987). According to Schütt (1988), such metabolic potential for degradation of recalcitrant DOM may be even genetically localized (extrachromosomal DNA; i.e. plasmids). Different microbial communities may be responsible for HS degradation (Wetzel, 1984). Carbon dating have revealed that formation and degradation of HS proceed slowly (Stuermer *et al.*, 1978). Therefore, the supply of energy and carbon via HS degradation to the rapid dynamic processes in aquatic microbial loops may only partially support the required merits for nutrients, carbon and energy.

[$^{14}$C]-glucose uptake experiments in Mekkojärvi water revealed that uptake velocity of glucose varied between 0.1–3.5 nmol $l^{-1}$ $h^{-1}$ (mean 1.65 nmol $l^{-1}$ $h^{-1}$); these values are in a similar range to those recorded in the eutrophic lake Plußsee (Chróst & Overbeck, 1989). Since UDOM pool sizes of DOM in Mekkojärvi are at least 5 to 10 times lower, and the bacterial biomass is in a similar range as in eutrophic lakes (e.g. lake Plußsee), a rapid turn-over of UDOM substrates is assumed. Unfortunately, we have not measured uptake kinetics to calculate the actual turnover e.g. for [$^{14}$C]-glucose. But, as we measured extracellular enzymatic substrate depolymerization, we could show that there might be significant regulatory mechanism in the water, which contributes to the processing of polymeric DOM. There was on average a release of about 15.6 nmol $l^{-1}$ $h^{-1}$ of glucose from added synthetic MUF-glucose by glucosidase activity (Fig. 3). In comparison to average glucose uptake, this means that theoretically there is an overflow of DF-glucose into the UDOM pool. Since such glucose overflow has been never reported, this may raise the question, whether the synthetic MUF-compounds represent a good natural substrate to the enzyme systems to calculate together with glucose uptake data the flow rates of natural substrates in the DOM pool. According to results in Fig. 8 and previous studies on extracellular substrate depolymerization and substrate uptake, we expect a close relationship and feed-back mechanism between substrate uptake and extracellular enzymatic release of glucose (Chróst *et al.*, 1989; Münster, 1991). According to our data, we may have a tenfold higher rate of inflow of DF-glucose by extracellular enzyme activity to the UDOM pool. Similar experiments by Hoppe *et al.* (1988) with the leu-MCA/leucine aminopeptidases system have shown that in the brackish water of Kiel Fjord, similar ratios for hydrolysis and uptake (6.8:1 and 13.4:1) of leucine occurred. We assume that there is an oscillating regulatory mechanism between substrate uptake systems and extracellular substrate depolymerization systems. This mechanism would have a much faster response of the uptake systems to the UDOM pool sizes, when compared to UDOM supply via enzymatic substrate depolymerization. This may be due to environmental parameters like pH, oxygen, HS (see Fig. 9), which influence more strongly the enzymatic hydrolytic processes than the uptake systems. Therefore, the actual rates of substrate uptake and enzymatic cleavage of polymers may be moderately coupled. We estimate from our studies of glycosidases and aminopeptidases that about 1–5% of DOC could be processed and utilized by the microorganisms. We consider our enzyme-substrate systems to be suitable tools to study detritus processing and UDOM pool size regulation on a biochemical level. This approach may provide deeper insight into ecophysiological processes and may improve our understanding of the dynamics and regulations of microbial food chain processes.

Microbial enzymatic substrate processing may certainly also exist for other, more recalcitrant substrates like HS, but we know less about the biochemical background of microbial HS depolymerization and utilization, which may explain the higher heterotrophy in humic compared to clear water lakes. Photochemical reactions may also be involved in the decomposition and (together with enzymatic attack by peroxidases of recalcitrant DOM) enhanced microbial utilization of HS may occur (Salonen & Tulonen, 1990; Münster, 1991).

## Acknowledgements

This work was supported by the Maj and Tor Nessling Foundation, Finland and the Max-Planck-Society, FRG. We also address our gratitude to the University of Helsinki for support with technical equipment and personal help during stay and experimental work at Lammi Biological Station. This work benefitted from stimulating discussions and valuable recommendations from Dr. K. Salonen, Dr. L. Arvola and Dr. T. Kairesalo. We further have to thank B. Albrecht for photographic assistance during preparation of the figures and Dr. L. Weider for correcting our English.

## References

Ammerman, J. W. & F. Azam, 1985. Bacterial 5′-nucleotidase in aquatic ecosystems: a novel mechanism of phosphorus regeneration. Science 277: 1338–1340.

Arvola, L., 1986. Spring phytoplankton of 54 small lakes in Southern Finland. Hydrobiologia 137: 125–134.

Arvola, L. & M. Rask, 1984. Relations between phytoplankton and environmental factors in a small, spring-meromictic lake in Southern Finland. Aqua fenn. 14: 129–138.

Azam, F. & B. C. Cho, 1987. Bacterial utilization of organic matter in the sea. In M. Fletcher, T. R. G. Gray & J. G. Jones (eds.), Ecology of Microbial Communities, Cambridge University Press, 261–281.

Azam, F., T. Fenchel, J. G. Field, J. S. Gray, L.-A. Meyer-Reil & F. Thingstad, 1983. The ecological role of water column microbes in the sea. Mar. Ecol. Prog. Ser. 10: 257–263.

Bergström, I., A. Heinänen & K. Salonen, 1986. Comparison of acridine orange, acriflavine, and bisbenzimide stains for enumeration of bacteria in clear and humic waters. Appl. envir. Microbiol. 51: 664–667.

Billén, G., 1984. Heterotrophic utilization and regeneration of nitrogen. In J. E. Hobbie & P. J. leB Williams, (eds), Heterotrophic activity in the sea. NATO SAD, Plenum Press, 1984, 313–355.

Cannell, R. J. P., S. J. Kellam, A. M. Owsianka & J. M. Walker, 1987. Microalgae and Cyanobacteria as a source of glycosidase inhibitors. J. gen. Microbiol. 133: 1701–1705.

Cannell, R. J. P., S. J. Kellam, A. M. Owsianka & J. M. Walker, 1988a. Results of a large scale screen of microalgae for the production of protease inhibitors. Planta Medica 54: 10–14.

Cannell, R. J. P., P. Farmer & J. M. Walker, 1988b. Purification and characterization of pentagalloylglucose, an α-glycosidase inhibitor/antibiotic from the freshwater gree alga Spirogyra varians. Biochem. J. 255: 937–941.

Christman, R. F. & E. T. Gjessing, 1983. Aquatic and Terrestrial Humic Materials. Ann Arbor Sci. Publ. 538 pp.

Chróst, R. J., 1989. Characterization and significance of $\beta$-glycosidase activity in lake water. Limnol. Oceanogr. 34: 660–672.

Chróst, R. J., 1990. Ectoenzymes in aquatic environments: origin, activity and ecological significance. In J. Overbeck & R. J. Chróst (eds.), Advanced biochemical and molecular approaches to Aquatic Microbial Ecology. Brock/Springer Series in Contemporary Bioscience, by Springer Verlag Berlin, 1990, p. 47–78.

Chróst, R. J., U. Münster, H. Rai, D. Albrecht, P. K. Witzel & J. Overbeck, 1989. Photosynthetic production and exoenzymatic degradation of organic matter in euphotic zone of an eutrophic lake. J. Plankton. Res. 11: 223–242.

Chróst, R. J. & J. Overbeck, 1989. Application of the isotope dilution principle to the determination of [$^{14}$C]-glucose incorporation by aquatic bacteria. Acta Microbiol. Polon. 38: 75–89.

De Haan, H., 1974. Effect of a Fulvic Acid fraction on the growth of a *Pseudomonas* from Tjeukemeer (the Netherlands). Freshwat. Biol. 4: 301–310.

De Haan, H., 1977. Effect of Benzoate on microbial decomposition of Fulvic Acids in Tjeukemeer (The Netherlands). Limnol. Oceanogr. 22: 38–44.

Fenchel, T., 1987. The ecology of heterotrophic microflagellates. Adv. microbial Ecol. 9: 57–97.

Haslam, E., 1974. Polyphenol-Protein Interactions. Biochem. J. 139: 285–288.

Hollibaugh, J. T. & F. Azam, 1983. Microbial degradation of dissolved proteins in seawater. Limnol. Oceanogr. 28: 1104–1116.

Hoppe, H.-G., 1983. Significance of exoenzymatic activities in the ecology of brackish water: measurements by means of methylumbelliferyl-substrates. Mar. Ecol. Progr. Ser. 11: 299–308.

Hoppe, H.-G., S. J. Kim & K. Gocke, 1988. Microbial decomposition in aquatic environments: combined processes of extracellular enzyme activity and substrate uptake. Appl. envir. Microbiol. 54: 784–790.

Ilmavirta, V., 1983. The role of flagellated phytoplankton in chains of small brown-water lakes in southern Finland. Ann. bot. fenn. 20: 187–195.

Ilmavirta, V., 1988. Phytoflagellates and their ecology in Finnish brown-water lakes. Hydrobiologia 161: 255–270.

Jannasch, H., 1970. Threshold concentration of carbon sources limiting bacterial growth in seawater. In D. W. Hood (ed), Symposium on organic matter in natural waters. Univ. Alaska, Mar. Sci. Occ. Publ. No. 1: 321–330.

Johansson, J. A., 1983. Seasonal development of bacterioplankton in two forest lakes in Central Sweden. Hydrobiologia 101: 71–87.

Jones, R. I. & L. Arvola, 1984. Light penetration and some related characteristics in small forest lakes in southern Finland. Verh. int. Ver. Limnol. 22: 811–816.

Jørgensen, N. O. G., 1982. Heterotrophic assimilation and occurrence of dissolved free amino acids in a shallow estuary. Mar. Ecol. 8: 145–159.

Jørgensen, N. O. G., 1987. Free amino acids in lakes: Concentrations and assimilation rates in relation to phytoplankton and bacterial production. Limnol. Oceanogr. 32: 97–111.

Lindroth, P. & K. Mopper, 1979. High performance liquid chromatographic determination of subpicomole amounts of amino acids by precolumn fluorescence derivatization with o-phthaldehyde. Analyt. Chem. 51: 1667–1674.

McManus, J. P., K. G. Davis, J. E. Beart, S. H. Gaffney, T. H. Lilley & E. Haslam, 1985. Polyphenol Interactions. Part 1. Introduction; Some observations on the reversible complexation of polyphenols with proteins and polysaccharides. J. Chem. Soc. Perkin Trans. II: 1419–1438.

Meyer-Reil, L.-A., 1986. Measurements of hydrolytic activity and incorporation of dissolved organic substrates by microorganisms in marine sediments. Mar. Ecol. Progr. Ser. 31: 143–149.

Münster, U., 1984. Distribution, dynamic and structure of free dissolved carbohydrates in the Plußsee, a North German eutrophic lake. Verh. int. Ver. Limnol. 22: 929–935.

Münster, U., 1985. Investigations about structure, distribution and dynamics of different organic substrates in the DOM of lake Plußsee. Arch. Hydrobiol., Suppl. 70: 429–480.

Münster, U., 1991. Extracellular Enzyme Activity in Eutrophic and Polyhumic Lakes. In R. J. Chróst (ed.) Microbial Enzymes in Aquatic Environments, Brock/Springer Series in Contemporary Bioscience, by Springer Verlag, 1991, in press.

Münster, U., P. Einiö & J. Nurminen, 1989. Evaluation of the measurements of extracellular enzyme activities in a polyhumic lake by means of studies with 4-methylumbelliferyl-substrates. Arch. Hydrobiol. 115: 321–337.

Münster, U. & R. J. Chróst, 1990. Dissolved organic matter (DOM) in aquatic environments: origin, distribution, composition and microbial utilization. In J. Overbeck & R. J. Chróst, (eds), Advanced Biochemical and Molecular Approaches to Aquatic Microbial Ecology, Brock/Springer Series in Contemporary Bioscience, by Springer Verlag 1990, p. 8–46.

Olsson, L. & O. Samuelson, 1974. Chromatography of aromatic acids and phenolics on cross-linked polyvinylpyrrolidone. J. Chromatogr. 93: 189–199.

Overbeck, J., 1979. Studies on the heterotrophic function and glucose metabolism of microplankton in Plußsee. Arch. Hydrobiol. Beih. Ergebn. Limnol. 13: 56–76.

Pigman, W. & E. F. L. J. Anet, 1972. Mutarotations and actions of acids and bases. In W. Pigman & D. Horton, D. (eds), The Carbohydrates, Chemistry and Biochemistry, Second Edition, Academic Press, New York, 1772, pp. 165–193.

Pollock, M. R., 1962. Exoenzymes. In I. C. Gunsalus & R. Y. Stanier, (eds), The Bacteria, Academic Press, New York, Vol. 2: 121–178.

Pomeroy, L. R., 1974. The ocean's food web, a changing paradigm. Bioscience 24: 499–504.

Pomeroy, L. R. & W. J. Wiebe, 1988. Energetics of microbial food webs. Hydrobiologia 156: 7–18.

Priest, F. G., 1984. Extracellular Enzymes, Aspects of Microbiology 9, Van Nostrand Reinhold (UK) Co. Ltd. pp. 79.

Rogers, H. J., 1961. The dissimilation of high molecular weight organic substrates. In I. C. Gunsalus & R. Y. Stanier (eds.), The Bacteria, Academic Press, New York, Vol. 2: 261–318.

Salonen, K., 1981. The ecosystem of the oligotrophic lake Pääjärvi. 2. Bacterioplankton. Verh. int. Ver. Limnol. 21: 448–453.

Salonen, K., 1984. Peculatities in the limnology of small polyhumic lakes. Lammi Notes 11: 5–7.

Salonen, K. & T. Hammar, 1986. On the importance of dissolved organic matter in the nutrition of zooplankton in some lakes waters. Oecologia 68: 246–253.

Salonen, K. & S. Jokinen, 1988. Flagellate grazing on bacteria in a small dystrophic lake. Hydrobiologia 161: 203–209.

Salonen, K., K. Kononen & L. Arvola, 1983. Respiration of plankton in two small polyhumic lakes. Hydrobiologia 101: 65–70.

Salonen, K. & T. Tulonen, 1990. Photochemical and biological transformation of dissolved humic substances. Verh. int. Ver. Limnol. 24: 294.

Schütt, C., 1988. Plasmid-DNA in natural bacterial populations of four brownwater lakes (South Sweden). Arch. Hydrobiol. Beih. 31: 133–139.

Sieburth McN, J., 1979. Sea Microbes. Oxford Univ. Press, New York. 420 pp.

Somville, M., 1984. Measurement and study od substrate specificity of exoglucosidase activity in eutrophic water. Appl. envir. Microbiol. 48: 1181–1185.

Sorokin, Y. I., 1977. The heterotrophic phase of plankton succession in the Japan Sea. Mar. Biol. 41: 107–117.

Steinberg, C. & U. Münster, 1985. Geochemistry and ecological role of humic substances in lakewater. In G. R. Aiken, D. M. McKnight, R. L. Wershaw & P. MacCarthy (eds), Humic substances in Soil, Sediment and Water. Geochemistry, Isolation, and Characterization. J. Wiley & Sons, N.Y., p. 105–145.

Stuermer, D. H., I. R. Kaplan & K. E. Peters, 1978. Source indicators of humic substances and proto-kerogen. Stable isotope ratios, elemental compositions and electron spin resonance spectra. Geochim. Cosmochim. Acta 42: 989–997.

Thurman, M., 1985. Organic Geochemistry of Natural Waters. Nijhoff/Junk Publ., Boston, 350 pp.

Tranvik, L., 1989. Bacteriaplankton growth, grazing mortality and quantitative relationship to primary production in a humic and clearwater lake. J. Plankton. Res. 11: 985–1000.

Tranvik, L. & M. Höfle, 1987. Bacterial growth in mixed cultures on dissolved organic carbon from humic and clear waters. Appl. envir. Microbiol. 53: 482–488.

Wetzel, R. G., 1983. Limnology, 2nd ed. Saunders College, Philadelphia, 830 pp.

Wetzel, R. G., 1984. Detrital dissolved and particulate organic carbon functions in aquatic ecosystems. Bull. Mar. Sci. 35: 503–509.

Wetzel, R. G., 1991. Extracellular enzymatic interactions in aquatic ecosystems: storage, redistribution, and interspecific communication. In: R. J. Chróst (ed.), Microbial Enzymes in Aquatic Environments, Brock/Springer Series in Contemporary Bioscience, by Springer Verlag, in press.

Wetzel, R. G. & G. E. Likens, 1979. Limnological Analyses, W. B. Saunders Company, Philadelphia, 151–156.

Williams, P. J. leB, 1981. Incorporation of microheterotrophic processes into the classical paradigm of the planktonic food web. Kieler Meeresforsch. Sonderh. 5: 1–28.

Winberg, G. G., 1980. General characteristics of freshwater ecosystems based on Soviet IBP studies. In: E. D. Le Cren & R. H. Lowe-McConnell (eds.), The functioning of Freshwater Ecosystems. IBP 22, Cambridge Univ. Press., p. 481–491.

*Hydrobiologia* **229**: 239–252, 1992.
*K. Salonen, T. Kairesalo & R.I. Jones (eds.),*
*Dissolved Organic Matter in Lacustrine Ecosystems: Energy Source and System Regulator.*




# Effects of different molecular weight fractions of dissolved organic matter on the growth of bacteria, algae and protozoa from a highly humic lake

T. Tulonen, K. Salonen & L. Arvola*
*University of Helsinki, Lammi Biological Station, SF-16900 Lammi, Finland (* author for correspondence)*

*Key words:* dissolved organic matter, humic substances, bacteria, algae, protozoa, growth

## Abstract

Effects of different molecular size fractions (< 1000 MW, < 10 000 MW, < 100 000 MW and < 0.1 $\mu$m) of dissolved organic matter (DOM) on the growth of bacteria, algae and protozoa from a highly humic lake were investigated. DOM from catchment drainage water as well as from the lake consisted mostly (59–63%) of high molecular weight (HMW) compounds (> 10 000 MW). With excess inorganic nutrients, the growth rate and yield of bacteria were almost identical in all size fractions. However, in < 1000 MW fractions and with glucose added, a longer lag phase occurred. Without added nutrients both the growth rates and biomasses of bacteria decreased towards the smaller size fractions and the percentage of dissolved organic carbon (DOC) used during the experiment and the growth efficiency of bacteria were lower than with excess nutrients. The growth efficiency of bacteria was estimated to vary between 3–66% in different MW fractions, largely depending on the nutrient concentrations, but the highest growth efficiencies were observed in HMW fractions and with glucose. The growth of algae was clearly lowest in the < 1000 MW fraction. In dim light no net growth of algae could be found. In contrast, added nutrients substantially enhanced algal growth and in deionized water with glucose, algae achieved almost the same growth rate and biomass as in higher MW fractions of DOM. The results suggested that bacteria and some algae were favoured by DOM, but protozoans seemed to benefit only indirectly, through bacterial grazing. The utilization of DOM by bacteria and algae was strongly affected by the availability of phosphorus and nitrogen.

## Introduction

Most dissolved organic matter (DOM) is considered to be relatively resistant to biodegradation (Wetzel, 1983; Geller, 1983; 1985; Münster, 1985), and so it is generally assumed that heterotrophic bacteria utilize mostly autochthonously produced, labile, low molecular weight (LMW) organic compounds (Jørgensen, 1986). However, in highly humic lakes where a high proportion of DOM is refractory humic compounds (De Haan, 1974; 1979; Steinberg & Münster, 1985; Thurman, 1985) of allochthonous origin (Arvola *et al.*, 1990), the high content of DOM apparently favours microbial growth (Salonen, 1981; Salonen *et al.*, 1983; Hessen, 1985). This indicates that DOM is at least partially degraded by bacteria and other microorganisms (Sederholm *et al.*, 1973; De Haan, 1974, 1979; Tranvik & Höfle, 1987; Benner *et al.*, 1988; see also Visser, 1985).

Jones & Salonen (1985) demonstrated in a humic lake that the extracellular products of algae could account for only a small proportion of total bacterial production, and so allochthonous DOM should be more important in the nutrition of heterotrophic microbes (see also Meyer *et al.*, 1987; Tranvik & Höfle, 1987; Tranvik, 1988; Benner *et al.*, 1988). The results of Meyer *et al.* (1987), who used different molecular weight fractions of DOM from a blackwater river, indicated that labile LMW substrates attached to more refractory high molecular weight (HMW) compounds were supporting bacterial growth. Also the amount of dissolved organic carbon (DOC) used was highest in the LMW fraction. Whether these compounds were predominantly allochthonously produced nonhumic substances, such as amino acids, carbohydrates and organic acids, typical components of humus (Stevenson, 1982), or substrates produced by algae, macrophytes and other organisms living in the river is not known, despite the authors' suggestion that there was no autochthonously produced DOC in the river. Similarly, Geller (1985) found that the smaller sized molecules contributed most of the DOC losses by biodegradation, and that photolysis was a prerequisite for the utilization of the inaccessible bulk of the macromolecules.

The aim of this study was to investigate the effects of different molecular weights of lakewater and allochthonous DOM on the growth of both prokaryotic and eukaryotic microorganisms from a highly humic lake. Special attention was paid to inorganic nutrient availability when estimating the bioavailability of different fractions of DOM.

## Material and methods

Experiments were carried out in batch cultures with water taken from the epilimnion (0–0.5 m) of Mekkojärvi, a highly humic lake in southern Finland (Experiment I, July, 1987), or with water taken from the small inflow to the lake (Experiment II, June, 1989). In experiment I (E I) the culture medium was made by filtering the lake water through a Whatman GF/C glass fibre filter and then through a 0.1 $\mu$m Nuclepore cartridge (No 720100) at a low vacuum (< 50 KPa). In experiment II (E II) the inflow water was sieved through a 10 $\mu$m plankton net and then filtered through a 0.1 $\mu$m Nuclepore cartridge to obtain bacteria-free water.

The 0.1 $\mu$m filtered water was fractionated by tangential flow ultrafiltration (Millipore Pellicon) into three molecular weight (MW) fractions: < 100 000 MW (filter type PCAC 000 05), < 10 000 MW (PTGC), and < 1000 MW (PTHK 000 05). Before use, the filters were rinsed with copious amounts of MQ water, and the first 100 ml of filtrate of the sample were discarded. 14 liters of water were first filtered through the 100 000 MW filter and 3 l of filtrate reserved for the experiments. Then the remaining filtrate was further filtered through the 10 000 MW filter and again 3 l filtrate was reserved for the experiments. The remaining filtrate was then filtered through the 1000 MW filter and the filtrate was collected for the experiments. The filtration pressure was kept as low as possible (< 70 KPa). The experiments were made in 500 ml (E I) or 250 ml (E II) glass bottles. In experiment I no nutrients were added, whereas in experiment II the filtrates were amended with inorganic nutrients and vitamins (Jaworski's medium except that the concentrations of $NaHCO_3$ and $KH_2PO_4$ were doubled; CCAP, 1988). Experiments were also made with < 0.1 $\mu$m water by adding Jaworski's medium (JM) without nitrogen. In addition, an experiment was made in MQ-water with glucose (2 mg C $l^{-1}$) and nutrient (JM) additions.

Experimental media were inoculated with lake water filtered through a 0.6 $\mu$m Nuclepore membrane (E I; dilution 1:10) or with 1.0 $\mu$m Nuclepore filtrate (E II; dilution 1:50). In experiment II phytoplankton was also inoculated (dilution 1:50) with lake water filtered through a 50 $\mu$m mesh plankton net.

In experiment I cultures were incubated in darkness at 20 °C for four days. Duplicate sample bottles were removed daily for measurements. In experiment II culture bottles were incubated under continuous light of 200 and 2 $\mu$E $m^{-2}$ $s^{-1}$ at 16.5 °C for nine days and sampled daily using a Finnpipette with sterile plastic tips.

All glassware was acid washed and rinsed with

MQ-water and then autoclaved (120 °C, 20 min). The Nuclepore cartridge was rinsed with MQ-water and dried at 60 °C > 12 h. Medium preparation, inoculation and sampling were all done inside a laminar flow hood to avoid contamination.

Dissolved organic carbon (DOC) was determined (Salonen, 1979) in 0.2 $\mu$m filtered (Millipore; first rinsed with MQ water and then with sample) samples stored frozen (−20 °C) in precombusted (450 °C, >4 h) glass vials. Before determination, inorganic carbon was purged from thawed samples by acidifying with 0.2 ml phosphoric acid to ca. pH 2.

Molecular size distributions of the DOM fractions used in experiment II were characterized using HPLC (Shimadzu LC-6A) with gel filtration column (Bio-Gel TSK 30-XL of 300 × 7.5 mm together with HPLC Guard column of 75 × 7.5 mm). The UV-VIS detector (Shimadzu SPD-6AV) was operated at 254 nm wavelength. The separation of humic material was performed with 0.01 M sodium acetate ($CH_3COONa$) solution (flow rate of 0.8 ml min$^{-1}$) at pH 7 (Vartiainen *et al.*, 1987).

Bacterial and algal samples were preserved with cold glutardialdehyde at a final concentration of 0.5% and samples were kept in a refrigerator until counted. Bacteria were stained with 0.2 $\mu$m filtered acriflavine solution (Bergström *et al.*, 1986) on 0.2 $\mu$m black Nuclepore membranes and counted with a Nikon Optiphot epifluorescent microscope at a magnification of 1200 × from at least 20 randomly selected fields or until the standard error of counted fields was < 10%. Volume of bacteria was determined from 100 randomly selected cells by measuring the length and width of each cell with a calibrated eyepiece graticule (Patterson Globe and Circle, G1). A carbon-to-volume conversion factor of 0.22 pg C $\mu$m$^{-3}$ (Bratbak & Dundas, 1984) was used rather than the recently more often cited factor of 0.38 pg C $\mu$m$^{-3}$ (Lee & Fuhrman, 1987), because of the rather large size of the cells. Algal and protozoan samples were counted with a Wild M 40 inverted microscope at a magnification of 600 × from 20–50 fields by the sedimentation chamber technique (Utermöhl 1958). A carbon-volume conversion factor of 0.2 pg C $\mu$m$^{-3}$ was used for algae (Rocha & Duncan, 1985) and protozoa (Børsheim & Bratbak, 1987).

The fluorescence of chlorophyll was determined directly from unfiltered samples using a Hitachi F4000 fluorescence spectrophotometer with 435 nm excitation and 686 nm emission wavelengths. The fluorescence values were corrected for the fluorescence of MQ and humic water.

## Results

### *Humic fractions*

According to the proportion of DOC in ultrafiltrates, the DOM pool consisted mostly of HMW compounds (Table 1). In experiment I compounds > 10000 MW represented 59% and in experiment II 63% of total DOC. The lowest MW fraction (< 1000 MW) represented 36% of the total DOC pool in both experiments. Thus, the carbon content between 1000 and 10000 MW represented only a few percent of total DOC. The HPLC chromatograms showed that ultrafiltration did make a rough separation between different molecular weights (Fig. 1). The peak areas of the chromatograms (Fig. 1) were directly proportional to the DOC concentrations (Table 1) of different fractions ($r^2 = 0.99$), and the decrease in humic matter was also reflected in water colour.

Ultrafiltration also affected nutrient concentrations (Table 1). Total nitrogen and total phosphorus decreased with the decrease in DOC concentration (E I). Inorganic N and P also decreased towards smaller size fractions and about 40% of total dissolved P was in the fraction > 10000 MW, consistent with the results obtained using Sephadex gel chromatography by Jones *et al.* (1988) from Mekkojärvi, indicating that part of organic or organic-absorbed inorganic nutrients are trapped by humic substances. Another explanation would be that P and N were present in inorganic colloids of various sizes, such as an association of P with iron hydroxides. Thus

*Table 1.* Some chemical characteristics of the culture media.

| Experiment | Fraction | DIC | DOC | Colour | pH | $NO_2 + NO_3$-N | $PO_4$-P | Tot. N | Tot. P |
|---|---|---|---|---|---|---|---|---|---|
| | | mg $l^{-1}$ | | mg Pt $l^{-1}$ | | $\mu g\ l^{-1}$ | | | |
| I | <0.1 μm | 0.7 | 19.2 | – | – | <10 | 6 | 540 | 15 |
| | <100000 MW | 0.7 | 15.9 | – | – | <10 | 4 | 460 | 14 |
| | <10000MW | 0.6 | 7.9 | – | – | <10 | 3 | 255 | 9 |
| | <1000 MW | 0.5 | 7.0 | – | – | <10 | 1 | 165 | 7 |
| II | <0.1 μm | 5.3 | 26.7 | 429 | 6.9 | *) | +) | | |
| | <100000 MW | 5.2 | 13.7 | 117 | 7.1 | *) | +) | | |
| | <10000 MW | 5.0 | 9.9 | 20 | 7.2 | *) | +) | | |
| | <1000 MW | 5.0 | 9.7 | 15 | 7.2 | *) | +) | | |
| | MQ with glucose | 3.8 | 8.1 | 8 | 7.3 | *) | +) | | |

*) Nitrate (as $Ca(NO_3)_2$ and $NaNO_3$) concentration was 15.6 mg N $l^{-1}$.
+) Phosphate (as $KH_2PO_4$) concentration was 5.7 mg P $l^{-1}$.

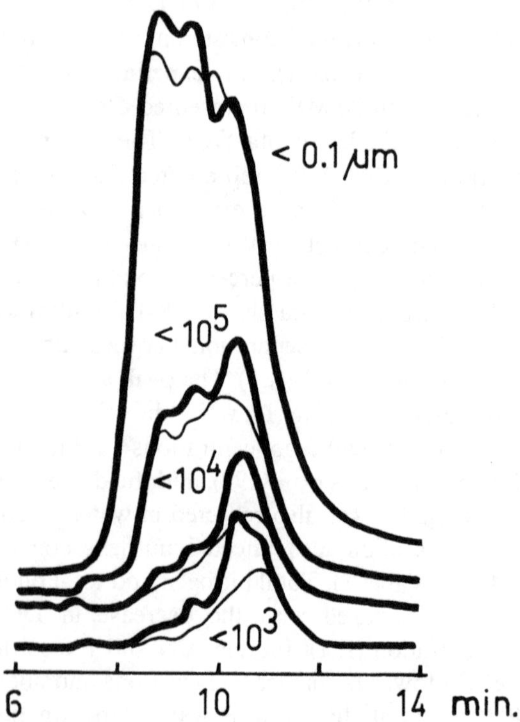


*Fig. 1.* The HPLC chromatograms of the MW fractions of allochthonous DOM represent the situation at the beginning (bold line) and at the end (thin lines) of the incubation with added nutrients and bacteria in bright light in experiment II.

routine chemical methods may overestimate inorganic nutrients. Small differences between the size fractions were also found in pH and DIC concentration, possibly due to escape of $CO_2$ during handling of samples.

During the experiments some changes took place in DOM. The height of HPLC chromatograms declined (Table 4) and in MW fractions there was also a change in the shape of MW distribution (Fig. 1). The most prominent alterations were observed in <10000 MW fractions, and when both algae and protozoa were present together with bacteria.

### *Cultures with bacteria*

Without added nutrients (E I) the highest numbers of bacteria were reached in the <0.1 μm and the highest MW fractions (Fig. 2). However, due to the increase of average cell volume in lower fractions (Table 2) the achieved biomasses were similar in all MW fractions except in <1000 MW.

With excess inorganic nutrients (E II) the numbers, biomass produced and growth of bacteria were almost identical in all size fractions, and also markedly higher (5–20-fold in biomass and 30–65% in growth rate) than in respective fractions in experiment I (Fig. 3, Table 2). However, in the <1000 MW fraction the lag phase in the growth of bacteria was longer than in other fractions. The decrease in the number of bacteria in the <0.1 μm fraction at the end of the experiment

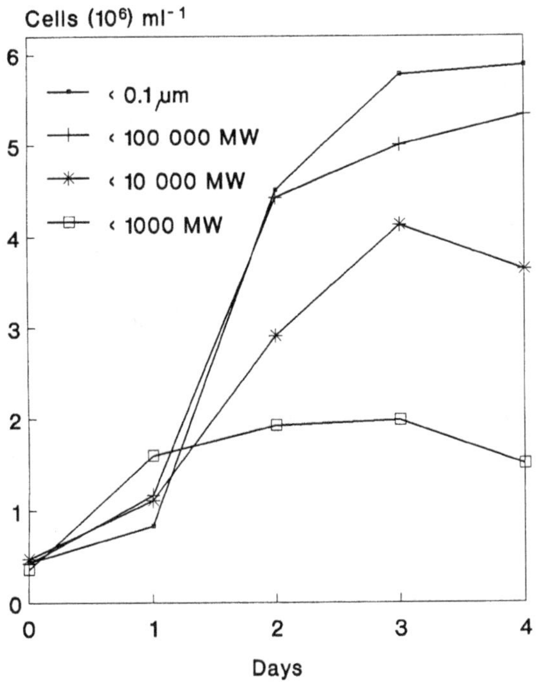


*Fig. 2.* The growth of bacteria in different MW fractions of lake water DOM. The standard error (SE) of the counted ($\geq 20$) fields was $< 10\%$ for all bacteria samples in this study.

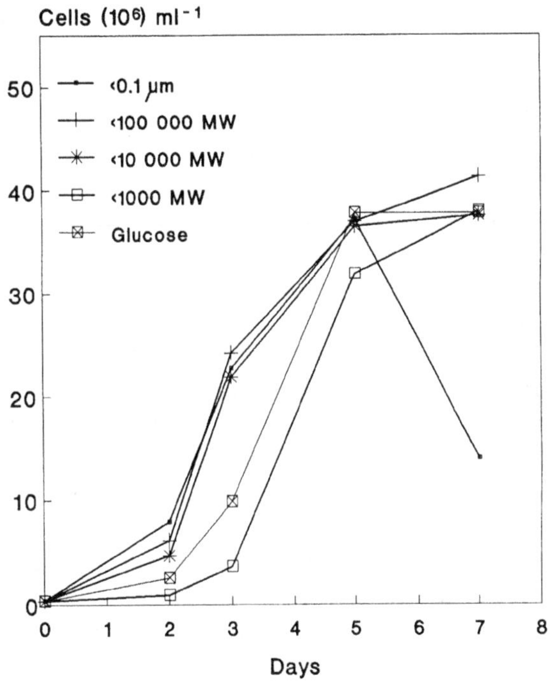


*Fig. 3.* The growth of bacteria in different MW fractions of allochthonous DOM and glucose when nutrients were added into the culture media.

was apparently due to grazing by small ($< 3$ µm) flagellates, which appeared in the culture after five days incubation. Because flagellates were observed only in $< 0.1$ µm fraction, the cultures may have become contaminated during filtration through the 0.1 µm Nuclepore cartridge (cf. Stockner *et al.*, 1989). Also, reduced nitrogen produced significantly lower bacterial biomass in

*Table 2.* The maximum density, biomass, and growth rate ($k$) of bacteria as well as their average cell volume in different cultures without algae.

| Fraction | Experiment I | | | | Experiment II | | | |
|---|---|---|---|---|---|---|---|---|
| | Density $10^6$ cells ml$^{-1}$ | Biomass mg C l$^{-1}$ | k Div. d$^{-1}$ | Volume µm$^3$ | Density $10^6$ cells ml$^{-1}$ | Biomass mg C l$^{-1}$ | k Div. d$^{-1}$ | Volume µm$^3$ |
| <0.1 µm | 5.9 | 0.19 | 0.94 | 0.14 | 37.2 | 1.12 | 1.36 | 0.14 |
| <100000 MW | 5.4 | 0.21 | 0.91 | 0.17 | 37.0 | 1.23 | 1.36 | 0.15 |
| <10000 MW | 3.6 | 0.24 | 0.74 | 0.30 | 36.5 | 1.24 | 1.36 | 0.15 |
| <1000 MW | 1.5 | 0.11 | 0.52 | 0.34 | 31.9 | 1.32 | 1.31 | 0.19 |
| <0.1 µm in dim light | – | – | – | – | 33.8 | 0.98 | 1.33 | 0.13 |
| <0.1 µm without N | – | – | – | – | 6.0 | 0.14 | 0.82 | 0.11 |
| Glucose | – | – | – | – | 37.8 | 1.61 | 1.36 | 0.19 |

this fraction (Fig. 4). In cultures where glucose was the only carbon source for bacteria, the growth of bacteria was similar to that in the < 1000 MW fraction, having a clear lag phase during the first few days, although a slightly higher biomass was achieved than in other cultures.

The decrease of DOC in cultures during the incubation was assumed to be due to bacterial consumption. In experiment I about 30% of DOC was consumed in the LMW fraction (< 1000 MW), but only 4–8% in higher MW fractions (Table 3, see also Table 4). Similarly, in cultures with excess nutrients (E II) the proportion of DOC consumed by bacteria was clearly less in high MW fractions compared to the low MW fraction (Table 4). The HPLC chromatograms at the end of the incubation (Fig. 2) also indicate that in low MW fractions proportionally more carbon was consumed than in high MW fractions. Under dim light the DOC consumed was twice that consumed in high light. Although light intensity did not affect the achieved biomass and growth rate of bacteria in the < 0.1 μm fraction, the growth efficiency of bacteria was markedly reduced in dim light (Table 4). With the addition of glucose, almost as much of the initial DOC was consumed as in the < 1000 MW fraction of lake water.

Bacterial growth efficiencies were estimated by comparing the bacterial biomass produced to the DOC consumed during the incubation up to the stationary phase. The highest efficiencies (42–66%) were observed in high MW fractions with added nutrients and also in cultures with glucose as the only carbon source for bacteria. Under dim light and nutrient depletion the growth efficiency of bacteria clearly decreased (Table 4). The growth rate of bacteria was strongly correlated with the amount of DOC in experiment I, whereas no correlation existed between these two parameters in experiment II.

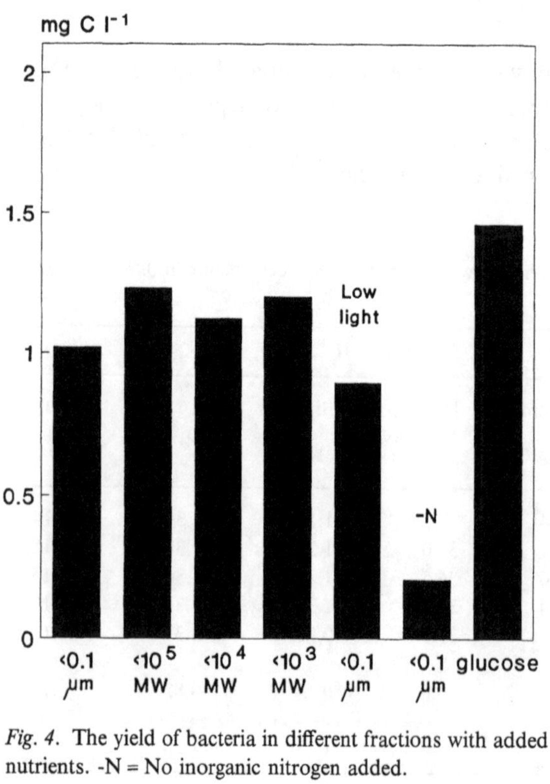


*Fig. 4.* The yield of bacteria in different fractions with added nutrients. -N = No inorganic nitrogen added.

*Cultures with algae*

In the presence of excess nutrients the growth of algae declined dramatically in the fraction < 1000 MW (Figs. 5 & 6, Table 5). Algae stopped growing at low densities before the end of the experiment, whereas in all other fractions the exponential growth continued to the end of the experiment (Fig. 5). Under nitrogen depletion the net growth was also reduced (biomass 17% of that in < 0.1 μm with N), and at dim light no net growth of algae was observed (Figs. 6 & 7, Table 5). With glucose, algae achieved almost the same growth rate and biomass as in higher MW fractions of allochthonous DOM.

Among the algae, *Chlamydomonas* spp. comprised 86–99% of total algal biomass at the end of the experiment. The biomass of *Chlamydomonas* increased with the MW fractions (i.e. with DOC in the MW fractions) while the biomasses of *Dictyosphaerium* spp., *Mallomonas heterospina* and *Cryptomonas* spp., seemed to vary irregularly in different fractions, except that in the < 1000 MW fraction biomasses were always much lower than in the others (Fig. 7).

*Table 3.* Final bacterial biomass, DOC consumed and the growth efficiency (biomass produced divided by DOC consumed) in experiment I.

| Fraction | Biomass produced* | | DOC consumed | | Growth efficiency |
|---|---|---|---|---|---|
| | mg C $l^{-1}$ | % of DOC | mg C $l^{-1}$ | % of DOC | % |
| <0.1 μm | 0.22 | 1.1 | 0.91 | 4.7 | 24 |
| <100000 MW | 0.18 | 1.1 | 0.67 | 4.2 | 26 |
| <10000 MW | 0.16 | 2.1 | 0.66 | 8.4 | 25 |
| <1000 MW | 0.08 | 1.1 | 2.30 | 32.9 | 3 |

* Mean of two days at the stationary phase of growth.

*Table 4.* The growth efficiency of bacteria (biomass produced divided by DOC consumed) in experiment II without algae. Values represent the third day of the incubation except cultures of <1000 MW and glucose, which are from the seventh day of incubation.

| Fraction | Biomass produced | | DOC consumed | | Growth efficiency | Decrease in HPLC peak % |
|---|---|---|---|---|---|---|
| | mg C $l^{-1}$ | % of DOC | mg C $l^{-1}$ | % of DOC | % | |
| <0.1 μm | 0.85 | 3.2 | 1.86 | 7.0 | 46 | 14 |
| <100000 MW | 1.12 | 8.2 | 1.70 | 12.4 | 66 | 18 |
| <10000 MW | 1.09 | 11.0 | 3.20 | 12.0 | 34 | 33 |
| <1000 MW | 1.15 | 11.9 | 3.96 | 40.7 | 29 | 30 |
| <0.1 μm in dim light | 0.83 | 3.1 | 3.70 | 13.9 | 22 | – |
| <0.1 μm without N | 0.23 | 0.8 | 1.37 | 5.1 | 16 | – |
| Glucose | 1.30 | 16.1 | 3.07 | 37.9 | 42 | 28 |

*Table 5.* Biomass, fluorescence, and growth rate (*k*) of algae after seven days incubation in experiment II.

| Fraction | Biomass mg C $l^{-1}$ | Fluorescence | *k* Div. $d^{-1}$ |
|---|---|---|---|
| <0.1 μm | 36.0 | 0.072 | 1.69 |
| <100000 MW | 30.3 | 0.064 | 1.65 |
| <10000 MW | 23.8 | 0.061 | 1.43 |
| <1000 MW | 1.1 | 0.003 | 1.17 |
| <0.1 μm in dim light | 0 | 0 | 0 |
| <0.1 μm without N | 6.0 | 0.012 | 1.29 |
| Glucose | 21.3 | 0.037 | 1.55 |

### *Protozoa*

The protozoan community consisted of few species, among which *Desmarella moniliformis* and an unidentified, small (diameter <3 μm) flagellate were the most common. Some *Monas* sp., ciliates and heliozoans were also present in low numbers. In general, the growth of protozoa had no clear trend in different MW fractions except that the growth was least in the lowest MW fractions (<1000 MW and glucose) and at low light (Figs. 8 & 9). However, *Desmarella* and the unidentified flagellate attained their highest biomass in water with glucose.

When protists were present in cultures, they grazed heavily on bacteria (Fig. 10). The observed

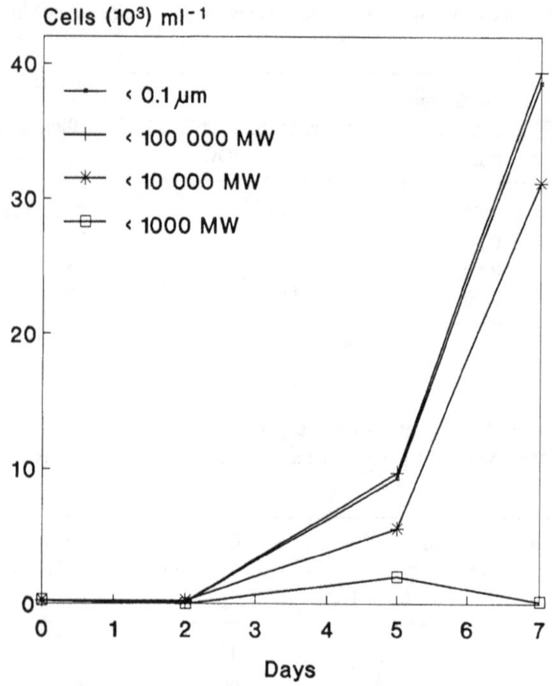


*Fig. 5.* The growth of algae in different MW fractions of allochthonous DOM when nutrients were added to the culture media. The points represent means of duplicate samples (Mean SE 16% at the end of the incubation).

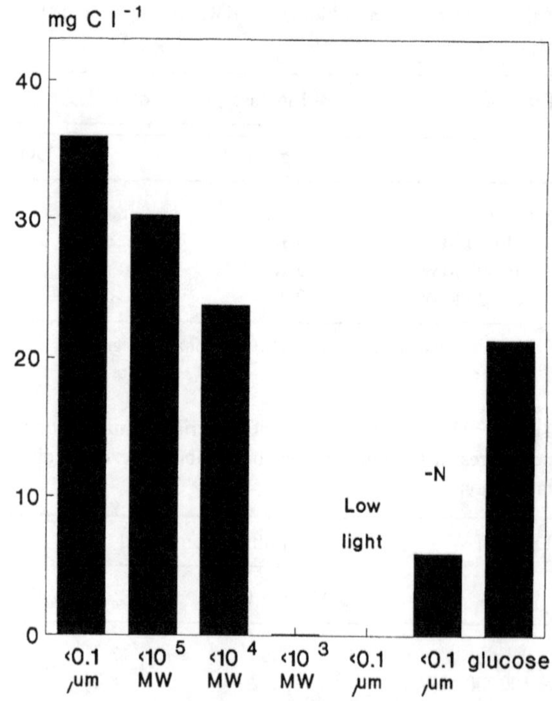


*Fig. 6.* The yield of algae in different fractions with added nutrients. The bars show the means of duplicate samples.

growth of protozoa commenced about two days after that of bacteria and, during the exponential growth of protozoa, a decrease in bacterial numbers was observed (Fig. 11). In cultures with delayed protozoan growth (< 1000 MW, glucose) bacteria achieved clearly higher numbers than in other cultures. Less than 500 protists per ml can markedly affect bacterial numbers in this type of experiment.

## Discussion

In the oceans DOM is mainly composed of small (< 1000 MW) molecular compounds (Carlson *et al.*, 1985). However, in lakes macromolecules (> 1500 MW) are usually prevalent (Geller, 1983; Münster, 1985). In Mekkojärvi and its inflow, the carbon content of the fraction between 1000 and 10000 MW was clearly lowest, representing only 1–5% of the DOC pool. This agrees with the observation of Allen (1976) that in natural waters this size fraction is quantitatively the least important. However, this does not necessarily mean that this fraction is less important for microorganisms than other fractions of DOM, because some smaller organic molecules (e.g. vitamins and some organic substrates) may have a signal or inducer function for microbial enzyme systems (U. Münster, pers. comm.). Our results are in agreement with those of De Haan *et al.* (1987) who, using Sephadex gel filtration, found that the bulk of DOM of Mekkojärvi has a nominal MW around 30000 and a small portion exceeded 100000 MW. Differences in the composition of DOM between the inflow and lakewater might occur because of both photooxidation (cf. Gjessing, 1970; 1976; Il'ln & Orlov, 1973; Chen *et al.*, 1978; Strome & Miller, 1978; Stewart & Wetzel, 1981), microbial degradation, algal exudates and different sampling time. For example, the results of Ward & Wetzel (1984) indicate that the proportions of molecular size fractions of DOM may differ markedly between lakes with similar phys-

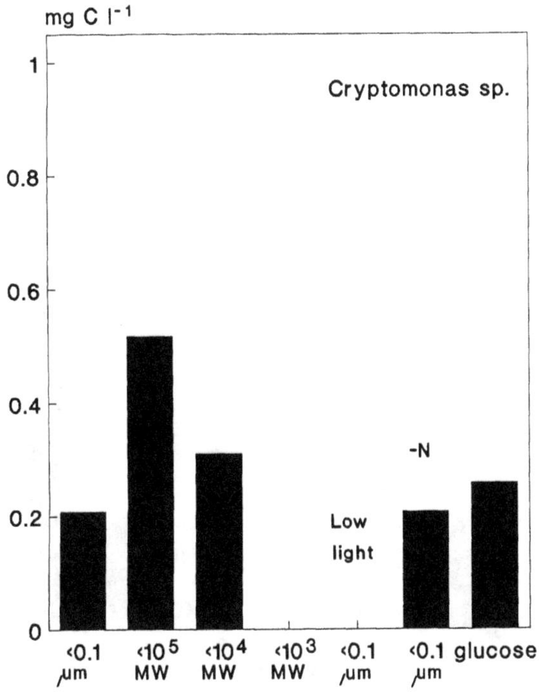


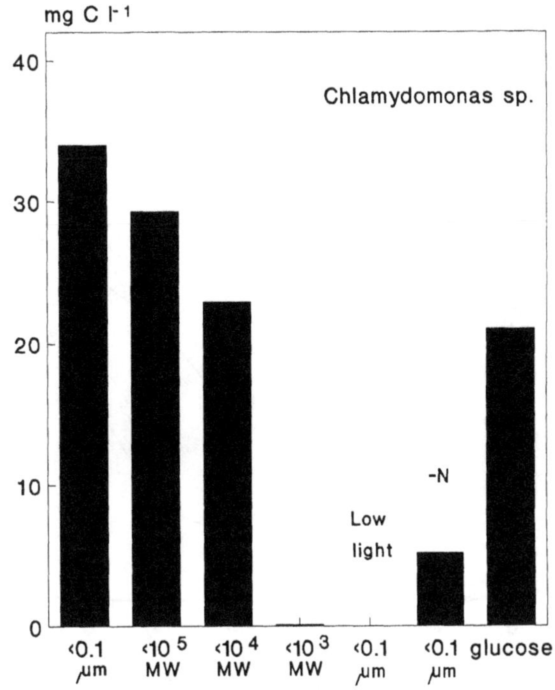


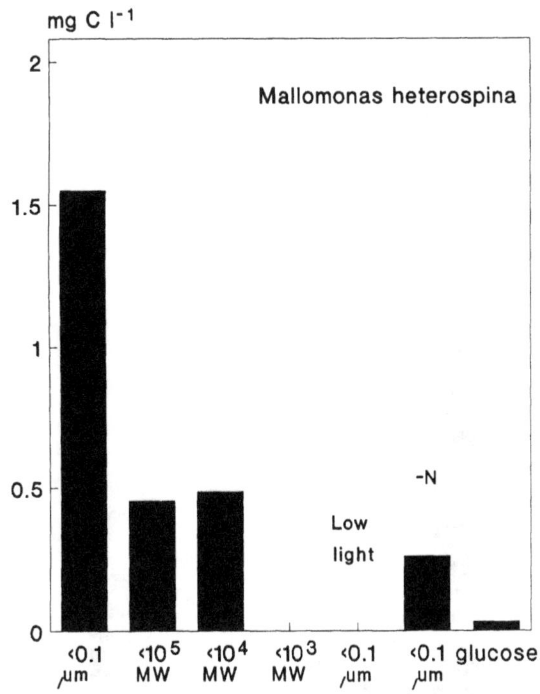


*Fig. 7.* The yields of the most common algal species in different fractions with added nutrients. The bars show the means of duplicate samples with a mean SE of 35%.

ico-chemical properties but different primary productivity.

Earlier studies on the availability of different MW fractions of DOM for bacteria have indicated that LMW compounds are generally more easily utilized than HMW substances (Ogura, 1975; Kaplan & Bott, 1983; Meyer *et al.*, 1987). Most cells are able to take up only LMW compounds (monomers and oligomers; e.g. peptides of $<1000$ MW, Coffin, 1989) through the cell membrane, whereas HMW compounds (larger proteins, carbohydrates and humic compounds) have first to be cleaved extracellularly by appropriate enzymes (see Chróst, 1989; Chróst *et al.*, 1989; Münster *et al.*, 1989). However, at least some heterotrophic flagellates are also apparently able to take up HMW ($>100\,000$ MW) carbohydrate substrates by ingestion (Sherr, 1988). Such an uptake mechanism would especially favour heterotrophic flagellates in polyhumic lakes where the DOM pool consists largely of HMW compounds, and since proteins and carbohydrates can be linked to the HMW humic compounds (Lytle & Perdue, 1981; Stevenson, 1982).

In this study bacteria consumed the highest

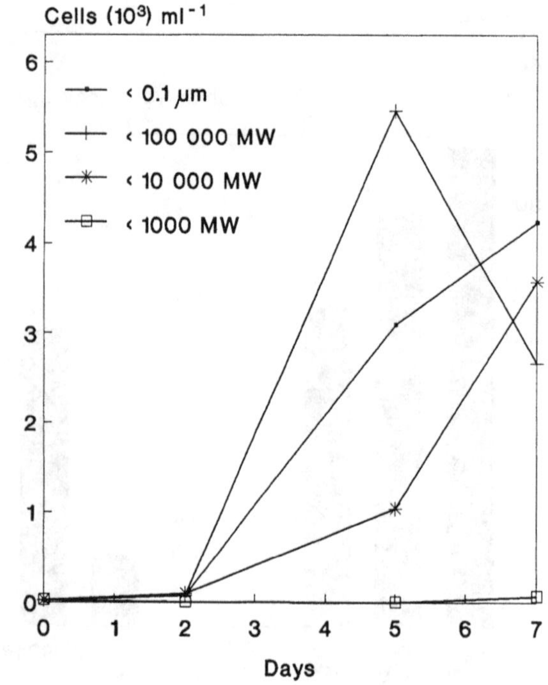


*Fig. 8.* The growth of protozoa in different MW fractions of allochthonous DOM when nutrients were added to the culture media. The average SE of duplicate samples was 6% of the mean.

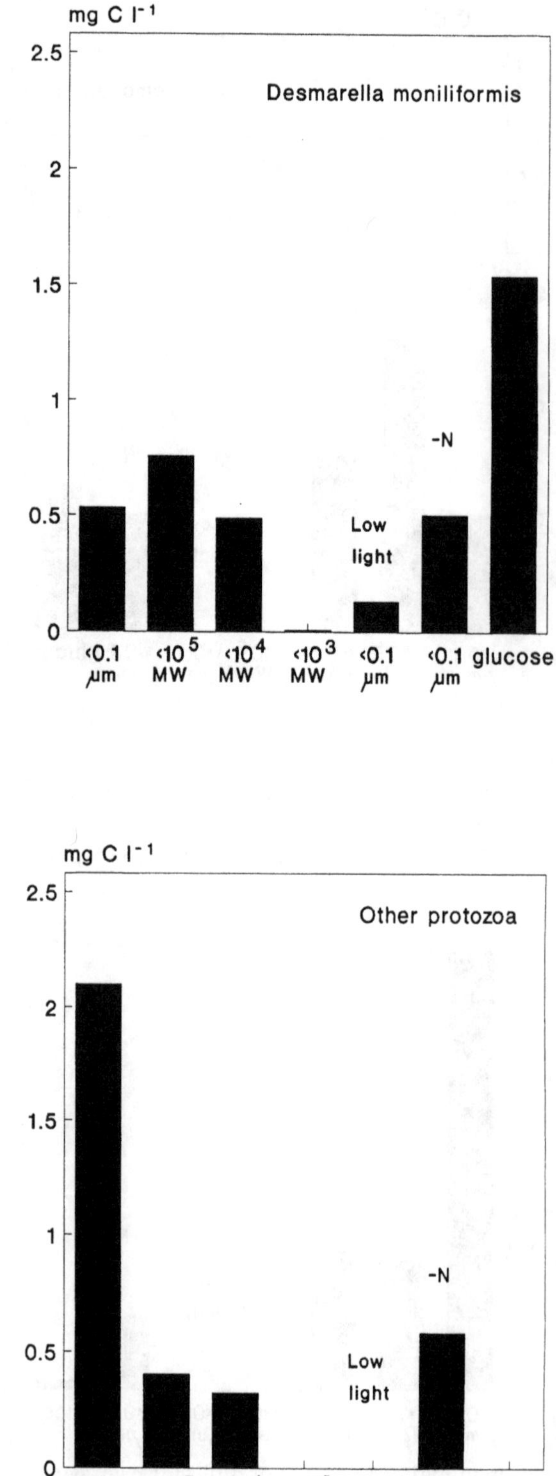


*Fig. 9.* The yields of protists in different fractions with added nutrients. The bars show means of duplicate samples.

proportion of DOC in the LMW fractions (< 1000 MW and glucose), although the growth efficiency of bacteria was highest with HMW compounds. The biomasses produced by bacteria were similar in all fractions, however. This indicates that bacteria used LMW fractions to a lesser extent for growth but rather more for energy requirements (Münster & Chróst, 1990). Tranvik (1990) also found that HMW compounds (> 10000 MW) of DOM were generally more available to bacteria than LMW ones in oligotrophic lakes with different humic content. The HMW compounds may better cover the nutritional and metabolic requirements of bacteria than the LMW compounds, because of higher diversity in composition (cf. Münster, 1991). The long lag phase and low growth efficiency of bacteria observed in the LMW fraction of DOM might indicate that the physiological state of bacteria was disturbed, perhaps because of the change in the quality of DOM. Bacteria might

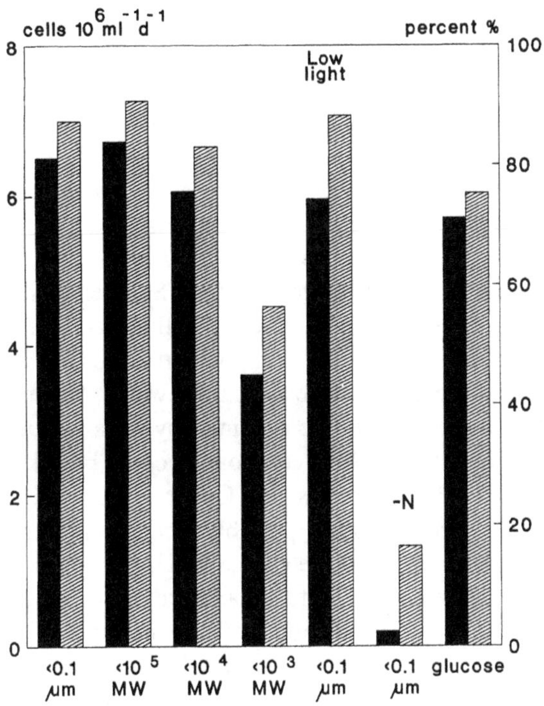


*Fig. 10.* Bacteria grazed by protists in different MW fractions of DOM with added nutrients. The black bars show the number of bacteria grazed daily and the hatched bars the difference (%) between the cultures with and without grazers.

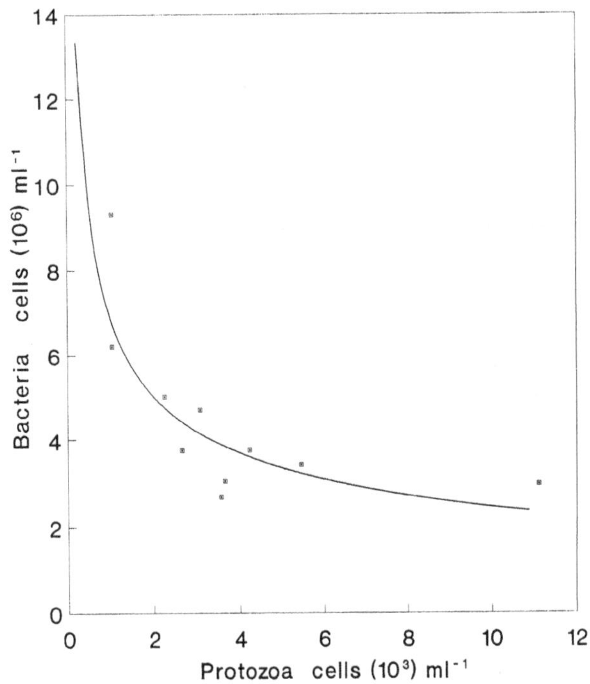


*Fig. 11.* The relationship between the numbers of bacteria and protozoa during the exponential growth of protists.

have had to reformulate their uptake mechanisms for DOM before they were able to use the LMW compounds of DOM efficiently. This would be in agreement with the observation of Geller (1985) that bacteria most efficiently degrade DOM from the parent lake. In contrast, Meyer *et al.* (1987) suggested that the LMW compounds from a blackwater river were most readily available, followed by HMW compounds, while intermediate MW compounds were least available. Regarding the similar growth of bacteria with all MW fractions of DOM when nutrients were added, the ultrafiltration procedure itself (i.e. water contact with the ultrafilters and tubes) did not cause any remarkable effect on bacteria. This was in agreement with the observation of Tranvik (1990), that the procedure itself had no effect on bacterial growth despite some contamination of DOC from the ultrafilter he used.

Besides the different nutritional and metabolic values of MW size fractions, inorganic nutrients, particularly phosphorus and nitrogen, also affected the decomposition of DOM by bacteria. The C:N ratio of DOM may largely dictate the organic carbon conversion efficiency of bacteria (Goldman *et al.*, 1987), and the addition of nitrogen can markedly enhance the availability of DOC of allochthonous origin and promote bacterial production (Benner *et al.*, 1988; see also Ryhänen, 1968). Particularly in allochthonous DOM, the C:N ratio is typically low (as in experiment I, where the C:N ratio was 35:1 even if lakewater was used). The effect of nitrogen on the growth of bacteria was clearly demonstrated in experiment II, when the growth efficiency of bacteria without added nitrogen was only one third of that with added nitrogen. Moreover, in the presence of extra nutrients bacteria consumed a higher proportion of the available DOC than under nutrient deficiency, and the growth efficiency of bacteria was also enhanced, thus supporting the results of Benner *et al.* (1988). When different MW fractions are considered our results also suggest that the availability of allochthonous

DOM would be strongly related to the phosphorus and nitrogen concentrations. In spite of the difficulties of reliably estimating the growth efficiency of bacteria (e.g. the carbon-volume conversion factors quoted in the literature vary more than 5-fold), the observed estimates in humic water without extra nutrients were comparable with other results from similar waters (Meyer *et al.*, 1987; Tranvik & Höfle, 1987; Tranvik, 1988; Arvola & Tulonen, ms.). These observations imply that in natural humic waters bacteria may be nutrient limited, which may reduce the decomposition of DOM. In addition to nutrients, another factor which clearly promoted the decomposition of DOM by bacteria was light. This supports the view that light could enhance bacterial growth by splitting larger molecules of DOM (e.g. Kieber *et al.*, 1990) and increasing their availability for bacteria (Strome & Miller, 1978; Salonen & Tulonen, 1990).

The results of earlier investigations with natural and laboratory strains of *Chlamydomonas* indicate that species of this genus may be able to utilize organic substrates as a carbon source (e.g. Bennett & Hobbie, 1972; Arvola & Tulonen, ms.). In this study a natural strain of *Chlamydomonas* grew well in the higher MW fractions, but in lower MW fractions the growth declined substantially in spite of abundant nutrients. In contrast, with glucose and extra nutrients it reached almost as high a biomass as with HMW fractions of DOM. Thus our results seem to demonstrate that DOM can have profound positive effect on some photosynthetic algae. This appeared only in the light, which indicates that the studied species were photoheterotrophic (Arvola & Tulonen, ms.). Further studies are clearly required to explain the particularly weak growth of algae in the < 1000 MW fraction.

Like algae, protozoans did not grow well with the < 1000 MW even if they were able to graze on bacteria. The results of this study show that protists were favoured indirectly, through grazing on bacteria, by allochthonous DOM, but no evidence was found for direct uptake of DOM. Thus, the results are consistent with other observations from humic waters (e.g. Salonen & Jokinen, 1988; Tranvik, 1989; Tranvik & Sieburth, 1989; Arvola & Tulonen, ms.) that protozoans are largely responsible for the decrease of bacteria in both experimental and natural conditions. Because of the rather high abundance of ciliates in the < 0.1 $\mu$m fraction, both *Chlamydomonas* and *Cryptomonas* as well as *Desmarella* were depressed markedly. This might also explain why *Mallomonas*, which is more resistant to grazing (e.g. Salonen & Arvola, 1988), was so abundant in this fraction.

In conclusion, the composition of DOM affected the growth of bacteria. However, because the molecular weights obtained by ultrafiltration are relative i.e. LMW compounds could be linked with HMW fractions (cf. Geller, 1983; Meyer *et al.*, 1987), it is not possible to draw any final conclusions on the preference for HMW fractions by bacteria until more detailed analyses of the biochemical composition of the nominal MW fractions of allochthonous DOM are available. Although in natural water fractions the HMW compounds seemed important, with added nutrients the difference in biomass produced between the size fractions of DOM almost disappeared. This might suggest that decomposition by bacteria was closely dependent on nutrients bound by organic matter. The HMW fractions of DOM clearly also favoured some algae, particularly *Chlamydomonas*. Further studies are needed to explain why the < 1000 MW fraction did not support the growth of algae although compounds of this size should be most available to them. Finally, the possible effects of the ultrafiltration procedure itself on the MW fractions of DOM and the growth of microorganisms should be evaluated in detail.

## Acknowledgements

This study is a part of the research program dealing with the structure and functioning of the food webs of highly humic lakes supported by the Academy of Finland and the Maj & Tor Nessling Foundation. We thank R. Ilola and J. Vainionpää for making most of the chemical measurements, U. Münster for valuable comments on the manuscript, and R. I. Jones for his comments and revision of the English of the manuscript.

## References

Allen, H. L., 1976. Dissolved organic matter in lakewater: Characteristics of molecular weight size-fractions and ecological implications. Oikos 27: 64–70.

Arvola, L. & T. Tulonen. Enhancement by dissolved organic matter of the growth of laboratory strains of bacteria and algae and the microorganisms of a highly humic lake. Manuscript.

Arvola, L., K. Salonen & M. Rask, 1990. Chemical budgets for a small dystrophic lake in southern Finland. Limnologica (Berlin) 20: 243–251.

Benner, R., J. Jay, E. K'nees & R. E. Hodson, 1988. Carbon conversion efficiency for bacterial growth on lignocellulose: Implications for detritus-based food webs. Limnol. Oceanogr. 33: 1514–1526.

Bennett, M. E. & J. E. Hobbie, 1972. The uptake of glucose by *Chlamydomonas* sp. J. Phycol. 8: 392–398.

Bergström, I., A. Heinänen & K. Salonen, 1986. Comparison of acridine organge, acriflavine, and bisbenzimide stains for enumeration of bacteria in clear and humic waters. Appl. envir. Microbiol. 51: 664–667.

Børsheim, K. Y. & G. Bratbak, 1987. Cell volume to cell carbon conversion factors for a bacteriovorous *Monas* sp. enriched from sea water. Mar. Ecol. Prog. Ser. 36: 171–175.

Bratbak, G. & I. Dundas, 1984. Bacterial dry matter content and biomass estimations. Appl. envir. Microbiol. 48: 755–757.

Carlson, D. J., M. L. Brann, T. H. Mague & L. M. Mayer, 1985. Molecular weight distribution of dissolved organic materials in seawater determined by ultrafiltration: A reexamination. Mar. Chem. 16: 155–171.

CCAP, 1988. Culture collection of algae and protozoa. Catalogue of strains. Natural Environmental Research Council, 164 pp.

Chen, Y., S. U. Khan & M. Schnitzer, 1978. Ultraviolet irradiation of dilute fulvic acid solutions. Soil Sci. Soc. Amer. J. 42: 292–296.

Chróst, R. J., 1989. Characterization and significance of $\beta$-glucosidase activity in lake water. Limnol. Oceanogr. 34: 660–672.

Chróst, R. J., U. Münster, H. Rai, D. Albrecht, P. K. Witzel & J. Overbeck, 1989. Photosynthetic production and exoenzymatic degradation of organic matter in euphotic zone of an eutrophic lake. J. Plankton Res. 11: 223–242.

Coffin, R. B., 1989. Bacterial uptake of dissolved free and combined amino acids in estuarine waters. Limnol. Oceanogr. 34: 531–542.

De Haan, H., 1974. Effects of a fulvic acid fraction on the growth of a *Pseudomonas* from Tjeukemeer (The Netherlands). Freshwat. Biol. 4: 301–310.

De Haan, H., 1979. Effect of benzoate on the microbial decomposition of fulvic acids in Tjeukemeer (The Netherlands). Limnol. Oceanogr. 22: 38–44.

De Haan, H., R. I. Jones & K. Salonen, 1987. Does ionic strength affect the configuration of aquatic humic substances, as indicated by gel filtration? Freshwat. Biol. 17: 453–459.

Geller, A., 1983. Degradability of dissolved organic lake water compounds in cultures of natural bacterial communities. Arch. Hydrobiol. 99: 60–79.

Geller, A., 1985. Degradation and formation of refractory DOM by bacteria during simultaneous growth on labile substrates and persistent lake water constituents. Schweiz. Z. Hydrol. 47: 27–44.

Gjessing, E. T., 1970. Reduction of aquatic humus in streams. Vatten 1: 14–23.

Gjessing, E. T., 1976. Physical and chemical characteristics of aquatic humus. Ann. Arbor Sci. Publ., Inc., Ann Arbor (Michigan): 1–20.

Goldman, J. C., D. A. Caron & M. R. Dennett, 1987. Regulation of gross growth efficiency and ammonium regeneration in bacteria by substrate C:N ratio. Limnol. Oceanogr. 32: 1239–1252.

Hessen, D. O., 1985. The relation between bacterial carbon and dissolved humic compounds in oligotrophic lakes. FEMS Microbiol. Ecol. 31: 215–223.

Il'ln, N. P. & D. S. Orlov, 1973. Photochemical destruction of humic acids. Soviet Soil Sci. 13: 75–83.

Jones, R. I. & K. Salonen, 1985. The importance of bacterial utilization of released phytoplankton photosynthate in two humic lakes in southern Finland. Holarct. Ecol. 8: 133–140.

Jones, R. I., K. Salonen & H. De Haan, 1988. Phosphorus transformations in the epilimnion of humic lakes: abiotic interactions between dissolved humic materials and phosphate. Freshwat. Biol. 19: 357–369.

Jørgensen, S. E., 1986. Fundamentals of ecological modelling. Elsevier.

Kaplan, L. A. & T. L. Bott, 1983. Microbial heterotrophic utilization of dissolved organic matter in a piedmont stream. Freshwat. Biol. 13: 363–377.

Kieber, R. J., X. Zhou & K. Mopper, 1990. Formation of carbonyl compounds from UV-induced photodegradation of humic substances in natural waters: Fate or riverine carbon in the sea. Limnol. Oceanogr. 35: 1503–1515.

Lee, S. & J. A. Fuhrman, 1987. Relationships between biovolume and biomass of naturally derived marine bacterioplankton. Appl. envir. Microbiol. 53: 1298–1303.

Lytle, C. R. & E. M. Perdue, 1981. Free, proteinaceous, and humic-bound amino acids in river water containing high concentrations of aquatic humus. Envir. Sci. Technol. 15: 224–228.

Meyer, J. L., R. T. Edwards & R. Risley, 1987. Bacterial growth on dissolved organic matter from a blackwater river. Microb. Ecol. 13: 13–29.

Münster, U., 1985. Investigations about structure, distribution and dynamics of different organic substancies in the DOM of lake Plu see. Arch. Hydrobiol. Suppl. 70: 429–480.

Münster, U., 1991. Extracellular enzyme activity in eutrophic and polyhumic lakes. In: R. J. Chróst (ed.), Microbial enzymes in aquatic environments, Brock/Springer Series, New York. (in Press).

Münster, U. & R. J. Chróst, 1990. Origin, composition, and microbial utilization of dissolved organic matter. In: J. Overbeck & R. J. Chróst (eds.), Aquatic microbial ecology, Brock/Springer Series, New York. pp. 8–46.

Münster, U., P. Einiö & J. Nurminen, 1989. Evaluation of the measurements of extracellular enzyme activities in a polyhumic lake by means of studies with 4-methyl-umbelliferyl-substrates. Arch. Hydrobiol. 115: 321–337.

Ogura, N., 1975. Further studies on decomposition of dissolved organic matter in coastal seawater. Mar. Biol. 31: 101–111.

Rocha, O. & A. Duncan, 1985. The relationship between cell carbon and cell volume in freshwater algal species used in zooplankton studies. J. Plankton Res. 7: 279–294.

Ryhänen, R., 1968. Die Bedeutung der Humussubstanzen im Stoffhaushalt der Gewässer Finnlands. Mitt. int. Ver. Limnol. 14: 168–178.

Salonen, K., 1979. A versatile method for the rapid and accurate determination of carbon by high temperature combustion. Limnol. Oceanogr. 24: 177–183.

Salonen, K., 1981. The ecosystem of the oligotrophic Lake Pääjärvi. 2. Bacterioplankton. Verh. int. Ver. Limnol. 21: 448–453.

Salonen, K. & L. Arvola, 1988. A radiotracer study of zooplankton grazing in two small humic lakes. Verh. int. Ver. Limnol. 23: 462–469.

Salonen, K. & S. Jokinen, 1988. Flagellate grazing on bacteria in a small dystrophic lake. Hydrobiologia 161: 203–209.

Salonen, K., K. Kononen & L. Arvola, 1983. Respiration of plankton in two small, polyhumic lakes. Hydrobiologia 101: 65–70.

Salonen, K. & T. Tulonen, 1990. Photochemical and biological transformation of dissolved humic substances (Abstract). Verh. int. Ver. Limnol. 24: 294.

Sederholm, H., A. Mauranen & L. Montonen, 1973. Some observations on the microbial degradation of humous substances in water. Verh. int. Ver. Limnol. 18: 1301–1305.

Sherr, E. B., 1988. Direct use of high molecular weight polysaccharide by heterotrophic flagellates. Nature 335: 348–351.

Steinberg, C. & U. Münster, 1985. Geochemistry and ecological role of humic substances in lake water. In: Aiken, G. R., D. M. McKnight, R. L. Wershaw & P. MacCarthy (Eds.): Humic substances in soil, sediment, and water. Geochemistry, isolation and characterization. J. Wiley & Sons, P. 105–145.

Stevenson, F. J., 1982. Humus chemistry. John Wiley & Sons, New York.

Stewart, A. J. & R. G. Wetzel, 1981. Dissolved humic materials: photodegradation, sediment effects and reactivity with phosphate and calcium carbonate precipitation. Arch. Hydrobiol. 92: 265–286.

Stockner, J. G., M. E. Klut & W. P. Cochlan, 1989. Leaky filters: a warning to aquatic ecologist. Can. J. Fish. aquat. Sci. 47: 16–23.

Strome, D. J. & M. C. Miller, 1978. Photolytic changes in dissolved humic substances. Verh. int. Ver. Limnol. 20: 1248–1254.

Thurman, E. M., 1985. Organic geochemistry of natural waters. Mertinus Nijhoff/Dr. W. Junk.

Tranvik, L. J., 1988. Availability of dissolved organic carbon for planktonic bacteria in oligotrophic lakes of differing humic content. Microb. Ecol. 16: 311–322.

Tranvik, L. J., 1989. Bacterioplankton growth, grazing mortality and quantitative relationship to primary production in a humic and a clearwater lake. J. Plankton Res. 11: 985–1000.

Tranvik, L. J., 1990. Bacterioplankton growth on fractions of dissolved organic carbon of different molecular weights from humic and clear waters. Appl. envir. Microbiol. 56: 1672–1677.

Tranvik, L. J. & M. G. Höfle, 1987. Bacterial growth in mixed cultures on dissolved organic carbon from humic and clear waters. Appl. envir. Microbiol. 53: 482–488.

Tranvik, L. J. & J. McN. Sieburth, 1989. Effects of flocculated humic matter on free and attached pelagic microorganisms. Limnol. Oceanogr. 34: 688–699.

Utermöhl, H., 1958. Zur Vervollkommung der quantitativen Phytoplanktonmethodik. Mitt. int. Ver. Limnol. 9: 1–38.

Vartiainen, T., A. Liimatainen & P. Kauranen, 1987. The use of TSK size exclusion columns in determination of the quality and quantity of humus in raw waters and drinking waters. Sci. Tot. Envir. 62: 75–84.

Visser, S. A., 1985. Effect of humic acids on numbers and activities of micro-organisms within physiological groups. Org. Geochem. 8: 81–85.

Ward, A. K. & R. G. Wetzel, 1984. Molecular weight fractionation of dissolved organic nitrogen and carbon compounds from two lakes of differing trophic status. Arch. Hydrobiol. 101: 481–488.

Wetzel, R. G., 1983. Limnology, 2nd. edn. W.B. Saunders Co., Philadelphia.

*Hydrobiologia* **229**: 253–269, 1992.
*K. Salonen, T. Kairesalo & R.I. Jones (eds.),*
*Dissolved Organic Matter in Lacustrine Ecosystems: Energy Source and System Regulator.*




# Vertical distributions of bacteria and algae in a steeply stratified humic lake under high grazing pressure from *Daphnia longispina*

L. Arvola, K. Salonen, P. Kankaala & A. Lehtovaara
*Lammi Biological Station, University of Helsinki, SF-16900 Lammi, Finland*

*Key words:* vertical distribution, photosynthetic bacteria, *Cryptomonas*, *Mallomonas*, *Chlamydomonas*, *Scourfieldia*, *Daphnia*, ciliates, humic lake

## Abstract

The vertical distributions of bacteria and algae in a steeply stratified, highly humic lake were studied during three 24 h periods in summer. The highest bacterial and algal densities and biomasses were recorded in the anoxic hypolimnion. The bacterial biomass in the hypolimnion was composed mainly of photosynthetic green sulphur bacteria (*Chlorobium*) which occurred at very low light intensity ($< 1.5\ \mu mol\ m^{-2}\ s^{-1}$). The numbers and biomasses of bacteria, both in the epilimnion and hypolimnion, were mostly higher at night than during the day, indicating possible asynchrony between the production and loss of bacteria. Because of vertical migration, the diurnal vertical distribution of algae was more variable than that of bacteria. Particularly in July and August, when cryptomonads were abundant, the biomass of algae was much higher in the epilimnion during the day than at night. The flagellated chlorophytes, *Chlamydomonas* spp. and *Scourfieldia cordiformis*, stayed mainly in the upper hypolimnion close to the oxic-anoxic boundary zone where only a small proportion of *Daphnia longispina* was continuously present. Unpalatable *Mallomonas* chrysophytes with silicified plates and bristles, and small, presumably heterotrophic, flagellates stayed in the oxic epilimnion together with a dense (up to $300\ ind\ l^{-1}$) population of *D. longispina*. The results indicated that, besides the physical and chemical properties of the water column, grazing pressure by *Daphnia longispina* strongly affected the vertical distribution of microorganisms in this polyhumic lake.

## Introduction

In humic lakes the uppermost few centimeters of water absorb most of the incoming solar radiation (Eloranta, 1978; Jones & Arvola, 1984). As a consequence, small humic lakes with high relative depth have a tendency to retain some stratification during the spring 'overturn period', because of the rapid warming of the thin surface water layer (Salonen *et al.*, 1984a). In summer, thermal stratification can be extremely steep, particularly in highly humic small lakes occupying a sheltered position. In such lakes the steep thermal stratification is generally accompanied by an equally steep oxygen gradient between the epilimnion and hypolimnion.

The steep stratification of physical and chemical conditions is reflected in the small scale vertical distributions of algae (Croome & Tyler, 1984; Arvola, 1985; Jones, 1988). Many flagellate algae migrate diurnally within the epilimnion (Ilmavirta, 1974; Jones & Ilmavirta, 1978), or even between epi- and hypolimnion (Arvola, 1984; Salonen *et al.*, 1984b; Arvola *et al.*, 1987; Jones, 1988;

Similä, 1988; Smolander & Arvola, 1988). Similarly, specific distribution patterns of bacteria and ciliates are found over the temperature and oxygen gradient (e.g. Anagnostidis & Overbeck, 1966; Sorokin, 1970; Goulder, 1972; Dubinina & Kuznetsov, 1976; Croome & Tyler, 1984; Baker *et al.*, 1985; Børsheim *et al.*, 1985; Berninger *et al.*, 1986; McDonough *et al.*, 1986).

In this paper we describe the vertical distributions of bacteria, algae, and protozoa under high grazing pressure from the cladoceran *Daphnia longispina* in a small, polyhumic lake.

## Material and methods

The study was performed at Mekkojärvi, southern Finland during three 24 h periods; 11–12 June, 15–16 July, and 20–21 August, 1986.

### *Study site*

Mekkojärvi (Fig. 1) is a small (0.0035 $km^2$) and shallow lake (mean depth 2 m, maximum depth 4 m) with high humic content (water colour 300–600 mg Pt $l^{-1}$) and acidic water (pH 5–6.5). The shoreline of the lake is mainly composed of

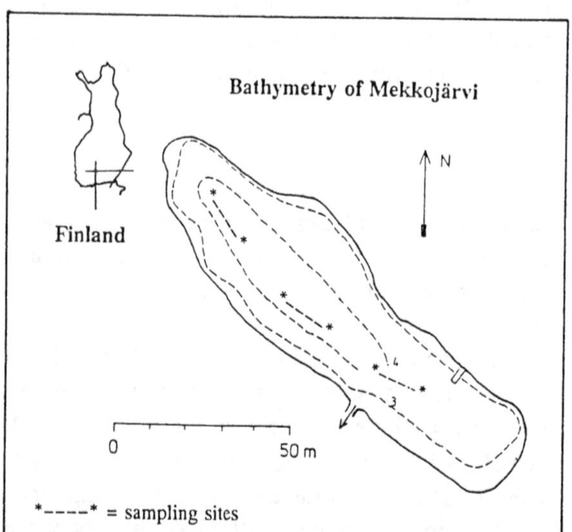


*Fig. 1.* Location and bathymetric map of Mekkojärvi with locations of the sampling areas.

*Sphagnum* moss. One small inlet brings water from the catchment area and water flows out of the lake via a shallow outlet. The average retention time for the epilimnion (ca. the uppermost 75 cm) varies from 3–4 days during high throughflow to ca. 150 days during dry periods. The ice-free period usually begins in early May and continues till the end of October.

### *Physical, chemical and biological measurements*

To reduce the influence of possible patchiness on the results three sets (Fig. 1) of vertical samples were taken twice a day (at solar noon and midnight) with a multicore sampler (Salonen & Lehtovaara, 1991), and every time the samples from each 0.1 m thick water layer were pooled.

Samples for pH, conductivity, nutrient analyses, and chlorophyll determinations were collected at 5.30 a.m. (solar time) for each study period. On each occasion water temperature and oxygen concentration were measured *in situ* with a combined probe (Yellow Springs Instrument) at the deepest part of the lake. Diel illumination (in July only at 5.15 a.m.) in the water column was measured with a quantum meter (Lambda Instruments, model Li 2190).

Dissolved nutrients were determined immediately upon transport to the laboratory. The samples were filtered through Whatman GF/C glass fibre filters, which were first rinsed with 0.5 l of deionized water. The first 200 ml of filtrate were discarded and only the last fraction was used for the determination of inorganic phosphorus (Murphy & Riley, 1962) and nitrate + nitrite (Wood *et al.*, 1967) with an AKEA autoanalyzer. Total nitrogen and total phosphorus were determined after persulphate digestion (Koroleff, 1979).

Samples for bacteria and phytoplankton were immediately fixed with formaldehyde (final concentration 3%). Every second vertical sample from the noon and midnight samples were counted. Bacteria were stained with acriflavine (Bergström *et al.*, 1986) on black Nuclepore (pore size 0.2 $\mu$m) membranes. At least 20 random fields per sample were counted (200–1000 cells) with an epifluorescence microscope at 1250 × magnification. For volume determinations length and width of at least 100 bacteria were measured with an eyepiece micrometer. The carbon-volume ratio of bacteria was assumed to be 0.38 pg C $\mu m^{-3}$ (Lee & Fuhrman, 1987; Salonen *et al.*, unpubl.).

The species composition and biomass of phytoplankton and protozoa were determined with an inverted microscope using a settling chamber technique (Utermöhl, 1958). Samples of 50 ml were sedimented for 48 hours and 25 random fields were counted in each sample at 400 × magnification (all cells) and another 25 fields at 200 × magnification (large cells). The biovolumes were estimated from the linear dimensions measured for each species. The carbon conversion factor used was 0.2 pg C $\mu m^{-3}$ (Rocha & Duncan, 1985).

Chlorophyll *a* (Chl *a*) and bacteriochlorophyll *d* (BChl *d*) were determined once for each sampling period (at 05.30 h). The pigments were extracted in ethanol (Arvola, 1981) and measured with a Shimadzu UV-240 spectrophotometer at 665 nm and 750 nm for Chl *a* and 654 nm and 750 nm for BChl *d*. The wavelength ratio between Chl *a*/BChl *d* was taken to indicate the shift from Chl *a* to BChl *d* in the hypolimnion. The concentration of Chl *a* was calculated using the equation of Lorenzen (1967) and BChl *d* using the equation of Takahashi & Ichimura (1970).

## Results

### *Physical and chemical conditions*

Because of the brown water colour and limited turbulent mixing, the water column was sharply stratified (Fig. 2). In June the maximal temperature difference between the surface and 1 m depth was 13.5 °C, recorded at 5 p.m., but next morning the respective difference was reduced to 7.5 °C. This demonstrates how effectively humic water absorbs solar radiation and how quickly this shallow, warm water layer can lose heat at night. Although during cooling some convection takes place in the epilimnion, in summer the temperature difference between the epi- and hypolimnion is high enough to maintain the stagnant stratification (cf. Arvola, 1984; Salonen *et al.*, 1984a; Bowling & Salonen, 1990). In June the boundary layer between the anoxic and oxic water of Mekkojärvi was located at around 0.9 m depth and in July at 1.1–1.2 m (Fig. 2). With the longer nights and lower air temperatures, thermal stratification weakened in late summer. In August the oxic water layer extended to 1.2–1.5 m depth.

Details of the vertical distribution of nutrients, conductivity, and pH are presented only for 16 July (Fig. 3), because their distribution patterns closely followed the gradients of temperature and oxygen. On each occasion the concentrations of total nitrogen and total phosphorus were much higher in the hypolimnion than in the epilimnion (Table 1). The same trend was seen in conductivity. The concentration of phosphate phosphorus declined from June to July in the epilimnion, but in the hypolimnion it increased. At the same time the concentration of nitrate + nitrite decreased in the whole 2 m water column. Water pH had a bimodal distribution with maxima near the epilimnion and sulphate reduction in the hypolimnion (Kelly *et al.*, 1982; Schindler, 1988). In the boundary layer $CO_2$ production (respiration) presumably greatly exceeded $CO_2$ consumption (primary production), thus decreasing pH.

The 10% light level was at a depth of around 0.45 m, the 1% light level at 1.05 m and 0.1% light level at 1.60 m. Thus at the top of the hypolimnion the maximum light intensity was less than 15 $\mu mol\ m^{-2}\ s^{-1}$ during a sunny day and at 160 cm depth it did not exceed 1.5 $\mu mol\ m^{-2}\ s^{-1}$.

### *Bacteria*

The numbers (Fig. 4) and biomasses (Fig. 5) of bacteria varied both within one day as well as between the sampling periods. In the oxic epilimnion the cell numbers fluctuated between 2.2–7.3 × $10^6$ cells $ml^{-1}$ and the biomasses between 30–195 $\mu g\ C\ l^{-1}$. Small cocci (vol $<0.04\ \mu m^3$) contributed 73–79% of the cell numbers and 32–

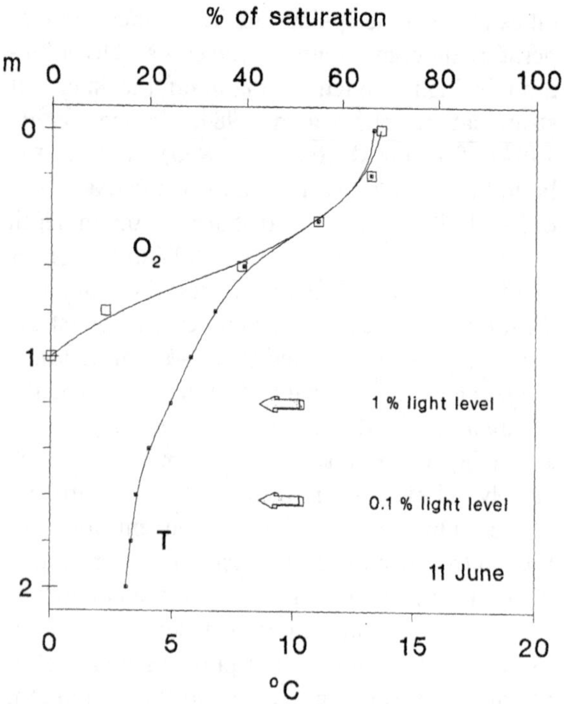


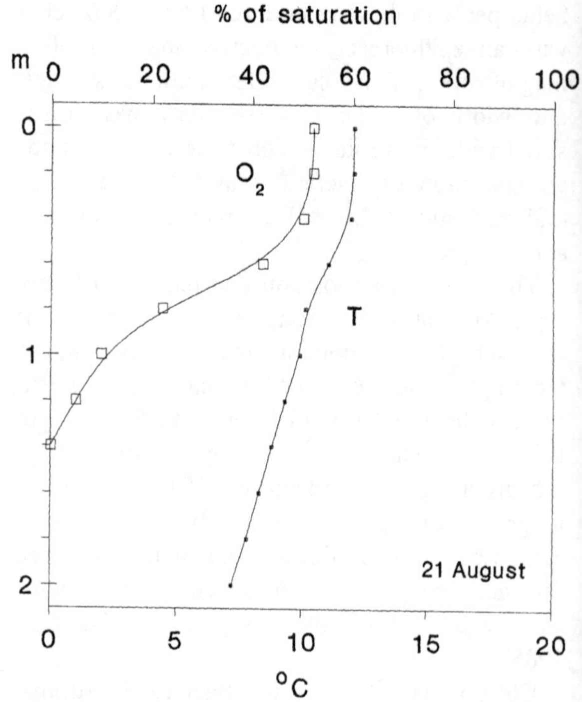


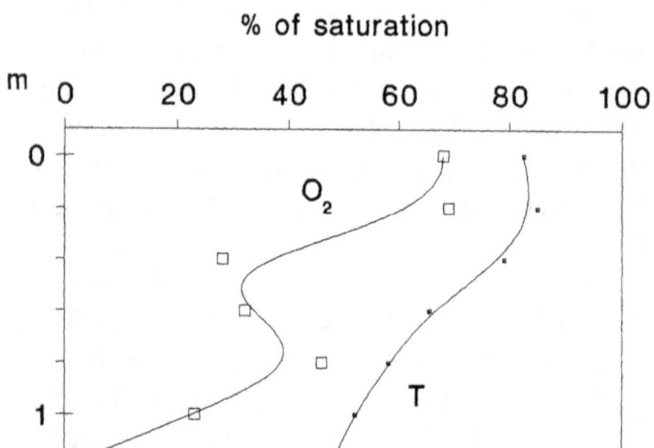


*Fig. 2.* Vertical distribution of temperature (T; °C) and oxygen ($O_2$; % saturation) at 5.30 a.m. in Mekkojärvi during each sampling period. The 1% and 0.1% light penetration levels for 11 June 1986 are given.

53% of the biomass. In the hypolimnion both the numbers and biomasses of bacteria were about an order of magnitude higher ($1.7–4.1 \times 10^7$ cells $ml^{-1}$ and 400–2300 $\mu g\ C\ l^{-1}$). Although again some 50% of the cells were small cocci ($<0.04\ \mu m^3$), larger rods and sickle shaped cells, mainly green sulphur bacteria (*Chlorobium*), contributed more to the biomass. According to the 665/654 wavelength ratio, the vertical distribution of photosynthetic bacteria coincided with the top of the anoxic hypolimnion (Fig. 6). The concentration of bacteriochlorophyll *d* reached its maximum (600–900 $\mu g$ BChl $d\ l^{-1}$) at a depth of 160–185 cm, where light intensity was less than 0.1% from that at the surface. Iron and manganese oxidizing bacteria (*Gallionella* sp., ?*Ferribacterium* sp., *Siderocapsa* sp.) and large purple sulphur bacteria (*Chromatium* sp.) were also present in the hypolimnion.

## Algae

The number of algal species recorded was slightly higher in June and July (> 20 taxa) than in August

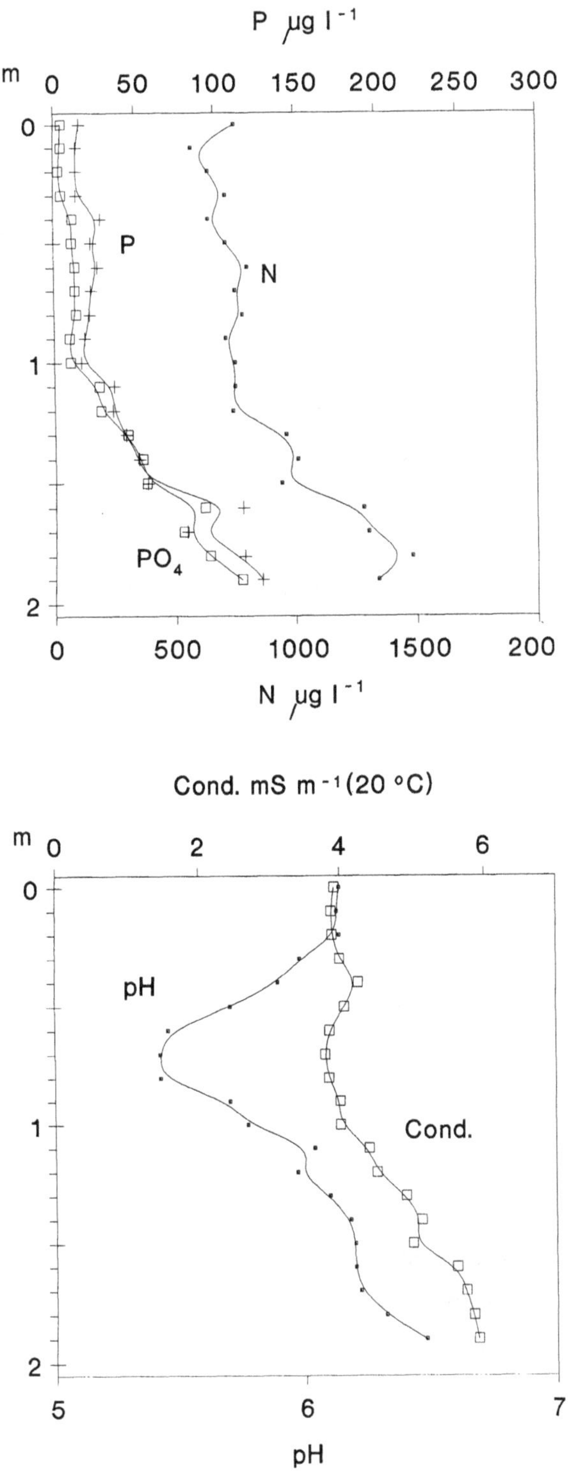


*Fig. 3.* Vertical distribution of total nitrogen (N), total phosphorus (P) and phosphate-phosphorus ($PO_4$) as well as pH and electrical conductivity on 16 July.

*Table 1.* The median concentrations ($\mu g\, l^{-1}$) and ranges (within the depth profile) of total phosphorus, phosphate phosphorus, total nitrogen, and nitrite + nitrate nitrogen in three different sections of the water column during the study.

| Depth | June | July | August |
|---|---|---|---|
| Total phosphorus | | | |
| 0–1 m | 31 | 21 | 26 |
| | 25–41 | 14–29 | 18–31 |
| 1–15 m | 45 | 37 | 38 |
| | 32–58 | 17–52 | 24–68 |
| 1.5–2 m | 91 | 117 | 167 |
| | 62–180 | 57–129 | 81–190 |
| Phosphate phosphorus | | | |
| 0–1 m | 20 | 10 | 13 |
| | 16–26 | 3–14 | 9–18 |
| 1–1.5 m | 25 | 29 | 38 |
| | 19–45 | 10–54 | 15–99 |
| 1.5–2 m | 58 | 93 | 191 |
| | 47–89 | 57–116 | 121–282 |
| Total nitrogen | | | |
| 0–1 m | 1190 | 710 | 770 |
| | 1070–1300 | 570–800 | 720–810 |
| 1–1.5 m | 1500 | 750 | 620 |
| | 1390–1700 | 740–1010 | 580–660 |
| 1.5–2 m | 1700 | 1300 | 950 |
| | 1640–1820 | 940–1480 | 600–1040 |
| Nitrite + nitrate nitrogen | | | |
| 0–1 m | 85 | <25 | 35 |
| | 60–90 | | |
| 1–1.5 m | 85 | <25 | 35 |
| | 50–110 | | |
| 1.5–2 m | 40 | <25 | 30 |
| | 40–70 | | 30–35 |

(< 15 taxa). The density of algae (together with small heterotrophic flagellates) also varied between the sampling periods; in June the density was ca. 8000 cells ml$^{-1}$ and in July 16000 cells ml$^{-1}$, whereas in August it was only 2700 cells ml$^{-1}$ (Fig. 7). The densities and biomasses were much lower in the epilimnion than beneath the chemocline (Figs. 7 and 8). *Scourfieldia cordiformis* Takeda was the most abundant algal species during the study. Because of the low density of most taxa and the difficulty of identifying the smallest flagellate cells, only the data concerning the most abundant and easily identified species are presented. Three different distribution patterns of the phytoplankton populations could be

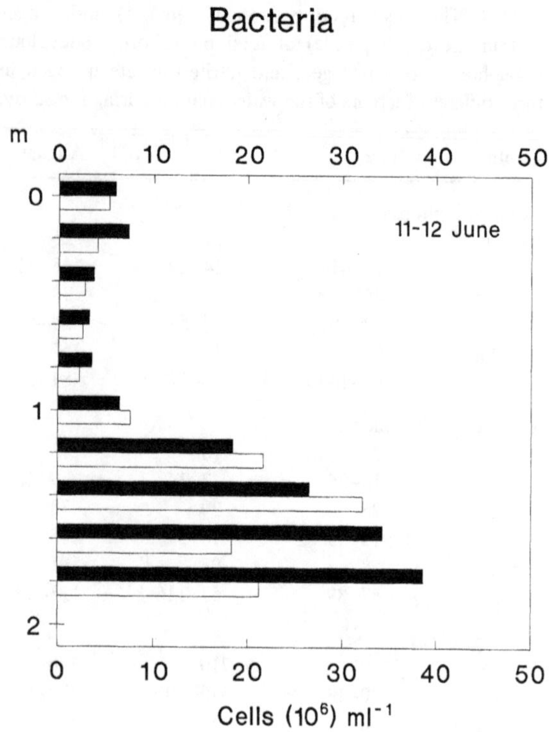
Bacteria
m
0
1
2
0 10 20 30 40 50
11-12 June
Cells (10⁶) ml⁻¹

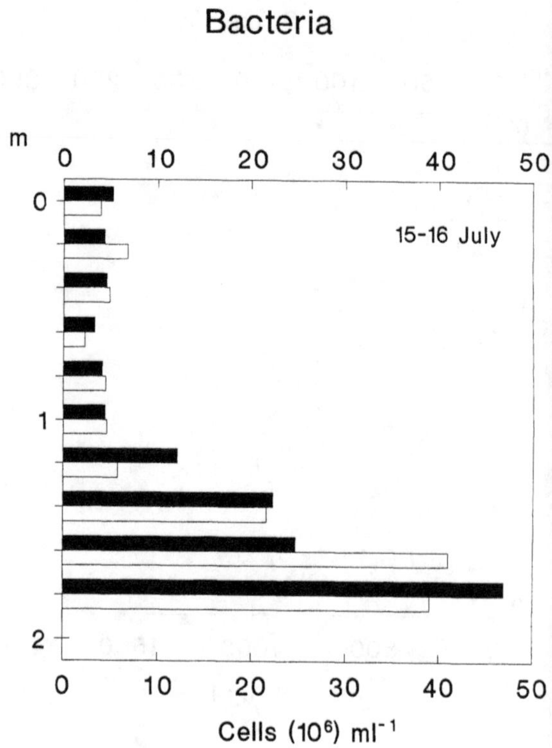
Bacteria
m
0
1
2
0 10 20 30 40 50
15-16 July
Cells (10⁶) ml⁻¹

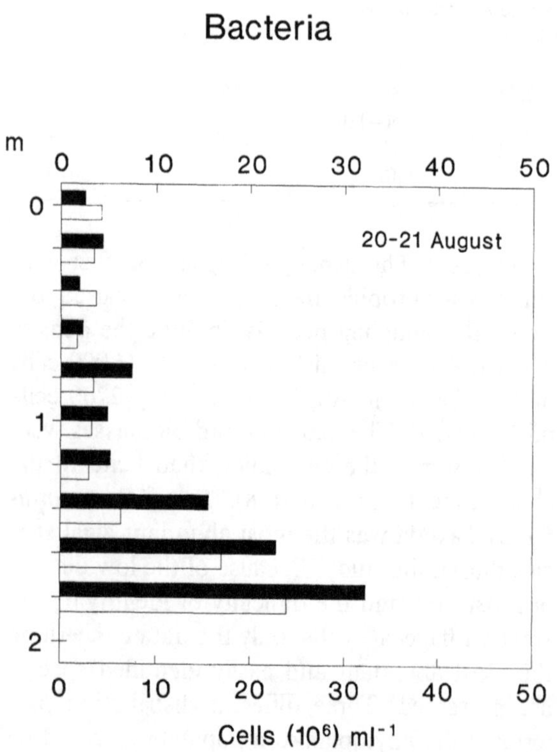
Bacteria
m
0
1
2
0 10 20 30 40 50
20-21 August
Cells (10⁶) ml⁻¹

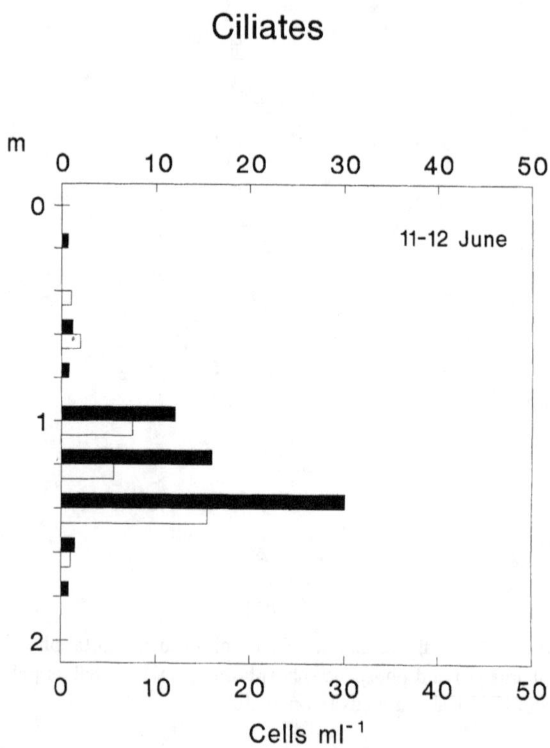
Ciliates
m
0
1
2
0 10 20 30 40 50
11-12 June
Cells ml⁻¹

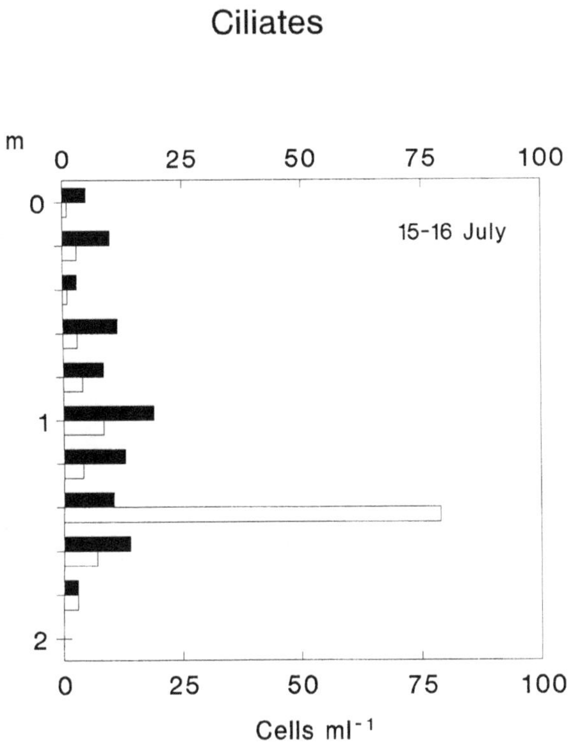


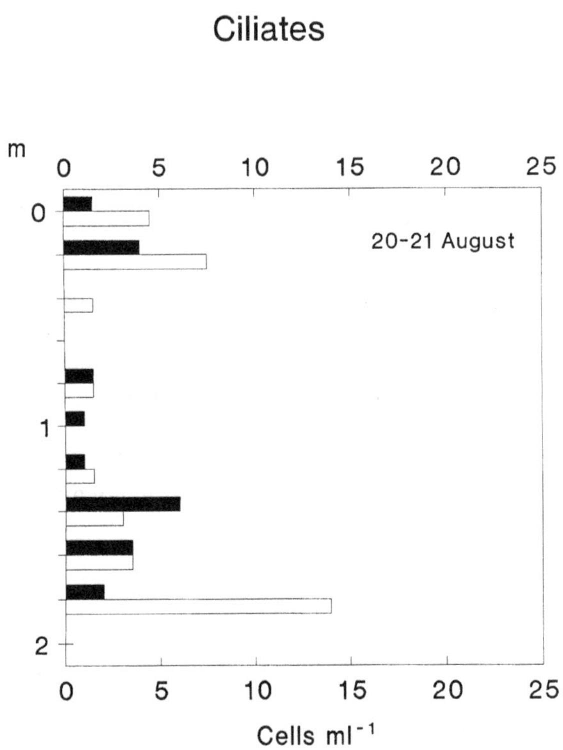


*Fig. 4.* Day (open bars) and night (black bars) profiles of the numbers of bacteria and ciliates during the study periods.

found: 1) strictly epilimnetic species; 2) species which occurred only in the anoxic hypolimnion; and 3) species which migrated diurnally between the epi- and hypolimnion.

Among the epilimnetic algae, chrysophycean *Mallomonas* species were most noticeable. *M. costata* Durrschmidt (determined by SEM), *M. heterospina* Lund emend. Asmund and *M. akrokomos* Ruttner were all restricted to the uppermost part of the water column (Fig. 8). Probably due to horizontal patchiness or incomplete vertical coverage of counted samples the numbers of these species varied considerably between day and night. In July and August small scale downward migration seemed to occur at night, although cells did not penetrate the chemocline into the anoxic hypolimnion.

Two algal species were particularly abundant in the hypolimnion. In June a volvocalean chlorophyte, *Chlamydomonas* spp., occurred beneath the thermocline (Fig. 8), although earlier in the spring it is usually abundant near the surface (Arvola & Kankaala, 1989; see also Similä, 1988). In August it also occurred predominantly in the anoxic hypolimnion, but some cells were found in the epilimnion at midday. The vertical distribution of *Scourfieldia cordiformis* (Prasinophyceae) was similar, but on all occasions the population of *Scourfieldia* was restricted to the anoxic water beneath the chemocline (Fig. 8).

Small cryptomonads, *Cryptomonas marssonii* Skuja and *C. gracilis* Skuja, were present in the lake in July and August, with *C. marssonii* always dominant. Although they migrated vertically in the water column (Fig. 8), it remained uncertain whether the cells passed through the chemocline. In the hypolimnion there might have been a subpopulation or else the vertical movements were not diurnally synchronised (see Smolander & Arvola, 1988). In August no small cryptomonads were observed in the hypolimnion. Particularly in July, a large *Cryptomonas* (*C. rostratiformis* Skuja) migrated through the chemocline into the anoxic hypolimnion (Fig. 8). However, one third of the population was continuously above the chemocline, and so possibly the cells stayed only a short time in the anoxic layer.

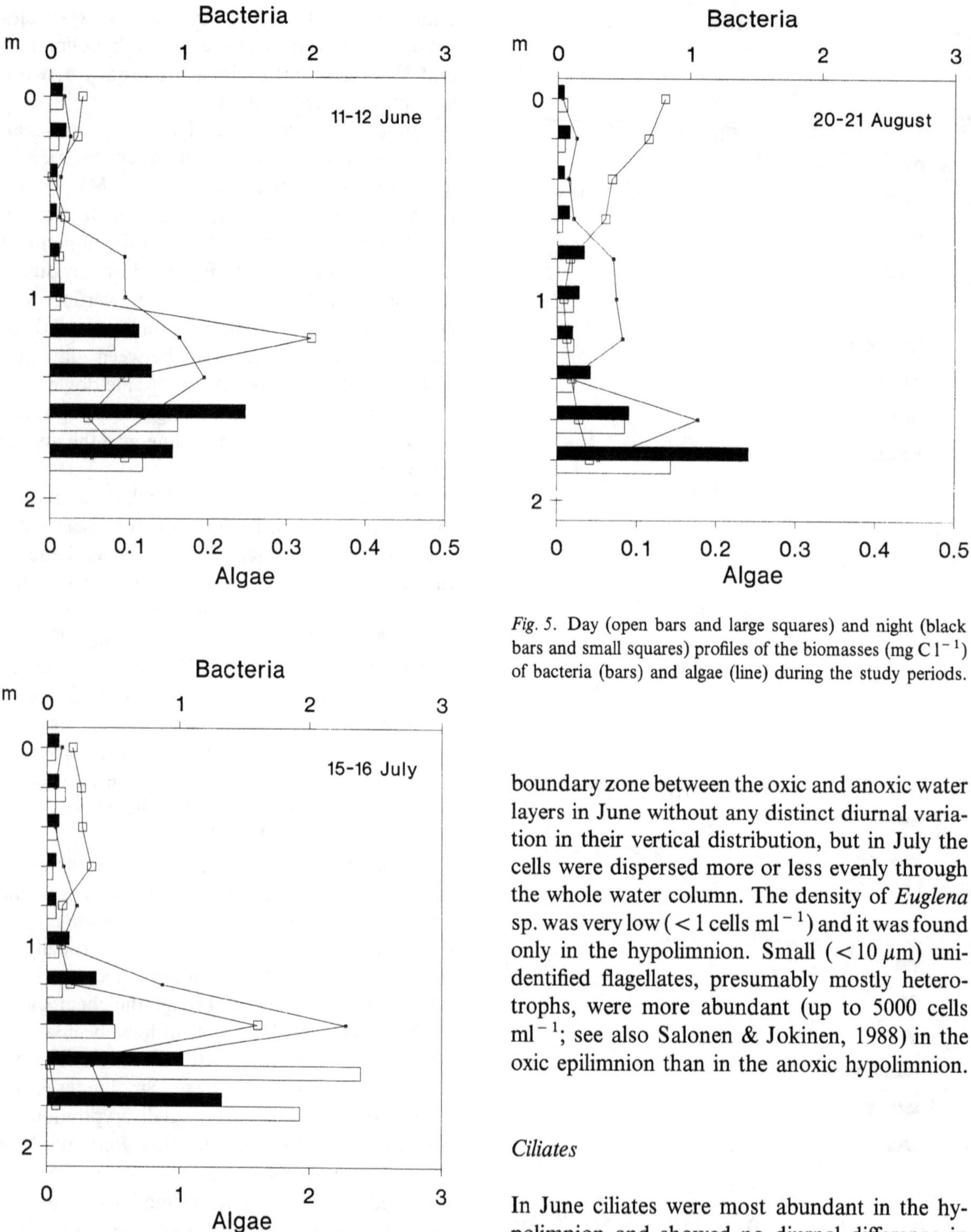


*Fig. 5.* Day (open bars and large squares) and night (black bars and small squares) profiles of the biomasses (mg C $l^{-1}$) of bacteria (bars) and algae (line) during the study periods.

Dinoflagellates (*Gymnodinium* sp. and *Glenodinium* sp.), which were present in low densities (<5 cells $ml^{-1}$), were concentrated into the boundary zone between the oxic and anoxic water layers in June without any distinct diurnal variation in their vertical distribution, but in July the cells were dispersed more or less evenly through the whole water column. The density of *Euglena* sp. was very low (<1 cells $ml^{-1}$) and it was found only in the hypolimnion. Small (<10 $\mu$m) unidentified flagellates, presumably mostly heterotrophs, were more abundant (up to 5000 cells $ml^{-1}$; see also Salonen & Jokinen, 1988) in the oxic epilimnion than in the anoxic hypolimnion.

## *Ciliates*

In June ciliates were most abundant in the hypolimnion and showed no diurnal difference in their vertical distribution (Fig. 4). In July they were dispersed in the water column during the night, but during the day there was a distinct peak

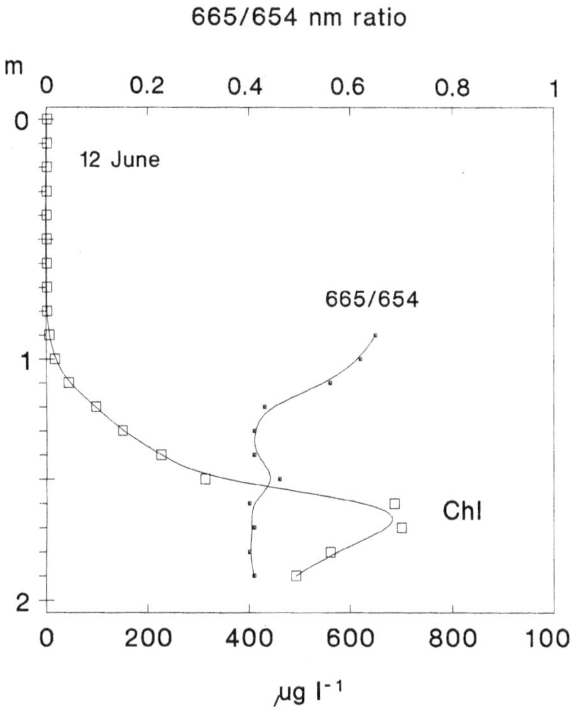


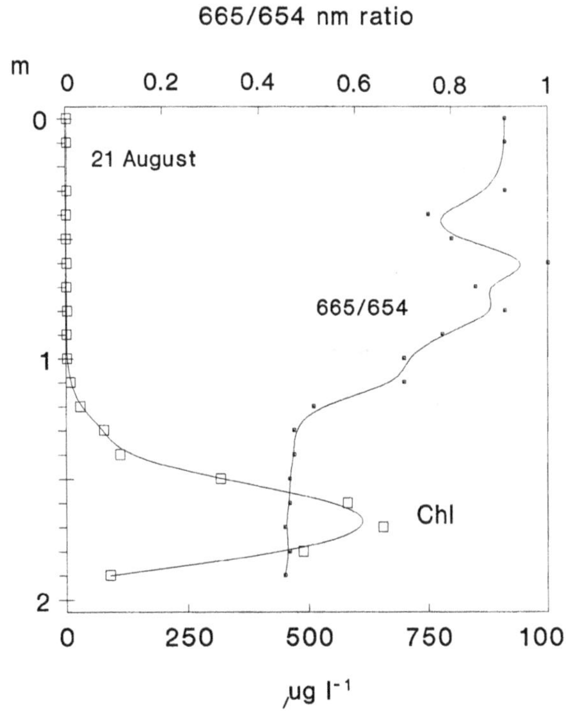


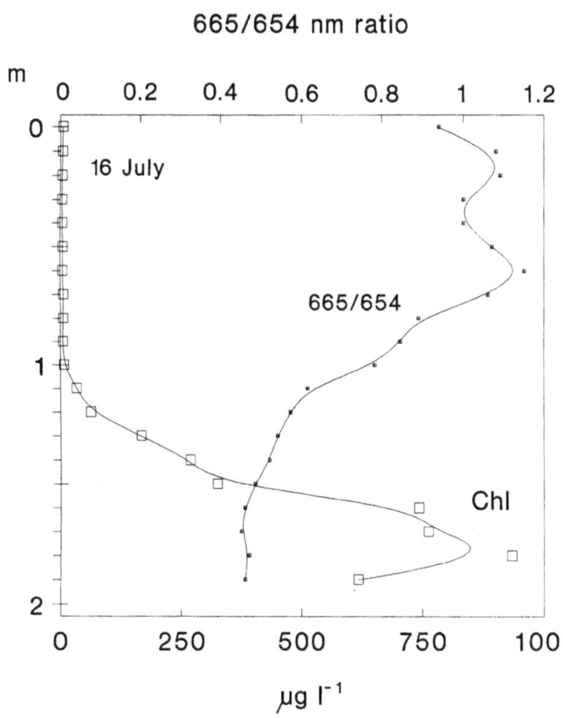


*Fig. 6.* Vertical distributions of total chlorophyll (Chl *a* + BChl *d*; $\mu g\, l^{-1}$) and the wavelength ratio of 665 nm/654 nm (665/654). Wavelength ratio of > 0.8 indicates the dominance of Chl *a*, and the ratio < 0.5 indicates the dominance of BChl *d*.

below the chemocline. In August their distribution was more or less even throughout the whole day. The cell size of ciliates varied from 10 to 75 µm (diameter or maximum length of the cell). Large cells belonging to the groups of Oligotrichida and Stichotrichida dominated the biomass.

## *Larger zooplankton*

During all the three study periods, *Daphnia longispina* dominated the macrozooplankton community in Mekkojärvi. However, because the details of the diel occurrence of *D. longispina* are described elsewhere in this volume (Salonen & Lehtovaara, 1991), the present paper gives only the vertical profiles of night and day for *Daphnia* (Fig. 7). During each occasion the individuals were mainly concentrated into the oxic water layer and the population did not overlap the maximum biomass of bacteria and algae (cf. Figs. 5, 6 & 7).

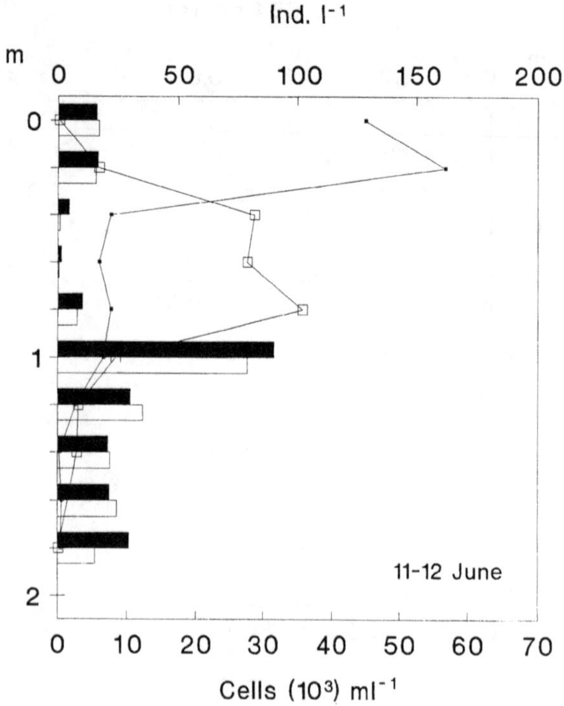


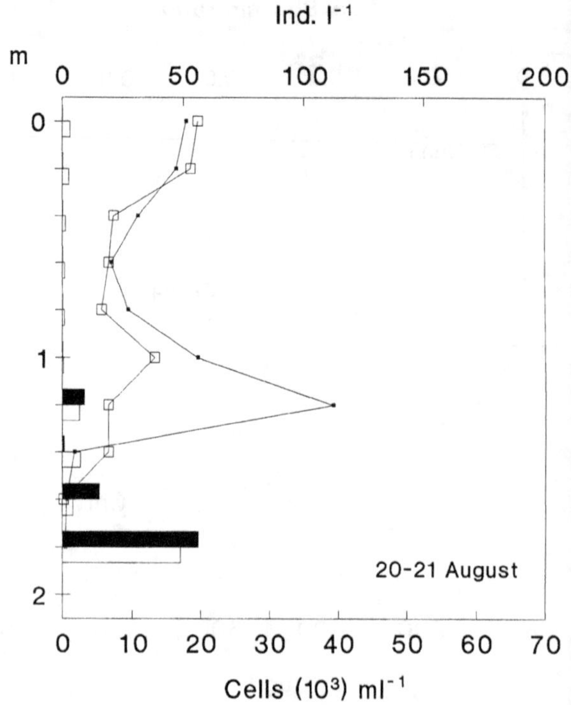


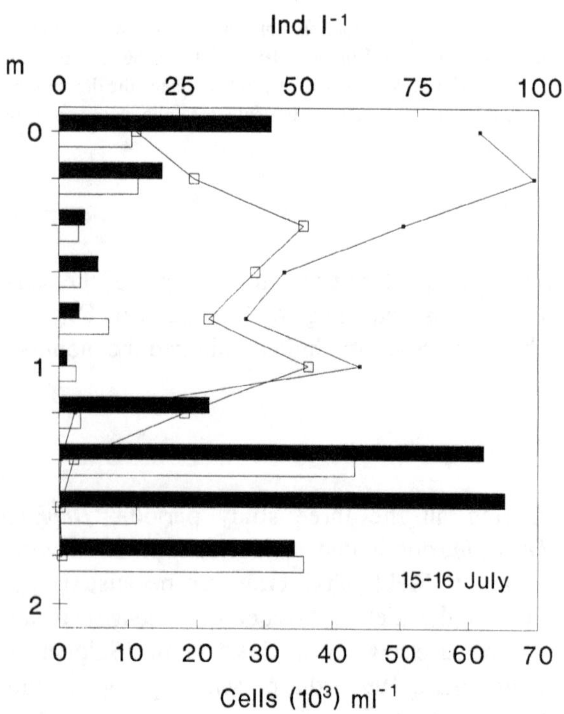


*Fig. 7.* Vertical distributions of phytoplankton (bars; cells $ml^{-1}$) and *Daphnia longispina* (solid line; ind $l^{-1}$) during day (open bars and large squares) and night (black bars and small squares).

## Discussion

In an undisturbed water column, the vertical distribution of microorganisms is primarily influenced by gradients of light, temperature, nutrients and other chemical factors like oxygen and hydrogen sulphide. In lakes with an anoxic hypolimnion, there can be a dense plate of phototrophic bacteria below the oxic zone, whereas in the epilimnion the density and biomass of bacteria may be similar to those in lakes with no anoxia (McDonough *et al.*, 1986). In highly humic lakes the occurrence of phototrophic bacteria is restricted to a very narrow anoxic zone where some light still penetrates. The light intensities at which the green sulphur bacteria exist in Mekkojärvi, $<0.1-1\%$ of the surface irradiance, is in agreement with observations made in other lakes (e.g. Parkin & Brock, 1981; Gorlenko *et al.*, 1983; Croome & Tyler, 1984; Steenbergen *et al.*, 1987; Burke & Burton, 1988; Overmann & Tilzer, 1989). The green sulphur bacteria (*Chlorobium*) are probably the chief consumers of $H_2S$ in Mekkojärvi as was the case in Knaack Lake (Parkin & Brock, 1981). According to Parking & Brock

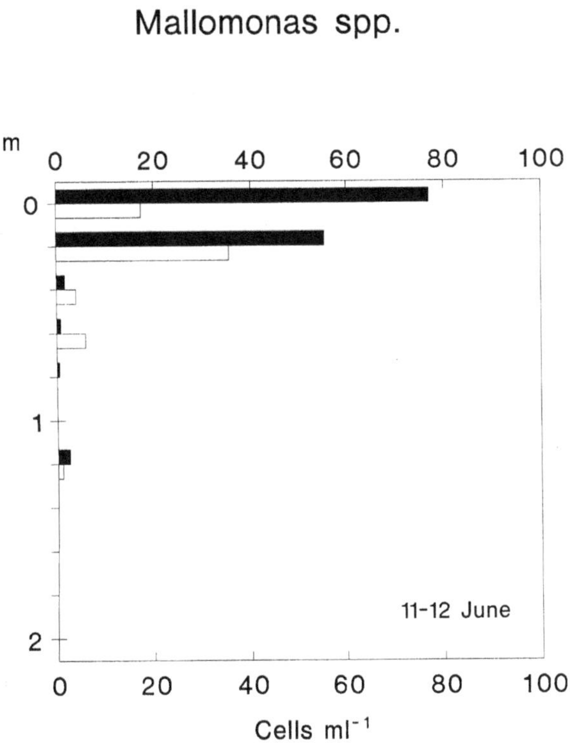
Mallomonas spp.
m
0
1
2
0 20 40 60 80 100
11-12 June
Cells ml⁻¹

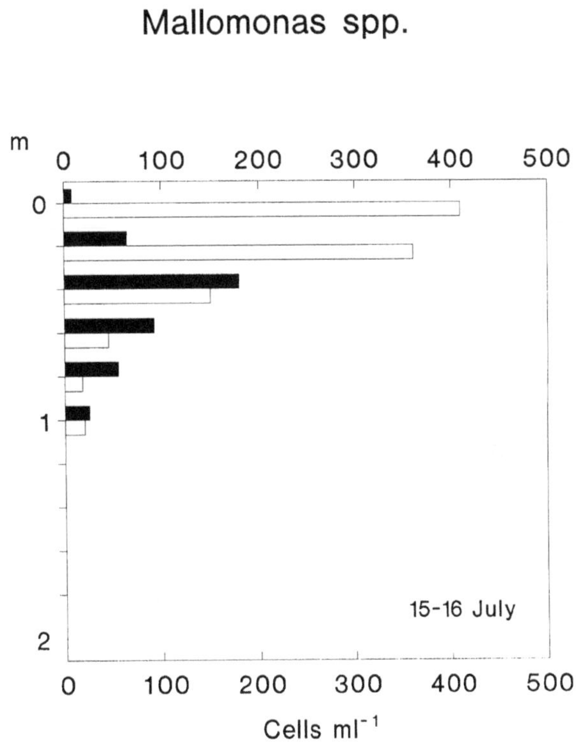
Mallomonas spp.
m
0
1
2
0 100 200 300 400 500
15-16 July
Cells ml⁻¹

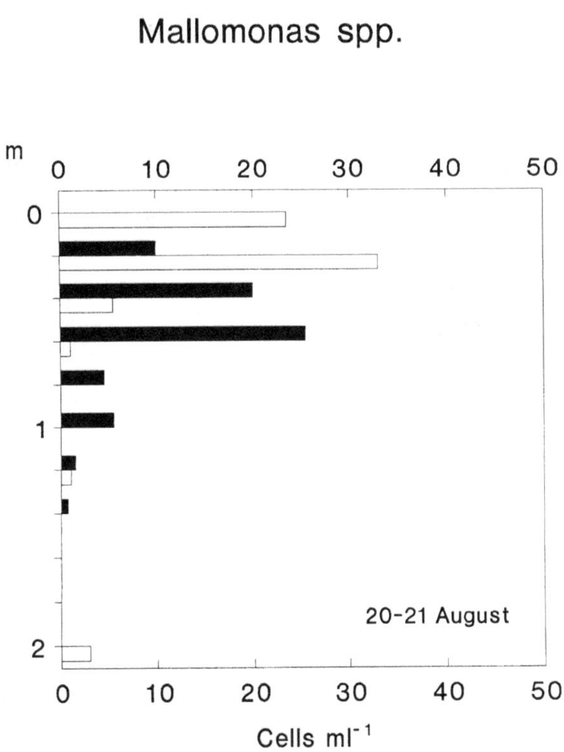
Mallomonas spp.
m
0
1
2
0 10 20 30 40 50
20-21 August
Cells ml⁻¹

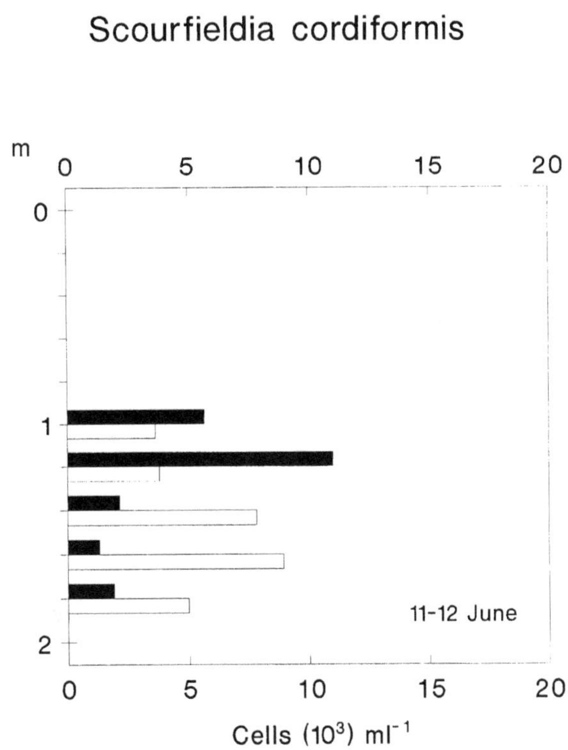
Scourfieldia cordiformis
m
0
1
2
0 5 10 15 20
11-12 June
Cells (10³) ml⁻¹

Scourfieldia cordiformis

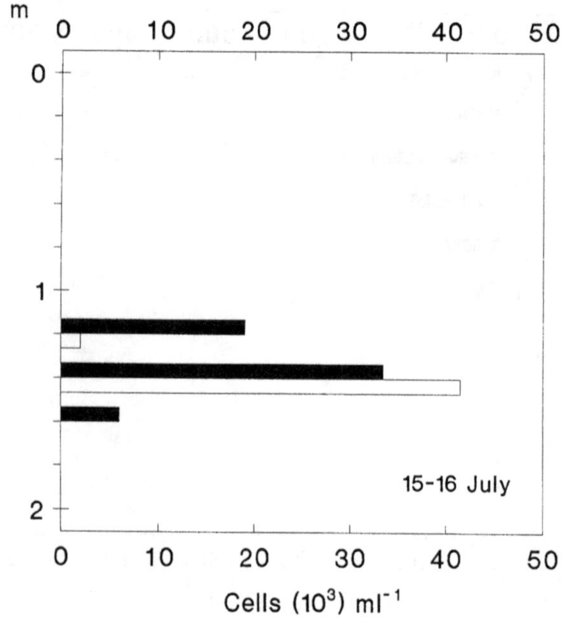


Scourfieldia cordiformis

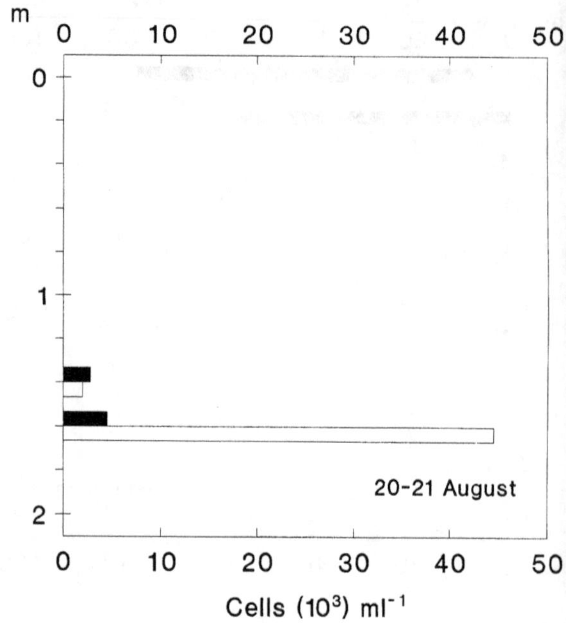


Cryptomonas small

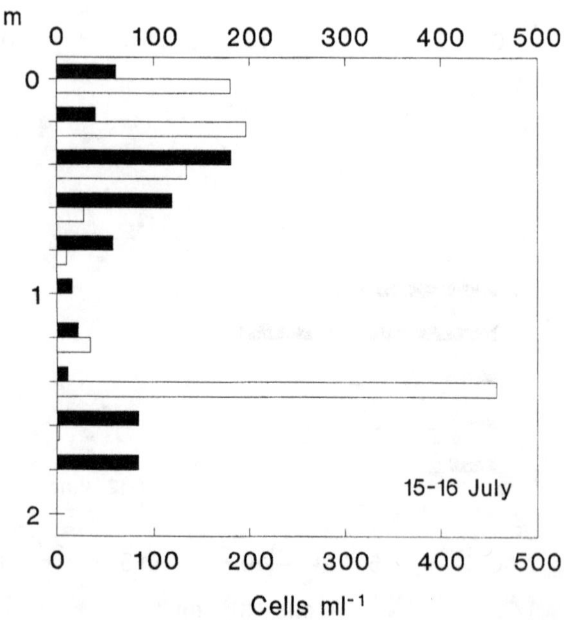


Cryptomonas small

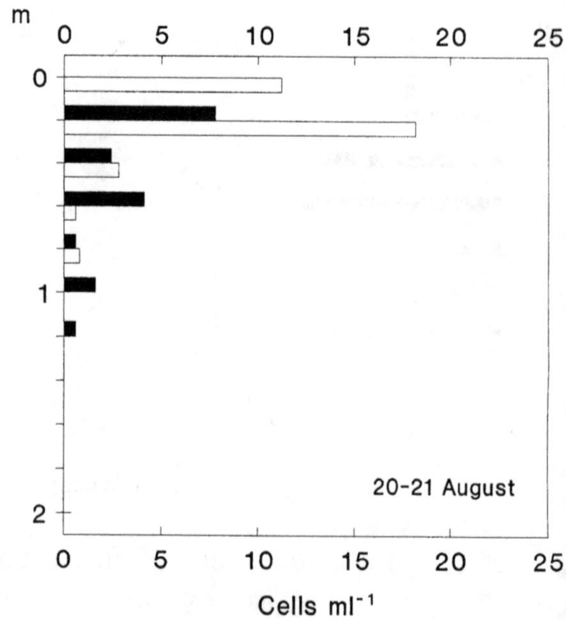

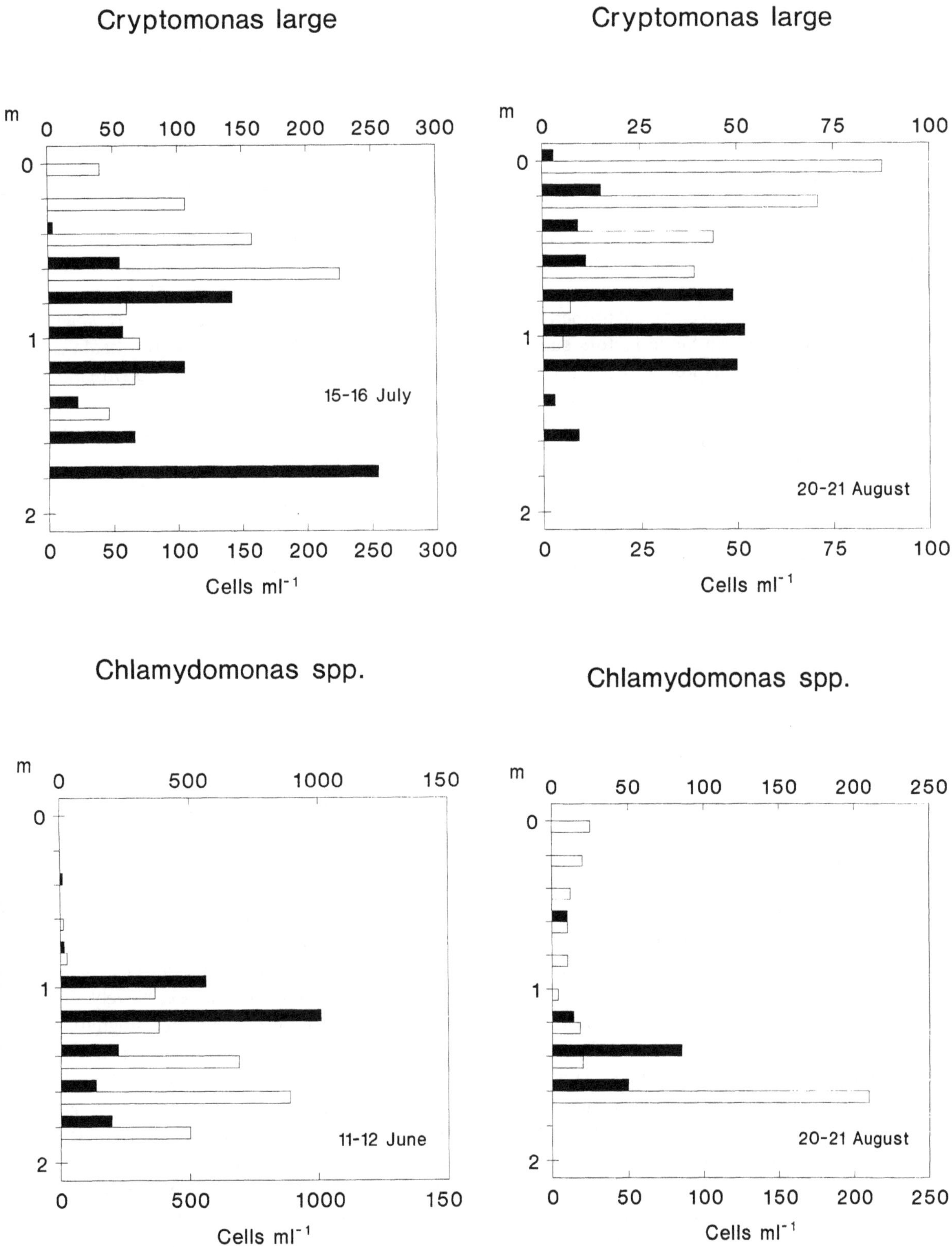


*Fig. 8.* Vertical distributions of some phytoplankton species during day (open bars) and night (black bars).

(1980) selective absorption of light by humic acids determines the presence of Chlorobiaceae and absence of Chromatiaceae; red light selects the former and green light the latter group. In addition, green sulphur bacteria are able to grow at lower light intensities (Biebl & Pfennig, 1978; Van Gemerden, 1980) and their affinity for sulphide is also higher than that of purple bacteria (Van Gemerden, 1987). Thus, in Mekkojärvi the growth rate of *Chromatium* is probably lower than that of *Chlorobium* (Matsuyama, 1987).

Besides the physical and chemical properties in the water column, several biotic factors such as competition and predator-prey relationships may also influence the vertical distribution of microorganisms (cf. Craig, 1987; Jones, 1991). In Mekkojärvi both bacteria and algae are intensively grazed by *Daphnia longispina* (Kankaala, 1988; Salonen *et al.*, 1991b). Mazumder & Dickman (1989) have shown that *Daphnia* can also consume phototrophic bacteria living in the anoxic hypolimnion. As a consequence we propose that the diurnal fluctuations of the biomass of bacteria in Mekkojärvi might be a result of asynchrony between the production and losses of bacteria.

The impact of grazing on the vertical distribution of phytoplankton seems evident in Mekkojärvi. *Chlamydomonas* and *Scourfieldia*, which in spring are in the epilimnion, were restricted to the hypolimnion during this study. Thus they were deeper than most of the larger individuals of *D. longispina*. In a mesocosm experiment we have shown that in the absence of *D. longispina*, these algal species grew well at temperatures of 10–16 °C (Salonen *et al.*, 1991a). However, in a similar experiment, but in the presence of *D. longispina*, grazing kept the densities of fast growing (generation times up to 0.4 days) phototrophic and heterotrophic nanoflagellates at a very low level (Salonen *et al.*, 1991b). Thus grazing by zooplankton is the most probable explanation for the changes in the vertical distribution of these two chlorophyte species. According to Salonen & Lehtovaara (1991), only the large adult females of *D. longispina* occasionally dive into the anoxic hypolimnion and graze on the photosynthetic bacteria and algae located there.

Among the algae of Mekkojärvi, only *Mallomonas* species were abundant through the whole day in the oxic epilimnion despite their small scale vertical movements. This observation is consistent with the results of Arvola (1984) and Jones (1988) who found, in two other small humic lakes in the same area, that the vertical movements of *M. caudata*, *M. reginae* (*M. punctifera*) Teiling and *M. akrokomos* took place mainly in the epilimnion. Although *Daphnia* is known as a nonselective filter feeder (cf. DeMott, 1986), in experiments carried out in Mekkojärvi *D. longispina* selected against *M. akrokomos* (Kankaala & Arvola, unpubl.). *M. costata* and *M. heterospina* were eaten by *Daphnia* but they were not such favoured food as small cryptomonads and softbodied flagellates. Its silicated bristles and plates make *Mallomonas* difficult for *Daphnia* to handle. Thus, *Mallomonas* is probably favoured by a high grazing pressure of *Daphnia* directed more against other algal species (Salonen & Arvola, 1988).

Both *Chlamydomonas* sp. and *Scourfieldia cordiformis* have been shown to grow rather quickly (generation time 1.3 days for *Chlamydomonas*, see also Happey-Wood, 1976; and 0.46 days for *Scourdieldia*) under a wide range of temperature (8–16 °C) in experimental conditions (Salonen *et al.*, 1991a). In Mekkojärvi these species are common under the ice and during the spring maximum of phytoplankton, when the density of *D. longispina* is still low (Arvola & Kankaala, 1989). In a stratified brackish water lake, Lindholm & Weppling (1987) found a maximum of *Chlamydomonas* in oxic water just above the layer of phototrophic purple and green sulphur bacteria. The appearance of *Chlamydomonas* and *S. cordiformis* almost exclusively in the hypolimnion during the summer stagnation period, consistent with the observation of Schiller (1954), may be due to better availability of inorganic nutrients, but in the epilimnion grazing pressure by *Daphnia* certainly limits their existence. In the presence of a high concentration of dissolved organic matter *Chlamydomonas* may be able to supplement its light energy demand by facultative mixotrophy (Salonen & Hammar, 1988; Arvola & Tulonen, ms.). Similar distributions for *Chlamydomonas* and *Scourfieldia* such as in Mekkojärvi have been described in another small humic lake in Finland (Salonen *et al.*, 1990) and in two humic lakes in Tasmania (Croome & Tyler, 1985, 1988).

In a small polyhumic lake, Smolander & Arvola (1988) have shown a prominent diurnal vertical migration of *Cryptomonas marssonii*. The depth of migration varied during the growing season without any clear relationship to the physical and chemical conditions in the lake. However, the distribution of *C. marssonii* was limited to the oxic water, and oxygen depletion may have restricted the downward migration. On the other hand, Pedrós-Alió *et al.* (1987) have discovered that *C. phaseolus* is able to grow when oxygen and $H_2S$ are present simultaneously at low concentrations. It has been shown, too, that *C. marssonii* and *C. rostratiformis* are able to take up phosphate at night in the cold hypolimnion, and then migrate into the epilimnion (Salonen *et al.*, 1984b; Arvola *et al.*, 1991). Thus, the diurnal vertical migration may have a high adaptive value for the species. It also diminishes the loss of cells by outflow (Arvola *et al.*, 1987; see also Smolander & Arvola, 1988), particularly in spring at ice-melt and in summer after heavy rains, although during dry periods the impact of hydraulic washout is small (Jones, 1991). If the migration of algae is reversed compared to the migration of zooplankton, the loss rate by grazing may also decrease (cf. Arvola *et al.*, 1987). Cryptomonads are high quality food (Ahlgren *et al.*, 1990) for Daphnia and other zooplankton species, and not selection against cryptomonads has been found (Arnold, 1971; Porter, 1973; Knisely & Geller, 1986; Skogstad *et al.*, 1987; Kankaala & Arvola, unpubl.).

Ciliates were concentrated into the anoxic hypolimnion, presumably because the densities and biomasses of bacteria and nanoplanktonic algae were much higher than in the oxic epilimnion (cf. Børsheim, 1984; Jonsson, 1986; Skogstad *et al.*, 1987; review by Beaver & Crisman, 1989). In this way the ciliates were also able to diminish competition with *Daphnia*.

In conclusion, it seems that those algal species, such as cryptomonads, which have an opposite diurnal migration pattern compared to zooplankton, gain a substantial benefit under high grazing pressure. On the other hand, edible algal species which do not have a synchronized diurnal vertical migration, such as *Chlamydomonas* and *Scourfieldia*, are not able to maintain high cell densities in the oxic epilimnion because the grazing loss apparently exceeds their growth rate. Thus, together with the physical and chemical properties in the water column, the predator-prey interactions may have a great influence on the vertical distribution of microorganisms in steeply stratified lakes such as Mekkojärvi. We suggest that in steeply stratified lakes with very small distances between drastically different physical, chemical, and biological conditions, vertical migrations of algae are an important adaptation which deserves more study.

## Acknowledgements

This study was supported by the Academy of Finland and the University of Helsinki. We also thank R. I. Jones for his valuable comments on the manuscript and the revision of the English, and P. Eloranta for the identification of *Mallomonas costata* and *M. heterospina*.

## References

Anagnostidis, K. & J. Overbeck, 1966. Methanoxydierer und hypolimnische schwefelbakterien. Studien zur ökologischen Biocönotik der Gewässermikroorganismen. Ber. Dtsch. Bot. Ges. 79: 163–174.

Ahlgren, G., L. Lundstedt, M. Brett & C. Forsberg, 1990. Lipid composition and food quality of some freshwater phytoplankton for cladoceran zooplankters. J. Plankton Res. 12: 809–818.

Arnold, D. E., 1971. Ingestion, assimilation, survival, and reproduction by *Daphnia pulex* fed seven species of blue-green algae. Limnol. Oceanogr. 16: 906–920.

Arvola, L., 1981. Spectrophotometric determination of chlorophyll *a* and phaeopigments in ethanol extractions. Ann. bot. fenn. 18: 221–227.

Arvola, L., 1984. Diel variation in primary production and the vertical distribution of phytoplankton in a polyhumic lake. Arch. Hydrobiol. 101: 503–519.

Arvola, L., 1985. On the factors affecting phytoplankton species composition in highly coloured small forest lakes. Lammi Notes 12: 5–9.

Arvola, L. & P. Kankaala, 1989. Winter and spring variability in phyto- and bacterioplankton in lakes with different water colour. Aqua fenn. 19: 29–39.

Arvola, L. & T. Tulonen. Enhancement by dissolved organic matter of the growth of laboratory strains of bacteria and algae and the micro-organisms of a highly humic lake. Manuscript.

Arvola, L., A. Ojala, F. Barbosa & S. I. Heaney, 1991. Migration behaviour of three cryptophytes in relation to environmental gradients: an experimental approach. Br. phycol. J. (in press).

Arvola, L., K. Salonen, R. I. Jones, A. Heinänen & I. Bergström, 1987. A three day study of the diel behaviour of plankton in a highly humic and stratified lake. Arch. Hydrobiol. 109: 89–106.

Baker, A. L., K. K. Baker & P. A. Tyler, 1985. Fine-layer depth relationships of lakewater chemistry, planktonic algae and photosynthetic bacteria in meromictic lake Fidler, Tasmania. Freshwat. Biol. 15: 735–747.

Beaver, J. & T. L. Crisman, 1989. The role of ciliated protozoa in pelagic freshwater ecosystems. Microb. Ecol. 17: 111–136.

Bergström, I., A. Heinänen & K. Salonen, 1986. Comparison of acridine orange, acriflavin, and bisbenzimide stains for enumeration of bacteria in clear and humic waters. Appl. envir. Microbiol. 51: 664–667.

Berninger, U.-G., B. J. Finlay & H. M. Canter, 1986. The spatial distribution and ecology of zoochlorellae-bearing ciliates in a productive pond. J. Protozool. 33: 557–563.

Biebl, H. & N. Pfennig, 1978. Growth yields of green sulfur bacteria in mixed culture with sulfur and sulfate reducing bacteria. Arch. Mikrobiol. 117: 9–16.

Børsheim, K. Y., 1984. Clearange rates of bacteria-sized particles by freshwater ciliates, measured with monodisperse latex beads. Oecologia (Berlin) 63: 286–288.

Børsheim, K. Y., J. G. Kuenen, J. Gottschal & I. Dundas, 1985. Microbial activities and chemical gradients in the chemocline of a meromictic lake in relation to the precision of the sampling procedure. FEMS Microb. Ecol. 31: 337–341.

Bowling, L. C. & K. Salonen, 1990. Heat uptake and resistance to mixing in small humic forest lakes in southern Finland. Aust. J. mar. Freshwat. Res. 41: 747–757.

Burke, C. M. & H. R. Burton, 1988. The ecology of photosynthetic bacteria in Burton Lake, Vestfold Hills, Antarctica. Hydrobiologia 165: 1–11.

Craig, S. R., 1987. The distribution and contribution of picoplankton to deep photosynthetic layers in some meromictic lakes. Acta Acad. Aboensis 47: 55–81.

Croome, R. L. & P. A. Tyler, 1984. Microbial microstratification and crepuscular photosynthesis in meromictic Tasmanian lakes. Verh. int. Ver. Limnol. 22: 1216–1223.

Croome, R. L. & P. A. Tyler, 1985. Structure and ecology of the flagellate *Scourfieldia caeca* (Korsh.) Belcher & Swale in two meromictic lakes in Tasmania. Aust. J. mar. Freshwat. Res. 36: 413–419.

Croome, R. L. & P. A. Tyler, 1988. Phytoflagellates and their ecology in Tasmanian polyhumic lakes. Hydrobiologia 161: 245–253.

DeMott, W. R., 1986. The role of taste in food selection by freshwater zooplankton. Oecologia (Berl.) 69: 334–340.

Dubinina, G. A. & S. I. Kuznetsov, 1976. The ecological and morphological characteristics of micro-organisms in Lesnaya Lamba (Karelia). Int. Revue ges. Hydrobiol. 61:1.

Eloranta, P., 1978. Light penetration in different types of lakes in Central Finland. Holarct. Ecol. 1: 362–366.

Gorlenko, V. M., G. A. Dubinina & S. I. Kuznetsov, 1983. The ecology of aquatic micro-organisms. Die Binnengewässer Vol. 28.

Goulder, R., 1972. The vertical distribution of some ciliated protozoa in the plankton of a eutrophic pond during summer stratification. Freshwat. Biol. 2: 163–176.

Happey-Wood, C. M., 1976. Vertical migration patterns in phytoplankton of mixed species composition. Br. phycol. J. 11: 355–369.

Ilmavirta, V., 1974. Diel periodicity in the phytoplankton community of the oligotrophic lake Pääjärvi, southern Finland. I. Phytoplanktonic primary production and related factors. Ann. bot. fenn. 11: 136–177.

Jones, R. I., 1988. Vertical distribution and diel migration of flagellated phytoplankton in a small humic lake. Hydrobiologia 161: 75–87.

Jones, R. I., 1991. Advantages of diurnal vertical migrations to phytoplankton in sharply stratified, humic forest lakes. Arch. Hydrobiol. 120: 257–266.

Jones, R. I. & L. Arvola, 1984. Light penetration and some related characteristics in small forest lakes in southern Finland. Verh. int. Ver. Limnol. 22: 811–816.

Jones, R. I. & V. Ilmavirta, 1978. A diurnal study of the phytoplankton in the eutrophic lake Lovojärvi, Southern Finland. Arch. Hydrobiol. 83: 494–514.

Jonsson, P. R., 1986. Particle size selection, feeding rates and growth dynamics of marine planktonic oligotrichous ciliates (Ciliophora: Oligotrichina). Mar. Ecol. Prog. Ser. 33: 265–277.

Kankaala, P., 1988. The relative importance of algae and bacteria as food for *Daphnia longispina* (Cladocera) in a polyhumic lake. Freshwat. Biol. 19: 285–296.

Kelly, C. A., J. W. M. Rudd, R. B. Cook & D. W. Schindler, 1982. The potential importance of bacterial processes in regulating rate of lake acidification. Limnol. Oceanogr. 27: 868–882.

Knisley, K. & W. Geller, 1986. Selective feeding of four zooplankton species on natural lake phytoplankton. Oecologia (Berl.) 69: 86–94.

Koroleff, F., 1979. Methods for chemical analysis of seawater. Meri 7: 1–60. (In Finnish).

Kuuppo-Leinikki, P. & K. Salonen, 1992. Bacterioplankton in a small polyhumic lake with an anoxic hypolimnion. Hydrobiologia 229: 159–168.

Lee, S. & J. A. Fuhrman, 1987. Relationships between biovolume and biomass of naturally derived marine bacterioplankton. Appl. envir. Microbiol. 53: 1298–1303.

Lindholm, T. & K. Weppling, 1987. Blooms of phototrophic bacteria and phytoplankton in a small brackish lake on Åland, SW Finland. Acta Acad. Aboensis 47: 45–53.

Lorenzen, C. J., 1967. Determination of chlorophyll and phaeopigments: spectrophotometric equations. Limnol. Oceanogr. 12: 343–346.

Matsuyama, M., 1987. A large phototrophic bacterium densely populating the $O_2$-$H_2S$ interface of lake Kaiike on Kamikoshiki island, southwest Japan. Acta Acad. Aboensis 47: 29–43.

Mazumder, A. & M. D. Dickman, 1989. Factors affecting the spatial and temporal distribution of phototrophic sulfur bacteria. Arch. Hydrobiol. 116: 209–226.

McDonough, R. J., R. W. Sanders, K. G. Porter & D. L. Kirchman, 1986. Depth distribution of bacterial production in a stratified lake with an anoxic hypolimnion. Appl. envir. Microbiol. 52: 992–1000.

Murphy, J. & J. P. Riley, 1962. A modified singlesolution method for the determination of phosphate in natural waters. Analyt. chim. Acta 27: 13–19.

Overmann, J. & M. Tilzer, 1989. Control of primary productivity and the significance of photosynthetic bacteria in a meromictic kettle lake. Mittlerer Buchensee, West-Germany. Aquat. Sci. 51: 261–278.

Parkin, T. B. & T. D. Brock, 1980. The effects of light quality on the growth of photosynthetic bacteria in lakes. Arch. Mikrobiol. 125: 19–27.

Parkin, T. B. & T. D. Brock, 1981. The role of phototrophic bacteria in the sulfur cycle of a meromictic lake. Limnol. Oceanogr. 26: 880–890.

Pedrós-Alió, C., J. M. Gasol & R. Guerrero, 1987. On the ecology of a *Cryptomonas phaseolus* population forming a metalimnetic bloom in Lake Cisó, Spain: Annual distribution and loss factors. Limnol. Oceanogr. 32: 285–298.

Porter, K. G., 1973. Selective grazing and differential digestion of algae by zooplankton. Nature 244: 179–180.

Rocha, O. & A. Duncan, 1985. The relationship between cell carbon and cell volume in freshwater algal species used in zooplankton studies. J. Plankton Res. 7: 279–294.

Salonen, K. & L. Arvola, 1988. A radiotracer study of zooplankton grazing in two small humic lakes. Verh. int. Ver. Limnol. 23: 462–469.

Salonen, K. & T. Hammar, 1988. On the importance of dissolved organic matter in the nutrition of zooplankton in some lake waters. Oecologia (Berl.) 68: 246–253.

Salonen, K. & S. Jokinen, 1988. Flagellate grazing on bacteria in a small dystrophic lake. Hydrobiologia 161: 203–239.

Salonen, K. & A. Lehtovaara, 1992. Migrations of haemoglobin-rich *Daphnia longispina* in a small, steeply stratified, humic lake with an anoxic hypolimnion. Hydrobiologia 229: 271–288.

Salonen, K., L. Arvola & M. Rask, 1984a. Autumnal and vernal circulation of small forest lakes in Southern Finland. Verh. int. Ver. Limnol. 22: 103–107.

Salonen, K., L. Arvola, T. Tulonen, T. Hammar, T.-R. Metsälä, P. Kankaala & U. Münster, 1992a. Planktonic food chains of a highly humic lake. I. A mesocosm experiment during the spring primary production maximum. Hydrobiologia 229: 125–142.

Salonen, K., M. Järvinen, K. Kuoppamäki & L. Arvola, 1990. Effects of liming on the chemistry and biology of a small acid humic lake. In: Kauppi, P., P. Anttila & K. Kenttämies (eds.), Acidification in Finland. Springer, Berlin. pp. 1145–1167.

Salonen, K., R. I. Jones & I. Arvola, 1984b. Hypolimnetic phosphorus retrieval by diel vertical migrations of lake phytoplankton. Freshwat. Biol. 14: 431–438.

Salonen, K., P. Kankaala, T. Tulonen, T. Hammar, M. James, T.-R. Metsälä & L. Arvola, 1992b. Planktonic food chains of a highly humic lake. II. A mesocosm experiment in summer during dominance of heterotrophic processes. Hydrobiologia 229: 143–157.

Schiller, J., 1954. Uber winterliche pflanzliche Bewohner des Wassers, Eises und des daraufliegenden Schneebreies. I. Österr. Bot. Z. 101.

Schindler, D. W., 1988. Confusion over the origin of alkalinity in lakes. Limnol. Oceanogr. 33: 1637–1640.

Similä, A., 1988. Spring development of a *Chlamydomonas* population in lake Nimetön, a small humic lake in southern Finland. Hydrobiologia 161: 149–157.

Skogstad, A., L. Granskog & D. Klaveness, 1987. Growth of freshwater ciliates offered planktonic algae as food. J. Plankton. Res. 9: 503–512.

Smolander, U. & L. Arvola, 1988. Seasonal variation in the diel vertical distribution of the migratory alga *Cryptomonas marssonii* (Cryptophyceae) in a small, highly humic lake. Hydrobiologia 161: 89–98.

Sorokin, Y. I., 1970. Interrelations between sulphur and carbon turnover in meromictic lakes. Arch. Hydrobiol. 66: 391–446.

Steenbergen, C. L. M., H. J. Korthals & M. van Nes, 1987. Ecological observations on phototrophic sulfur bacteria and the role of these bacteria in the sulfur cycle of monomictic Lake Vechten (The Netherlands). Acta Acad. Aboensis 47: 97–115.

Takahashi, M. & S. Ichimura, 1970. Photosynthetic properties and growth of photosynthetic sulfur bacteria in lakes. Limnol. Oceanogr. 15: 929–944.

Utermöhl, H., 1958. Zur Vervollkommung der quantitativen Phytoplanktonmethodik. Mitt. int. Ver. Limnol. 9: 1–38.

Van Gemerden, H., 1980. Survival of *Chromatium vinosum* at low light intensities. Arch. Mikrobiol. 125: 115–121.

Van Gemerden, H., 1987. Competition between purple sulfur bacteria and green sulfur bacteria: Role of sulfide, sulfur and polysulfides. Acta Acad. Aboensis 47: 13–27.

Wood, E. D., F. A. J. Armstrong & F. A. Richards, 1967. Determination of nitrate in sea water by cadmium-copper reduction to nitrite. J. mar. biol. Ass. U.K. 47: 23–31.

*Hydrobiologia* **229**: 271–288, 1992.
*K. Salonen, T. Kairesalo & R.I. Jones (eds.),*
*Dissolved Organic Matter in Lacustrine Ecosystems: Energy Source and System Regulator.*



# Migrations of haemoglobin-rich *Daphnia longispina* in a small, steeply stratified, humic lake with an anoxic hypolimnion

Kalevi Salonen & Anja Lehtovaara
*Lammi Biological Station, University of Helsinki, SF-16900 Lammi, Finland*

*Key words: Daphnia longispina*, haemoglobin, vertical migration, humic lake

## Abstract

Migrations of *Daphnia longispina* were studied in a small humic lake with an exceptionally shallow oxic epilimnion. Horizontal distributions showed clear avoidance of the shoreline, which might be explained by the lower density of predators (*Chaoborus* sp. and *Notonecta* sp.) in the central parts of the lake. In early summer all size classes of *D. longispina* exhibited upward nocturnal vertical migration, descending to the upper hypolimnion in daytime. Later in summer, when the nocturnally migrating *Chaoborus* sp. had grown large enough to graze on small *Daphnia*, the latter seemed to shift towards twilight migration. However, large *Daphnia* individuals showed no synchronized migration; rather their bimodal vertical distributions suggested asynchronous vertical migration. Large individuals showed a particular tendency to concentrate near to the oxycline, close to the dense phytoplankton and bacteria populations in the upper part of the anoxic hypolimnion. According to vertical trap experiments, large *D. longispina* visited the anoxic hypolimnion and might harvest its abundant food resources. The high haemoglobin content of large individuals seems a specific adaptation to allow access to low oxygen water and hence to maximize grazing potential, in both epi- and hypolimnion, and minimize predation pressure. By staying predominantly in cooler water near the oxycline, *Daphnia* might also minimize its energy consumption to adjust to low food availability while sustaining a sufficiently high population density to exploit those unpredictable short periods with abundant food which are common in small headwater lakes. It is suggested that migrations of zooplankton are a complex behavioural adaptation which may not be explained by any single factor. In humic lakes with shallow stratification, vertical migrations seem to offer particularly high potential advantages, because of the short distances between dramatically different environments in the water column. In further studies more emphasis should be placed on migrations of individuals rather than populations, and migrations should be considered as a dynamic part of the structure and function of the whole planktonic ecosystem.

## Introduction

Vertical migration is a widespread behaviour among planktonic invertebrates, and a variety of hypotheses has been proposed to explain selective advantage(s) of this behaviour. However, with the exception of the predation hypothesis, postulating that feeding in the food rich surface waters during night but returning to dark, deep water in daytime minimizes the risk of being eaten by visual predators (Zaret & Suffern, 1976; Wright *et al.*, 1980; Gliwicz, 1986; Lampert,

1989), there is quite little agreement. Vertical migrations are found in all kinds of water bodies ranging from small ponds to the ocean. Although the basic behaviour of animals may be similar in different environments, local conditions are likely to modify patterns of vertical migration.

We have studied the migrations of a cladoceran, *Daphnia longispina*, in a small, humic, forest lake. In such lakes wind contributes little to the vertical mixing of water and the vertical distributions of light and temperature are largely determined by water colour (e.g. Jones & Arvola, 1984). High humic concentration typically leads to an anoxic hypolimnion and organisms with a variety of metabolic types may develop thin layers with high density within the oxygen gradient. In such cases zooplankton is often abundant around the oxycline (e.g. Sorokin, 1966; Hanazato *et al.*, 1989; Mazumder & Dickman, 1989). Thus, unlike in many large lakes, migration into deeper water does not mean reduced food availability. We paid particular attention to studying the adaptations of *Daphnia longispina* to cope with low or no oxygen which enable it to obtain food from the upper part of the anoxic hypolimnion or to escape predation by *Chaoborus* sp. and *Notonecta* sp. predators.

## Study site

Mekkojärvi is a small and shallow (area 0.35 ha, mean depth 3 m), highly humic lake in southern Finland, surrounded by coniferous forest. The shoreline is composed of floating *Sphagnum* and floating Warnstortia moss and hence there is no shallow littoral. The small size and very brown water (colour 300–600 mg Pt $l^{-1}$) result in steep thermal and oxygen stratification. Annual primary production is below 10 g C $m^{-2}$. Some physical, chemical and other biological properties of the lake are described elsewhere in this volume (Salonen *et al.*, 1992a; Salonen *et al.*, 1992b; Kuuppo-Leinikki & Salonen, 1992; Münster *et al.*, 1992).

*Daphnia longispina* is very abundant in Mekkojärvi throughout the summer and dominates the zooplankton biomass almost exclusively. The few other species occur only occasionally and in summer their biomass is low compared with *D. longispina*. With the exception of very young pike, there are no planktivorous fish in the lake. Therefore the most important invertebrate predators are *Notonecta* sp. and larvae of *Chaoborus* sp.

## Material and methods

In the first stage of the study, vertical zooplankton samples were collected at 6 h intervals over a 24 h period on 5 sites along the lake (Fig. 1). Water was sampled from the surface to 2 m with a 0.5 m vertical tube sampler (volume 7.7 l) with lids opening so that water can freely pass through the sampler when it is lowered. The samples from each site were counted separately. Zooplankton was sieved through a 50 μm net, washed into bottles and preserved in sugar-formaldehyde solution (Haney & Hall, 1973). In the laboratory all species, except rotifers, were counted under a Wild M5 dissection microscope.

Because of large variation in the total numbers of individuals between the sampling sites and times, and high numbers of *Daphnia* found in the night samples, we also studied the horizontal dis-

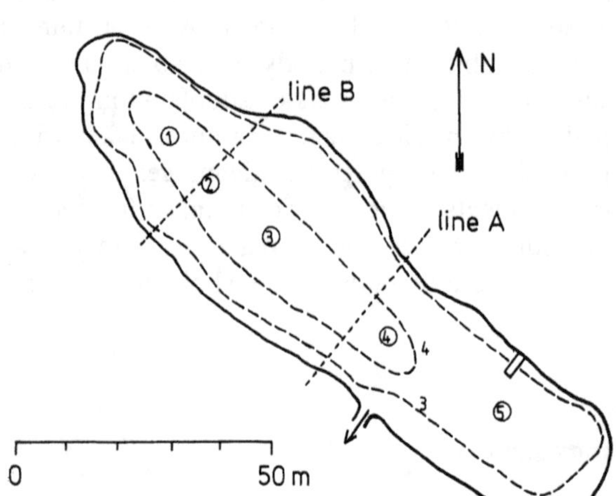


*Fig. 1.* Bathymetric map of Mekkojärvi (61° 17′ N, 25° 7′ E). The sampling sites (circled) of the first vertical study and the transects used in the horizontal study are shown.

tribution of *Daphnia* over one 24 h period in July 1985. Two lines (A: 26 m with 7 sampling points and B: 34 m with 8 sampling points; Fig. 1) were marked by string stretched across the lake ca. 1 m up from the surface. The first and last marks were 1–2 m from the shoreline. We sampled the depth zone 0–2 m (but at the shoreline 0–1 m) at 4 hour intervals, using a 1 m long (volume 6.2 l) tube sampler as described above.

In 1986 we built a new sampler (Fig. 2) which allowed more detailed vertical sampling of plankton. It has four 0.3 m long square section tubes with 100 × 100 mm opening (volume 4.5 l) stacked on each other. The tube assemblage is fixed to a vertical metal rod which can be immersed to the desired depth. The vertical rod is connected to a horizontal one so that the sampler can be moved so far (ca. 2 m) in front of the boat that water movements caused by the boat do not disturb the vertical distribution. The boat was rowed slowly forward and water moved freely through the open lids. After some time the closing mechanism was released by pulling a thread. Then rubber bands closed all the lids simultaneously and the tube assemblage was taken into the boat. Water from each tube was drained through a 100 $\mu$m plankton net and preserved. Care was taken always to sample water where the boat had not moved. Therefore water was sampled along longitudinal lines within the 4 m depth curve (Fig. 1). Three samples from each 10 cm depth zone taken at different localities were pooled. Altogether 20 pooled samples were taken between the surface and 2 m. Sampling of one set lasted 1–1.5 hours.

In the laboratory all animals in the zooplankton samples were counted. In addition, 100–150

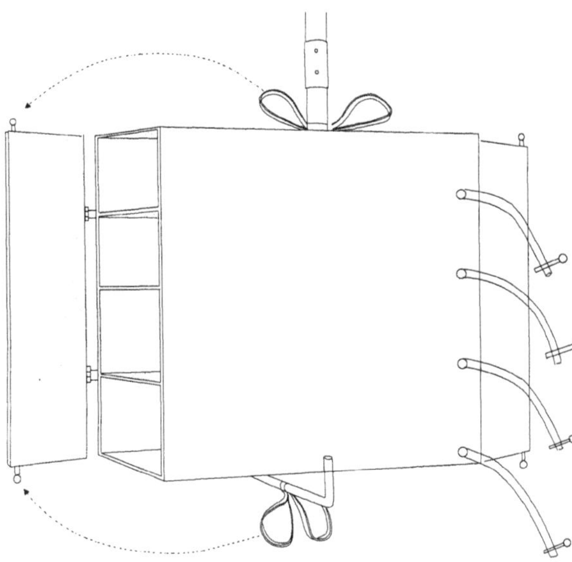

*Fig. 2.* Horizontal tube sampler. The trigger mechanism behind the box releases the threads which keep the sampler loaded and the rubber bands close to lids.

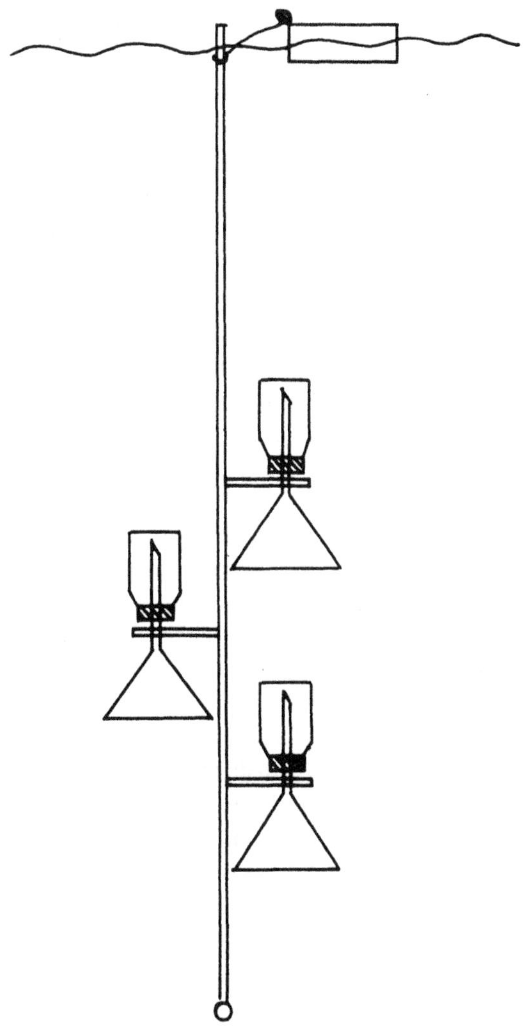

*Fig. 3.* Schematic view of the traps used to collect upswimming zooplankton.

animals selected at random from each sample were measured under a dissection microscope at 25× magnification and with a precision of 0.04 mm. The total length was measured from the top of the head to the base of the carapace spine.

To test if *Daphnia* visit the anoxic hypolimnion, we used vertical traps to collect upswimming individuals. These traps consisted of inverted glass funnels anchored to a suspending rod system (resembling the funnels used by Landon & Stasiak, 1983). Small (100 ml transparent plastic bottles were attached on the top of the funnel tube (Fig. 3). Three funnels were fixed to a rod at 0.1 or 0.2 m intervals alternately on different sides of the rod. Two rods were suspended in the lake so that traps covered a 0.5 or 1 m depth zone. The bottom of the bottles was of 300 $\mu$m plankton net, which was dense enough to retain young individuals. The funnels were filled with filtered water from the lake surface and closed with a plastic film kept in place with rubber bands. The rods with funnels full of water were lowered to the desired depth and the plastic films were drawn away after releasing the rubber bands. Thus we excluded the possibility that zooplankton could enter the traps during their installation. The funnels were fixed to a buoy at the desired depth and kept in the lake 1–2 days. At the end of the experiment the traps were removed and zooplankton in the bottles were washed out and preserved in sugar-formaldehyde. Whenever traps were suspended in the lake we also collected (as described above for the sampling of along the horizontal lines across the lake) at 5 sites altogether 62 l pooled zooplankton samples covering the depth 0–2 m.

Haemoglobin in *Daphnia* was determined according to Landon & Stasiak (1983). The animals were frozen in small plastic tubes in lake water. Haemoglobin of 30–100 individuals of about the same size was measured with a Shimadzu UV 240 spectrophotometer. Crystallized rabbit haemoglobin (H-7255, Sigma Chemical Company) was used as a standard. The haemoglobin content of *Daphnia* was calculated per individual dry mass, which was obtained as a mean of 15–40 individuals, dried for 24 hours at 60 °C and weighed on an electronic microbalance at a precision of 0.001 mg.

Temperature and oxygen concentrations were measured at every sampling with a combined probe (Yellow Spring Instruments). Light penetration into water was measured with a quantum sensor and integrator (Lambda Instruments, Li 2190). Samples for chemical determinations and bacterio- and phytoplankton were collected from water filtrate obtained in sampling zooplankton (Arvola *et al.*, 1991).

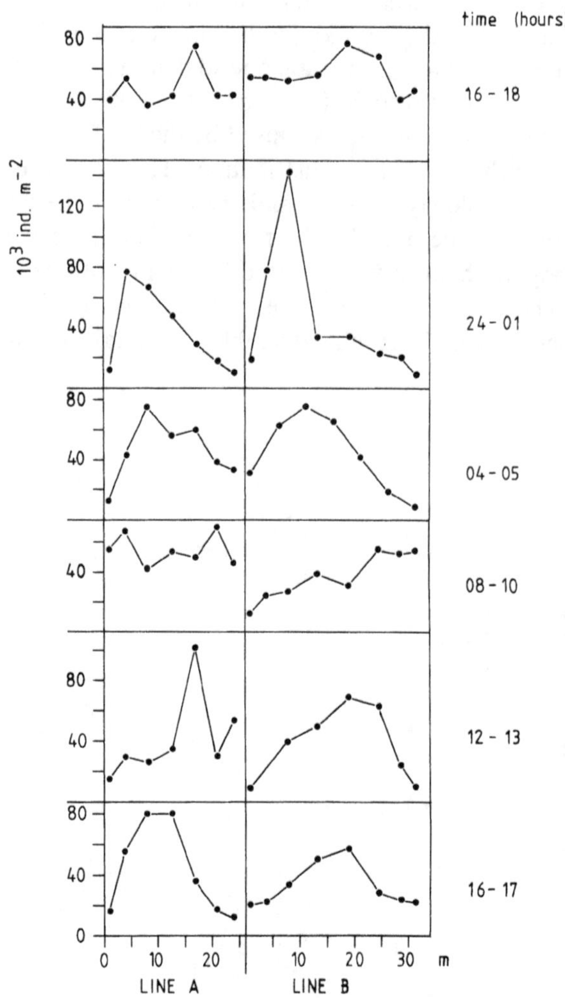


*Fig. 4.* Diel horizontal distribution of *Daphnia longispina* in July 1985 along two transects over Mekkojärvi. The distances are measured from the northeastern side of the lake.

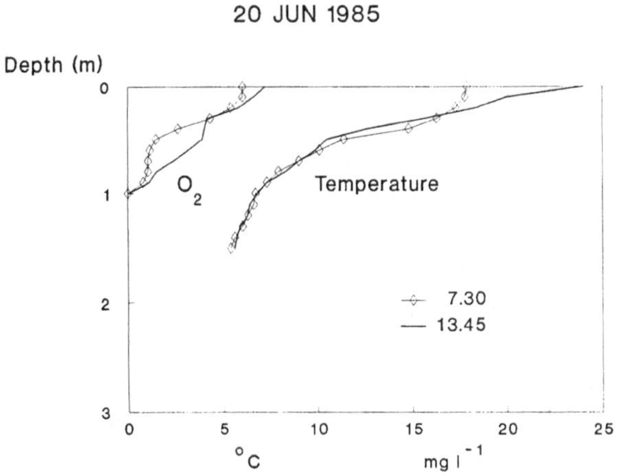


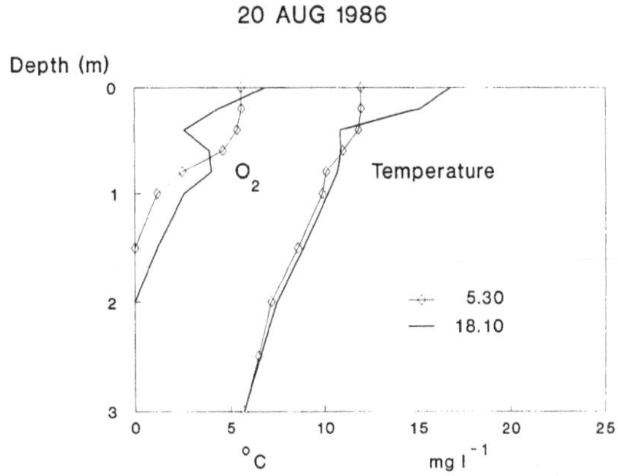


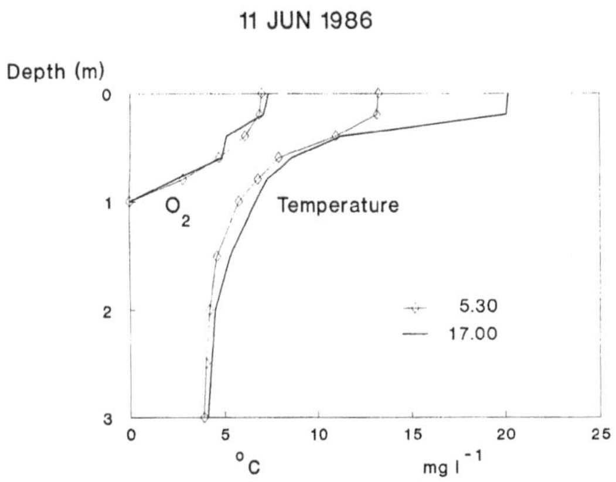


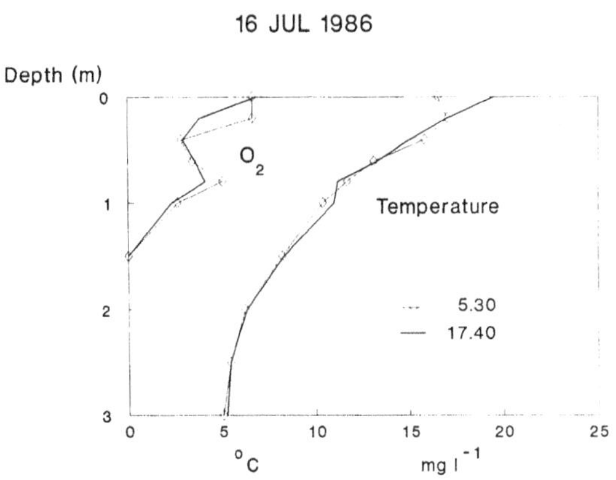


*Fig. 5.* Vertical distributions of temperature and oxygen in the water column during the vertical migration studies. Only distributions during the minimum (diamonds) and maximum surface temperature are shown with the time of their occurrence.

## Results

### *Horizontal distribution*

During the time of the horizontal migration study in July 1985, *Daphnia longispina* was by far the most abundant zooplankton species in terms of biomass. Among other cladocerans *Polyphemus pediculus* was present at low density. *Chaoborus* larvae and *Notonecta* sp. were probably the main invertebrate predators. Most *D. longispina* were in the 0–1 m water layer (mean 92% and minimum 87% of that in the 0–2 m water layer). Therefore although the sampling points nearest to the shoreline extended only to the depth of 1 m rather than the depth 2 m at other sampling points, this does not significantly bias the observed horizontal distributions. The mean densities of *D. longispina* along both transects over the lake were similar (line A: $43.3 \pm 5.6 \times 10^3$ ind. $m^{-2}$, line B: $42.3 \pm 8.6 \times 10^3$ ind. $m^{-2}$; both with SD). With the exception of samplings at 16–18 and 8–10, the horizontal distributions (Fig. 4) demonstrated a tendency for *D. longispina* to avoid water near to the shoreline. The two sample sets taken in the evening on successive days (Fig. 4) showed quite a marked difference, which might have been caused by weather conditions. The results along the transects were different at each sampling time ($P < 0.01$; Kolgomorov-Smirnov test, equalising the number of observation points arbitrarily by omitting the fifth observation from line B)), but the distributions do suggest some agreement and possibly diel horizontal migration which might be explained by the development of illumination and the direction of sunlight. The verification of diel horizontal migration in Mekkojärvi would, however, have required several successive sampling days. During the experiment wind was low and at night and in the morning conditions were completely calm and cloudless, but in spite of this the movements of water masses in the shallow epilimnion might still have been significant and led to differences in the observed horizontal distributions.

*Vertical distribution*

In June 1985 Mekkojärvi was so steeply stratified (Fig. 5) that at the warmest time of the day there was no homothermic epilimnion at all. Although the concentration of oxygen was low below the uppermost 0.3 m (Fig. 5), *Daphnia longispina* inhabited the whole oxygenated zone down to the upper anoxic hypolimnion and probably migrated towards the surface at night (Fig. 6). However, the high variability of the results and rather crude resolution of 0.5 m layer samples makes the interpretation difficult. Further difficulties arose from the considerably higher total number of animals in the vertical column at night ($99 \times 10^3$ ind. $m^{-2}$) than at other sampling times ($36–67 \times 10^3$ ind. $m^{-2}$), indicating either horizontal migration or differences in the efficiency of the sampling under different illuminations.

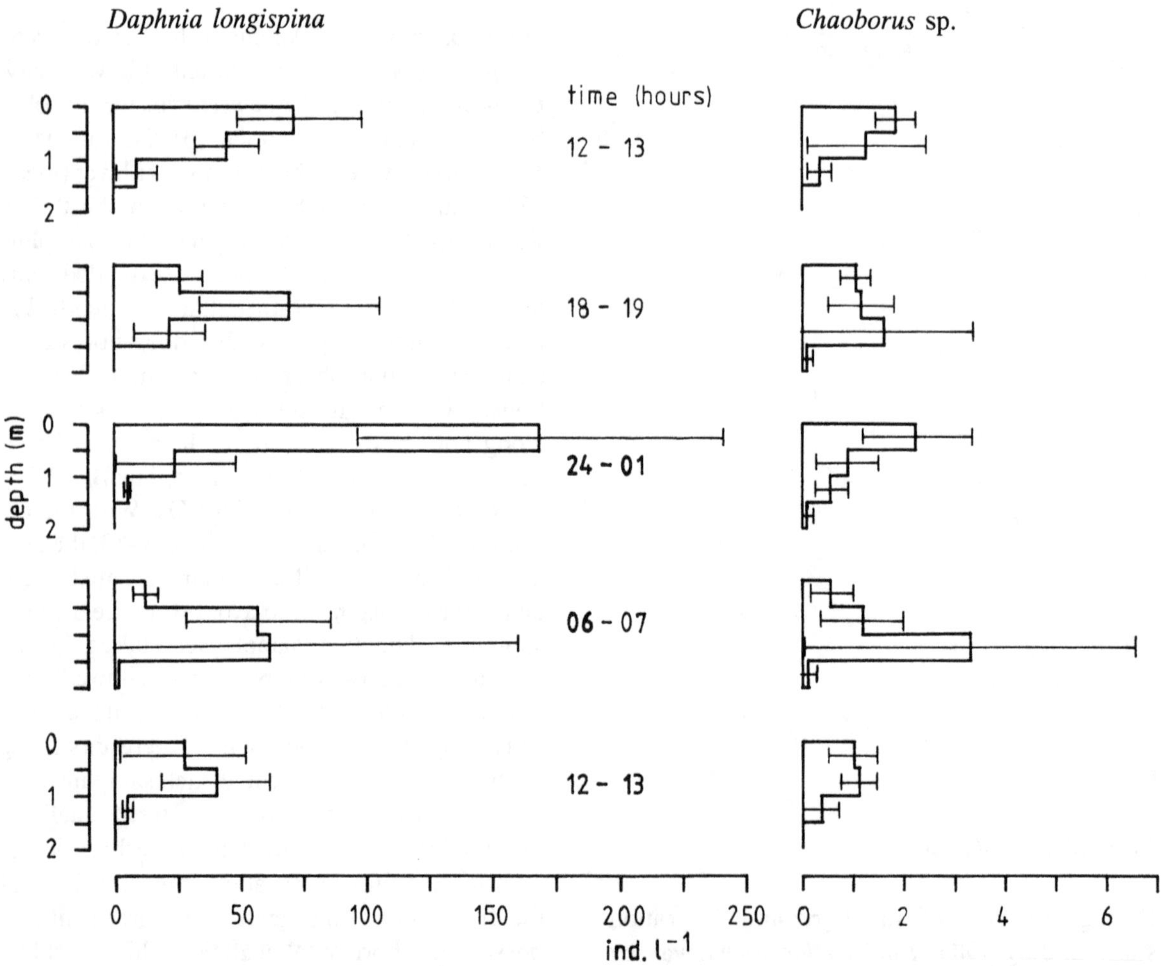


*Fig. 6.* Diel vertical distributions (with SD) of *Daphnia longispina* and *Chaoborus* sp. in June 1985.

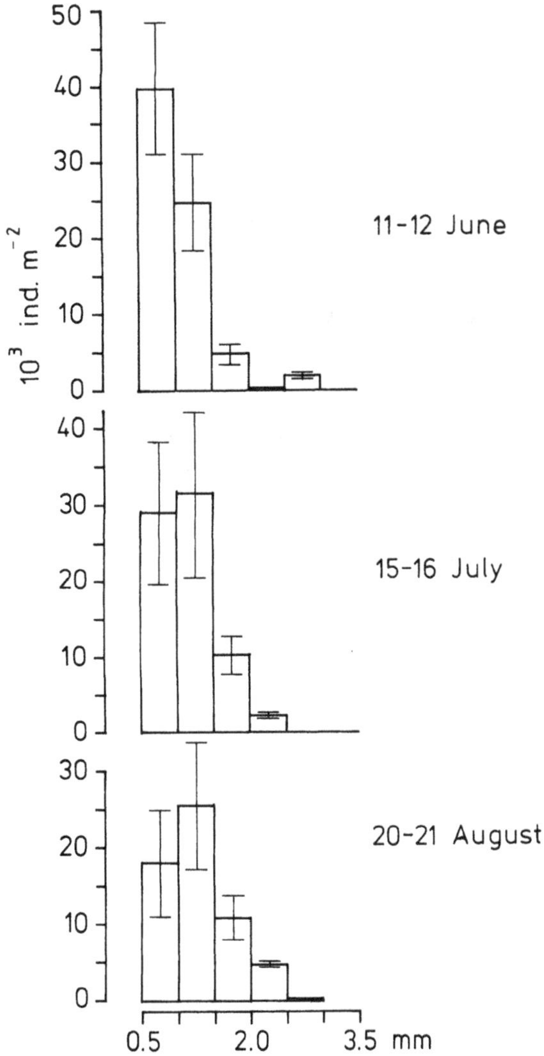


*Fig. 7.* Mean size distributions (with SD) of *Daphnia longispina* populations in diel samples in 1986.

In June 1986 *Daphnia longispina* again dominated the biomass, but nonmigrating *Cyclops* sp. was also abundant, consisting of naupliar and C III–IV copepodite stages with a maximum at ca. 1 m (110–220 ind. $l^{-1}$ for both). The *D. longispina* population was mainly young females, but there were also large females hatched from ephippia (Fig. 7).

The smoothness of vertical distributions obtained with our high resolution horizontal tube sampler in 1986 (Fig. 8) indicates a high reliability for the results. This may be explained by the smoothing effect of pooled samples and minimal disturbance of plankton by boat during sampling, unlike during sampling by the 0.5 m vertical tube sampler. The variation (CV 14–29%) between the diel sampling times was moderate, but again samples taken at solar noon generally yielded the lowest densities (Table 1). The sampling in June was an exception; then the minimum was found in the evening, but the maximum was still at midnight. The differences were probably caused by different horizontal distributions at different times of the day. For example, the diel changes in the horizontal distributions observed in late July 1985 would have resulted in similar variation in observed numbers as was found in June 1986.

*Table 1.* The observed relative abundances (%) of *Daphnia longispina* at different times of day in 1986. Densities at noon were taken as a reference.

| Date | Evening | Night | Morning | Noon | Evening | Night |
|---|---|---|---|---|---|---|
| 11–12 June | – | 131 | 89 | 100 | 81 | 132 |
| 15–16 July | 116 | 157 | 216 | 100 | 156 | – |
| 20–21 August | 110 | 131 | 131 | 100 | 135 | – |

In June, when daylight hours were nearly at their longest and solar radiation most intensive (Fig. 9), *Daphnia longispina* exhibited a typical pattern of diel vertical migration. At night >70% of the individuals were concentrated at the depth of 0–0.3 m and their maximum density was very high, more than 300 ind. $l^{-1}$. At other times only 1–3% of the individuals appeared at this zone while the maximum density was observed at 0.7–1.0 m, immediately above anoxic water (Fig. 5), and corresponding to ca. 1% of the surface light intensity at noon (Fig. 9). Large females (>1.5 mm) migrated similarly, but occurred in deeper water layers (Fig. 8). Although the observed vertical migration was pronounced, the results might still have suffered from possible diel horizontal migration (Fig. 4) or efficiency of sampling. All size classes clearly avoided the uppermost water layers, but it was most striking among ovigerous females (Figs. 10–11). The difference in temperature between the surface and 1 m was high (ca. 14 °C) but because of the ca. 7 °C diurnal variation of temperature, the daily temper-

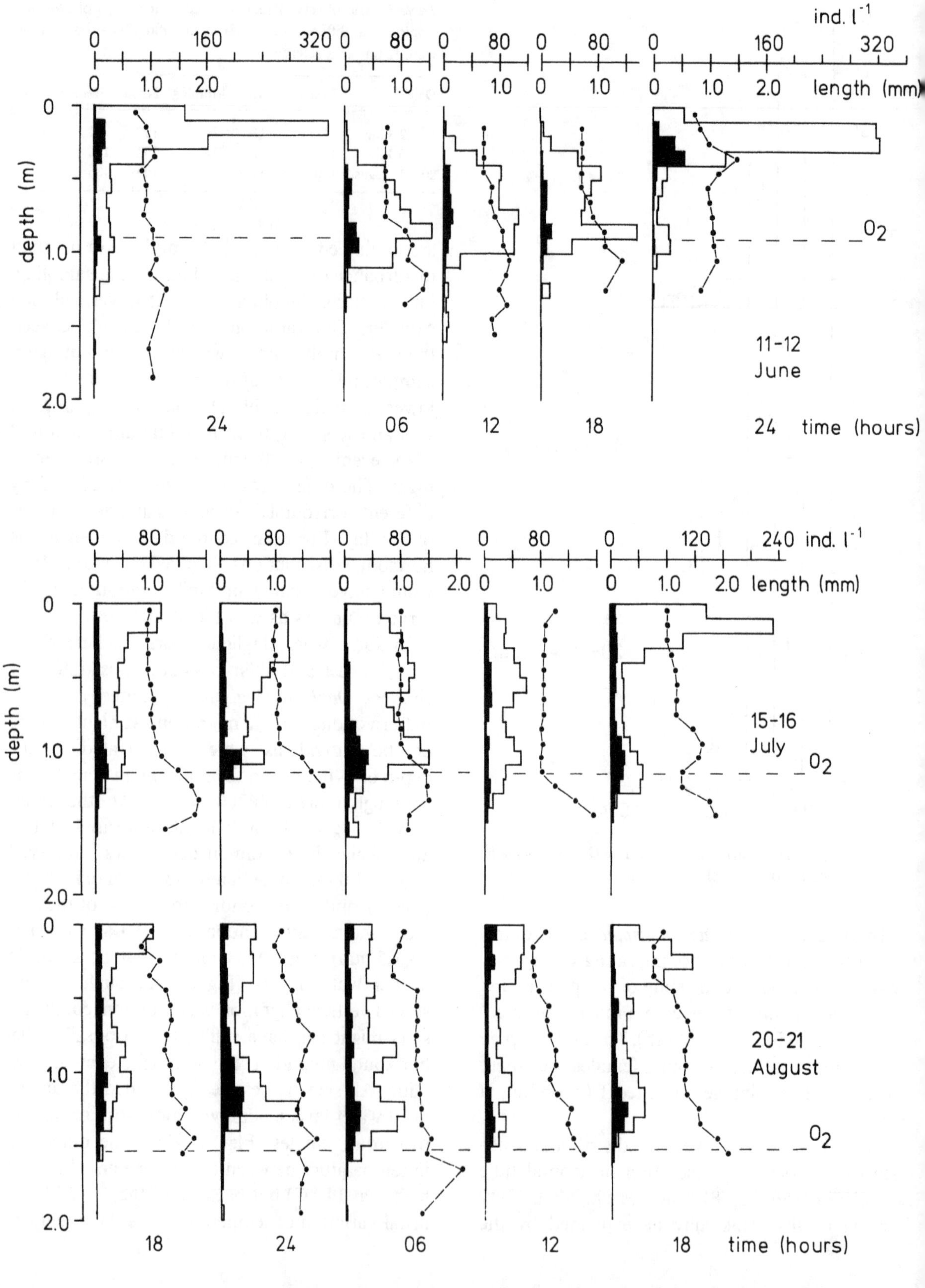

ind. l⁻¹
0 160 320
0 80
0 80
0 80
0 160 320
0 1.0 2.0
length (mm)
depth (m)
0
1.0
2.0
O2
11-12
June
24 06 12 18 24 time (hours)
0 80
0 80
0 80
0 80
0 120 240 ind. l⁻¹
0 1.0 2.0 length (mm)
15-16
July
20-21
August
18 24 06 12 18 time (hours)

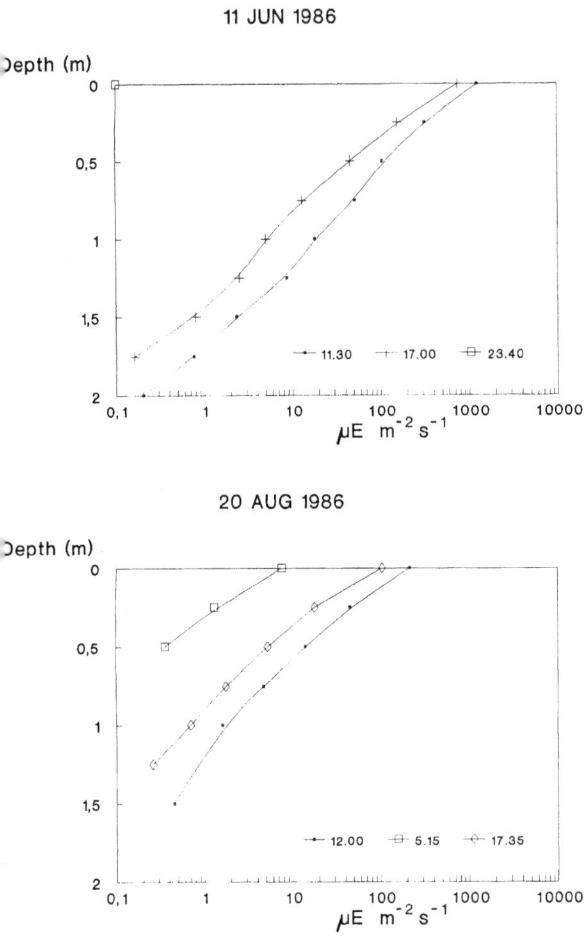


*Fig. 9.* Some vertical distributions of light intensity at different times of day in the water column during the vertical migration studies.

ature gradient calculated from the modal depths in the distributions of animals was in practise only ca. 10 °C.

In August *Daphnia longispina* exclusively dominated the community. Later in summer the nights got longer and light intensity decreased. During the sampling periods in July and August the weather was colder leading to an increase in the depth of the epilimnion and decrease in its temperature (Fig. 5). Thus the temperature gradient between the surface and 1 m was only 8 °C in

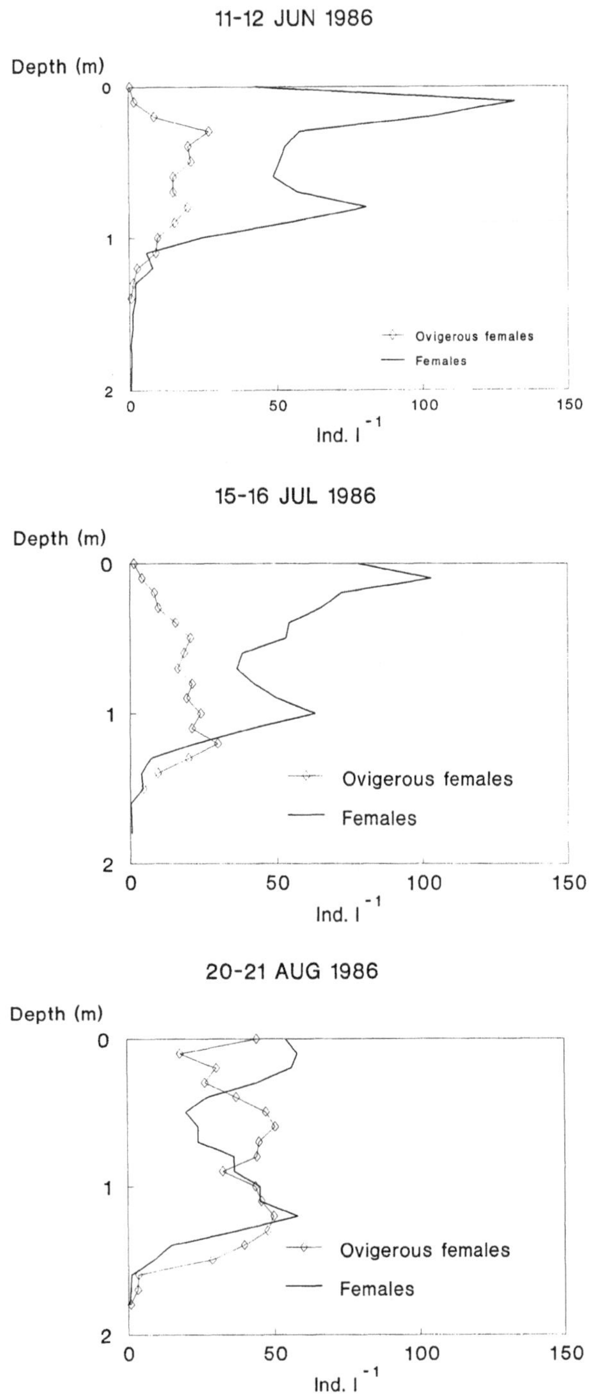


*Fig. 10.* Mean daily vertical distributions of *Daphnia longispina* and its ovigerous females. The arrows at the time scale denote sunrise and sunset.

*Fig. 8.* Diel vertical distributions of *Daphnia longispina* in the summer 1986. The broken horizontal lines indicate the oxygen limit. Open bars – Whole population; Black bars – > 1.5 mm long animals; Dots – Mean length of animals.

*Table 2.* Some mean population parameters for *Daphnia longispina* (with SD for examples taken at different times of the day).

| | 11–12 June | 15–16 July | 20–21 August |
|---|---|---|---|
| $10^3$ Females $m^{-2}$ | 72 ± 16 | 73 ± 21 | 59 ± 8 |
| $10^3$ Males $m^{-2}$ | 0 | 0.22 ± 0.11 | 5.2 ± 1.3 |
| $10^3$ Eggs $m^{-2}$ | 16 ± 11 | 9 ± 3 | 27 ± 2 |
| Eggs per ovigerous female | 7.6 ± 1.4 | 3.4 ± 0.2 | 4.3 ± 0.1 |

July and 6 °C in August. The population structure was no longer dominated by smallest individuals (Fig. 7). Few males were present in July, but in August they formed 8% of the population (Table 2). In July and August, diel vertical migration of *D. longispina* was still evident (Fig. 8), but less striking than in June. The vertical distributions of juveniles suggested a shift towards twilight migration. Large females did not show any synchronized vertical migration and again they occurred in deeper water than juveniles. Particularly large females were found near the oxycline (Fig. 8) and many of them were ovigerous (Fig. 11). The difference in temperature between the epilimnetic and deep water maxima of *D. longispina* was 6–7 °C in July and 2.5–4.5 °C in August depending on the time of the day.

All vertical distributions of *Daphnia longispina* showed a bimodality (although slight in June) (Fig. 8) indicating asynchronus vertical migration (Pearre, 1979). Bimodality was strikingly evident in the mean daily vertical distribution (Fig. 11) irrespective of different vertical migrations between June and late summer. However, it could not be observed in the vertical distribution of ovigerous females which showed a tendency to avoid surface water layers. In spite of that, in June most ovigerous females were in the epilimnion. Later in summer the situation was more or less reversed and most ovigerous females occurred in the lower part of the oxic water column.

In June in 1985 (Fig. 6) and 1986 (Figs. 8, 12) the vertical migrations of *Daphnia longispina* and *Chaoborus* larvae were similar. Taking into account the large variations and some inconsistency in the diel distributions (Fig. 6), this was particularly striking. The larvae were small (only 2–3 mm long) and their density was high (800 ind. $m^{-2}$). In July *Chaoborus* larvae were mostly large (8–11 mm long), and their density had decreased to about 270 ind. $m^{-2}$. Although *D. longispina* in late summer changed its migration behaviour (Fig. 8), *Chaoborus* continued nocturnal vertical migration. In late summer they seemed to continue the same behaviour, but their density was so low (170 ind. $m^{-2}$) that the number of animals counted in samples makes the interpretation uncertain.

Vertical migrations and distributions of bacte-

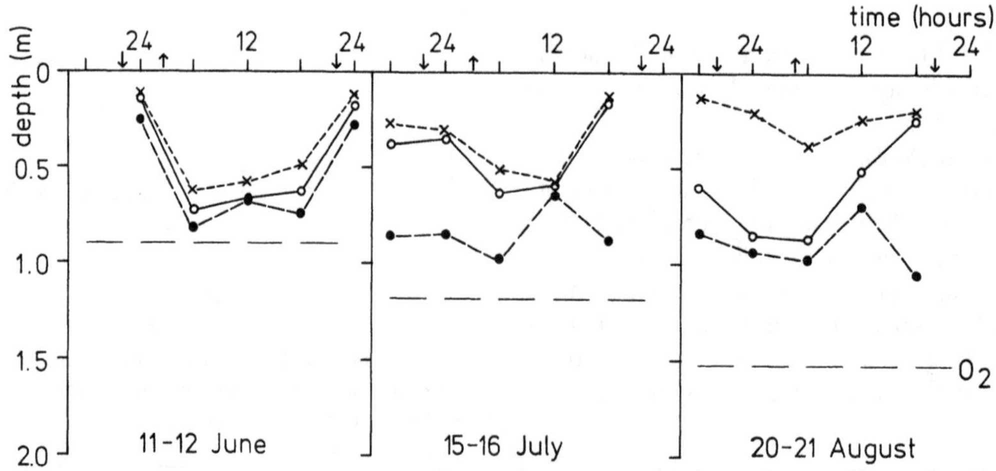


*Fig. 11.* The mean vertical positions of different size classes of *Daphnia longispina* in diel samples in 1986. Circles – All animals; Dots – > 1.5 mm long animals; Crosses – < 1 mm long animals. The broken line shows the oxygen limit.

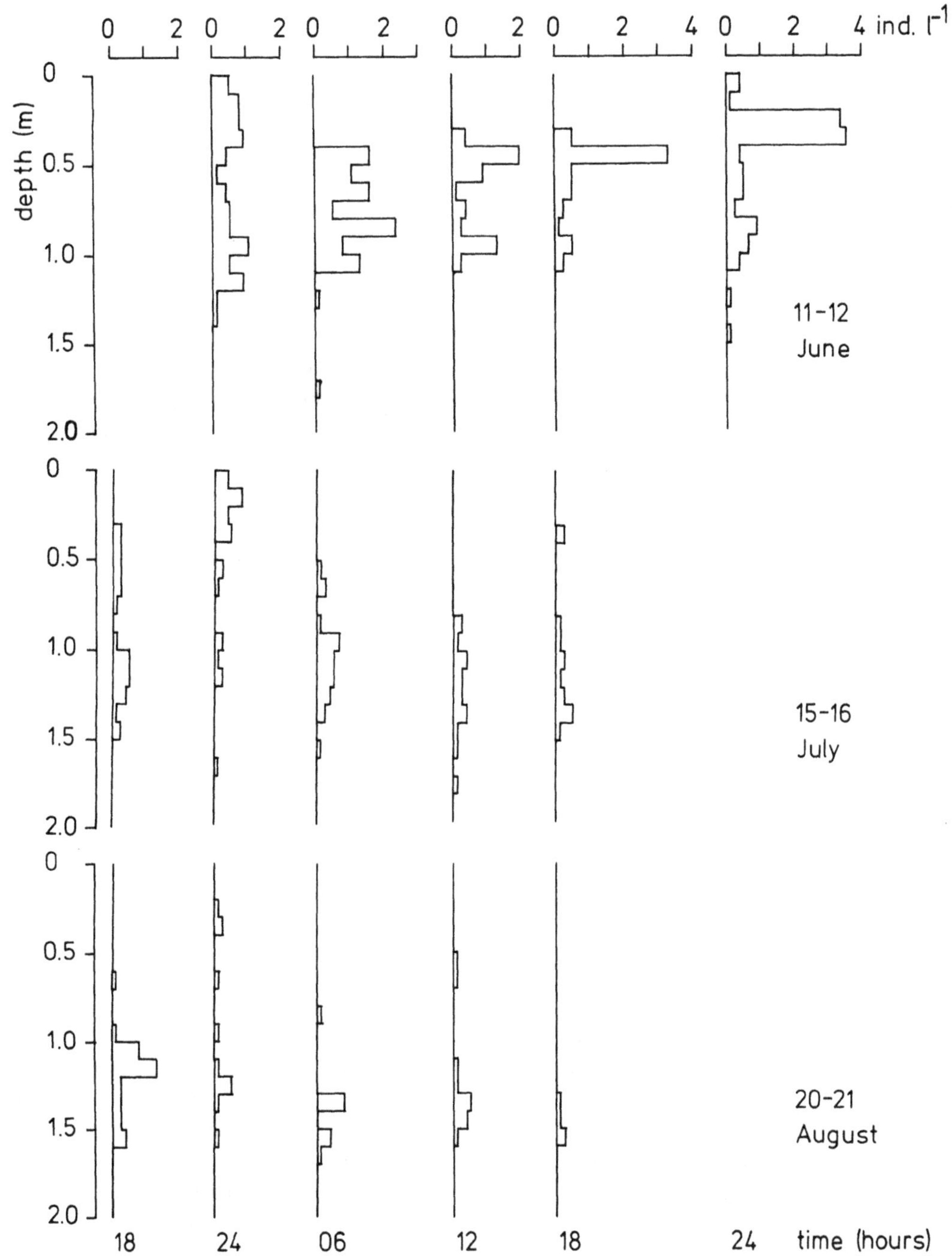


*Fig. 12.* Diel vertical distributions of *Chaoborus* sp. in the summer 1986. The broken horizontal lines indicate the oxygen limit.

ria and phytoplankton were studied using the same samples (Arvola *et al.*, 1991). During all three sampling periods in 1986, bacteria and algae were most abundant throughout the day in the upper hypolimnion (Fig. 13; Table 3). Small algae, *Chlamydomonas* sp. and *Scourfieldia cordiformis*, were particularly abundant near the deep water maximum of *Daphnia longispina*. The vertical migrations of algae are reported in more detail by Arvola *et al.* (1992).

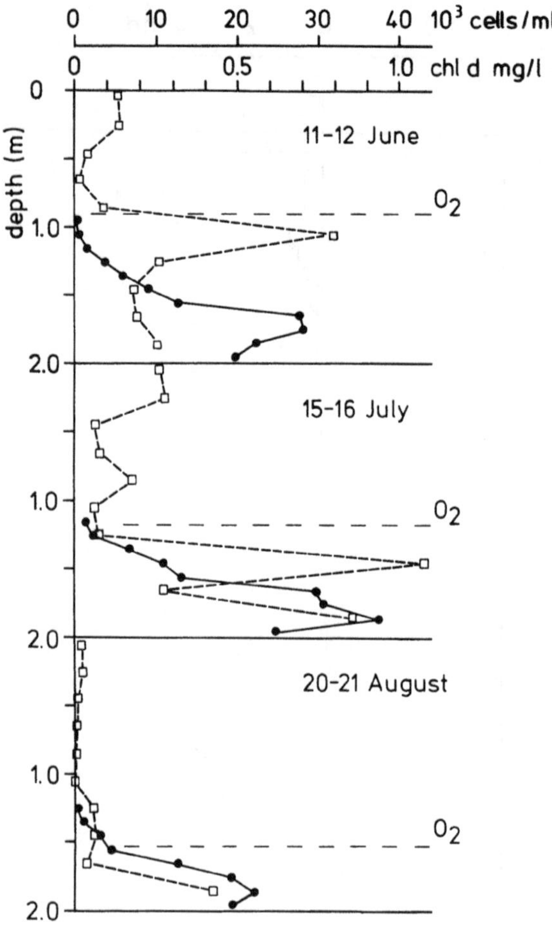


*Fig. 13.* Vertical distributions of phyto- and bacterioplankton at noon in Mekkojärvi in 1986. Squares – Density of phytoplankton; Dots – Concentration of bacteriochlorophyll *d*.

Vertical traps showed that *Daphnia longispina* can visit the upper anoxic hypolimnion at least 0.5 m below the oxycline. The results of successive periods showed high variation in the numbers of animals caught (Fig. 14) and hence the visit rates of animals at various depths cannot be reliably estimated. Most of the animals caught were large females while in the whole population most individuals were small (Fig. 15). The latter looked pale and contained less than 10% haemoglobin in their dry mass (Fig. 16), but large animals were dark brown and their haemoglobin concentration was up to one third of dry mass.

In August 1985 we determined chlorophyll extracted from *Daphnia longispina* collected from

*Table 3.* Mean daily biomass (g $m^{-3}$) of algae and bacteria in the upper water layers of Mekkojärvi in 1986. Cell volumes were converted to carbon using factors 0.2 and 0.38 pg $\mu m^{-3}$ for algae and bacteria, respectively.

| | Algae | | Bacteria | | Alga + bacteria | |
|---|---|---|---|---|---|---|
| | 0–1 m | 1–2 m | 0–1 m | 1–2 m | 0–1 m | 1–2 m |
| 11–12 June | 0.19 | 0.55 | 0.44 | 3.21 | 0.63 | 3.77 |
| 15–16 July | 0.69 | 2.91 | 0.69 | 3.92 | 1.67 | 6.83 |
| 20–21 August | 0.31 | 0.21 | 0.56 | 1.99 | 0.86 | 2.20 |

0.5 m depth to assess possible feeding on photosynthetic bacteria. Although located on the tail of the high haemoglobin absorbtion curve, a clear absorption peak was observed at 653 nm, which coincides with the absorption maximum of bacteriochlorophyll *d* and indicates that the animals had ingested photosynthetic bacteria.

## Discussion

Diel vertical distributions observed in nature are often difficult to interpret. Water masses at the sampling site may move, the distribution of organisms is generally patchy and the efficiency of sampling may vary according to weather conditions. In addition to sampling problems, the vertical migration behaviour itself may change due to rather small changes in the environment. Hence it is very tedious to collect and count enough samples to get statistically unambiguous results. In Mekkojärvi we demonstrated horizontal variation in the distribution of *Daphnia longispina*, and the results indicated some regularity in the distribution rather than random patchiness. This probably resulted from the low turbulence and the absence of waves in this small and sheltered lake. Similarly the diel vertical samples taken with the high resolution sampler showed no major irregularities in the distributions of *D. longispina*. Thus, although no replicate samples were counted, the smooth vertical distributions shown by the pooled samples seem to provide a reasonable basis for the interpretation of the results.

In very small lakes horizontal migrations are certainly realistic in terms of the swimming speed

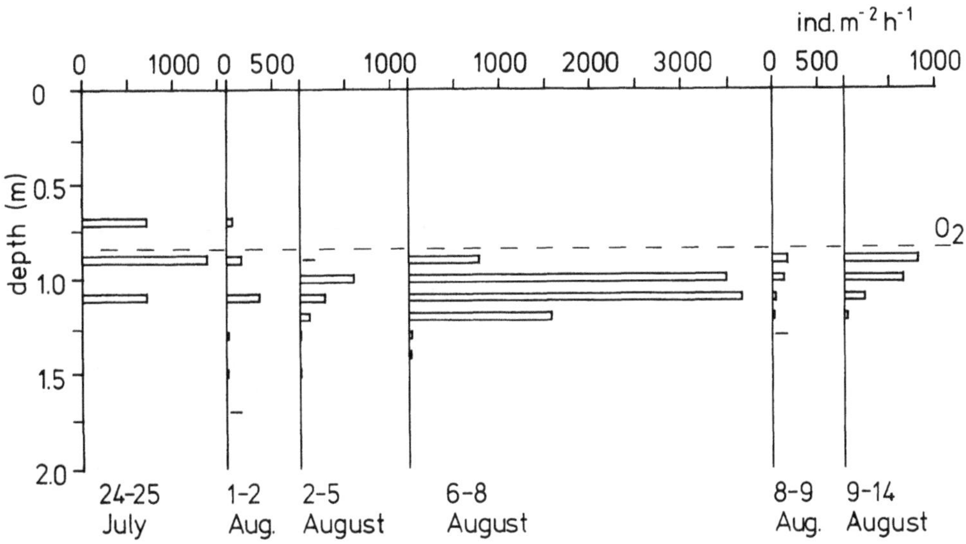


*Fig. 14.* The catches of *Daphnia longispina* in traps suspended near or below the oxycline.

of *Daphnia* (e.g. Hutchinson, 1967; Gerritsen, 1982). In Mekkojärvi the horizontal migration of *Daphnia longispina* was not as clear as found by Davies (1985) for *D. hyalina lacustris* which, in daytime, moved towards the shoreline. He explained horizontal migration as an avoidance of fish predation in open water. Although the shoreline avoidance observed in Mekkojärvi is seemingly opposite to the results of Davies, it can also be explained by predation, because in the middle of the lake there was no fish predation, whereas invertebrate beetle as well as heteropteran (*Notonecta* sp.) predators were visibly much more abundant near the shoreline.

*Daphnia* populations frequently occur in anoxic or nearly anoxic water (e.g. Sorokin, 1966; Mazumder & Dickman, 1989; Prepas & Rigler, 1978; Murtaugh, 1985) and their ability to synthesize haemoglobin is common. The presence of haemoglobin has been proposed as an adaptation to unpredictable environments where the oxygen concentration may suddenly drop to lethally low levels (e.g. Heisey & Porter, 1977; Weider & Lampert, 1985). This explanation is unlikely for

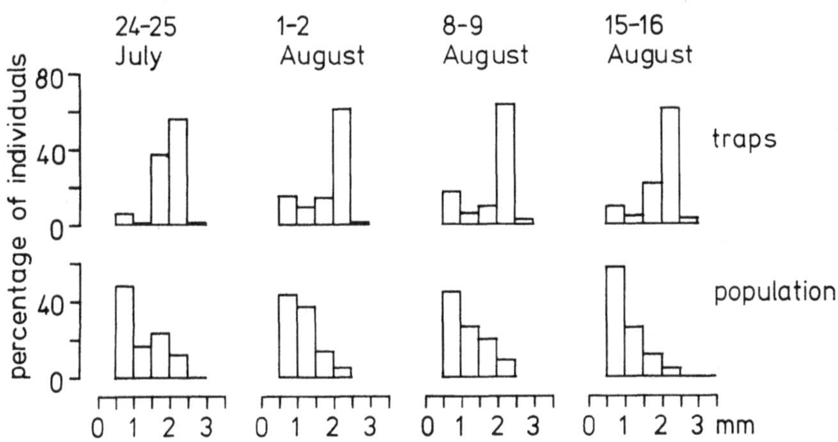


*Fig. 15.* Size distributions of *Daphnia longispina* population in Mekkojärvi and in the catches caught by the vertical traps.

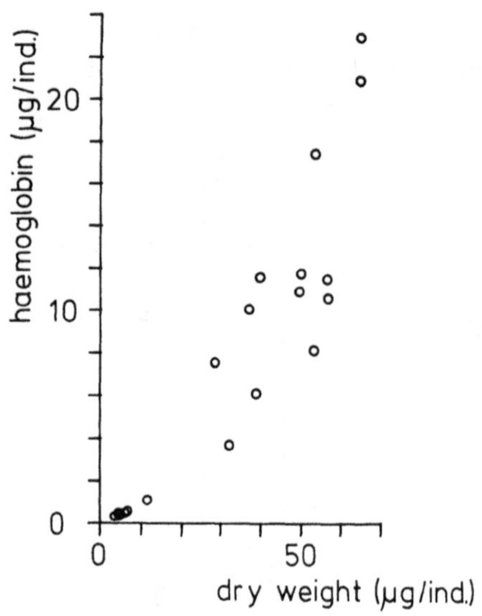


*Fig. 16.* The haemoglobin content as a function of the size of *Daphnia longispina.*

Mekkojärvi where the summer stratification of oxygen is stable. During 6 summers of study we have never seen any sign of mixing of water that would turn the epilimnion anoxic. In Mekkojärvi the haemoglobin concentrations were about ten times higher than those observed in the experiments of Weider & Lampert (1985) and Kring & O'Brien (1976), manifesting the importance of the ability to tolerate low oxygen concentration. Similarly high values have been reported only for *Daphnia* overwintering in very poorly oxygenated water (Landon & Stasiak, 1983) and for *D. pulex* (Engle, 1985) in a pond without vertebrate predators.

High haemoglobin concentration may yield several advantages for *Daphnia*. Weider & Lampert (1985) found *Daphnia pulex* to be able to regulate oxygen consumption across a wide range of oxygen concentrations down to a level of 0.5–1 mg $O_2$ $l^{-1}$. In Mekkojärvi, *D. longispina*, with an order of magnitude higher haemoglobin concentration, was probably able to tolerate even lower concentrations. Evidently a low oxygen level provides an even better refuge from predators than darkness alone. Analogous to Mekkojärvi, Hanazato *et al.* (1989) found *D. longispina* to escape fish predation by retreat into the anoxic water of Lake Yonoko. With high haemoglobin filtering rates are also higher (Fox *et al.*, 1951; Kring & O'Brien, 1976). Thus even in the aerobic epilimnion the feeding rate of haemoglobin rich *D. longispina* can be 2–3 times higher than in the absence of haemoglobin. Therefore animals remaining a long time in deep water without feeding might still be able to collect as much food as their counterparts, without haemoglobin, living continuously in the epilimnion. Haney (1985) observed higher filtering rates of several cladoceran species at night than in daytime and this was most pronounced in large individuals. Although there is no information about the haemoglobin level in those species, his results suggest that vertically migrating animals may in general be able to compensate for the time spent in deeper water, perhaps with low food concentration, by more intensive filtering. Combined with high feeding rate, vertical migration, irrespective of whether it is synchronous or not, may then provide a scanning mechanism through the water column to detect and rapidly utilize the most favourable food patches. If haemoglobin-containing animals are such superior filter feeders and they do not have essential costs from vertical migration, any advantages of vertical migration are amplified. This raises the question of why cladocerans everywhere do not take advantage of the benefits of high haemoglobin concentration. Kring & O'Brien (1976) and Engle (1985) suggested the higher visual predation of pigmented animals as the most probable explanation. In the very brown water of Mekkojärvi, the high visibility of strongly pigmented animals is less disadvantageous.

Lampert (1989) concluded that vertical migration may not evolve without light dependent mortality (visual predation) of zooplankton. Ohman (1990) has extended the predation hypothesis by differentiating between visual and nonvisual predation and was able to explain large variation in diel migration behaviour of the same species ranging from reversed vertical migration to the typical nocturnal one. When mortality by visually hunting predators is dominating, nocturnal migration is adopted, but when mortality is mainly caused by nonvisually hunting predators (e.g.

predatory zooplankton) reversed migration may be adopted. In Mekkojärvi, *D. longispina* had to adjust its behaviour to meet mortality caused by two predators with different behaviour, *Notonecta* sp. and *Chaoborus* sp. larvae. Dodson (1988) found that the chemical stimulus from these predators caused completely different behaviour of several *Daphnia* species. Contact with the stimulus from the visual predator, *Notonecta* (e.g. Dodson & Havel, 1988; Dodson, 1989) leads into the sinking of *Daphnia* within less than one hour, while the presence of *Chaoborus* causes escape upwards. In June, *Notonecta* sp. was likely the main predator, because *Chaoborus* larvae were probably too small to prey on *D. longispina* (Moore, 1988; Vanni, 1988). Therefore the identical diel migration patterns of *Chaoborus* and *D. longispina* do not conflict with the predation avoidance hypothesis. Later in summer *Chaoborus*, which continued to migrate nocturnally, got larger and, in spite of its lower abundance, became an important predator of small *D. longispina* and may have forced them to shift from nocturnal migration towards twilight and asynchronous migration. Large individuals were probably not affected by *Chaoborus* larvae. The increase in the size of *Notonecta* and increase in temperature may explain the change in their behaviour. The lack of information of the numbers and feeding of *Notonecta* sp. in Mekkojärvi does not allow more detailed explanation of the results.

In addition to predation, food (e.g. Bohrer, 1980; Walls *et al.*, 1990) is another probable ultimate cause of vertical migration. Among modifying factors, temperature and light are certainly important (Hutchinson, 1967). At high latitudes the seasonal course of illumination and the length of the day might explain the apparent shift from typical nocturnal vertical migration in Mekkojärvi in June towards twilight migration in late summer. In midsummer of high latitudes, midnight would be the most favourable time to avoid visual predation in the epilimnion and hence nocturnal migration would be a reasonable strategy. However, later in summer when nights are long, the time to be able to avoid visual predators is longer and allows avoidance of nonvisual predators without sacrifying too much feeding time. Thus twilight migration would have been the optimum behaviour to minimize temporal overlap with nocturnally migrating *Chaoborus* as well as visual predators. However, because the change in the light climate between June and July sampling was slight, light cannot alone explain the change in the migration behaviour. In late summer the need for more frequent feeding in the epilimnion and asynchronous migration, as indicated by the bimodal distributions of *Daphnia longispina* (Pearre, 1979) might have followed the ca. 5 °C rise in hypolimnetic temperature at the depth of the deep population maximum. This may have increased the metabolism of *D. longispina* so much that at least part of the population may not have been able to follow the diel light cycle in its vertical migration, but was forced to have more frequent feeding forays in the epilimnion. Asynchronous vertical migration is, particularly when combined with such short distances as present in Mekkojärvi, also effective to reduce epilimnetic predation.

In the conditions prevailing in Mekkojärvi, asynchronous migration might be particularly probable, because the migration over short distances requires insignificant energetic cost (Hutchinson, 1967: 785). That would also allow larger scale of variation in the behaviour of animals. Short scale asynchronous migration can take place at a frequency allowed by the swimming speed of animals. For *D. longispina* the time required for one vertical cycle is only of the order of minutes which is of the same order as the reaction time of *Daphnia* to the stimulus from its predators (Dodson, 1988). In large lakes the respective distance is much larger and hence the time required to swim between two vertical localities might dictate that the possible frequency of visits to near surface layers is not far from the diel solar cycle. Under such conditions any other factors coupled with the light cycle can more easily lead to synchronous diel vertical migration.

Pearre (1979) suggested that asynchronous migration is most probable with surplus food. The biomasses of algae and bacteria in the epilimnion of Mekkojärvi seem to support this hypothesis

(Table 2), but the high variation between the day and night results of algae and bacteria (Arvola *et al.*, 1991) and the uncertainty of volume-to-carbon conversion factors for reliable summation of algal and bacterial biomasses do not allow unambiguous conclusions. On the other hand, small and large *D. longispina* may differ in their ability to eat bacteria (Kankaala, 1988) and the availability of food may be highly size specific. The size distribution and egg ratio of *D. longispina* in fact indicate that in late summer the food availability was somewhat worse. However, in June large individuals were derived from ephippia and their egg ratio may not be comparable with individuals derived from parthenogenetic eggs. Under high grazing pressure, such as in Mekkojärvi, biomasses of food organisms might be misleading in the estimation of food availability, but knowledge of food production would be needed. It has been shown that with enough food *Daphnia longispina* may migrate vertically, but when food is scarce hunger forces it to feed all the time in the epilimnion irrespective of predation (Dagg, 1985; Johnsen & Jakobsen, 1987). However, De Meester & Dumont (1989) obtained opposite results and concluded that the vertical daytime distribution cannot be ascribed to a physiological adaptation alone, but that genetic differences are also significant.

Mazumder & Dickman (1989) made direct determinations of the feeding rate of *Daphnia* spp. on photosynthetic bacteria in meromictic Crawford lake in Ontario and found good evidence that *Daphnia* do feed on those organisms. In contrast, Murtaugh (1985) suggested that in the Canadian Roi lake the organisms at the chemocline and in deeper water were not heavily exploited by zooplankton. In Mekkojärvi the deep living *Daphnia longispina* population was able to penetrate into anoxic water and was potentially able to utilize the dense community of food organisms living near the oxycline. Populations of *Scourfieldia cordiformis* and *Chlamydomonas* sp. (Fig. 13; Arvola *et al.*, 1992) and photosynthetic bacteria (Arvola *et al.*, 1992; Kuuppo-Leinikki & Salonen, 1992) represented, together with heterotrophic bacteria, higher biomass per volume than is available in the epilimnion (Table 2). In support of the hypothesis of utilization of organisms in anaerobic layers, we were able to detect bacteriochlorophyll *d* in *D. longispina* caught in the epilimnion at 0.5 m depth. Although we could probably avoid any passive contamination of animals at sampling, our method did not yield any quantitative estimate about the importance of photosynthetic bacteria as a food. Similarly the importance of algae living in anoxic water cannot be quantified, but the near proximity of the vertical distributions of algae and *D. longispina* and the ability of the latter to visit anoxic water might allow grazing on the deep water algal population.

Among other alternatives to explain the diel vertical migration of *Daphnia longispina*, metabolic advantages (McLaren, 1963) are often considered inconclusive (Stich & Lampert, 1984; Kerfoot, 1985). However, although vertical migration would yield not instantaneous growth advantage, the lifetime reproduction may become higher and make it possible to maintain higher population density than in nonmigrating, but faster growing populations. In Mekkojärvi, large *D. longispina*, often carrying eggs, clearly preferred deeper and colder water than did the small ones (for other *Daphnia* species see also Murtaugh, 1985; Buchanan & Haney, 1980). With vertical migration between cold and warm water the population would also be able, to some extent, to regulate its food consumption and reduce the effects of starvation (Geller, 1986; 1989) under fluctuating food level. *Daphnia* in general seems to possess life history parameters which allow rapid exploitation of favourable periods although its responses may not always be optimal over a short time scale (Porter *et al.*, 1983; Threlkeld, 1987). Increased growth potential and regulation of food consumption might be particularly advantageous in small headwater lakes where the terrestrial influence is not buffered by the large volume of the epilimnion, but cause frequent nutrient input pulses markedly affecting the nutrient status. As a consequence algal populations, generally composed of small, fast growing species, also develop temporal maxima. The duration of such events are typically not compatible with the length of the life cycle of large zooplankton species such as *Daphnia longispina*. In strongly fluctuating environments, a sufficiently dense population of large reproducing *D. longispina* individuals can also obtain an advantage over competing herbivores.

Migrations represent a versatile behaviour by animals to cope with a multitude of different factors distributed unevenly in time and space. Consequently diel migration is a complex behaviour which combines optimised reactions to various constraints encountered by zooplankton. Along with temperature, feeding and avoidance of predation are certainly among the most important factors affecting vertical migrations. Because rather small differences in some parameters – e.g. about 10% in mortality (Vuorinen, 1987; Ohman, 1990) – may have a significant effect on the success of species depending on its behaviour, all ultimate reasons for vertical migration are seldom self evident. A few aspects seem to deserve more emphasis in future studies of vertical migrations. First, more efforts should be taken to study vertical migration of individuals rather than populations. Although that is very difficult, the results should be more unequivocally explainable. Such studies would also yield a better basis to take into account possible individual solutions and genetic shifts in the population along with varying environmental conditions. Second, one should take into account that diel migrations of plankton probably play an essential role in the functioning of the whole ecosystem. Therefore large and intensive efforts with many factors including not only information about the numbers of single migrating species, but also the metabolic rates of animals as well as their food, are needed to more thoroughly understand the migrations of zooplankton. Migrations are dynamic responses to the environment (Ohman, 1990) and such behaviour is an essential part in studying the structure and function of the ecosystems.

## Acknowledgements

This study was supported by the Academy of Finland and the Maj and Tor Nessling Foundation. Our thanks are due to R. Ilola and J. Vainionpää for making most of chemical determinations, S. Hällfors for the identification of *Paraphysomonas* species and Roger Jones for the revision of English of the manuscript.

## References

Arvola, L., K. Salonen, P. Kankaala & A. Lehtovaara, 1992. Vertical distributions of bacteria and algae in a steeply stratified humic lake under high grazing pressure from *Daphnia longispina*. Hydrobiologia 229: 253–269.

Bohrer, R. N., 1980. Experimental studies on diel vertical migration. In W. C. Kerfoot (ed.), Evolution and ecology of zooplankton communities. The University Press of New England, Hanover (N.H.); Lond.: 111–121.

Buchanan, C. & J. F. Haney, 1980. Vertical migrations of zooplankton in the Arctic: A test of the environmental controls. In W. C. Kerfoot (ed.), Evolution and ecology of zooplankton communities. The University Press of New England, Hanover (N.H.); Lond.: 69–79.

Dagg, M. J., 1985. The effects of food limitation of diel migratory behaviour in marine zooplankton. Arch. Hydrobiol. Beih. Ergebn. Limnol. 21: 247–255.

Davies, J., 1985. Evidence for a diurnal horizontal migration in *Daphnia hyalina lacustris* Sars. Hydrobiologia 120: 103–105.

De Meester, L. & H. J. Dumont, 1989. Phototaxis in *Daphnia*: Interaction of hunger and genotype. Limnol. Oceanogr. 34: 1322–1325.

Dodson, S., 1988. The ecological role of chemical stimuli for the zooplankton: Predator-avoidance behaviour in *Daphnia*. Limnol. Oceanogr. 33: 1431–1439.

Dodson, S. I., 1989. The ecological role of chemical stimuli for the zooplankton: predator-induced morphology in *Daphnia*. Oecologia (Berlin) 78: 361–367.

Dodson, S. I. & J. E. Havel, 1988. Indirect prey effects: Some morphological and liffe history responses of *Daphnia pulex* exposed to *Notonecta undulata*. Limnol. Oceanogr. 33: 1274–1285.

Engle, D. L., 1985. The production of haemoglobin by small pond *Daphnia pulex*: intraspecific variation and its relation to habitat. Freshwat. Biol. 15: 631–638.

Fox, H. M., B. M. Gilchrist & E. A. Phear, 1951. Functions of haemoglobin in *Daphnia*. Proc. Royal Soc. London Ser. B. 138: 514–528.

Geller, W., 1986. Diurnal vertical migration of zooplankton in a temperate great lake (L. Constance): a starvation avoidance mechanism? Arch. Hydrobiol. Suppl. 74: 1–60.

Geller, W., 1989. The energy budget of two sympatric *Daphnia* species in lake Constance: productivity and energy residence times. Oecologia (Berlin) 78: 242–250.

Gerritsen, 1982. Behavioural response of *Daphnia* to rate of temperature change: Possible enhancement of vertical migration. Limnol. Oceanogr. 27: 254–261.

Gliwicz, Z. M., 1986. Predation and the evolution of vertical migration in zooplankton. Nature 320: 746–748.

Hanazato, T., M. Yasuno & M. Hosomi, 1989. Significance of a low oxygen layer for *Daphnia* population in Lake Yonoko, Japan. Hydrobiologia 185: 19–27.

Haney, J. F., 1985. Regulation of cladoceran filtering rates in nature by body size, food concentration, and diel feeding patterns. Limnol. Oceanogr. 30: 397–411.

Haney, J. F. & D. J. Hall, 1973. Sugar-coated *Daphnia*: A preservation technique for Cladocera. Limnol. Oceanogr. 18: 331–333.

Heisey, D. & K. G. Porter, 1977. The effect of ambient oxygen concentration in filtering and respiration rates of *Daphnia galeata mendotae* and *Daphnia magna*. Limnol. Oceanogr. 22: 839–845.

Hutchinson, G. E., 1967. A treatise on limnology. 2. Introduction to lake biology and the limnoplankton. John Wiley & Sons, N.Y., 1115 pp.

Johnsen, G. H. & P. J. Jakobsen, 1987. the effect of food limitation on vertical migration in *Daphnia longispina*. Limnol. Oceanogr. 32: 873–880.

Jones, R. & L. Arvola, 1984. Light penetration and some related characteristics in small forest lakes in southern Finland. Verh. int. Ver. Limnol. 22: 811–816.

Kankaala, P., 1988. The relative importance of algae and bacteria as food for *Daphnia longispina* (Cladocera) in a polyhumic lake. Freshwat. Biol. 19: 285–296.

Kerfoot, W. C., 1985. Adaptive value of vertical migration: comments on the predation hypothesis and some alternatives. In M. A. Rankin (Ed.), Migration: Mechanisms and adaptive significance. University of Texas, Port Aransas: 91–113.

Kring, R. L. & W. J. O'Brien, 1976. Effect of varying oxygen concentrations on the filtering rate of *Daphnia pulex*. Ecology 57: 808–814.

Kuuppo-Leinikki, P. & K. Salonen, 1992. Bacterioplankton in a small polyhumic lake with anoxic hypolimnion. Hydrobiologia 229: 159–168.

Lampert, W., 1989. The adaptive significance of diel vertical migration of zooplankton. Functional Ecology 3: 21–27.

Landon, M. S. & R. H. Stasiak, 1983. *Daphnia* hemoglobin concentration as a function of depth and oxygen availability in Arco Lake, Minnesota. Limnol. Oceanogr. 28: 731–737.

Mazumder, A. & M. D. Dickman, 1989. Factors affecting the spatial and temporal distribution of phototrophic sulfur bacteria. Arch. Hydrobiol. 116: 209–226.

McLaren, I. A., 1963. Effects of temperature on growth of zooplankton and the adaptive value of vertical migration. J. Fish. Res. Bd Can. 20: 685–727.

Moore, M. V., 1988. Differential use of food resources by the instars of Chaoborus punctipennis. Freshwat. Biol. 19: 249–268.

Münster, U., 1991. Extracellular enzymes in a polyhumic lake: Important regulators in detritus processing. Hydrobiologia 00: 00–00.

Murtaugh, P. A., 1985. Vertical distributions of zooplankton and population dynamics of *Daphnia* in a meromictic lake. Hydrobiologia 123: 47–57.

Ohman, M. D., 1990. The demographic benefits of diel vertical migration by zooplankton. Ecol. Monogr. 60: 257–281.

Pearre, S. Jr., 1979. Problems in the detection and interpretation of vertical migration. J. Plankton Res. 1: 29–44.

Porter, K. G., J. D. Jr. Orcutt & J. Gerritsen, 1983. Functional response and fitness in a generalist filter feeder, *Daphnia magna* (Cladocera: Crustacea). Ecology 64: 735–742.

Prepas, E. & F. H. Rigler, 1978. The enigma of *Daphnia* death rates. Limnol. Oceanogr. 23: 970–988.

Salonen, K., L. Arvola, T. Tulonen, T. Hammar, T.-R. Metsälä, P. Kankaala & U. Münster, 1992a. Planktonic food chains of a highly humic lake. I. A mesocosm experiment during the spring primary production maximum. Hydrobiologia 229: 125–142.

Salonen, K., P. Kankaala, T. Tulonen, T. Hammar, M. James, T.-R. Metsälä & L. Arvola, 1992. Planktonic food chains of a highly humic lake. II. A mesocosm experiment in summer during dominance of heterotrophic processes. Hydrobiologia 229: 143–157.

Sorokin, Ju. I., 1966. On the trophic role of chemosyntesis and bacterial biosynthesis in water bodies. In C. R. Goldman (Ed.), Primary productivity in aquatic environments. University of California Press, Berkeley: 187–205.

Stich, H.-B. & W. Lampert, 1984. Growth and reproduction of migrating and non-migrating *Daphnia* species under simulated food and temperature conditions of diurnal vertical migration. Oecologia (Berlin) 61: 192–196.

Threlkeld, S. T., 1987. *Daphnia* life history strategies and resource allocation patterns. Mem. Ist. ital. Idrobiol. 45: 353–366.

Vanni, M. J., 1988. Freshwater zooplankton community structure: Introduction of large predators and large herbivores to a small-species community. Can. J. Fish. aquat. Sci. 45: 1758–1770.

Vuorinen, I., 1987. Vertical migration of *Eurytemora* (Crustacea, Copepoda): a compromise between the risks of predation and decreased fecundity. J. Plankton Res. 9: 1037–1046.

Walls, M., I. Kortelainen & J. Sarvala, 1990. Prey responses to fish predation in freshwater communities. Ann. zool. fenn. 27: 183–199.

Weider, L. J. & W. Lampert, 1985. Differential response of *Daphnia* genotypes to oxygen stress: respiration rates, hemoglobin content and low-oxygen tolerance. Oecologia (Berlin) 65: 487–491.

Wright, D., W. J. O'Brien & G. L. Vinyard, 1980. Adaptive value of vertical migration: A simulation model argument for the predation hypothesis. In W. C. Kerfoot (ed.), Evolution and ecology of zooplankton communities. The University Press of New England, Hanover (N.H.); Lond.: 111–121.

Zaret, T. M. & J. S. Suffern, 1976. Vertical migration in zooplankton as a predator avoidance mechanism. Limnol. Oceanogr. 21: 804–813.

*Hydrobiologia* **229**: 289–291, 1992.
*K. Salonen, T. Kairesalo & R.I. Jones (eds.),*
*Dissolved Organic Matter in Lacustrine Ecosystems: Energy Source and System Regulator.*




# Concluding remarks*

Robert G. Wetzel
*The University of Alabama, Tuscaloosa, Alabama 35487, USA*

When asked to reflect upon the essence of our understanding of dissolved organic matter (DOM) in lacustrine ecosystems, as encapsulated in the presentations and discussions of the past three days, I obtained the same feeling incurred during a well-executed experiment – the results enhanced our understanding of the questions being addressed but they represented only a step in a long trail and generate many more questions for every answer obtained. To be emphatically emphasized, however, is the united conclusion that this long neglected subject is not only a major focus for study in freshwater ecosystems, but that DOM serves both a primary energy source and as a metabolic regulator within lacustrine ecosystems. Quantitative resolution of these functions still eludes us, but definite progress was apparent.

A number of questions have been addressed: What are the sources of DOM and regulatory mechanisms of loading to fresh waters? How do the chemical properties of DOM influence their efficacy as an energy source? And of great importance, how do these compounds regulate both the environment and the metabolism of biota, directly and indirectly. These questions are not new – the insightful pioneering studies of Birge and Juday (1926, 1927, 1934) demonstrated the large amounts of DOM, approximately an order of magnitude more than all particulate organic matter in the water. The complex direct and indirect *functional* significance of the great quantities of the DOM were championed long ago by Saunders (1957), Wetzel (1968, 1972, 1975), and others. The analytical tools available at the present time have led to a spate of renewed interest because the chemical complexity is now more tangible (e.g., recent compendia on the subject include Seki, 1982; Aiken *et al.*, 1985; Thurman, 1985; Suffet & MacCarthy, 1989; Hayes *et al.*, 1989; Perdue & Gjessing, 1990). That interest is particularly high in Finland, of course, because of the large quantities of DOM in many of the fresh waters of this country. I am adamant, however, that the principles we are discussing apply to all inland waters; the functional differences in observed concentrations are largely in the extent of influence, not whether it occurs. Certain major generalizations about the functional significance of DOM are possible.

Two primary directions were being addressed. Firstly, how is DOM in general utilized as an energy source? With what efficiencies is energy acquired from the diverse organic molecules? How is energy obtained from the more recalcitrant molecules coupled to that energy acquired from DOM released from more 'conventional' autochthonous phytoplanktonic productivity? All of these questions relate to the importance of the sources of DOM and chemical reactive, physical, and biological exchange mechanisms as DOM is transported to and circulated within the lacustrine ecosystem. Secondly, dissolved organic matter (DOM) was being evaluated as a metabolic regulator within the lacustrine ecosystem.

* International Symposium: Dissolved organic matter in lacustrine ecosystems: energy source and system regulator. Lammi Biological Station, University of Helsinki, Finland, 29–31 May 1990.

Physical factors were considered in a general sense only, such as the importance of drainage basin characteristics, hydrology, and precipitation events on the loading of DOM to recipient lakes. From a more functional viewpoint, the important role of ultraviolet radiation in the decoupling of humic acid-enzyme complexes was demonstrated. Other important physical aspects were addressed here only in general terms, but should not be neglected in future analyses and interpretations. For example, the selective spectral attenuation of light can certainly affect photosynthetic efficiencies and influence physiological regulators such as phytochromes or possibly migratory behavior of planktonic algae.

The importance of allochthonous and the wetland-littoral sources of DOM were repeatedly emphasized. In particular, the land-water interface macrophytes were shown to be often much more significant than the phytoplankton for DOM loading for many lacustrine ecosystems. Enhanced loading of humic compounds from terrestrial sources was evaluated in relation to increased precipitation and thermal enrichments associated with global climate changes. Altered rates of decomposition of DOM as a result of temperature increases are potential counteractive responses. The qualitative differences in DOM from different sources and how the compounds are altered during transport and distribution within the lake ecosystems represent important questions yet to be resolved.

Chemical definition of DOM is a perennial problem. Obviously the bulk of DOM consists of a complex composite of many compounds, nearly all of plant origins and many in various stages of chemical alteration and biological degradation (e.g., Conn, 1981; Haslam, 1989). Identification of chemical constituents is essential if detailed assimilation rates and physiological effects are to be evaluated and understood completely. Identification is difficult at best and, even with constant evolution and improvement of methodology, rapid and consistent chemical identification is distant in the future. The question then is whether we can progress significantly in ecological understanding without complete or even partial identification of the chemical constituents of DOM. Molecular size fractionation of DOM coupled to biological utilization experiments and uptake kinetics of the organic fractions, for example, can assist in interpretation of availability of DOM as an energy and carbon source to higher trophic levels. I argue that we can progress most effectively, and indeed it is essential to proceed on both of these chemical and biological fronts simultaneously. Although resolution of interactive effects of specific compounds of DOM may not be resolved at the species and population levels, important community and ecosystem interactions can be. Analyses here that coupled DOM loadings with physical parameters, changes in nutrient loadings, and other effects were numerous. All stressed the potential and realized major effects of DOM from both allochthonous, land-water interface, and autochthonous sources.

Much needed emphasis of the Symposium was directed to the utilization of DOM by bacteria directly, and what indirect effects to higher trophic levels result from this productivity. The importance of accurate metabolic methodology was repeatedly emphasized; clearly many methods should be adopted more extensively from related disciplines, including molecular biology, than has been the case in the past. Application of different methods should be applied to natural communities simultaneously in order to maximize understanding of the processes and mechanisms involved in metabolic regulation of bacteria by different natural dissolved organic compounds.

Bacterial utilization rates and efficiencies of different types of source materials of humic-type DOM were examined by many researchers. Certain studies were particularly effective in not only demonstrating very significant utilization of the relatively recalcitrant DOM pool by bacteria but also that carbon budgets cannot balance without this source of organic carbon inputs to the ecosystems.

The biogeochemistry of humic substances of aquatic ecosystems is an extraordinarily interactive subject. The complex and highly dynamic interactions of chemical and biological processes were evaluated and coupled to observed system variations. Interactions of humic compounds with metal and nutrient availability were also presented. Of great interest were interactions among extracellular enzymes and dissolved organic compounds, particularly humic substances. Although we have only an inkling of the many variegated and diverse interactions of these compounds, the information is adequate to indicate a multitude of important regulatory mechanisms.

This symposium reiterated the need for enhanced communication and collaboration among the organic chemists, biogeochemists, and biologists to resolve the functional interactive mechanisms of DOM in aquatic ecosystems. Certainly the biologists can give guidance to the most probably fruitful avenues for chemical analyses from community and ecosystem perspectives. Similarly, needs exist for biologists, particularly those involved with organic regulators and enzymological interactions, to collaborate more rigorously with biochemical expertise. The many chemical and biological interactions of DOM with the environment and the metabolism of many types of microbiota operate in dynamic and diverse ways. Dissolved organic matter provides a major energetic source to aquatic ecosystems. In addition, the complex interactions of DOM with the metabolism of organisms at many levels in the ecosystem structure provide boundless enticement and fascination. These functions are foundational controls of the whole ecosystem operation. This symposium was a small, but important step in our understanding and collaboration. As Bede said 'It is better never to begin a good work than, having begun it, to stop'. Many more of such interactive meetings are needed in the future.

## References


Aiken, G. R., D. M. McKnight, R. L. Wershaw & P. MacCarthy, Editors, 1985. Humic substances in soil, sediment, and water. Geochemistry, isolation, and characterization. John Wiley & Sons, New York. 692 pp.

Birge, E. A. & C. Juday, 1926. Organic content of lake water. Bull. U.S. Bur. Fish. 42: 185–205.

Birge, E. A. & C. Juday, 1927. The organic content of the water of small lakes. Proc. Amer. Phil. Soc. 66: 357–372.

Birge, E. A. & C. Juday, 1934. Particulate and dissolved organic matter in inland lakes. Ecol. Monogr. 4: 440–474.

Conn, E. E., Editor, 1981. The biochemistry of plants. Vol. 7. Secondary plant products. Academic Press, New York. 791 pp.

Haslam, E., 1989. Plant polyphenols. Vegetable tannins revisited. Cambridge Univ. Press, Cambridge. 230 pp.

Hayes, M. H. B., P. MacCarthy, R. L. Malcolm & R. S. Swift, 1989. Humic substances II. In search of structure. John Wiley & Sons.

Perdue, E. M. & E. T. Gjessing, Editors, 1990. Organic acids in aquatic ecosystems. John Wiley & Sons, Chichester, 345 pp.

Saunders, G. W., 1957. Interrelations of dissolved organic matter and phytoplankton. Bot. Rev. 23: 389–410.

Seki, H., 1982. Dynamics of organic materials in aquatic environments. CRC Press, Boca Raton, Florida, 201 pp.

Suffet, I. H. & P. MacCarthy, Editors, 1989. Aquatic humic substances. Influence on fate and treatment of pollutants. Amer. Chemical Soc., Washington, DC, 964 pp.

Thurman, E. M., 1985. Organic geochemistry of natural waters. Martinus Nijhoff/Dr. W. Junk Publ., Dordrecht, 497 pp.

Wetzel, R. G., 1968. Dissolved organic matter and phytoplanktonic productivity in marl lakes. Mitt. int. Ver. Limnol. 14: 261–270.

Wetzel, R. G., P. H. Rich, M. C. Miller & H. L. Allen, 1972. Metabolism of dissolved and particulate detrital carbon in a temperate hard-water lake. Mem. Ist. ital. Idrobiol. 29 (Suppl.): 185–243.

Wetzel, R. G., 1975. Limnology. W.B. Saunders Company, Philadelphia, 743 pp.

www.ingramcontent.com/pod-product-compliance
Lightning Source LLC
LaVergne TN
LVHW062118200926
848484LV00007B/1706
* 9 7 8 0 7 9 2 3 1 6 5 2 7 *